CATALYTIC NAPHTHA REFORMING

CHEMICAL INDUSTRIES

A Series of Reference Books and Textbooks

Founding Editor

HEINZ HEINEMANN

CATALYTIC NAPHTHA REFORMING
SECOND EDITION, REVISED and EXPANDED

edited by
George J. Antos
UOP, LLC
Des Plaines, Illinois, U.S.A.

Abdullah M. Aitani
King Fahd University of Petroleum and Minerals
Dhahran, Saudi Arabia

CRC Press
Taylor & Francis Group
Boca Raton London New York

CRC Press is an imprint of the
Taylor & Francis Group, an **informa** business

The first edition of this book was published as *Catalytic Naphtha Reforming: Science and Technology*, edited by George J. Antos, Abdullah M. Aitani, and José M. Parera (Marcel Dekker, Inc., 1995).

First published 2004 by Marcel Dekker, Inc.

Published 2020 by CRC Press
Taylor & Francis Group
6000 Broken Sound Parkway NW, Suite 300
Boca Raton, FL 33487-2742

First issued in paperback 2020

© 2004 by Taylor & Francis Group, LLC
CRC Press is an imprint of Taylor & Francis Group, an Informa business

No claim to original U.S. Government works

ISBN 13: 978-0-367-57842-8 (pbk)
ISBN 13: 978-0-8247-5058-9 (hbk)

Visit the Taylor & Francis Web site at
http://www.taylorandfrancis.com

and the CRC Press Web site at
http://www.crcpress.com

Library of Congress Cataloging-in-Publication Data
A catalog record for this book is available from the Library of Congress.

Preface to the Second Edition

Nearly a decade has passed since the publication of the first edition of *Catalytic Naphtha Reforming*. That book was a survey of the technology encompassing the first 45 years of the use of this process in the refining industry. In preparing the second edition, perspective on this refining process was again considered. It is still true that catalytic reforming is the primary process in the refinery for producing high-octane gasoline to be blended into the gasoline pool. As needs for gasoline have risen, the demands on the reformer have also increased. What has changed is that additional drivers have surfaced, which have added to the demands on the process.

The first of these demands are the new environment-based regulations for fuel quality parameters. In particular, the targeted reduced-sulfur content for gasoline and diesel fuel has had an impact on the catalytic reformer. Although sulfur in the gasoline pool does not originate with the reformer, sulfur content of naphtha from the fluid catalytic cracker does require significant treatment in order to continue inclusion in the pool. Most of the schemes to deal with this sulfur involve some level of hydrodesulfurization. Hydrogen is required, and the reformer is one of the few units to provide hydrogen in the refinery. The result is an increased demand on the catalytic reformer. Although many of these hydrotreating schemes attempt to minimize octane loss, any loss will need to be countered with more output from the octane machine—the reformer. Environmental regulations aimed at lowering sulfur in diesel fuel also increase the need for hydrogen in the refinery. Hydrogen demand has increased overall in the refinery, and the catalytic reformer is under pressure to produce more hydrogen by an increased severity of operation or by improved selectivity to aromatics.

In addition, in the United States, the drive to eliminate the use of MTBE as an oxygenate component in the gasoline pool will impact the reformer situation. Octane barrels are lost when MTBE is replaced by ethanol. The catalytic reformer will need to replace these lost octane barrels, largely through an increased severity of operation or through higher yields of high octane gasoline.

Over the past decade, refiners have been forced to maximize their existing asset utilization. With capital at a premium, refiners must deliver more from the units they already have. These twin pressures from environmental regulations and asset utilization have impacted the catalytic reformer. New technology, in the form of new catalysts or a minimal revamp of process improvements, was required. The catalyst vendors and process licensors have responded to these needs, thereby fulfilling predictions in the first edition and providing the basis for this book.

For this edition, prominent authors were again invited to either update an existing chapter or write a new chapter. The layout of the book is logical and similar to that of the first edition. Part I covers the chemistry of naphtha reforming, emphasizing basic reforming reactions, metal/acid catalysis, and naphtha hydrotreatment. Part II is a detailed review of reforming catalysts. The chapter on catalyst preparation has been extensively enhanced with an in-depth treatment of platinum impregnation chemistry, a topic that has been extensively investigated over the past decade. When combined with the updated chapter on catalyst characterization, this section serves as a reference source for anyone involved in the preparation of or research on platinum-containing catalysts. Included in this section is a completely updated discussion of the commercial reforming catalysts available from vendors today. Two chapters that are more experimental have been included on the future direction of catalyst technology in pore structure optimization and zeolite-hybrid catalysts.

Part III focuses on catalyst deactivation by coking and regeneration. Added to this is a discussion on some of the issues that are important to continuous reformer operations involving catalyst movement and continuous regeneration as experienced by refinery personnel. A separate chapter is dedicated to the recovery of the precious metals from the reforming catalyst.

In Part IV, commercial process technology is covered. The licensed processes are reviewed in conjunction with chapters on control systems and modeling for commercial reformer units.

Once again it has been our pleasure to work with the contributors of this book. They paid much attention to reviewing the literature in the area, and then skillfully combined it with their own work and insights. It has been an extensive effort and has taken time to bring it to completion. We give special thanks to the contributors and the publisher for their patience. Our intent was to place our combined experience and knowledge of the technology of catalytic naphtha reforming into one book in order to share it with all those who need this information. We hope that this second edition will be recognized as a valuable resource for those involved in the reforming or related catalysis areas, whether as academics, graduate students, industrial researchers, chemical engineers, or refinery personnel. Knowledge and the time to gain it are two assets that we have attempted to help you manage to your advantage with this volume.

George J. Antos
Abdullah M. Aitani

Preface to the First Edition

The use of catalytic naphtha reforming as a process to produce high-octane gasoline is as important now as it has been for over the 45 years of its commercial use. The catalytic reformer occupies a key position in a refinery, providing high value-added reformate for the gasoline pool; hydrogen for feedstock improvement by the hydrogen-consuming hydrotreatment processes; and frequently benzene, toluene, and xylene aromatics for petrochemical uses. The technology has even further impact in the refinery complex. The processes of hydrogenation, dehydrogenation, and isomerization have all benefited from the catalyst, reactor, and feed treatment technologies invented for catalytic reforming processes. The long-term outlook for the reforming catalyst market remains strong. The conditions of operation of catalytic reforming units are harsh and there is an increasing need for reformate. Presently, the catalytic reforming process is currently operated to produce research octane numbers of 100 and more.

Since its introduction, catalytic reforming has been studied extensively in order to understand the catalytic chemistry of the process. The workhorse for this process is typically a catalyst composed of minor amounts of several components, including platinum supported on an oxide material such as alumina. This simplification masks the absolute beauty of the chemistry involved in combining these components in just the proper manner to yield a high-performance, modern reforming catalyst. The difficulty in mastering this chemistry and of characterizing the catalyst to know what has been wrought is the driving force behind the many industrial and academic studies in reforming catalysis available today.

Several questions come to mind. Why are scientists continuing to research this area of catalysis? What have all the preceding studies taught us about these catalysts, and what remains unknown? Given the numerous studies reported in the patent literature and in technical journals, it is surprising that a survey aimed at answering these questions summarizing the preceding experiences is not readily found. All the editors and contributors of this book are experienced in the study of reforming catalysts, and each one of them would have employed such a survey in his own research program. This volume provides information not currently available from one single literature source. The chapters are written by well-known authorities in the fields encompassed by catalytic reforming, starting with the process chemistry and focusing on the preparation, characterization, evaluation, and operation of the catalyst itself. The unknown aspects of catalyst chemistry and fundamental studies attempting to provide an understanding are also presented. Some attempt is made to predict the future for this catalyst technology, a task made complicated by the conflicting demand for more transportation fuels and petrochemicals, and the resolution to reduce the pollution resulting from their use.

It has been our pleasure to work with the contributors involved in this book. Their effort in combining their own research with the recent literature in the field of catalytic naphtha reforming is highly appreciated. This effort would not have been possible without their willingness to share valuable knowledge and experience. Moreover, we express our gratitude for their responsiveness to deadlines and review comments.

The editors hope that veteran industrial researchers will recognize this volume as an important resource and that novice researchers in the field of reforming and related catalysts—industrial chemists assigned to their first major catalysis project, graduate students embarking on the study of catalysis, and chemical engineers in the refinery responsible for full-scale commercial catalytic reforming—will find this a valuable reference volume and tool for their future endeavors in this exciting area.

George J. Antos
Abdullah M. Aitani
José M. Parera

Contents

Contributors

Abdullah M. Aitani *King Fahd University of Petroleum and Minerals, Dhahran, Saudi Arabia*

Syed Ahmed Ali *King Fahd University of Petroleum and Minerals, Dhahran, Saudi Arabia*

George J. Antos *UOP, LLC, Des Plaines, Illinois, U.S.A.*

Raimundo Arvelo *University of La Laguna, Laguna, Spain*

Burtron H. Davis *University of Kentucky, Lexington, Kentucky, U.S.A.*

Patricia K. Doolin *Marathon Ashland Petroleum, LLC, Catlettsburg, Kentucky, U.S.A.*

Nora S. Fígoli *Instituto de Investigaciones en Catálisis y Petroquímica (INCAPE), Santa Fe, Argentina*

Knut Grande *STATOIL Research Centre, Trondheim, Norway*

Matthias Grehl *W.C. Heraeus GmbH & Co., Hanau, Germany*

Anders Holmen *Norwegian University of Science and Technology, Trondheim, Norway*

Mark P. Lapinski *UOP, LLC, Des Plaines, Illinois, U.S.A.*

Rafael Larraz* *University of La Laguna, Laguna, Spain*

Cheng-Lie Li[†] *National University of Mexico, Mexico City, Mexico*

Horst Meyer *W.C. Heraeus GmbH & Co., Hanau, Germany*

Kjell Moljord *STATOIL Research Centre, Trondheim, Norway*

Mark D. Moser *UOP, LLC, Des Plaines, Illinois, U.S.A.*

**Current affiliation*: CEPSA, Madrid, Spain
[†]*Current affiliation*: East China University of Science and Technology, Shanghai, China

Octavio Novaro   *National University of Mexico, Mexico City, Mexico*

Soni O. Oyekan   *Marathon Ashland Petroleum, LLC, Catlettsburg, Kentucky, U.S.A.*

Zoltán Paál   *Hungarian Academy of Sciences, Budapest, Hungary*

José M. Parera   *Instituto de Investigaciones en Catálisis y Petroquímica (INCAPE), Santa Fe, Argentina*

Grigore Pop   *S.C. Zecasin S.A., Bucharest, Romania*

Rune Prestvik   *SINTEF Applied Chemistry, Trondheim, Norway*

J. R. Regalbuto   *University of Illinois at Chicago, Chicago, Illinois, U.S.A.*

Jerzy Szczygieł   *Wroclaw University of Technology, Wroclaw, Poland*

Lee Turpin   *Aspen Technology Inc., Bothell, Washington, U.S.A.*

Jin-An Wang   *National Polytechnic Institute, Mexico City, Mexico*

David J. Zalewski   *Marathon Ashland Petroleum, LLC, Catlettsburg, Kentucky, U.S.A.*

CATALYTIC NAPHTHA REFORMING

1

Compositional Analysis of Naphtha and Reformate

Rune Prestvik
SINTEF Applied Chemistry
Trondheim, Norway

Kjell Moljord and Knut Grande
STATOIL Research Centre
Trondheim, Norway

Anders Holmen
Norwegian University of Science and Technology
Trondheim, Norway

1 INTRODUCTION

Naphtha is transformed into reformate by catalytic reforming. This process involves the reconstruction of low-octane hydrocarbons in the naphtha into more valuable high-octane gasoline components without changing the boiling point range. Naphtha and reformate are complex mixtures of paraffins, naphthenes, and aromatics in the C_5-C_{12} range. Naphthas from catalytic or thermal cracking also contain olefins. Naphthas of different origin contain small amounts of additional compounds containing elements such as sulfur and nitrogen. These elements affect the performance of the bifunctional noble metal catalyst used in catalytic reforming and must be removed to low levels prior to entering the reformer unit. The composition of hydrocarbons and the concentration of additional elements determine the quality as reforming feedstock or as a gasoline blending component.

This chapter describes the chemistry of naphtha and reformate. It includes the origin from crude oil, the overall composition, and key parameters with respect to processing ability and product quality. Finally, analytical methods available for performing a complete compositional analysis and parameter detection are described.

2 THE NAPHTHA FRACTION

2.1 Origin from Crude Oil Distillation and Processing

Hydrocarbons are the major constituents of crude oil, or petroleum, and account for up to 97% of the total mass.[1] These are paraffinic, naphthenic, or aromatic structures ranging from light gaseous molecules (C_1–C_4 alkanes) to heavy waxes or asphaltenic matter. The rest are organic compounds of sulfur, nitrogen, and oxygen, as well as water, salt, and a number of metal containing constituents such as vanadium, nickel, and sodium. Although elemental concentrations of carbon and hydrogen vary only slightly within narrow limits, typically 82–87 wt % C and 10–14 wt % H, the individual concentrations of the different compounds that determine the physical properties are highly variable and depend on the crude oil origin.

Full-range naphtha is the fraction of the crude oil boiling between 30°C and 200°C, and constitutes typically 15–30% by weight of the crude oil. This includes hydrocarbons ranging from C_5 to C_{12}, some sulfur, and small amounts of nitrogen. Metal containing compounds are usually not present. The naphtha obtained directly from the atmospheric crude distillation column is termed *straight run* (SR). However, naphtha is also produced during processing of heavier parts of the crude oil (e.g., catalytic cracker naphtha, visbreaker naphtha, coker naphtha). As opposed to the straight-run streams, these naphthas also contain olefinic hydrocarbons. *Light naphtha* is the fraction boiling from 30°C to 90°C, containing the C_5 and C_6 hydrocarbons. *Heavy naphtha* is the fraction boiling from 90°C to 200°C. The term '*medium naphtha*' is sometimes used for the fraction of this heavy cut that boils below 150°C and includes mostly C_7–C_9 hydrocarbons. Table 1 illustrates how naphtha fractions can range from highly

Table 1 Composition of Medium Naphtha Cuts from Different Crude Oils[2]

Oil field	Paraffins (wt %)	Naphthenes (wt %)	Aromatics (wt %)	Sulfur (wt ppm)	Nitrogen (wt ppm)
Troll (Norway)	13.9	**75.2**	10.8	20	<1
Norne (Norway)	27.7	34.8	**37.5**	10	<1
Heidrun (Norway)	35.4	51.2	13.5	10	<1
Lufeng (China)	**69.5**	27.5	2.9	<10	1

paraffinic to highly naphthenic and from low in sulfur to high in sulfur, depending on the crude oil.

Hydrotreated (desulfurized) medium naphtha is the favored feedstock for catalytic reforming, although full-range stocks are sometimes processed if benzene is a desired product. The light naphtha is preferentially upgraded by isomerization whilst the heaviest part of the naphtha is often included in the light gas oil fraction (jet fuel/diesel). Figure 1 gives an example of a processing scheme for refinery gasoline production with catalytic reforming.

2.2 Naphtha Composition

Hydrocarbons

Paraffins or *alkanes* are saturated aliphatic hydrocarbons with the general formula C_nH_{2n+2}. They are either straight-chain (*n*-paraffins) or branched structures (*i*-paraffins). The boiling point increases by about 25–30°C for each carbon

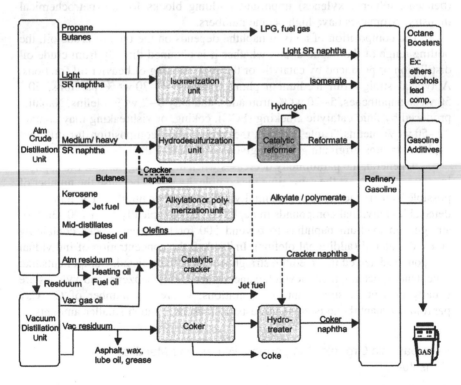

Figure 1 Example of a processing scheme for refinery gasoline production with catalytic reforming.

atom in the molecule, and the boiling point of an n-paraffin is always higher than that of the i-paraffin with the same carbon number. The density increases with increasing carbon number as well. **Olefins** or *alkenes* are unsaturated aliphatic hydrocarbons. Like the paraffins, they are either straight chains or branched structures, but contain one or more double bonds. Monoolefins have the general formula C_nH_{2n}. **Naphthenes** or *cycloalkanes* are saturated cyclic hydrocarbons that contain at least one ring structure. The general formula for mononaphthenes is C_nH_{2n}. The most abundant naphthenes in petroleum have a ring of either five or six carbon atoms. The rings can have paraffinic side chains attached to them. The boiling point and the density is higher than for any paraffin with the same number of carbon atoms. **Aromatics** have the general formula C_nH_{2n-6} and contain one or more polyunsaturated rings (conjugated double bonds). These benzene rings can have paraffinic side chains or be coupled with other naphthenic or aromatic rings. The boiling points and the densities of these polyunsaturated compounds are higher than that of both paraffins and naphthenes with the same carbon number. The reactivity of the unsaturated bonds make the C_6, C_7, and C_8 aromatics or *BTX* (benzene, toluene, xylenes) important building blocks for the petrochemical industry. Aromatics have high octane numbers.

The composition of a given naphtha depends on the type of crude oil, the boiling range of the naphtha, and whether it is obtained directly from crude oil distillation or produced by catalytic or thermal cracking of heavier oil fractions. A typical straight-run medium naphtha contains 40–70 wt % paraffins, 20–50 wt % naphthenes, 5–20 wt % aromatics, and only 0–2 wt % olefins. Naphtha produced by fluid catalytic cracking (FCC), coking, or visbreaking may contain 30–50 wt % olefins. Table 2 shows the hydrocarbon composition for different naphtha streams originating from a given crude.

In general, the paraffinicity decreases when the boiling point of the naphtha increases (Fig. 2). At the same time the complexity grows because the number of possible isomers increases exponentially with the carbon number. The number of detectable individual compounds in naphthas ranges typically from 100–300 for straight-run medium naphthas to beyond 500 for full-range stocks containing cracked material (additional olefins). In Table 3 the concentration of individual compounds detected in a medium straight-run naphtha is listed. Components like n-heptane, n-octane, methylcyclohexane, toluene, ethylbenzene, and xylenes are usually present in significant concentrations, whereas a number of C_7–C_{9+} paraffin and naphthene isomers are usually present in much smaller amounts.

Heteroatomic Organic Compounds, Water, and Metallic Constituents

Sulfur is an important heteroatomic constituent in petroleum. The concentration is highly dependent on the type of crude oil and may range from virtually zero to

Table 2 Typical Compositions and Characteristics of Refinery Naphtha Streams Originating from the Same Crude Oil

Stream	Paraffins (wt %)	Olefins (wt %)	Naphth. (wt %)	Aromatics (wt %)	Density (g/ml)	IBP–FBP (°C)	Crude (wt %)
Light SR	55	—	40	5	0.664	C_5–90	3.2
Medium SR	31	—	50	19	0.771	90–150	8.6
Heavy SR	30	—	44	26	0.797	150–180	4.7
FCC	34	23	11	32	0.752	C_5–220	20
Light VB	64	10	25	1	0.667	C_5–90	—
Heavy VB	46	30	16	8	0.750	90–150	—

SR, straight-run; FCC, fluid catalytic cracker; VB, visbreaker; IBP, initial boiling point; FBP, final boiling point.

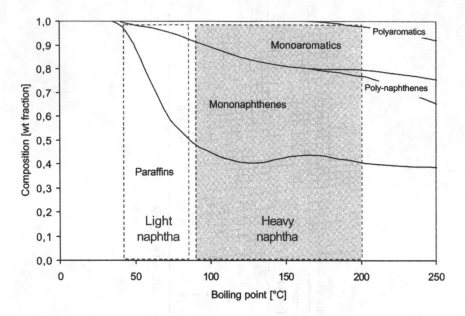

Figure 2 Hydrocarbon composition as a function of boiling point upon distillation of a North Sea crude.

more than 5% by weight. The sulfur tends to be more concentrated in the heavy end of the crude oil, which means that only ppm levels of sulfur are found in straight-run naphtha fractions. Still, even small concentrations are of great importance when it comes to processing the feedstock or using it directly as fuel. Sulfur poisons the noble-metal catalyst used in reforming and also promotes formation of undesirable SO_x during combustion. Cracker and coker naphthas originating from heavier oil fractions often contain much more sulfur, up to a few thousand ppm. Sulfur is removed from naphtha by hydrotreating, which means conversion to H_2S over a hydrotreating catalyst under hydrogen pressure. Hydrotreating is described more extensively in Chapter 4. The types of sulfur compounds found in crude oil are many: mercaptans, sulfides, disulfides, cyclic sulfides, alkylthiophenes, benzothiophenes, sulfates, traces of sulfuric acid, and sulfur oxides. In the naphtha boiling range thiophenes, noncyclic mercaptans and sulfides are the major groups. Identified sulfur compounds in naphtha are shown in Figure 3.

Organic **nitrogen** is present in even smaller concentrations than sulfur in the crude oil (<1.0 wt %) and mostly in the higher boiling point fractions. The compounds are usually classified as basic or nonbasic. Basic compounds are

Table 3 Hydrocarbon Composition[a] in a Straight-Run Naphtha from North Sea Crude, Identified by GC

Compound	Wt %	Compound	Wt %	Compound	Wt %	Compound	Wt %
2,4-Dm-Pentane	0.018	c-1,4-Dm-CyC6	0.914	C_9 naphthene 16	0.784	1-Me-2-Et-Benz	0.225
3,3-Dm-Pentane	0.078	n-Octane	5.263	C_9 naphthene 18	0.152	3-Et-Octane	0.094
2-Me-Hexane	2.287	iPr-CyC5	0.065	C_9 naphthene 20	0.269	C_{10} naphthene 10	0.014
2,3-Dm-Pentane	1.140	C_8 naphthene 6	0.074	C_9 naphthene 22	0.013	C_{10} naphthene 11	0.030
1,1-Dm-CyC5	0.716	c-2-Octane	0.066	C_9 naphthene 23	0.039	C_{10} paraffin 8	0.040
3-Me-Hexane	3.216	c-1,2-Et-Me-CyC5	0.154	C_9 naphthene 24	0.079	3-Me-Nonane	0.073
c-1,3-Dm-CyC5	1.742	2,2-Dm-Heptane	0.089	C_9 naphthene 26	0.052	C_{10} paraffin 9	0.021
t-1,3-Dm-CyC5	1.650	c-1,2-Dm-CyC6	0.279	C_9 naphthene 29	0.036	1,2,4-Tm-Benz	0.281
t-1,2-Dm-CyC5	3.328	2,2,3-Tm-Hexane	0.106	C_9 naphthene 31	0.070	C_{10} naphthene 14	0.067
C_7 Olefin 7	0.017	2,4-Dm-Heptane	0.276	n-Nonane	2.226	C_{10} naphthene 15	0.081
n-Heptane	7.885	4,4-Dm-Heptane	0.035	C_9 naphthene 32	0.062	i-But-CyC6	0.010
Me-CyC6	17.38	Et-CyC6	3.052	C_9 naphthene 33	0.046	C_{10} naphthene 16	0.012
1,1,3-Tm-CyC5	0.866	2-Me-4-Et-Hexane	0.038	iPr-Benzene	0.205	C_{10} naphthene 17	0.013
2,2-Dm-Hexane	0.105	2,6-Dm-Heptane	0.719	C_9 olefin 13	0.342	C_{10} naphthene 18	0.013
Et-CyC5	1.056	1,1,3-Tm-CyC6	0.918	C_9 naphthene 35	0.206	i-But-Benzene	0.037
2,2,3-Tm-Pentane	0.409	1,1,4-Tm-CyC6	0.136	iPr-CyC6	0.009	s-But-Benzene	0.055
2,4-Dm-Hexane	0.595	2,5-Dm-Heptane	0.394	2,2-Dm-Octane	0.067	n-Decane	0.258
ct-124-Tm-CyC5	0.990	3,5-Dm-Heptane	0.205	C_{10} paraffin 1	0.109	C_{10} naphthene 20	0.013
3,3-Dm-Hexane	0.137	C_9 naphthene 3	0.179	C_{10} paraffin 2	0.015	1,2,3-Tm-Benz	0.079
tc-123-Tm-Pentane	1.051	C_9 naphthene 4	0.078	C_9 naphthene 36	0.053	1,3-Me-iPr-Benz	0.087
2,3,4-Tm-Pentane	0.162	Ethylbenzene	1.265	n-Pr-CyC6	0.519	1,4-Me-iPr-Benz	0.132
Toluene	6.765	C_9 naphthene 5	0.226	C_{10} paraffin 3	0.090	C_{10} naphthene 22	0.119
1,1,2-Tm-CyC5	0.308	tt-1,2,4-Tm-CyC6	0.472	n-But-CyC5	0.096	Indane	0.070
2,3-Dm-Hexane	0.452	C_9 naphthene 7	0.050	C_{10} naphthene 2	0.074	C_{10} naphthene 24	0.016
2-Me-3-Et-Pentane	0.180	C_9 naphthene 8	0.031	C_{10} naphthene 3	0.020	C_{10} naphthene 25	0.030
2-Me-Heptane	2.741	m-Xylene	3.039	C_{10} naphthene 4	0.041	C_{11} paraffin 2	0.040

(Table continues)

Table 3 *Continued*

Compound	Wt %	Compound	Wt %	Compound	Wt %	Compound	Wt %
4-Me-Heptane	0.888	p-Xylene	0.927	3,3-Dm-Octane	0.250	n-But-CyC6	0.036
3,4-Dm-Hexane	0.123	2,3-Dm-Heptane	0.860	C10 paraffin 4	0.059	C10 naphthene 30	0.011
C8 naphthene 1	0.064	C9 naphthene 9	0.048	n-Pr-Benzene	0.278	1,3-De-Benzene	0.012
C8 naphthene 2	0.065	3,3-Dm-Heptane	0.090	C10 naphthene 5	0.065	1,3-Me-nPr-Benz	0.026
c-1,3-Dm-CyC6	2.904	4-Et-Heptane	0.105	2,6-Dm-Octane	0.145	1,4-Me-nPr-Benz	0.009
3-Me-Heptane	1.699	4-Me-Octane	0.433	C10 naphthene 7	0.039	n-But-Benzene	0.012
3-Et-Hexane	1.664	2-Me-Octane	0.570	1-Me-3-Et-Benz	0.383	13-Dm-5Et-Benz	0.009
1,1-Dm-CyC6	0.454	C9 naphthene 11	0.124	1-Me-4-Et-Benz	0.150	C10 naphthene 31	0.014
t-13-Et-Me-CyC5	0.374	3-Et-Heptane	0.157	C10 naphthene 9	0.063	1,2-Me-nPr-Benz	0.015
c13-Et-Me-CyC5	0.413	3-Me-Octane	0.571	1,3,5-Tm-Benz	0.166	14-Dm-2Et-Benz	0.015
t-12-Et-Me-CyC5	0.733	C9 naphthene 11	0.050	C10 paraffin 5	0.076	12-Dm-4Et-Benz	0.013
1-Me-1-Et-CyC5	0.107	o-Xylene	1.260	C10 paraffin 6	0.042	n-Undecane	0.018
t-1,2-Dm-CyC6	1.649	C9 naphthene 12	0.061	C10 paraffin 7	0.030		
cc-123-Tm-CyC5	0.023	C9 naphthene 14	0.023	4-Me-Nonane	0.025		

[a]Structures not fully identified are numbered according to type of compound and carbon number.

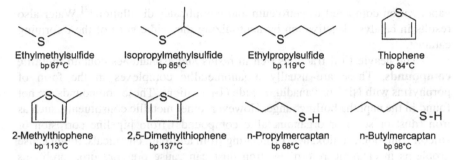

Figure 3 Identified sulfur compounds in naphtha.

pyridine, piperidine, or indoline derivatives whereas the nonbasic are pyrrole derivatives. Straight-run naphtha fractions usually contain sub-ppm concentrations of nitrogen, whereas cracker and coker naphthas may contain typically 10–100 ppm by weight. Nitrogen is poisonous to the reforming catalyst as it adsorbs strongly on its acidic sites. Common N-containing components in the naphtha boiling range are shown in Figure 4.

Oxygen-containing organic compounds are normally present only in the heavy fractions of the crude. These are phenols, furanes, carboxylic acids, or esters. The different acids account for the petroleum's acidity. High acidity can cause serious corrosion problems in the refinery. Little or no organic oxygen is found in the naphtha fractions.

Water is normally present in crude oil to some extent, partly dissolved in the oil and possibly as a separate water phase. Naphtha fractions will to some extent dissolve moisture during handling and storage. Water has a high heat of

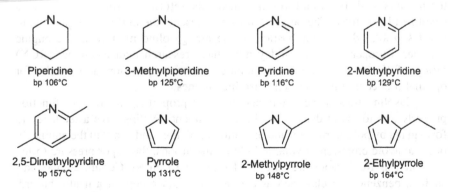

Figure 4 Identified nitrogen compounds in straight-run naphtha.

vaporization compared to petroleum and complicates distillation.[3] Water also results in catalyst deactivation by neutralizing the acidic sites of the reforming catalyst.

The heaviest oil fractions rich in resins and asphaltenes contain **metallic compounds**. These are usually organometallic complexes in the form of porphyrins with Ni^{2+} or vanadium oxide $(1+)$ cations. These compounds are not found in the naphtha boiling range. However, other metallic constituents, such as iron (dust or scale or organometallic compounds) from pipeline corrosion or silicon compounds (siloxanes) originating from antifoam chemicals, might cause problems in catalytic reforming. Iron dust can cause pressure drop problems whereas the silicon compounds adsorb onto and deactivate the reforming catalyst.

3 EFFECT OF NAPHTHA COMPOSITION ON PROCESS PERFORMANCE AND PRODUCT QUALITY IN CATALYTIC REFORMING

The hydrocarbon composition, the naphtha boiling range, and the concentration of *impurities* affect the quality of the reformate product. The same feedstock characteristics also influence the reforming process, including the performance and lifetime of the catalyst. In order to understand these relationships it is useful first to define some quality requirements of the product (gasoline specifications, octane ratings) and to describe briefly the reactions involved in the catalytic reforming process.

3.1 Gasoline Quality Requirements

The purpose of catalytic reforming is primarily to increase the octane number of the naphtha feedstock to a level that makes the reformate product suitable as a gasoline blend stock. The octane number represents the ability of a gasoline to resist *knocking* during combustion of the air–gasoline mixture in the engine cylinder. European gasoline today must have research octane number (RON) ratings of 95–98. Such high octane numbers allow compression ratios needed for optimal fuel economy of present gasoline engines.

Gasoline must have a number of other properties in order to function properly and to avoid damage to the environment. Olefins have a tendency to form gums by polymerization and oxidation of olefins, and can foul the engine. In order to avoid emission of volatile light hydrocarbons, the vapor pressure (often measured as Reid vapor pressure, RVP) must be limited. Certain compounds, such as benzene, are classified as carcinogenic and represent a health hazard. Tetraalkyllead has long been used as an octane booster, but will accumulate in

Table 4 Present Gasoline Specifications for the United States, Europe, and Japan[4-6]

Max values	USA	EU	Japan
RVP (kPa)	—	60	78
Sulfur (wppm)	50	150[a]	100
Oxygen (wppm)	2.2	2.7	—
Benzene (vol %)	1.0	1	—
Aromatics (vol %)	35	45	—
Olefins (vol %)	15	18	—
Lead (g/L)	—	0.005	—

[a]50 wppm from 2005.

nature, and is today strictly regulated and largely eliminated. Combustion of carbon leads to CO_2 (global warming problem) and poisonous CO. Combustion of sulfur and nitrogen (from air) leads to production of SO_x and NO_x that cause acid rain pollution. The volatile organic compounds produced during combustion of heavy aromatics are toxic in nature and are involved also in the photochemical reaction with NO_x to form ground-level ozone (smog). Exhaust catalysts have reduced emissions of NO_x to some extent, but present catalysts are sensitive to sulfur. Stringent regulations on the sulfur level of gasoline are therefore being developed. The present gasoline specifications (Table 4) set upper limits for the allowable concentrations of sulfur, benzene, olefins, and aromatics. Some countries have tax incentives for 50 or 10 ppm sulfur.

3.2 The Octane Number

In practice two octane ratings are measured, the research octane number (RON) and the motor octane number (MON), which differ in test procedure used. RON represents the engine performance at low speed whereas MON is representative for high-speed driving. By definition, the octane number of n-heptane is zero and the octane number of isooctane (2,2,4-trimethylpentane) is 100. The octane number for a gasoline is defined as the volume percent of isooctane in blending with n-heptane that equals the knocking performance of the gasoline being tested. Some gasoline components have octane numbers exceeding 100 and have to be characterized by use of mixtures. A common mixture contains 20% of the actual compound and 80% of an n-heptane/isooctane (40 : 60) mixture. A hypothetical blending octane number is then obtained by extrapolating from 20% to 100% concentration. The blending octane number is specific for the mixture and usually different from the octane number of the pure component, as seen for a range of different hydrocarbons with octane numbers less than 100 in Table 5.

Table 5 Pure and Blending[a] Research Octane Numbers of Hydrocarbons[7]

Hydrocarbon	RON pure	RON blending	Hydrocarbon	RON pure	RON blending
Paraffins			*Naphthenes*		
n-Butane	94.0	113	Cyclopentane	>100	141
Isobutane	>100	122	Cyclohexane	83.0	110
n-Pentane	61.8	62	Methylcyclopentane	91.3	107
2-Methylbutane	92.3	100	Methylcyclohexane	74.8	104
n-Hexane	24.8	19	t-1,3-Dimethylcyclopentane	80.6	90
2-Methylpentane	73.4	82	1,1,3-Trimethylcyclopentane	87.7	94
2,2-Dimethylbutane	91.8	89	Ethylcyclohexane	45.6	43
n-Heptane	0.0	0	Isobutylcyclohexane	33.7	38
3-Methylhexane	52	56	*Aromatics*		
2,3-Dimethylpentane	91.1	88	Benzene	—	98
2,2,3-Trimethylbutane	>100	112	Toluene	>100	124
n-Octane	<0	−18	Ethylbenzene	>100	124
3,3-Dimethylhexane	75.5	72	o-Xylene	—	120
2,2,4-Trimethylpentane	100.0	100	m-Xylene	>100	145
n-Nonane	<0	−18	p-Xylene	>100	146
2,2,3,3-Tetramethylpentane	>100	122	n-Propylbenzene	>100	127
n-Decane	<0	−41	Isopropylbenzene	>100	132
Olefins			1-Methyl-3-ethylbenzene	>100	162
1-Hexene	76.4	96	1,3,5-Trimethylbenzene	>100	170
1-Heptene	54.5	65	n-Butylbenzene	>100	114
2-Methyl-2-hexene	90.4	129	1-Methyl-3-isopropylbenzene	—	154
2,3-Dimethyl-1-pentene	99.3	139	1,2,3,4-Tetramethylbenzene	>100	146

[a]Obtained using a 20% hydrocarbon – 80% 60:40 mixture of isooctane and n-heptane.

Table 5 shows that aromatics generally have much higher octane numbers than naphthenes, olefins, and paraffins and are therefore desired reformate hydrocarbon components. The octane number of the aromatics (except for benzene) is always above 100. Straight-chain paraffins have very low octane numbers (RON < 0 for *n*-octane and *n*-nonane), but the octane number increases markedly with the degree of branching (RON > 100 for 2,2,3-trimethylbutane). Light olefins and naphthenes generally have higher RON than the paraffins, but as for the *n*-paraffins the octane number declines as the number of carbon atoms increases. This decline is much less pronounced for the isoparaffins. Considering the boiling range of gasoline (C_5-C_{12} hydrocarbons) and the above comparison, visualized in Figure 5, an increase in the octane number of the reformate can best be obtained by transformation of naphthenes into aromatics and of linear paraffins into branched paraffins or aromatics. These transformations are the key reactions of the catalytic reforming process.

3.3 Catalytic Reforming Process

Catalytic reforming is carried out at elevated temperature (450–520°C) and moderate pressure (4–30 bar). By use of a proper catalyst in three or four serial reactors and in the presence of hydrogen (H_2/oil equal to 4–6 mol/

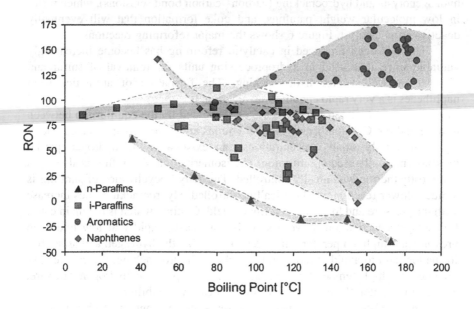

Figure 5 Octane numbers vs. boiling point for hydrocarbon families.[7,8]

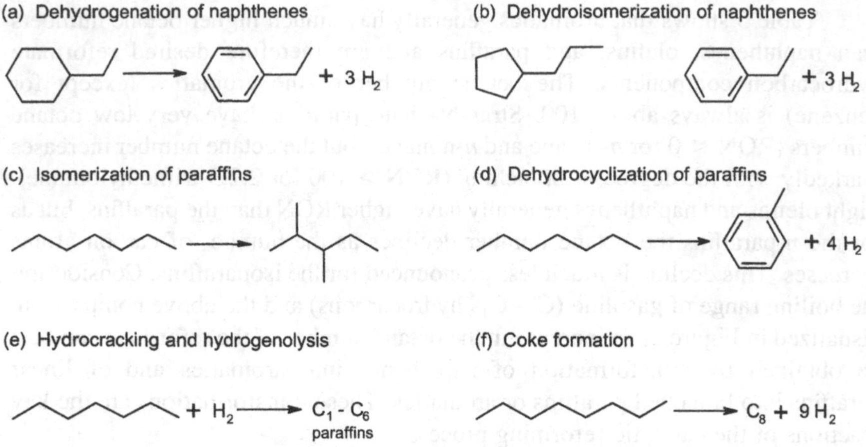

(a) Dehydrogenation of naphthenes

(b) Dehydroisomerization of naphthenes

(c) Isomerization of paraffins

(d) Dehydrocyclization of paraffins

(e) Hydrocracking and hydrogenolysis

(f) Coke formation

Figure 6   Major reactions in catalytic reforming of naphtha.

mol), naphthenes are transformed into aromatics by dehydrogenation and straight-chain paraffins into branched paraffins by isomerization. Paraffins also undergo dehydrocyclization to form aromatics. Other important reactions are hydrogenolysis and hydrocracking (carbon–carbon bond scissions), which result in low molecular weight paraffins, and coke formation that will eventually deactivate the catalyst. Figure 6 shows the major reforming reactions.

The hydrogen produced in catalytic reforming has become increasingly valuable since it is used in hydroprocessing units for removal of sulfur and nitrogen as well as for hydrocracking. The formation of aromatics from naphthenes is a very rapid endothermic reaction. It is thermodynamically favored by high temperature and low pressure, as illustrated by the equilibrium between toluene and the C_7 naphthenes (Fig. 7). Olefins are readily hydrogenated and at equilibrium only small concentrations can exist with the hydrogen partial pressures normally used in reforming. The isomerization of paraffins is also rapid and mostly thermodynamically controlled. The dehydrocyclization of paraffins is a much slower reaction and kinetically controlled. Hydrocracking rates increase with the pressure and lower the reformate yield. Coking, which is the main cause for catalyst deactivation, is very slow but increases rapidly at low hydrogen pressure and high temperature. In order to optimize the hydrogen and aromatics formation, and to avoid severe yield loss due to hydrocracking, the choice is to operate at a high temperature and at the lowest possible hydrogen pressure, although the latter always is a trade-off with catalyst stability.

The catalyst is bifunctional in the sense that it contains both a metallic function (platinum) that catalyzes dehydrogenation reactions and an acidic

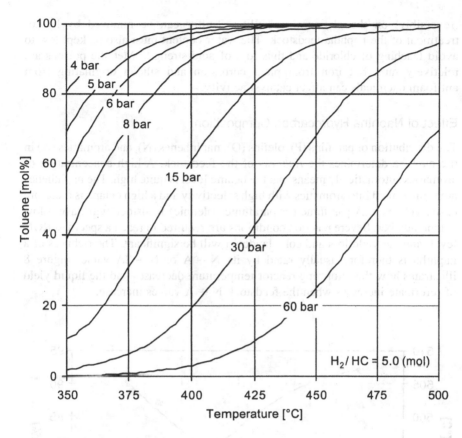

Figure 7 Effect of temperature and pressure on the concentration of toluene in thermodynamic equilibrium with H_2 and C_7 naphthenes.[9]

function (chlorided alumina) that catalyzes isomerization reactions. Platinum, which is usually used with a second metal, needs to be highly dispersed on the acidic carrier in order to maintain high activity and selectivity throughout a commercial cycle. In units designed for periodic regeneration of the catalyst (semiregenerative reforming), a cycle typically lasts 1–2 years. Most new units are designed with continuous catalyst regeneration implying that each catalyst particle has a cycle time of typically 6–8 days between regenerations. Two catalyst formulations prevail commercially: $Pt-Re/Al_2O_3$ and $Pt-Sn/Al_2O_3$. The former is the most stable and is preferred in semiregenerative units, whereas the latter has the highest selectivity at low pressure and is the best choice in continuous reforming units. These catalysts are sensitive to sulfur which adsorbs (reversibly) on the platinum crystallites. Sulfur can be removed by hydro-treatment of the naphtha feedstock. The water content must also be kept low to avoid leaching of chloride and thus loss of acid strength. Metallic poisons are relatively rare, but iron from plant corrosion and silicon originating from antifoam chemicals can affect catalyst activity.

Effect of Naphtha Hydrocarbon Composition

The distribution of paraffins (P), olefins (O), naphthenes (N), and aromatics (A) in the naphtha determines the *richness* of the feedstock. A high concentration of aromatics automatically means that the octane level is quite high. The naphthenes are transformed into aromatics with high selectivity and a high octane is therefore easily achieved. A paraffinic (or paraffinic–olefinic) feedstock will have a low octane number. Severe reaction conditions are required to reach a specified RON level, and the yield loss and coke laydown will be significant. The richness of a naphtha is therefore usually rated by its N + A or N + 2A value. Figure 8 illustrates how the reforming reactor temperature decreases and the liquid yield of reformate increases when the feedstock N + A values increase.

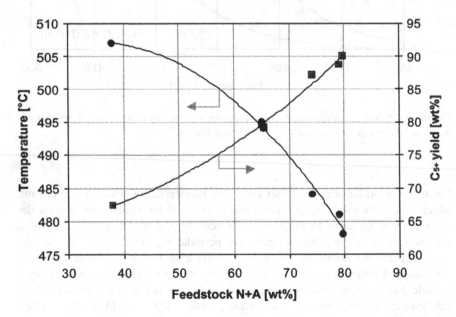

Figure 8 Reactor temperature and reformate yield as a function of naphtha N + A (naphthenes + aromatics) value. Reaction conditions: 100 RON, $P = 30$ bar, WHSV = 2.0 h^{-1}, and H$_2$/HC = 4.5.

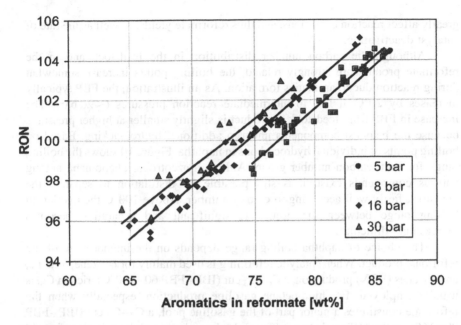

Figure 9 RON as a function of aromatics concentration from a number of pilot experiments using a range of different naphthas and variable reaction conditions.

The hydrocarbon composition in the naphtha does not affect the reformate composition much. The reformate consists mainly of paraffinic and aromatic hydrocarbons since the large part of the naphthenes is consumed in the reaction. There is a near-linear relationship between the RON value and the concentration of aromatics (Fig. 9). Thus, regardless of feedstock composition, when operating with a constant RON level in the product, the aromatic and paraffin concentrations are usually fixed within narrow limits. However, as shown in Figure 9, the RON–aromatics relationship changes somewhat with reaction pressure. At elevated pressures the concentration of high-octane cracked products (C_5 and C_6 isoparaffins) increases, and subsequently less aromatics are required to reach a specified RON in the product.

Effect of Naphtha Boiling Range

The boiling range of the naphtha feedstock is a key factor in catalytic reforming. The initial and final boiling points (IBPs and FBPs) and the boiling point distribution not only determine the carbon number distribution of the product but greatly affect reaction conditions, and thus reformate yields, as well as the rate of catalyst deactivation.

Although the carbon number distribution in the feedstock and in the reformate product are strongly related, the boiling points increase somewhat during reaction due to aromatics formation. As an illustration, the FBP typically increases by 20°C at low to intermediate reaction pressures (<20 bar). The increase in FBP from feedstock to product is slightly smaller at higher pressures because the heaviest components undergo additional hydrocracking. Based on boiling points of individual hydrocarbons in naphtha, Figure 10 shows the boiling range for each carbon number group. Although azeotropic phenomena among various compounds exist, it is still possible by distillation to separate the feedstock fairly well according to carbon number. Above 100°C the overlap in boiling range between the groups is significant and separation becomes increasingly difficult.

The choice of naphtha boiling range depends on the intended use of the reformate product. When catalytic reforming is used mainly for benzene, toluene, and xylenes (BTX) production, a C_6-C_8 cut (IBP–FBP 60–140°C), rich in C_6, is usually employed. For high-octane gasoline production, especially when the reformate constitutes a major part of the gasoline pool, a C_7-C_9 cut (IBP–FBP 90–160°) is the preferred choice. The C_6 hydrocarbons may be removed to avoid the benzene in the naphtha and to avoid further benzene formation from the C_6 naphthenes. The benzene yield from cyclohexane and methylcyclohexane (primary production) is significant, as illustrated in Figure 11. These reactions are controlled by thermodynamics and favored by low pressure. Benzene is also formed by dealkylation of heavier aromatics (secondary production). This reaction is kinetically controlled and favored by high temperature and low space velocity (Fig. 12).

The benzene selectivity from substituted aromatics increases with the length of the side chain (*n*-butylbenzene > *n*-propylbenzene) and with the degree

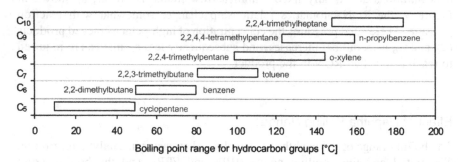

Figure 10 Boiling range of naphtha hydrocarbons grouped by carbon number.[8]

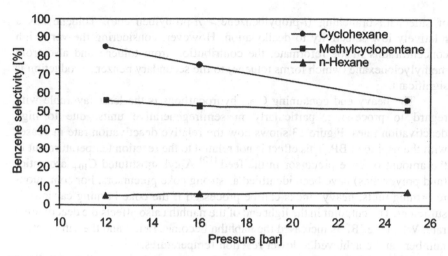

Figure 11 Benzene selectivity (percentage of the components feed concentration found as wt % benzene yield) vs. pressure in semiregenerative reforming with PtRe catalyst. RON = 101.

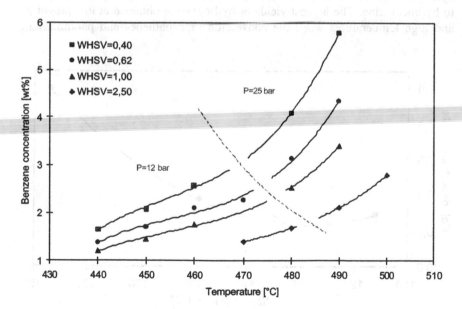

Figure 12 Benzene in reformate as a function of reaction temperature and space velocity using a feedstock with 0.23 wt % C_6 hydrocarbons.

of sidechain branching (*i*-propylbenzene > *n*-propylbenzene). Toluene has a relatively low selectivity to dealkylation. However, considering the very high concentrations in the reformate, the contribution from toluene and also from methylyclohexane (which forms toluene) to the secondary benzene production is significant.

The heavy end containing C_{10+} hydrocarbons is the least favorable with regard to processing, particularly in semiregenerative units, due to high deactivation rates. Figure 13 shows how the relative deactivation rate increases with the naphtha FBP. This effect is not related to the reaction temperature but to the amount of coke precursor in the feed.[10] Alkyl-substituted C_{10+} aromatics (and polycyclics) have been identified as strong coke precursors. For continuous reforming units, heavy stocks can be processed if the coke burning capacity is sufficient. The cutpoint in the light end of the naphtha also affects the deactivation rate. When the IBP is increased the naphtha becomes richer and the same octane number can be achieved at lower reaction temperatures.

The legislative requirements for sulfur removal from gasoline and diesel have increased hydrogen use in the refineries; hence, refiners are looking for ways to maximize their hydrogen yields. The optimal feedstock to a reformer with respect to hydrogen yield is a C_6–C_9 cut that contains the highest naphthene concentrations. No hydrogen can be produced from the C_5 fraction and little hydrogen is produced from the C_{10+} hydrocarbons, which are highly susceptible to hydrocracking. The highest yields of hydrogen are obtained at low pressures and high temperatures when the conversion of naphthenes and paraffins into

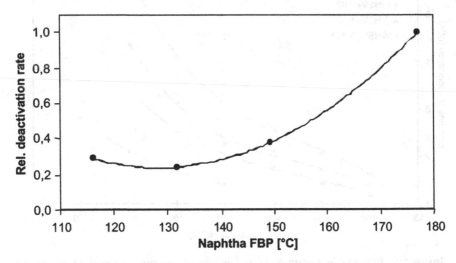

Figure 13   The deactivation rate (measured as the temperature rise needed to maintain 102.4 RON relative to a base naphtha) as a function of final boiling point (FBP).[10]

aromatics is high. The temperature must, however, be kept below a point when hydrocracking becomes important, which would lower the yield of both hydrogen and reformate. The octane number during maximized hydrogen production is typically in the order of 102–105 RON, and it follows that the deactivation rate is high.

Effect of Naphtha Sulfur Content

Reforming catalysts are sensitive to sulfur impurities in the naphtha feedstock. The surface platinum atoms of the catalyst convert the sulfur compounds into H_2S molecules that readily adsorb onto the surface metal atoms. The poisoned platinum atoms are no longer active and the temperature must be increased to maintain RON (i.e., produce aromatics by dehydrogenation). Reformate yield decreases somewhat due to the temperature rise but not as much as normally observed. This is due to the reduced methane production by the metal-catalyzed hydrogenolysis reaction. However, the rate of deactivation increases according to the temperature increase. The adsorption of sulfur is strong but reversible. A given sulfur concentration in the feedstock results in a specific sulfur coverage. However, if the sulfur is removed from the naphtha, the activity will eventually return to very near the initial level as shown in Figure 14.

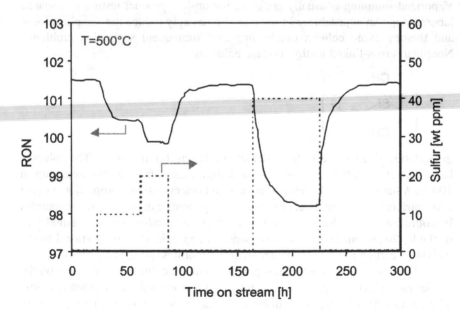

Figure 14 Effect of sulfur upset on RON level at $T = 500°C$, $P = 16$ bar, WHSV = 2.0 h^{-1}, and $H_2/HC = 4.3$.

4 ANALYSIS METHODS

The high complexity of naphtha and reformate fractions requires advanced techniques to obtain a complete compositional analysis and to determine the chemical and physical parameters needed for the refiner. Many different approaches exist, and the choice of analytical method depends on the needed resolution, the analysis time, and the cost. For industrial products that must meet defined specifications, refiners are required to follow standardized analysis procedures. The American Society for Testing and Materials (ASTM) is one of several recognized organizations for standardization. This chapter will concentrate on the most common methods for determining hydrocarbon composition, distillation range, octane numbers, and sulfur/nitrogen contents. Examples of both ASTM and nonstandardized methods are included.

4.1 Hydrocarbon Composition

The most powerful and widely used technique for analysis of hydrocarbons in naphthas or reformates is gas chromatography (GC). This is a separation method in which the sample is injected into a carrier gas stream, usually helium, and brought through a dedicated capillary column allowing transport of the different molecules at different rates (Fig. 15). The samples may be gaseous or liquid. Vaporized sampling is usually preferred for on-line product testing in research laboratories. An adjustable split injector can strongly reduce the sample amount and thereby avoid column overloading and subsequent separation problems. Nonpolar, cross-linked methylsiloxane columns

$$\left(\begin{array}{c} CH_3 \\ | \\ -Si-O- \\ | \\ CH_3 \end{array} \right)_n$$

give elution times close to the order of increasing boiling point. The columns have diameters of 0.1–0.5 mm and the length ranges from a few meters up to 100 m. A flame ionization detector creates and detects a signal proportional to the concentration of each hydrocarbon as the components exit the column. It operates by collecting (by an electrode) the ions of the flame produced during combustion of the hydrocarbon. The detector response is approximately proportional to the weight of carbon present,[11] which greatly simplifies quantitative analysis.

The rate of hydrocarbon transport through the column is dependent on the carrier gas velocity, adjusted for the injector pressure and the oven temperature. The lightest hydrocarbons (methane and ethane) are transported very quickly through the column and separation requires low temperature (ambient). On the

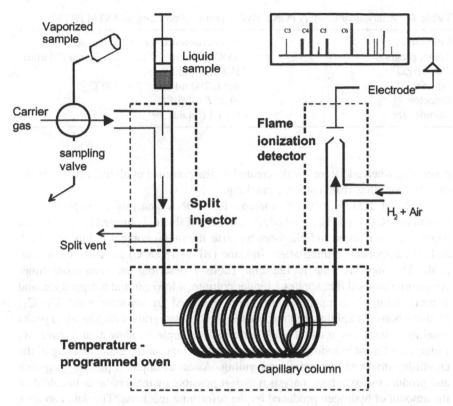

Figure 15 Schematic illustration of gas chromatography (GC) with gas/liquid sampling, split injection, and flame ionization detection (FID).

other hand, the heaviest aromatics need a temperature of 200°C or more in order not to adsorb strongly at the column front. Thus, an advanced temperature program and column pressure selection is required to optimize separation and time consumption of a GC analysis. The column material and length, the detector temperature, the carrier gas type, and the split flow rate also affect the separation.

Gas chromatography is not an identification method. In order to identify the large number of peaks in the chromatogram, the system must be calibrated. This can best be obtained by coupling a mass spectrometer to the column exit of an identical chromatographic setup (gas chromatography–mass spectrometry, GC-MS). Most of the resolved peaks are identified from MS spectra libraries. The equipment is costly and such an analysis is time consuming, but a good 'peak library' for the GC user is obtained given that the column separation is good. In practice, the 'heavy region' of the chromatogram is never fully resolved,

Table 6   Analysis of C_5-C_{12} PONA Hydrocarbons According to ASTM D5134

Column	50 m cross-linked methylsiloxane
Temp. program	35°C (30 min) → 200°C, 2°C/min (20 min)
Carrier gas	Helium, 215 kPa
Injector	Split, 200 ml/min; $T = 200$°C
Detector	FID; $T = 250$°C
Sample size	0.1 μl (liquid)

especially when additional peaks created by the presence of olefins exist, as is the case for naphthas from catalytic cracking.

ASTM D5134 is a GC method for PONA analysis in naphthas and reformates (C_5-C_{12}). The method, described in Table 6, is limited to straight-run naphthas, reformates, and alkylates because the olefin content is limited to 2% and all components eluting after *n*-nonane (BP > 150.8°C) are collected as one peak. The analysis time is 122 min. Table 7 describes an even more time-consuming method that applies a longer column, a lower initial temperature, and a more complex temperature program designed to separate most C_1-C_{12} hydrocarbons in naphthas and reformates. A chromatogram with identified peaks obtained using this method on a reformate sample is shown in Figure 16. Detection of most individual compounds is important for the understanding of the chemistry involved in catalytic reforming. As an example, a precise feedstock and product hydrocarbon analysis makes it possible by mass balance to calculate the amount of hydrogen produced by the reforming reactions. The data can also, based on simple models, be used to calculate density, vapor pressure, carbon and hydrogen content, and octane numbers. For the process engineer it is often sufficient to know the PONA group concentrations in order to verify the feedstock or product qualities, and the least time-consuming GC methods are chosen. Specialized methods for more precise analysis of single compounds are available.

Table 7   Comprhensive Laboratory Analysis for C_1-C_{12} PONA Hydrocarbons

Column	100 m cross-linked methylsiloxane
Temp. program	30°C (30 min) → 50°C, 1°C/min (10 min) → 140°C, 2°C/min (0 min) → 250°C, 10°C/min, (30 min)
Carrier gas	Helium, 300 kPa
Injector	Split, 800 ml/min; $T = 250$°C
Detector	FID; $T = 280$°C
Sample size	0.2 μl (liquid)

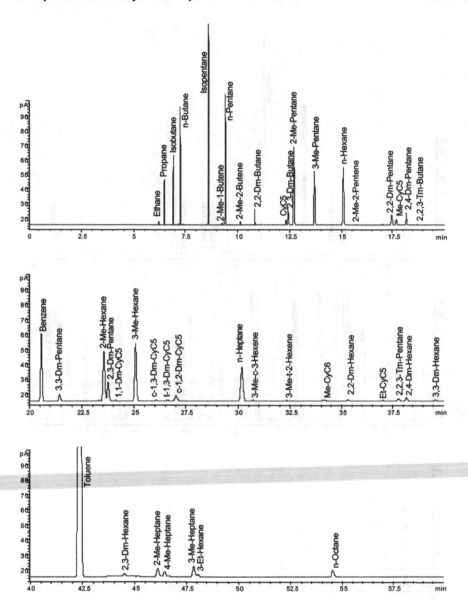

Figure 16 Chromatogram of reformate (liquid sample) using the GC method listed in Table 7. Page 1 of 2.

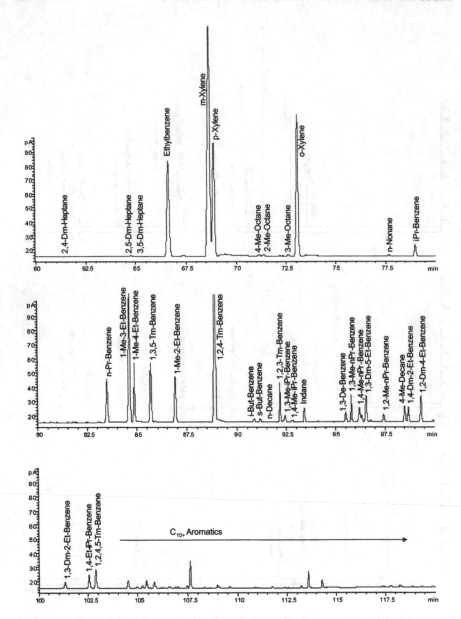

Figure 16 Continued.

4.2 Distillation Range

Knowledge about the boiling point distribution of gasolines is most frequently obtained by distillation according to ASTM D86. A batch distillation is conducted at atmospheric pressure and the resulting curve shows the temperature as a function of percent volume distilled. Automated instruments perform the measurement. Figure 17 shows an ASTM D86 distillation curve and tabulated values for a reformate sample.

Another way of analyzing the boiling range characteristics is to simulate the distillation by use of GC. By using an inert column stationary phase, the components elute in order of their boiling points. The ASTM D3710 method is specialized for gasoline fractions and gives the result within 15 min.

4.3 Sulfur and Nitrogen Analysis

The fact that only small concentrations of sulfur and nitrogen poison reforming catalysts calls for highly accurate analysis methods, capable of measuring down to sub-ppm levels. Non-hydrotreated naphthas from thermal or catalytic cracking

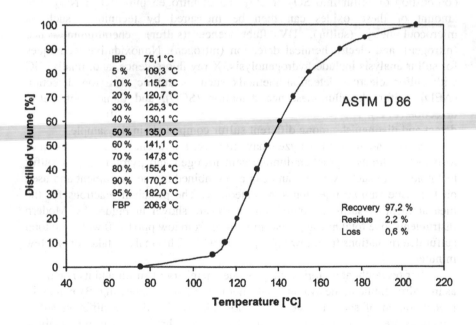

IBP	75,1 °C
5 %	109,3 °C
10 %	115,2 °C
20 %	120,7 °C
30 %	125,3 °C
40 %	130,1 °C
50 %	135,0 °C
60 %	141,1 °C
70 %	147,8 °C
80 %	155,4 °C
90 %	170,2 °C
95 %	182,0 °C
FBP	206,9 °C

ASTM D 86

Recovery 97,2 %
Residue 2,2 %
Loss 0,6 %

Figure 17 Results for an ASTM D86 distillation of a reformate sample.

Table 8 Sulfur and Nitrogen Analysis Methods

Target	Technique	ASTM	Range (wppm)[a]
N	OC/chemiluminescence	D4629	0.3–100
N	OC/electrochemical detection	D6366	0.05–100
S	OC/microcoulometry	D3120	3–100
S	OC/UV fluorescence	D5453	1–8000
S	Hydrogenolysis	D4045	0.02–10
S	X-ray fluorescence	D4294	>1000
S	GC/selective sulfur detector	D5623	0.1–100[b]

[a]Analytical range suggested by ASTM method.
[b]Concentration range of each individual sulfur compound.
OC, oxidative combustion.

processes may reach percent levels of sulfur and 100 ppm levels of nitrogen. Thus, versatile analysis methods covering sulfur and nitrogen from ppb up to percent levels are needed. A large number of methods and instrument types are available as shown in Table 8. Most analysis techniques are based on initial combustion of sulfur into SO_2 or SO_3 and of nitrogen into NO or NO_2. The amount of these oxides can then be measured by techniques such as microcoulometry (sulfur), UV fluorescence (sulfur), chemiluminescence (nitrogen), and electrochemical detection (nitrogen). Nonoxidative techniques for sulfur analysis include hydrogenolysis, X-ray fluorescence, and, finally, GC with sulfur-selective detection methods such as atomic emission detection (AED), sulfur chemiluminescence detection (SCD), and flame photometric detection (FPD). The GC technique not only measures total sulfur but may also detect and distinguish among different sulfur compounds in the sample.

Sulfur and nitrogen analyzers have improved in recent years when it comes to detection limits. Pyrochemiluminescent nitrogen and pyrofluorescent sulfur technology are such examples and can be combined in one instrument and used on the same sample injection simultaneously. The principal reactions for the measurement of sulfur by pyrofluorescence are shown in Figure 18. Modern instruments give total nitrogen determinations from low ppb to 20 wt % and total sulfur determinations from low ppb up to 40 wt %. The analysis takes only a few minutes.

For research laboratories studying the chemistry of sulfur and its reactions, as in hydrotreatment, the available GC methods are most appealing. By extensive precalibration of such a system it is possible to identify the different sulfur structures present in the sample. Figure 19 shows a chromatogram from analysis of a cracker naphtha using AED. Integration of all peaks in the chromatogram

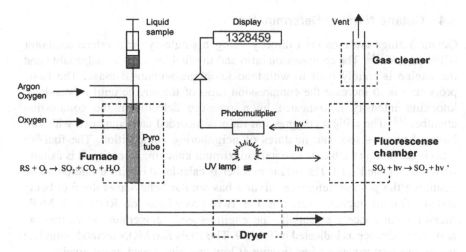

Figure 18 Schematic illustration of total sulfur analysis by pyrofluorescence method.

yields the total sulfur concentration. Very good comparisons have been measured between total sulfur analysis by GC/AED and sulfur analysis by pyrofluorescence.[12] The new SCD instruments[13] have extremely high sensitivity and are the choice for low-sulfur samples.[14]

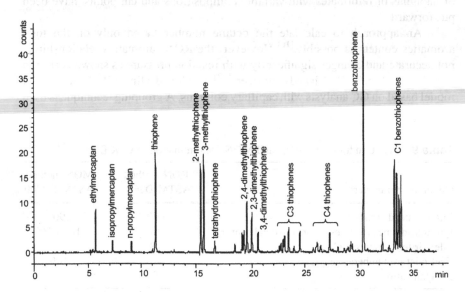

Figure 19 GC/AED chromatogram showing the sulfur distribution in full-range catalytic cracker naphtha.

4.4   Octane Number Determination

Octane ratings are measured directly using a single-cylinder reference motor (CFR engine).[15] The compression ratio and the fuel/air ratio are adjustable and the engine is solidly built to withstand knocking without damage. The basic procedure is to increase the compression ratio of the engine until a 'standard' knocking intensity is indicated by a pressure detector in the combustion chamber.[15] The critical compression ratio is recorded and compared with two binary heptane–isooctane mixtures of neighboring composition. The fuel/air ratio is adapted in each case to obtain maximum knocking intensity; it is usually between 1.05 and 1.10. The octane number is calculated by linear interpolation, assuming the primary reference mixture has similar behavior as the fuel being tested. The distinctions between the two procedures of RON and MON measurement concern essentially the engine speed, temperature of admission, and spark advance as indicated in Table 9. The RON and MON methods simulate the engine performance when driving at low and high speed, respectively.

An alternative method for determination of the octane number of a gasoline is by means of calculation, using the hydrocarbon composition from GC analysis as input data. It is not a straightforward task to develop such a model because blending of different individual hydrocarbons does not result in an engine knocking performance as expected from the octane numbers of the individual components. Advanced models, both linear and nonlinear and based on a number of naphthas or reformates with variable compositions and cut points, have been put forward.

An approach to calculate the octane number based only on the total aromatics content is possible.[16] However, the RON–aromatics relationship is not accurate and changes significantly with reaction pressure as shown earlier in this chapter (Fig. 9). Walsh and coworkers[17] developed a linear RON calculation model based on GC analysis with capillary columns. A grouping technique is used

Table 9   Test Conditions for RON and MON Determination in CFR Engines[15]

Operating parameters	RON method ASTM D2699	MON method ASTM D2700
Engine speed (rpm)	600	90
Ignition advance (degrees before top dead center)	13	14 to 26[a]
Inlet air temperature (°C)	48	—
Fuel mixture temperature (°C)	—	149
Fuel/air ratio	b	b

[a]Variable with the compression ratio.
[b]Adapted in each case to obtain maximum knocking intensity.

to produce a manageable number of pseudocompounds. Thirty-one groups were defined by the order of elution in the GC chromatogram and given a regression coefficient (b_r) for calculation of RON after the simple equation $RON = \sum (b_r W_r)$, where W_r is the weight fraction of group r. Durand and coworkers[18] have demonstrated the versatility of this RON model using 60 different gasoline samples that were analyzed by GC and rated by ASTM engine tests. The difference in RON values turns out to be less than 1 RON unit in most cases. The defined model groups with regression coefficients are listed in Table 10.

Table 10 Group Definition and Regression Coefficients of Linear RON Model Developed by Walsh and Coworkers[17]

Group no.	Group definition by GC elution times	Regression coefficient
1	Components eluting before *n*-butane	103.9
2	*n*-Butane	88.1
3	Components eluting between *n*-butane and isopentane	144.3
4	Isopentane	84.0
5	Components eluting between isopentane and *n*-pentane	198.2
6	*n*-Pentane	67.9
7	Components eluting between *n*-pentane and 2-methylpentane	95.2
8	2- and 3-Methylpentane and components eluting between these	86.6
9	Components eluting between 3-methylpentane and *n*-hexane	95.9
10	*n*-Hexane	20.9
11	Components eluting between *n*-hexane and benzene	94.9
12	Benzene	105.2
13	Components eluting between benzene and 2-methylhexane	113.6
14	2- and 3-Methylhexane and components eluting between these	80.0
15	Components eluting between 3-methylhexane and *n*-heptane	97.8
16	*n*-Heptane	−47.8
17	Components eluting between *n*-heptane and toluene	62.3
18	Toluene	113.9
19	Components eluting between toluene and 2-methylheptane	115.1
20	2- and 3-Methylheptane and components eluting between these	81.7
21	Components eluting between 3-methylheptane and *n*-octane	109.7
22	*n*-Octane	10.5
23	Components eluting between *n*-octane and ethylbenzene	96.1
24	Ethylbenzene	122.6
25	Components eluting between ethylbenzene and *p*-xylene	45.4
26	*p*-xylene + *m*-xylene	102.0
27	Components eluting between *m*-xylene and *o*-xylene	73.3
28	*o*-Xylene	123.6
29	Components eluting after *o*-xylene up to and including *n*-nonane	35.0
30	Components eluting between *n*-nonane and *n*-decane	112.0
31	*n*-Decane and components eluting after *n*-decane	85.6

Complex, nonlinear models in which the deviation from ideality (as expressed by the regression coefficients) of each component or component group is set as a function of the concentrations of the different hydrocarbon families can reduce the error of calculation to less than 0.5 RON unit. Such models will be especially useful for more complex gasolines in which the concentration of nonreformate material (alkylates, isomerates, cracker naphtha, polymerate, alcohols, and ethers) is high.

A fast and simple alternative to the previously described methods for octane number determination was proposed by BP[19] and involves the use of infrared (IR) spectroscopy. The near-IR region of the spectrum (wavelength: 800–2500 nm) contains many bands that result from overtones and combinations of carbon–hydrogen stretching vibrations, which are particularly useful for analyzing gasoline (Fig. 20). The variations in IR spectra can be coupled to a range of gasoline properties including RON and MON numbers. Automated and computerized instruments offer fast (1 min) analysis and have the possibility of

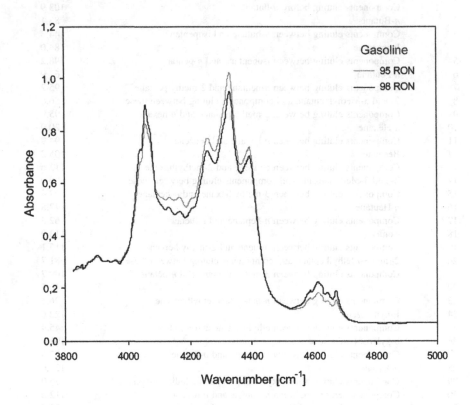

Figure 20   Near-IR absorbance spectra of two different gasolines.

on-site sampling. The error of calculation is not significantly higher than for the composition–octane models derived from GC analysis.

REFERENCES

1. Speight, J.G. *The Chemistry and Technology of Petroleum*, 3rd Ed.; Chemical Industries Vol. 3; Marcel Dekker: New York, 1999.
2. http://www.statoil.com (Products and Services/Crude Oil and Condensate).
3. Parera, J.M.; Figoli, N.S. In *Catalytic Naphtha Reforming*, 1st Ed.; Chemical Industries Vol. 61; Marcel Dekker: New York, 1995.
4. Martino, G. Catalysis for oil refining and petrochemistry: recent developments and future trends. In *Studies in Surface Science and Catalysis*; Corma, A., Melo, F.V., Mendioroz, S., Fierro, J.L.G., Eds.; *Proceedings of the 12th ICC*, Granada, Spain, July 9–14, 2000; Vol. 130A; Elsevier: Amsterdam, 2000; 83–103.
5. Hartman, E.L.; Hanson, D.W.; Weber, B. Hydrocarbon Proc. **1998**, 77.
6. http://www.paj.gr.jp/html/english/index.html (Petroleum Association of Japan, Annual Review 1999).
7. American Institute Research Project 45, 16th annual report, 1954.
8. Weast, Ed. *Handbook of Chemistry and Physics*, 58th ed.; CRC Press: Boca Raton, 1978.
9. Gjervan, T.; Prestvik, R.; Holmen, A. In *Basic Principles of Applied Catalysis*; Baerns, M., Ed.; *in press*.
10. Moljord, K.; Grande, K.; Tanem, I.; Holmen, A. In *Deactivation and Testing of Hydrocarbon-Processing Catalysts*; O'Connor, P., Takatsuka, T., Woolery, G.L., Eds.; ACS Symposium Series No. 634, 1995; 268–282.
11. Dietz, W.A. J. Gas Chromatogr. **1967**, 5, 68.
12. Steiner, P.; Myrstad, R.; Thorvaldsen, B.; Blekkan, E., *in preparation*.
13. Benner, R.L.; Stedman, D.H. Anal. Chem. **1989**, 61, 1268.
14. Adlard, E.R. Ed. *Chromotography in the refining industry*. J. Chromatogr. Lib., Vol. 56; Amsterdam, 1995.
15. Wauquier, J.-P. Ed. *Petroleum Refining 1, Crude Oil, Petroleum Products, Process Flowsheets*; IFP Publications, Editions Technip: Paris, 1994.
16. McCoy, R.D. ISA AID 73442, 187, 1973.
17. Anderson, P.C.; Sharkey, J.M.; Walsh, R.P. J. Inst. Petr. **1972**, 58 (560), 83.
18. Durand, J.P.; Boscher, Y.; Petroff, N. J. Chromatrogr. **1987**, 395, 229.
19. Descales, B.; Lambert, D.; Martens, A. *Determination des nombres d'octane RON et MON des essences par la technique proche infrarouge*, Revue de l'Association Francaise des Techniciens du Petrole, No 349, 1989.

2
Basic Reactions of Reforming on Metal Catalysts

Zoltán Paál
Hungarian Academy of Sciences, Budapest, Hungary

1 INTRODUCTION

Since the first industrial application of reforming for fuel upgrading using supported Pt catalysts, this large-scale commercial process has proved to be a driving force for research of metal-catalyzed hydrocarbon reactions. Laboratory studies, which frequently employed conditions vastly different from industrial ones, provided a scientific background for catalytic reforming, and these apparently remote investigations prepared the ground for several industrially important innovations in the past, and will do so in the future, too. This chapter concentrates on a few points of laboratory-scale studies that might be of value for industry.

Several catalytic reactions of reforming involve the rearrangement of the hydrocarbon skeleton; hence, they can be termed as 'skeletal reactions': aromatization, isomerization, C_5 cyclization, and hydrogenolysis. The first three reactions are 'useful' or value enhancing, the last one 'disadvantageous' for operation of a reforming plant, since products of lower value are produced.

This chapter concentrates on metal catalysts and mechanisms of reactions catalyzed by them. Relevant problems and the numerous hypotheses suggested for their solution will be pointed out rather than by presenting ready and apparently finalized theories. Interactions between metallic and support sites will also be mentioned. The diversity of ideas, methods, approaches, etc., reflects truly the present situation, where the experimental results as a function of several parameters lack well-established and generally valid interpretations. This is the reason why a relatively high number of references has been included; still, the literature covered is far from being comprehensive. Most of the basic information included in the first edition of this book[1] has been retained, although several recent references have been added.

2 POSSIBLE MECHANISMS OF THE REACTIONS

The chemistry of the industrial reforming process has been extensively reviewed.[2] All the valuable information from results obtained in the 1960s and 1970s will not be repeated here. Another, more concise review dealing with both chemistry and industrial aspects was published in 1991.[3] The excellent book by Olah and Molnár[4] summarized all relevant hydrocarbon reactions. *Every* reaction important in reforming (aromatization, C_5 cyclization, isomerization, and fragmentation) can also proceed with catalysts possessing metallic activity only. This feature will be stressed in the present chapter. Laboratory measurements are often carried out in the temperature range of 500–650 K and pressures up to 1 bar, being much lower than the conditions of industrial reforming. Yet these studies will be useful in understanding underlying phenomena.

Aromatization (or C_6 dehydrocyclization) was first observed by a Russian group as the formation of a second aromatic ring from an alkylbenzene on monofunctional Pt/C catalyst; the same group reported also the formation of an aromatic C_6 ring from alkanes.[5] Later they described the metal-catalyzed C_5 *cyclization* of alkanes to alkylcyclopentanes.[6] The aromatic ring is very stable under these conditions but C_5 cyclization is reversible: a ring opening of the C_5 ring to alkanes also takes place.[7] Metal-catalyzed *isomerization* [8] may occur (1) via the formation and splitting of the C_5 ring;[9] (2) in the case of hydrocarbons whose structure does not allow the formation of C_5 cyclic intermediate, by a so-called 'bond shift' mechanism.[10,11] The former isomerization route is often termed as 'cyclic mechanism'.[12] The present author prefers the name 'C_5 cyclic mechanism',[13] which will be used throughout this chapter, in agreement with de Jongste and Ponec who pointed out[14] that 'bond shift' may also involve a 'C_3 cyclic' intermediate. *Hydrogenolysis* of alkanes has also been a well-known and widely studied reaction.[15] The reaction mechanisms of these reactions and their relative importance over various catalysts have been comprehensively reviewed.[13,14,16–20]

Early ideas for *aromatization* [21] assumed the dehydrogenation of an open-chain hydrocarbon and the subsequent ring closure of the olefin directly to give a six-membered ring. Aromatization on carbon-supported metals was interpreted in terms of a direct 1,6 ring closure of the alkane molecule without its preliminary dehydrogenation.[5b] Past and present state of the art has been discussed in the excellent review by Davis.[22] With the appearance of bifunctional catalysts, the concept of this 1,6 ring closure has fallen temporarily into the background in favor of the two-dimensional mechanism.[2] This described very satisfactorily the reactions observed under industrial conditions. Still, the possibility of the 1,6 ring closure has again surfaced due to new evidence. The stepwise dehydrogenation of heptanes to heptenes, heptadienes, and heptatriene followed by cyclization has been shown over oxidic catalysts.[23] This idea was confirmed recently with n-octane aromatization over CrO_x clusters or Cr^{3+} ions as the catalyst, stabilized by La_2O_3.[24] Another novel catalyst family included Zr, Ti, and Hf oxides on carbon support.[25,26] These oxides were claimed to decompose upon pretreatment in Ar at 1273 K and were described as nonacidic 'Zr/C, Hf/C, Ti/C', producing aromatics with selectivities up to 67% from n-hexane[25] and 80–92% from n-octane, likely *via* the 'triene route'.

Hexatriene as an intermediate has been shown also on unsupported Pt catalysts, partly by using ^{14}C radiotracer.[27,28] This triene mechanism has also been regarded as one of the possible reaction pathways over Pt/Al_2O_3, together with another, direct C_6 ring closure.[29] The assumption of dienes and trienes does not mean that these intermediates should appear in the gas phase. It is more likely that a 'hydrocarbon pool' is produced on the catalyst surface upon reactive chemisorption of the reactant(s). As long as sufficient hydrogen is present, all of the chemisorbed species are reactive and may undergo dehydrogenation, rehydrogenation, and, if they have reached the stage of surface olefins, double bond or cis–trans isomerization may also occur.[13,28] Their desorption is possible in either stage; hence, hexenes, hexadienes, etc., may appear as intermediates. The true intermediates of aromatization are *surface* unsaturated species;[30] those appearing in the gas phase are the products of surface dehydrogenation *and* desorption process. Desorption should be less and less likely with increasing unsaturation of the surface intermediates. The loss of hydrogen produces either *cis* or *trans* isomers. The *cis* isomer of hexatriene is expected to aromatize rapidly, the chance of its desorption being practically zero. The *trans* isomer, on the other hand, has to isomerize prior to cyclization and, during this process, it has also a minor chance to desorb to the gas phase.[13,28] It is also a misunderstanding to suggest that *thermal* cyclization of triene intermediates would have any noticeable importance in heterogeneous reactions[30] just because a gas-phase hexatriene molecule would cyclize spontaneously and very rapidly at or above about 400 K.[28] The temperatures in any *catalytic* reaction exceed this value.

The 1–6 cyclization of hexane proceeds between two primary C atoms, but at least one secondary carbon must be involved with alkanes having seven or more C atoms in their main chain.[22] Tracer studies using ^{14}C labeled n-heptane indicated that, over Pt supported on a nonacidic Al_2O_3, 1,6 ring closure was the main reaction of aromatization; in addition, 1,5 ring closure, opening, and repeated cyclization might also occur.[31] The aromatization of $[1-{}^{13}C]$n-heptane on Te/NaX resulted in toluene with 93% of the label in the methyl group.[32]

Davis[22] summarized the results of tracer methods as well as the peculiarities of 'monofunctional' aromatization. Random C_6 ring closure of n-octane would give o-xylene (oX) and ethylbenzene (EB) in a ratio of oX/EB = 0.5. Isomerization of these primary alkylaromatics on acidic sites would shift this value while producing p- and m-xylene. The distribution of primary ring closure products is, however, different on various nonacidic catalysts: oX/EB $\cong$ 1 on Hf/C, Zr/C, Ti/C[26] as well as on Pt/nonacidic alumina but $\sim$0.7 on Pt/SiO$_2$.[33a] At 1 bar pressure the oX/EB values were about 1.3–1.5 on PtSn on both silica and alumina support.[33b] Adding Sn to a Pt/SiO$_2$ catalyst increased the oX/EB ratio from about 0.6 to about 1.6 when the experiment was carried out at 7.8 bar, with eight-fold H$_2$ excess,[33c] and the same trend was seen on Pt/Al$_2$O$_3$ as well.[33d] These results indicate that the activation of the primary and secondary C atoms is strongly dependent on the nature of the catalyst, and the reasons are still to be clarified.

The role of hydrogen in the scheme is twofold. First, the metal catalyzed *trans-cis* isomerization through half-hydrogenated surface intermediates[20,34] requires hydrogen. Second, the degree of dehydrogenation of the surface entities is often too deep, and thus their removal to the gas phase is a hydrogenative process, e.g.:

$$C_6H_4[ads] + 2H[ads] \rightarrow C_6H_6[gas]$$

The reality of such processes has been confirmed experimentally by at least two independent methods: temperature-programmed reaction (TPR)[35] and transient response method.[36] Another TPR study, combined with infrared [RAIRS] and near-edge X-ray absorption fine structure (NEXAFS), confirmed the possible intermediate character of hexatriene in aromatization of 1-hexene on the Cu$_3$Pt(111) single-crystal surface[37] (see also Sec. 5.2). Benzene formation from *trans*-3-hexene—requiring geometric isomerization—has also been confirmed. Only a fraction of unsaturated adsorbates formed benzene, with this fraction being higher starting with cyclohexene and cyclohexadiene (about 100% vs. about 70% from both 1-hexene and 3-hexene). The E_a values were 24–25 kJ mol^{-1} with open-chain and 14–17 kJ mol^{-1} with cyclic feeds. The availability of surface hydrogen and its activation may also be important in activating primary and secondary C atoms in the direct C_6 ring closure.

The conformation of *cis*-hexatriene would permit an easy ring closure.[13] *Trans* isomers, in turn, may be coke precursors.[13,28] The *trans* $\rightarrow$ *cis* isomerization involves half-hydrogenated species,[20] and thus the transition can be promoted even by small amounts of hydrogen. When 1-hexene reacted on unsupported Pt precovered with HT,[38] the unreacted hexene fraction exchanged on average one of its H atoms with T (Table 1). The relative radioactivity of n-hexane points to the uptake of two labeled H atoms from the retained surface hydrogen pool. Analogous values for other products showed incorporation of

Table 1 Selectivity and Relative Molar Radioactivity of Products from 1-Hexene on Tritiated Pt[a]

Effluent component	Composition (%)	Relative molar radioactivity (a.u.)
$<C_6$	4	11
n-Hexane	12	2.0
Unreacted 1-hexene	62	0.2
2-Hexenes	14.5	1.0
Hexadienes	4.0	1.2
Benzene	3.5	3.4

[a]$T = 633$ K, 0.76 g Pt black pretreated with 3×0.5 ml tritiated hydrogen, 3 µl pulses into He carrier gas. Adapted after Ref. 38.

about one H atom per hexadiene and about three H atoms per benzene molecule.[38] Thus, in addition to benzene as a π complex [produced, perhaps, directly from cis-hexatriene], more deeply dehydrogenated species could also been produced. Their hydrogenative desorption, such as

$$C_6H_3[ads] + 3H[ads] \rightarrow C_6H_6[gas]$$

may belong to the slow steps of the aromatization.[35]

C_5 cyclization and C_5 ring opening are closely related and likely have a common surface intermediate.[7,39] An 'alkene-alkyl insertion' [like that mentioned for C_6 cyclization][40] as well as a 'dicarbene' and also a 'dicarbyne' mechanism involving surface intermediates attached by two double or two triple bonds to the surface[12] were proposed for ring closure. An alternative pathway would involve a much less dehydrogenated intermediate where the position of the cycle would be roughly parallel to the catalyst surface.[41] The relatively low degree of dehydrogenation of this latter intermediate has been shown by deuterium tracer studies.[42] This type of intermediate of the C_5 cycle, which is to be formed or to be split, and its preferred 1,3 attachment to two sites of the catalyst has been suggested,[13] and supported experimentally by comparing several open-chain[43] and C_5 cyclic hydrocarbons.[44] At the same time, a 'dehydrogenative' C_5 cyclization of unsaturated molecules [hexenes, methyl-pentenes] is also possible.[13,45] The C_5 ring opening can be hindered to various extents in the vicinity of an alkyl substituent [position **a**], with this hindrance depending on the nature of the metal and its dispersion.[12,13,44,46,47] The selectivity of ring opening of methylcyclopentane [MCP] in this position can vary from the statistical value of 40% down to 1–2% depending on the nature of the metal and its support.[22,46] Metal particles in narrow zeolite pores selectively catalyze the opening of MCP in position **c**.[48]

The parallel occurrence of C_5 *cyclic* and *bond shift isomerization* has been shown by the use of ^{13}C tracer.[12,39] Whenever the C_5 cyclic mechanism was possible, it was usually predominant; at the same time, strong sensitivity to the structure of the reactant and to the catalyst has been observed[18] (see also later). Bond shift means a transfer of a C—C bond to the next carbon atoms ('1,2 bond shift'). This route has been demonstrated on Pt and Pd. These metals interact in different ways with the reactant.[12] The product composition pointed to the interaction of two methyl groups of 2,2-dimethylbutane with Pt while one methyl group and the secondary C atom formed the preferred surface intermediate on Pd.[49] A 1,3 bond shift of methylpentanes over Ir at 493 K was also shown by ^{13}C tracer studies.[50]

Hydrogenolysis is related to isomerization reactions.[11,20,49,51] The Anderson–Avery mechanism[19] assumed a rather deeply dehydrogenated surface intermediate. If the starting molecule has at least three carbon atoms, a 1,1,3 interaction with the surface is preferred. Pt, Pd, Ir, and Rh were found to split an alkane molecule predominantly into two fragments.[52] A multiple hydrogenolysis prevailed over other metals. A fragmentation factor [ζ] has been defined as the average number of fragments per decomposed C_n hydrocarbon molecule. Its value is around 2 in the case of single hydrogenolysis and can go up to n with multiple hydrogenolysis. The variation of ζ with conversion was discussed in chapter 2 of Ref. 1. Hardly any change was seen in the value of ζ factor as a function of the degree of conversion with single hydrogenolyzing metals while it increased with higher conversions up to n in the case of multiple hydrogenolysis.[53] Hydrogenolysis *activity* [15] can be quite high in the case of single hydrogenolysis, e.g., with Rh or Ir. The probability of the rupture of individual C—C bonds of an alkane molecule is not equal either. An ω factor has been defined as the ratio of actual rupture and random rupture at a given C—C bond.[54] Its calculation is strictly valid for the case of single hydrogenolysis only; still, ω values provide useful information when the value of ζ is between 2 and 3 by considering the amounts of the larger fragments, i.e., the products C_5, C_4, and C_3 from a C_6 feed and disregarding the slight C_1 and C_2 excess. Such a calculation was provided, e.g., for fragmentation of 2,2-dimethylbutane.[55] Multiple fragmentation can occur via a true disruption of the molecule during one sojourn on the catalyst (e.g., with Os, Ref. 53) or, alternatively, by subsequent end-demethylation (e.g., with Ni). A way of distinguishing between these cases is to calculate the relative amounts of higher than C_1 fragments and methane. Their ratio, the so-called fission parameter [M_f], permits one to distinguish between terminal, multiple, or random hydrogenolysis.[56,57] Both factors were strongly dependent on the nature of the metal (M_f was determined for Pt, Ir, Ni, Cu, Pd;[49] ω for Pt, Pd, Ir, Rh[58]). The increase of temperature on Ir/Al_2O_3 increased the ζ value from n-hexane; the dropping of M_f below unity indicated a true multiple splitting at 603 K.[59] Decreasing H_2 pressures had the same effect. As for more detailed discussion of the suggested surface intermediates and proposed mechanisms of metal catalyzed reactions, the reader is referred to the literature.[3,12–20,22,39]

3 CATALYSTS AND THEIR ACTIVE SITES

Platinum is the most important catalyst capable of catalyzing all types of hydrocarbon reactions mentioned. Metals of group VIII of the periodic system [except for Fe and Os], as well as Re and Cu, have been found to catalyze aromatization to at least a slight extent.[13] Hydrogenolysis and bond shift isomerization seem to be an inherent property of several metals.[60] On the other hand, C_5 cyclization and related isomerization of 3-methylpentane was found[52] to be restricted to a few metals—Pt, Pd, Ir, and Rh—concomitant with their propensity to catalyze single hydrogenolysis. All these catalysts form face-centered cubic crystals and their atomic diameters are not far from each other. Figure 1 shows the nondegradative and fragmentation activity of these metals in the reaction of 3-methylpentane and methylcyclopentane, demonstrating the

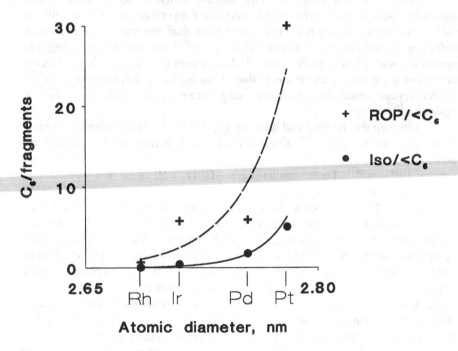

Figure 1 The ratio of activity of four metals in nondegradative reaction (isomerization of 3-methylpentane and ring opening of MCP) and in fragmentation as a function of their atomic diameter. (Adapted after Ref. 16.)

superiority of Pt for the former class of processes. Theoretical calculations on the correlation between adsorption properties, surface reactivity, and inherent properties of metals have reached a stage where the theoretical information can be useful for explaining reactions on real-world catalysts and, in particular, for explaining the difference between various transition metals.[61] Bond[62] pointed out the possible role of relativistic effects in determining structure, bond strength, and other properties of different metals. The increased 'speed' of the core $1s$ electrons leads to the shrinking of these orbitals above atomic number of about 50 [Sn] and the outer s orbitals will also contract. This will affect the atomic diameters and also the reactivities of the valence electrons in forming and splitting surface bonds. This may have been one of the reasons for the differences of chemisorption and catalytic properties of the Os, Ir, Pt triad as opposed to the group VIII–X metals of lower atomic number. The differences between Pd and Pt (e.g., prevailing π or σ adsorption) can perhaps also be attributed to such effects. Fundamental studies in the fields of solid-state physics and materials science will certainly contribute to better understanding of the inherent catalytic properties of different metals.

Various skeletal reactions may require different active sites. Those reactions whose active intermediate forms more than one chemical bond with the surface require active centers consisting of more than one atom. One Pt atom is sufficient to dehydrogenate propane to propene,[63] but aromatization requires ensembles with up to three Pt atoms.[64] Active sites for C_5 cyclic isomerization consist of more surface metal atoms than those for bond shift isomerization.[57] Under certain conditions, however, single-atom active sites may also be operative.[65]

Although the material and pressure gap between surface science and real-world catalysis still exists,[66] single-crystal studies permit one to elucidate the possible role of surfaces of various geometries by artificially creating various crystal surfaces.[67] These results agree fairly well with the conclusions mentioned in the preceding paragraph: platinum crystal faces with sixfold (111) symmetry—where active ensembles of three atoms are abundant—exhibited an enhanced aromatization activity as compared with those with fourfold symmetry (100) planes. As a rule, single crystals with steps and corner sites were more active than flat planes.[67,68] Much lower crystal plane sensitivity was observed in methylcyclopentane ring opening:[69] two-atom ensembles can be present on any crystal plane configuration. Iridium single-crystal surfaces were less active than corresponding Pt surfaces in cyclohexane dehydrogenation and n-heptane dehydrocyclization at low pressures.[70] Both C–H and C–C breaking occurred more easily on Ir than on Pt, which can explain the higher activity of Ir.[15,52] Present-day developments of surface science—among others, scanning tunneling microscopy (STM), in situ infrared (IR) spectroscopy,[71] and sum-frequency generation vibrational spectroscopy (SFG)[72]—contributed to bridge the pressure and material gap. However, the model systems are less complex than those involved in naphtha-reforming reactions, such as CO adsorption and oxidation[71] or hydrogenation–dehydrogenation reactions.[72] The most relevant result revealed so far is the role of both 1,3- and 1,4-cyclohexadiene in both hydrogenation and dehydrogenation of cyclohexane.[72,73]

The catalytic properties of various single-crystal planes have been compared directly with those of Pt/Al_2O_3 catalysts with various metal loadings and different crystallite sizes.[74,75] The importance of the so-called B_5 sites containing five atoms along a step, with (220) or (311) Miller index in most cases, termed sometimes also as 'ledge structures', has been pointed out on both single-crystal and 10% Pt/Al_2O_3 catalysts for bond shift isomerization and related hydrogenolysis. A sample containing 0.2% Pt/Al_2O_3 and very small crystallites was active in catalyzing C_5 cyclic reactions only. No exact single-crystal analogy was found for the catalytic behavior of this very disperse catalyst; its behavior was closest to (311) crystal planes.[75]

Ledge structures were found to be indispensable for alkane or cycloalkane hydrogenolysis on Pt single crystals.[76] Rounded crystallites—like a field emission tip[77]—must contain a large number of such high Miller index sites; hence, rounded crystallites should promote hydrogenolysis. Indeed, supported Pt catalysts exhibiting large rounded crystallites showed higher hydrogenolysis selectivity than 6% Pt/SiO_2 (EUROPT-1).[78] This latter catalyst has smaller but more perfect crystallites, and is approximated well with 55-atom cubooctahedra, as indicated by X ray diffraction (XRD) investigations.[79] Aromatization has been attributed to the six-atom (111) facets of the cubooctahedra whereas other skeletal reactions have been ascribed to (100) facets.[80] The hydrogenolysis activity of single-crystal and dispersed Ni catalysts also correlated well with the abundance of their ledge structures.[81]

Not only the atoms comprising the active sites but also the adjacent atoms may be important for catalysis. All hydrocarbon reactions mentioned involve the dissociation of at least one C-H bond, which can occur with or without participation of hydrogen adsorbed on the catalyst.[82] The kinetics of skeletal rearrangements of alkanes have been discussed in these terms where the initial step of all reactions of, say, a heptane isomer would be '*reactive* chemisorption':[83]

$$C_7H_{14} + H_a \rightarrow (C_7H_{15})_a \rightarrow (C_7H_{13})_a + H_2$$

as opposed to a '*dissociative* chemisorption':

$$C_7H_{14} \rightarrow (C_7H_{13})_a + H_a$$

Not all sites of the metal surface proper are active catalytically; 'chemisorption sites' (or 'landing sites'[84]) and 'reactive sites' have been distinguished. One approach attributes the former ones to flat planes while the reactions would likely proceed[75] on the reactive sites such as steps, kinks, and 'B_5 sites' (Fig. 2). The

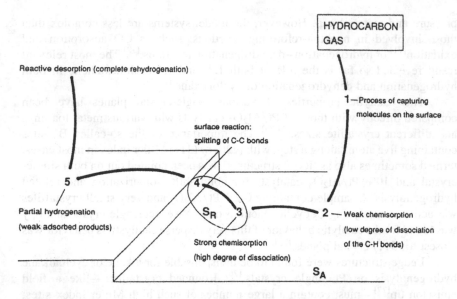

Figure 2 Possible adsorption sites and reactive sites and migration of surface species on a stepped single crystal surface. (Reproduced by permission after Ref. 75, Copyright Academic Press, Inc., 1987.)

underlying ideas are valid not only for large single-crystal planes but also for areas comprising terraces of a few atom widths and adjacent steps in the case of dispersed catalysts.

A strong kinetic isotope effect was reported reacting an equimolar mixture of n-octane (C_8H_{18}) and its fully deuterated counterpart, C_8D_{18}, on Pt/SiO_2.[22,85] The aromatization rate of C_8H_{18} was more rapid by a factor of 3–4 than that of the deuterated octane. This kinetic isotope effect points to the rate-limiting character of the initial adsorption step.[22,86a] Little H-D isotope exchange took place between n-octane and methylcyclohexane on Pt/SiO_2.[86b] The H-D distribution in the ethylbenzene and o-xylene products indicated that their unsaturated precursors remained long enough on the surface to reach equilibrium. The free migration of surface intermediate between 'chemisorption' and 'reactive' sites may not be applied without further consideration to those systems.

Somorjai[87] considered the 'rigidity' of crystal surfaces as an important factor in determining the catalytic activity. As opposed to the 'most rigid' flat planes, the 'most flexible' small particles which are assumedly able to change their structure under the effect of chemisorptive bonds were regarded as being the most active catalysts.[87] This idea may have its roots in the original suggestion of Taylor who postulated that the reactant itself may find its active site and at the same time represents a step further: the reactant may have an active role in *creating* the proper active site. Thus, the behavior of single-crystal surfaces may be different from that of disperse catalysts. A recent XRD study[88] of Pt black indicated an increase of the (220) and (311) reflections upon hydrogen treatment, both representing 'ledge structures'. Deactivation during hexane reaction not only caused carbon accumulation but also suppressed the above structures. This study demonstrated that individual Pt particles of about 20 nm were small enough to exhibit a 'flexible' behavior and, at the same time, are large enough that this effect would be detected by a bulk method: XRD.

The environment of the active metal atom can modify its catalytic properties, especially in the case of single-atom active sites.[65,67] The hypothesis states that the solid catalyst, the reactant, and possible other components all participate (by just being present as coadsorbed entity or by inducing solid-state transformations) in creating an active 'catalytic system'. This idea was postulated in 1974 largely on an intuitive basis.[89] One factor may be the interaction of a support atom in the vicinity of the one active atom.[65] This may be especially remarkable with TiO_2-supported metals where so-called strong metal–support interactions may occur.[90] This phenomenon has minor importance in most hydrocarbon reactions discussed. One important significant exception is the ring opening of cyclopentanes (being a very important step of metal-catalyzed isomerization, too). 'Nonselective' C_5 ring opening—which may occur in any position of the ring producing *n*-hexane and methylpentanes from methylcyclopentane— has been attributed to the so-called adlineation sites at the metal–support borderline. The 'selective' reaction occurring far from the alkyl substituent(s)—producing only methylpentanes from methylcyclopentane—in turn, was attributed to sites consisting of metal atoms only.[91] This suggestion was experimentally demonstrated by creating artificial metal–support borderlines by vacuum depositing Al_2O_3 on small Pt particles.[92] The abundance of *n*-hexane in the ring-opening products increased on these samples.[92] The same effect was observed with disperse Pt: admixing fine oxidized Al powder to Pt black also created 'adlineation sites'.[93] Hardly any adlineation effect was observed with Rh catalysts.[46,94] Pd, in turn, could reduce its support (e.g., SiO_2 or Al_2O_3) in such a way that a two-dimensional 'rim' exhibiting particular catalytic properties formed around the metal particle.[95]

One of the most important support effects in reforming reactions is the contribution of the acidic sites to the final product composition. A further description is covered in Chapter 3 of this edition. The classical picture of reforming on so-called bifunctional catalysts containing both metallic and acidic sites attributes dehydrogenation reactions to the metallic sites, ring closure, and $C_5 \rightarrow C_6$ ring enlargement reactions to acidic active centers.[96] The two types of active centers need not be situated on the same catalyst particles; a gas-phase transfer of reactive intermediates (stable alkene or cycloalkene molecules) in physical mixtures of nonacidic-supported metal and acidic catalyst grains would also result in almost the same product composition.[2,97] An intermediate situation can arise in a way analogous to the 'adlineation effect':[91] Cl added to the Al_2O_3 support to enhance its acidity can accumulate on the metal–support borderline. Isomerization selectivity vs. hydrogenolysis of 2,2-dimethylbutane increased after adding CH_2Cl_2 to Pt/Al_2O_3.[55] Cl can form Pt—O—Cl entities and, by so doing, can increase the oxidation state of Pt up to Pt^{+4}. This may be the reason for the higher aromatization selectivity.[98]

The fact that a nonnegligible fraction of the products can be produced over the metallic sites 'makes it necessary to alter the reaction scheme...'—i.e., the original two-dimensional scheme of Ref. 96—'...to include the additional reactions (particularly cyclization and isomerization) which occur on the metal surface alone, without involving acidic centers'.[2] Attempts to construct such alternative, more complete reaction schemes have been put forward by Parera et al.[99] as well as by the author of the present chapter.[100] Figure 3 includes the stepwise aromatization pathway on the metallic sites; the metal-catalyzed C_5 ring closure, ring opening, and isomerization are also shown. In our view, the reactions included in this scheme should be kept in mind even if the cooperation of metallic and acidic centers, as suggested originally by Mills et al.,[96] describes fairly well the processes occurring under the conditions of commercial reforming.[1,99]

The amorphous silica–alumina support can be replaced by acidic zeolite supports in reforming reactions.[101] The product composition on Pt/HY and on mechanical mixtures of Pt/SiO_2 and HY were close to each other.[102] However, smaller Pt particles formed in cavities of zeolites were less likely to undergo sintering and coking. Small metal particles in the zeolite framework and acidic sites in their close vicinity may form combined metal–proton adducts. These can arise during reduction of Pt precursor by H_2.[103] These entities can be called 'collapsed' (or 'compressed') bifunctional sites,[104] and these are claimed to be active first of all in the enlargement reaction of the five-membered ring. The existence of such sites was proven for Pd,[104] and they seem to be present also on $Rh^{[104]}$ and $Pt^{[105]}$. The commencing collapse of the zeolite structure upon heating to 833 K increased the Pt crystallites and brought the properties of Pt/NaY zeolites closer to those of nonacidic Pt. Product composition may serve as 'fingerprint-like' characterization for the prevailing active sites. Table 2 shows three characteristic product ratios obtained on Pt supported on zeolites with different acidity.[106] Hexane isomerization to dimethylbutanes decreased with decreasing acidity. A similar drop of dimethylbutanes was observed on larger Pt particles created by high-temperature calcination of 8% Pt/NaY.[105] The C_5/C_6 cyclic ratio (Bz/MCP) was higher on more acidic catalysts (Table 2). The high amount of skeletal isomers (Table 2) indicated a favored isomerization on acidic sites whereas the inherent C_5 cyclization activity was more pronounced on

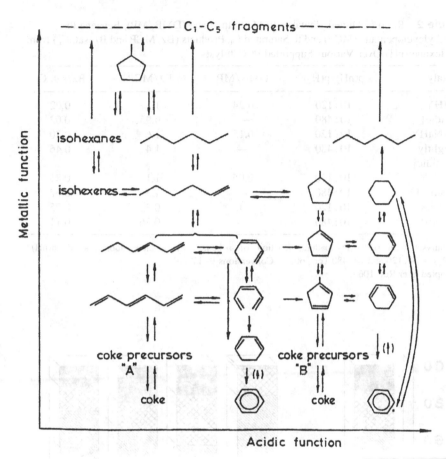

Figure 3 A modified two-dimensional reaction scheme of reactions over bifunctional catalysts. (Adapted from Ref. 100, copyright Academic Press, Inc., 1987.)

nonacidic catalysts. The neutral Pt/NaY sample was most active and exhibited the highest aromatic selectivities. A greater hydrogen excess during reaction enhanced metal-catalyzed reactions. Figure 4 compares fragment composition obtained on four supported Pt catalysts: 0.5% Pt/ZSM-5,[107] 0.8% Pt/HY,[106] 0.8% Pt/KL,[108] and 6% Pt/SiO$_2$: EUROPT-1.[109] The amounts of methane and ethane in the fragments were also negligible with acidic supports,[105,106] the composition being different with the two zeolite supports. Introduction of Pt hardly affected the fragment abundance as seen from the results obtained with HY alone. The splitting of the C$_6$ chain occurred in a nearly random way on both nonacidic catalysts, with smaller ω values in position C$_2$–C$_3$.[110]

Table 2 Ratio of Dimethylbutanes to Methylpentane (DMB/MP), Benzene to Methylcyclopentane (MCP) and to Saturated C_6 Products (Bz/MCP and Bz/sat. C_6) from n-Hexane (nH) Over Various Supported Pt Catalysts[a]

Catalyst	p(nH)/p(H_2)	DMB/MP	Bz/MCP	Bz/sat. C_6
Pt/HY	10 : 120	0.24	1.87	0.02
[acidic]	10 : 480	—	0.95	0.02
Pt/NaHY	10 : 120	0.15	1.6	0.20
[slightly acidic]	10 : 480	—	1.4	0.46
Pt/NaY	10 : 120	0.05	1.1	0.81
[neutral]	10 : 480	—	1.3	0.62
Pt/NaX	10 : 120	0	0.35	0.25
[basic]	10 : 480	—	0.36	0.11

[a]Catalysts with 0.8% Pt loading, reaction of n-hexane in a closed-loop reactor, p(nH)/p(H_2) = 10:120 and 10:480 Torr, 603 K. Conversions 9–11%.
Adapted after Ref. 106.

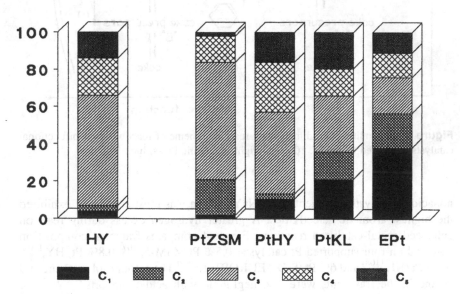

Figure 4 The abundance of fragments from hexane on HY zeolite and various Pt catalysts: 0.5% Pt on H-ZSM;[107] 0.8% Pt on HY;[105] 0.8% Pt on KL;[108] 'EPt': 6.3% Pt on SiO_2. Typical conversions were $X \approx 5\%$. (The fragment distribution on Pt/ZSM was nearly constant from $X \approx 4$–60%; $T = 603$ K; p(nH):p(H_2) = 10:120.)

Pt/KL zeolites represent a family of rather active monofunctional catalysts for aromatization of, first of all, hexane.[111] These contain Pt on alkali or alkali-earth-modified zeolites with negligible acidity. Counterclaims have been proposed to explain their activity.[101] The 'confinement model'[112] stressed the geometrical effect of narrow zeolite channel, representing a geometrical constraint for the reactive chemisorption of the alkane reactant. Accordingly, the reactant would form a six-ring pseudocycle facilitating C_6 dehydrocyclization on small Pt clusters exposed in the windows of zeolite cages of Pt/KL.[112] This hypothesis represents an interesting example of C_6 ring closure on single-atom active sites. A direct six-ring closure of *n*-hexane over nonacidic Pt/KL between 600 and 690 K was suggested and the confinement model found to be valid by comparing this catalyst with Pt/KY.[113] Pt/KL catalyst reduced by H_2 can still contain some acidic sites.[103,108a] These samples produced more isomers and less benzene than a neutralized Pt/KL at 603 K.[108b] Benzene formation (via hexenes) predominated at 693 K. The confinement model has been claimed to be unsatisfactory since high benzene selectivity from *n*-hexane was observed over Pt on various micro- and mesoporous supports,[114] although it fitted better the results obtained over Pt/KL. Incorporation of larger alkali cations, such as Cs, into β-zeolite (having also a tubular pore structure) resulted in smaller Pt particles, exhibiting activity similar to Pt/K/LTL zeolite.[115]

Another explanation suggested an electron transfer from the basic zeolite to Pt sites, and this was supported by infrared spectroscopy of CO.[116] More 'electron-rich' Pt zeolites produced more benzene and also more MCP within the C_6-saturated fraction. The role of support basicity was confirmed also by the similar behavior of Pt/KL and Pt/MgO,[117] the latter also showing a fraction of small negatively charged clusters.[117b] These 'electron-rich' Pt particles would, accordingly, favor direct C_6 ring closure, in agreement with the stepwise aromatization pathway.[118] The binding energy of the Pd 3d in X-ray photoelectron spectroscopy (XPS) decreased with increasing alkalinity of the L-zeolite support, parallel with the activity in neopentane hydrogenolysis. However, 'electron transfer' from the support to Pt was disclaimed, and the changes of electronic properties of Pt on highly alkaline catalyst supports were attributed to coulombic forces through space between metal atoms and surrounding O ions.[119a,b] The reverse interaction would occur between Pt and acidic sites. Apart from Pt–Pt bonds, Pt–O bonds are also present in small supported particles. A recent X-ray absorption fine structure (XAFS) study combined with molecular orbital calculations assumed a Pt_3O_4 cluster as a good model for the strong electronic interaction and charge rearrangement between O and Pt atoms. The electron density of support oxygen can influence the *monofunctional* activity of Pt.[119c] A further possible reason for the enhanced dehydrocyclization selectivity over Pt/KL catalysts may be the hindered deactivation of metal particles within the zeolite channels.[120,121] At the same time, even a slight sintering of Pt particles could grow them to an extent sufficient for pore blocking.[122]

4 RELATIONSHIP BETWEEN REACTANT STRUCTURE AND REACTIVITY

It can be easily understood that not all hydrocarbons have the same reactivity: the reactant structure permits or excludes certain catalytic processes (e.g., alkylbutanes are not able to participate in C_5 cyclic reactions). Furthermore, even if the reactions are theoretically possible, some structures must be more favorable for certain reactions.

The species in the partly dehydrogenated hydrocarbon pool responsible for stepwise aromatization apparently 'remember' the structure of the hydrocarbon feed. For example, the structure of trans-2-hexene is more favorable for C_5 cyclization than for aromatization; more hexadienes appear from 1-hexene than from both 2-hexene isomers.[123] Hardly any hexatriene was formed from 2,4-hexadiene whereas 1,4-hexadiene produced rather high amounts of trans-1,3,5-hexatriene.[28] A possible aromatization route may involve the reaction 2-hexene → 2,4-hexadiene followed by cyclization of the latter, bypassing the stage of hexatrienes.[13,28] This route seems to be likely on the basis of temperature-programmed reaction studies where a reactant (such as 1-hexene or 2-hexene) was first adsorbed on Pt-black and Pt/Al_2O_3 catalysts and the removal of product(s) of surface reaction(s) followed during programmed heating.[35] These measurements also indicate that a cyclization step hexatriene → cyclohexadiene belongs to the rapid elementary reactions as compared with dehydrogenation and/or trans-cis isomerization processes.

As far as skeletal rearrangement reactions are concerned, hydrocarbons were assumed[124] to react according to their structure as 'C_2 units' (consisting of a two-atom molecular fraction containing primary and/or secondary C atoms only) or 'iso-units' (where one of the reacting C atoms is tertiary or quaternary). These structures react in a different way in skeletal isomerization and their hydrogenolysis pattern is also different. Garin and Maire[125] defined an 'agostic entity' as the part of the molecule interacting directly with the catalyst surface during reaction. They considered the electronic state of the catalyst (i.e., the nature and the actual state of the metal) as the main factor deciding the prevailing pathway of the reaction, based on the variation of the ratio of iso-unit mode to C_2 unit mode reactions. This ratio was between 0.85 and 1.2 for various branched alkanes reacting on Pt single-crystal planes as well as over Pt/Al_2O_3, the dispersion of the latter being between 0.04 and 1. If the predominating surface interaction is secondary alkyl, the iso-unit mode reactions will prevail, whereas multiple carbon–metal bonds favor C_2 unit reactions. The type of this primary interaction was claimed to be a sensitive indicator for the different catalytic properties of various metals. If the electronic factor has already decided the way of the primary reactive adsorption, geometrical effects will control whether bond shift or C_5 cyclic mechanism (with or without hydrogenolysis) follows and in what ratio.

There are experimental evidences that a hydrocarbon with C_5 main chain will undergo predominantly C_5 cyclization (and C_5 cyclic isomerization). At the same time, the reactivities of 3-ethylpentane and 3-methylhexane (= '2-ethylpentane') were different over both Pt-black[43] and 6.3% Pt/SiO_2 (EUROPT-1),[126] simply because of the steric hindrance of the latter molecule in forming a flat-lying intermediate. The same conclusion can be drawn from the different reactivity of *cis*- and *trans*-dimethylcyclopentanes.[44] The formation of aromatics from alkanes with fewer than six C atoms in their main chain (e.g., methylpentanes) is also possible but proceeds much less readily than direct C_6 dehydrocyclization.[13,127] This mechanism should involve a sort of dehydrogenative bond shift isomerization. The same is valid for the $C_5 \rightarrow C_6$ ring enlargement. This can take place over metal catalysts but is not a favored process.[127] The presence of a quaternary C atom enhances this reaction and subsequent aromatization.[13,42,44,128,129] More toluene was formed from 3,3-dimethylpentane than from other alkylpentanes with no quaternary C atom.[126]

Bearing in mind the different energies of activation for different bond ruptures,[13,39] a totally random C–C bond rupture is an exception rather than the rule. Bond rupture position showed also a reactant structure sensitivity on supported Pt[130] as well as on unsupported Pt, Pd, Ir, and Rh catalysts.[58] The preferential position of hydrogenolysis of hexane, heptane, and octane isomers indicated that the splitting of a C_5 unit was facilitated from all of those molecules to give $C_1 + C_5$, $C_2 + C_5$, and $C_3 + C_5$ fragments, respectively.[110] 2,2-Dimethylbutane over various metals[55] produced predominantly isobutane plus ethane on Pt and Pd (along with isomerization), while mostly neopentane plus methane was formed on Rh, Ru, Ir, and Ni (no isomerization).

5   EFFECT OF SECOND COMPONENTS ON THE ACTIVE SITES OF THE CATALYSTS

Modern reforming catalysts often contain at least one second component in addition to Pt. The industrial aspects of the multimetallic reforming catalysts of the early 21st century are described in Chapter 8 of this volume. A short summary of the effects of metallic and nonmetallic components obtained by laboratory studies is presented here.

5.1 Metallic Components

High hydrogen pressures are thermodynamically not favorable for benzene formation[2,3,101] but are necessary in practice to slow down coking of the catalyst. The development of bimetallic catalysts[20,22,131,132] permitted this hydrogen excess to decrease considerably. Their preparation, properties, and catalytic behavior were discussed in the first edition of this volume.[1] Some of the added metals have catalytic properties on their own [Ir, Rh, Re], while others, such as Sn, Ge are catalytically inactive. For example, Ir added to Pt utilizes the higher hydrogenolysis activity of Ir in breaking up carbonaceous deposits. The selectivity pattern of hexane reactions on a Pt-Ir/Al$_2$O$_3$ with a 1 : 1 ratio was closer to that observed on Ir/Al$_2$O$_3$. The use of a bimetallic precursor resulted in somewhat lower degradation and more stable aromatization and isomerization selectivities.[133] Hydrogenolysis was much more marked on Pt-Rh/Al$_2$O$_3$ than on Pt-Ir/Al$_2$O$_3$.[134] Adding an unusual component, U, to Pt resulted in activity drop. However, Pt-U containing 3% F was superior in isomerization,[134] due perhaps to maintaining fluorine on metal–support borderline sites, as discussed earlier (Sec. 3).

Bimetallic or multimetallic catalysts are often regarded as alloys, although alloying would mean an intimate mixing of all components, which would create new phases. In another approach, two suggestions have been put forward in the literature to explain the peculiar properties of bi- and multimetallic catalysts: the 'electronic' and the 'geometrical' theories.[20,135] The shifts in the infrared band of chemisorbed CO indicated an electronic interaction between Sn and Pt[136] as a result of formation of intermetallic compounds such as Pt$_3$Sn or PtSn.[137] These unsupported compounds started to be active above about 700 K only in hydrogenolysis and aromatization.[138] Even if their formation is possible, it is not certain that they are really present under catalytic conditions. Davis et al.[139] distinguished three regions of Sn on reforming activity; tin added to a bifunctional Pt eliminated support acidity, thus decreasing the overall activity. Adding tin to nonacidic Pt had the same effect as introducing additional Sn to the bifunctional one: aromatization by monofunctional pathway was promoted. Excess tin [as Sn^0] gradually poisoned both types of catalysts. Of all possible phases, PtSn and PtSn$_2$ were identified on SiO$_2$ support.[140] Hexenes and benzene were the main products (at 753 K with fourfold H$_2$ excess), with the selectivity of the former increasing from 53% to 85% between 5Pt1Sn and 1Pt1Sn. The stability of the samples with Pt/Sn > 1 was attributed to the PtSn phase.

The geometrical theory may be closer to reality with catalytically inactive additives such as Cu or Au.[49] Here the second metal is totally inert and its effect is to dilute multiatomic active ensembles. This inactive second component can divide the active surface to smaller units ('ensemble' effect). 'Structure–insensitive' reactions (like cyclohexane dehydrogenation) would then exhibit almost the same rate up to rather high additive concentration, whereas 'structure–sensitive' processes are hampered at lower amounts of the second, inactive metal. Adding either Sn or sulfided Re to Pt may hamper coke formation by this geometrical effect.[64,141] Although several conclusions on ensemble effects and sizes of active centers have been drawn by this method,[49] in reality, the geometrical effect as a sole explanation may not be sufficient. Burch[142] postulated that no geometrical effect can exist without electronic interaction, as shown for example, by infrared studies of chemisorbed CO on Pt-Sn/Al$_2$O$_3$.[143]

More and more studies point to the importance of the method of preparation of bimetallic catalysts.[135] Introducing tin from the complex precursor [Pt(NH$_3$)$_4$] [SnCl$_6$] should result in smaller contiguous Pt islands in the resulting PtSn/Al$_2$O$_3$ catalysts.[137] Hydrogenolysis of *n*-hexane on larger Pt ensembles was not favored on this sample. The selectivities for isomerization and C$_5$ cyclization, in turn, were higher and so was the selectivity of hexene production, as in Ref. 140. The benzene selectivity was practically independent of the method of preparation.

One novel and successful method of preparing bimetallic catalysts is by reacting the corresponding alkyl compounds with hydrogen adsorbed on Pt. Applying typically tetraethyltin precursor, Margitfalvi et al. produced catalysts with a controlled dispersion of tin on Pt.[144] The changes of the surface Pt-H bond strength on PtSn catalysts may be due to electronic interaction. It was also possible to control the site of Sn as attached to the support and/or to the Pt metal particles.[145] With a large excess of tetraethyltin, multilayer organometallic complexes were formed on the surface, enhanced by excess hydrogen. Their decomposition in H$_2$ resulted in a PtSn alloy whereas decomposition in O$_2$ yielded an Sn oxide situated in the vicinity of Pt particles, representing a peculiar type of Lewis acid, and decreasing aromatization and increasing isomerization selectivities.[145] Tin oxide interacted with the more reactive Al$_2$O$_3$ as opposed to SiO$_2$. Such an interaction was also reported with added Re which—as Re$_2$O$_7$—interacted with the Al$_2$O$_3$ support.[146] This metal–additive–support interaction can also influence the acidic function.

Pt-Ge catalysts prepared by anchoring tetrabutylgermanium to the parent Pt/Al$_2$O$_3$ exhibited rather unusual catalytic properties.[146a] Benzene formation from *n*-hexane was suppressed on the sample containing one-eighth of a monolayer of Ge, especially in higher H$_2$ excess. Instead, cyclohexane was produced. Since the same sample was inactive in benzene hydrogenation, the small amount of Ge has most likely selectively poisoned sites with threefold symmetry where an aromatic ring could be formed or reacted.[64]

The number and variety of second (and/or third) metals added to Pt is very high. Examples for bimetallic systems are mentioned throughout this chapter but a detailed discussion of bimetallic catalysts is beyond the scope of the present work. The reader is referred to review papers[18,20,132,147] as well as to Chapters 4 and 5 of the first edition.[1]

5.2   Nonmetallic Components

These components either are deliberately added (e.g., sulfur or hydrogen) or represent hardly avoidable impurities (e.g., carbon, oxygen, or nitrogen). Their effect can be harmful or beneficial.[148] No important new information has been generated on the effect of silicon since the first edition of this book (chapter 2, Ref. 1); therefore, it will not be discussed further.

Sulfur and Nitrogen

It is a customary industrial practice to pretreat Pt-reforming catalysts by a 'selective, controlled poisoning procedure with sulfur to reduce its initial hydrogenolysis activity' after their regeneration with oxygen.[2, pp. 291–293] Under industrial conditions, Pt-Re/Al$_2$O$_3$ operates also in the sulfided state.[99,149] These catalysts were found to have a much lower sulfur tolerance than Pt/Al$_2$O$_3$.[148,150] The primarily adsorbed, irreversibly held sulfur was responsible for suppression of fragmentation reactions and consequent enhancement of aromatization over Pt-Re/Al$_2$O$_3$ while an additional, reversibly held sulfur was necessary to this end with Pt/Al$_2$O$_3$.[150]

A comparative study[151] of irreversibly chemisorbed sulfur by using H$_2$ ^{35}S radiotracer for catalyst sulfidation shows that its amount increases in the order Pt/Al$_2$O$_3$ < Pt-Re/Al$_2$O$_3$ < Re/Al$_2$O$_3$. The behavior of Pt-Re/Al$_2$O$_3$ in comparison with Pt/Al$_2$O$_3$ and Re/Al$_2$O$_3$ in sulfur desorption and exchange was nonadditive.[99,151] This indicates the formation of mixed active sites in Pt-Re/Al$_2$O$_3$.

The effect of sulfur on supported Pt, Ir, Pt-Ir, Ru, and Pt-Re catalysts was compared.[99,152,153] One of the main effects of sulfur was to suppress selectively the hydrogenolysis activity. The remaining activity in the sulfided state was 1–6% of that observed in the sulfur-free state. The activity for skeletal isomerization, in turn, decreased to a much lesser extent (residual activity, 25–75%); in two cases (with Ir and Pt-Re catalysts), the relative isomerization rates were even *higher* after sulfidation (Fig. 5). A basic additive, pyridine, suppressed mainly isomerization rates and thus must have been selectively attached to acidic sites. Hence, bifunctional activity could manifest itself even at a temperature as low as 620 K.

Adding S even in very small quantities [about 1% of a monolayer] to a stepped Pt(557) single crystal surface induced its reconstruction.[154] The selectivity of hydrogenolysis of 2-methylpentane decreased gradually by a factor of 2 when the S content increased from 1% to 18% of a monolayer. The overall rate per unit free surface showed a similar increase. A Pt-Re single crystal showed a highly enhanced hydrogenolysis activity with *n*-hexane model reactant as compared with pure Pt. Adding about one monolayer of sulfur suppressed this

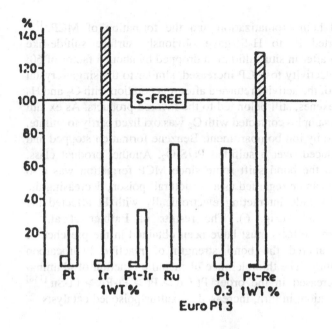

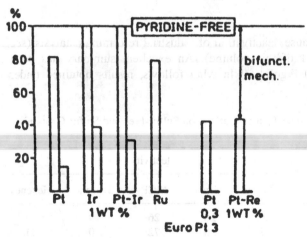

Figure 5 Relative rates of hydrogenolysis (left bar of each pair) and skeletal isomerization (right bar of each pair) of *n*-hexane at 620 K over various Al_2O_3-supported catalysts after sulfidation (upper panel) and pyridine addition (lower panel). The sulfur-free and nitrogen-free states are taken as reference (100%). (Reproduced by permission after Ref. 153, copyright Elsevier Science Publishers, 1991.)

reaction and enhanced both aromatization and the formation of MCP.[155] Exposing an unsupported Pt to H_2S gave obviously surface sulfide-like species.[156a] Its activity after in situ sulfidation dropped by about a factor of 60; at the same time, the selectivity to MCP increased, similar to the single-crystal results (Table 3). Some of the activity returned after regeneration with O_2 and H_2 but MCP, along with hexenes, still belonged to the favored products. As ex situ XPS indicated, sulfur in samples contacted with O_2 was oxidized partly to sulfate, which could be removed by ion bombardment. Benzene formation stopped and much hexene was produced over a sulfided Pt/SiO_2. Another product class, isomers, likely involved the bond shift route since MCP formation was also suppressed.[156b] Sulfate can be regarded as a 'structural' poison, decreasing the active surface whereas sulfide interacting electronically with Pt affected the reaction routes considerably (Fig. 6). The results of Barbier et al.[157] emphasizing the electronic effects must have been obtained in the presence of sulfidic sulfur, which altered the bond strength of reactive hydrocarbon intermediates in its vicinity. The thiotolerance of metallic function of alumina-supported catalysts decreased in the order Pt-Ge > Pt > Pt-Re > Pt-Sn.[158] Acid-catalyzed isomerization, in turn, increased on sulfur-poisoned catalysts.

Carbon

Carbonaceous deposits cause deactivation of industrial reforming catalysts (see also Chapters 10 and 11 of this volume). An excellent summary has been published by Parera and Figoli.[159] In what follows, results obtained under

Table 3 Selectivity of Hexane Transformations on Sulfur-Free and Sulfur-Containing Pt Catalysts[a]

Catalyst	Conv. (%)	Selectivity (%)				
		$<C_6$	Iso	MCP	Benzene	Hexenes
Reg. O_2/H_2	16	26	26	26	19	3
Sulf.[b]	0.2	7	0	72	0	21
O_2[c]	1.5	11	12	31	14	32
EUROPT-1	⌣8	9	21	45	8	7
EUROPT-1/S[d]	⌣0.5	8	13	5	0	74

[a]Pt black, $p[nH]/p[H_2] = 10 : 120$ Torr, 603 K.
[b]Sulfidation with H_2S; sulfur present as S^{2-}
[c]Sulfur partly oxidized to sulfate.
[d]Both sulfide and sulfate were present.
Adapted after Refs. 156a and 156b.

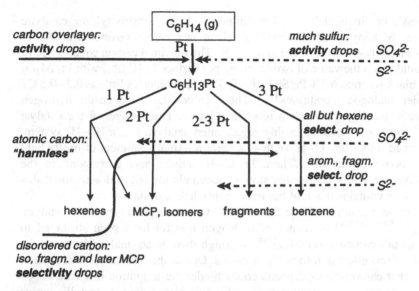

Figure 6 Possible modifier effects of different types of sulfur and carbon chemisorbed on Pt as observed in the skeletal reactions of hexane. Atomic carbon ('Pt/C' species) is shown as 'harmless': they may correspond to the residues of the 'Pt-C-H' state. (Adapted after results of Refs. 156 and 180.)

laboratory conditions and with model catalytic systems, especially concerning the metallic function, are to be discussed.

A rather marked carbon coverage has been claimed to be the normal state of Pt single crystals during hydrocarbon reactions.[160] The surface analysis necessarily involved ultrahigh vacuum (UHV) conditions. A hydrogen flushing of the catalyst before evacuation removed a large fraction of the residual hydrocarbons.[161] Upon direct evacuation, the chemisorbed hydrocarbons lost hydrogen, increasing the amount of surface carbonaceous deposits. The total carbon content of a Pt single-crystal or polycrystalline Pt foil as measured by Auger electron spectroscopy (AES) was divided into three groups, namely (1) a constant amount of 'residual carbon' (up to 4%), (2) 'reversible carbon' formed instantaneously during working conditions, and (3) 'irreversibly adsorbed carbon' accumulating during reaction (length of run, up to 10 h) and ultimately forming coke. Carbon coverages calculated from the C and Pt AES intensity ratio were 0.21–0.34 when the hydrocarbon reactant was just evacuated after run; the maximal coverage was, however, as low as 0.09 when the run was concluded by a hydrogen flushing.[161] Menon[162] classified various types of surface carbon on different catalysts, including industrial ones, as being 'beneficial', 'harmful', 'harmless', or 'invisible'. He also defined a 'coke sensitivity' for catalytic reactions. Sárkány[163] determined the clean and carbon-covered fraction of various Pt catalysts between 453 and 593 K. The maximal carbon coverage of Pt sites could reach the value of two C atoms per surface Pt [C/Pt_s] with Pt/Al_2O_3 and Pt-black whereas 6.3% Pt/SiO_2 [EUROPT-1] retained as little as 0.3–0.5 C/Pt_s under analogous conditions.[164] These values depended on the hydrogen pressure in the H_2/hydrocarbon reactant mixture. Three stages of the catalyst life[163] were postulated. A freshly regenerated catalyst, i.e., a 'Pt–H' system, transformed into a 'Pt–C–H' state as a certain—small—amount of firmly held hydrocarbons accumulated. These Pt–C–H entities may correspond to the 'reversible' carbon.[161] A further slow C accumulation led to deactivation: this 'Pt–C' state contains much of the irreversibly held carbon.

All these results indicate that the Pt+C system possesses some catalytic activity.[147,160,163,164] Reactions of hydrogen transfer have been attributed to mixed metal–carbon ensembles,[165] although there is no final evidence on the real role of this effect in reforming reactions. One of the reasons for deactivation may be that carbonaceous deposits could hinder the migration of chemisorbed hydrocarbon species to reactive sites and would also deactivate metallic active sites.[160] The changes of carbon coverages with $p(H_2)$ were accompanied by marked selectivity changes in skeletal reactions of n-hexane. The Pt–H state exhibited a strong hydrogenolysis activity.[166] This period lasted for a few minutes on Pt black[167] or on 6% Pt/SiO_2[168] at 603 K in a closed-loop reactor. With the development of the 'Pt–C–H' stage, an almost steady-state nondegradative activity was reached. During deactivation, first the isomerization activity dropped and that of C_5 cyclization increased; as the 'Pt–C' state was approached, more and more olefins were produced at 600 K. Single C atoms on valley positions of close-packed metal surface were regarded as equivalent to inert metal additives [such as Cu or Au], both shifting the selectivity toward processes requiring 1,3 interaction with the surface.[49] At constant hydrogen pressure, a more marked activity change at 693 K indicated more rapid carbon accumulation. This was true for both 6.3% Pt/SiO_2 [EUROPT-1][109,168] and Pt/ KL zeolite.[108] A nonuniform deactivation was reported at higher temperatures [773 K] with an industrial reforming catalyst[169] as well. Hydrogenolysis activity was deactivated first, followed by the loss of isomerization and aromatization ability.

At least two types of coke formation routes have been reported on metallic sites, one involving C_1 units and, the other involving polymerization of surface polyenes.[170] The C_1 route was discovered by using methane as the carbonizing agent. Theoretical calculations indicated the relative stability of C atoms, as opposed to CH_x species.[171] The polyene route may result in three-dimensional carbon islands observed by electron spectroscopy over Pt-black catalysts,[172] containing also some graphitized carbon. It is not excluded that these entities correspond to 'ordered' or 'disordered' carbonaceous deposits, the interconversion of which having been reported on Pt single crystals.[67] Precursors of carbonaceous deposits may also migrate from metallic sites to support sites and back. However, coke can also be formed on acidic sites of bifunctional catalysts involving a third route: cationic polymerization. Recent excellent reviews describe the chemistry of polymeric 'coke' formation on acidic catalysts.[173,174] One of the secrets of preparing high-performance bifunctional catalysts can be how to facilitate migration of coke precursors to liberate more valuable metallic sites, while the prevention of their subsequent polymerization is important for avoiding catalyst fouling and pore blocking.

Several methods are suitable and useful to detect carbon on catalysts but each has its own drawback. Temperature-programmed oxidation[148,175] is excellent for distinguishing C on metal and support sites but gives little information on their structure. When the retained hydrocarbonaceous entities are removed by hydrogen, some actual surface intermediates may appear. The detection of MCP after hexane reaction[163,176] confirmed its actual role in surface reactions. Heptane isomers appeared upon hydrogen treatment of Pt/SiO$_2$ after reacting propylcyclobutane in the presence of H$_2$.[177] Sárkány determined the amount of retained C by gravimetry.[178] A subsequent hydrogen flushing did *not* remove all of this residue. This was kept in mind when circulating hydrogen was applied to remove carbonaceous residues after alkane reaction in a closed loop.[179] Mostly methane appeared in the fraction removed after hexane reaction from Pt-black. This may originate from C$_1$ units present on the surface or from the hydrogenolytic disruption of more massive deposits. The predominance of CD$_4$ after deuterative removal points to the former possibility,[180] and so does the enhanced tritium uptake by fragments from the surface hydrogen pool (Sec. 2, Table 1). The absence of removed hydrocarbons following propylcyclobutane reaction *without* H$_2$ also indicates the difficult disruption of massive surface C.[177] The hydrocarbon fraction removed from 6% Pt/SiO$_2$ [EUROPT-1] contained about 20% benzene. More benzene was removed from an industrial 0.6% Pt/Al$_2$O$_3$ (Fig. 7). MCP was detected in one case only, after reaction on EUROPT-1 in excess H$_2$.[179a,b] The amount of hydrocarbon transformed during catalysis was about 100 times greater than the removable residue. The aromatic fraction removed by H$_2$ after reaction of nonane also contained more benzene than C$_6$ aromatics. Thus, hydrogen could be active in removing mostly C$_6$ units.[179b] These may have formed by partial dehydrogenation during evacuation.[161] The number of hydrocarbon molecules removed from Pt/KL was highest.[179a] Hardly any C$_2$–C$_5$ fragments appeared in the fraction hydrogenated from monometallic Pt, but they were abundant from PtSn/Al$_2$O$_3$ catalysts.[179c]

Electron microscopy can detect only thicker carbon layers. Graphitic entities—with crystal planes almost perpendicular to the Pt surface—were

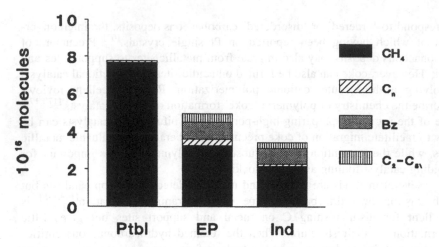

Figure 7 Hydrocarbons removed from three Pt catalysts: Pt black, 6% Pt/SiO$_2$ ('EPT'), industrial 0.6% Pt/Al$_2$O$_3$ ('Ind') with nearly the same metallic surface in the reactor after a reaction with hexane + H$_2$, p(nH) : p(H$_2$) = 3 : 120 Torr, T = 633 K, conversion ∼10%.

discovered on heavily carbonized Pt.[181] Deactivating deposits outside the porous structure of molecular sieve–supported catalysts showed a similar picture but with carbon layers parallel to the particle surface.[174]

Various metals retain different amounts of carbon. A spectacular difference was observed between Rh- and Pt-blacks: While about 8% C remained from the initial C impurity of about 12% of a (freshly reduced) Pt black, Rh lost more than 90% of its C impurity after hydrogen treatment at 483 K (about 16% → 1.5%). Purely graphitic carbon remained on Rh-black,[182] but XPS revealed several components in the residual C 1s component on Pt. Carbon on Pt usually contained 'Pt/C' (single C entities on Pt), a 'disordered' ('chain') carbon likely containing multiple C–Pt bonds, graphite or graphite precursors, some aliphatic polymer, and oxidized carbon entities. After intentional carbonization,[181] the loss of overall activity could be attributed to the accumulation of three-dimensional carbon whereas the disordered deposits poisoned selectively isomerization and C$_5$ cyclization reactions (Table 4). The same conclusion was reached when small carbocycles reacted on Pt/SiO$_2$;[177] here the importance of the coke–metal interface was also stressed. The Pt 4f peaks[172,181]—as well as UPS spectra[183]—indicated that, in spite of the presence of 'Pt/C', Pt remained in Pt^0 metallic state.

The poisoning effects of carbon and sulfur are compared in Figure 6. Both of the 'structural' poisons—C overlayer and oxidized sulfur—affected the activity. The 'electronic' poisons—sulfide and 'disordered' carbon—affected the selectivities in a different way. Both enhanced hexene selectivities but while

Table 4 Abundance of C 1s Components on Pt black, and Selectivity of Hexane Transformations After Various Carbonizing Treatments[a]

A. *Catalytic Properties*

Catalyst	TOF (h^{-1})	Selectivity (%)				
		$<C_6$	Hexenes	Iso	MCP	Benzene
1. Reg. O_2/H_2	17.6	33	2	20	16	29
2. HD[b]	1.2	5	50	2	1	43
3. HD + H_2[c]	1.9	23	9	13	18	37

B. *Distribution of C 1s Components*

Catalyst	Total C (%)[d]	BE (eV)				
		$\backsim283$ "Pt/C"	$\backsim284.1$ Disord. C	284.7 Graph.	285.6 Poly-C_x-H_y	287–288 Ox.C
1. O_2/H_2	28	3	14	6	3	2
2. HD[b]	51	2	25	14	6	3
3. HD + H_2[c]	44	2	12	18	9	3

[a]Pt black, p[nH] : p[H_2] = 10 : 120 Torr, 603 K.
[b]Exposed for 40 min to 53 mbar *t,t*-hexa-2,4-diene, 603 K.
[c]Exposed for 20 min to 13 mbar *t,t*-hexa-2,4-diene plus 210 mbar H_2, 603 K.
[d]Percent of the total surface (Pt4f + C1s + O1s, the latter being between 3 and 5%).
Adapted after Ref. 181.

sulfur deactivated aromatization, carbon poisoned mostly isomerization and fragmentation (Tables 3 and 4). The selectivity of C_5 cyclization first increased, but predominantly only hexenes were formed on Pt strongly deactivated by either sulfur or carbon. The opposite effect was reported on industrial Pt, Pt-Ir, Pt-Re, and Pt-Re-Cr catalysts supported on Al_2O_3. Heptane and MCP produced more aromatics after S treatment. Coke accumulation also increased in the latter case.[184] Sulfur obviously could not prevent polymerization of methylcyclopentadiene, representing a peculiar polyene route of coking.[175] The Pt-Ge/Al_2O_3 combination was most resistant to the combined effect of coke and sulfur, its sulfur tolerance being the decisive factor.[158]

Oxygen

Pt-black pretreated with oxygen showed enhanced fragmentation selectivity from *n*-hexane,[185] and so did Pt single-crystal catalysts.[67] The rate of ethane, propane, and n-butane hydrogenolysis was reported to be higher on SiO_2-supported Ni, Ru, Rh, Pd, Ir, and Pt catalysts when they were pretreated in O_2 at 773 K.[186] Reconstruction upon H_2/O_2 cycling was demonstrated by electron microscopy and was considered as a reason for these phenomena. Another possible reason may be the actual presence of surface oxygen. UPS of Pt-black taken a few minutes after H_2 treatment at 600 K showed the presence of intensive peaks attributed to chemisorbed O species. Keeping the sample overnight at 603 K in UHV, these transformed into surface OH/H_2O species.[167] Hydrogen treatment before the catalytic run did not remove all surface oxygen, its value varying from about 3%[88] to about 15%,[172] depending on previous treatments. The main reaction of n-hexane over Pt-black[167] and Pt/SiO_2[168] containing presumably Pt–O entities (very short contact times after regeneration) was predominantly hydrogenolysis to methane with nearly 100% selectivity. A few minutes was sufficient to produce a stable catalytic surface where active oxygenated species may have transformed into OH/H_2O and/or were replaced by 'C–H' deposits.[78,167] Sintering or particle size redistribution of this Pt-black sample of rather large crystallite size is not likely. It is not excluded that the reconstruction observed by electron microscopy[186] and the activity changes are somehow related to the amount and chemical state of surface oxygen.

Hydrogen

Both industrial and laboratory experiments are usually carried out in the presence of hydrogen. Hydrogen may induce surface reconstruction, including sintering, particle migration, growth or coalescence, and particle shape changes with various supported and unsupported metals.[186–188] Once a catalyst has a likely stable morphology, hydrogen effects in the catalytic reactions manifest themselves. Maximal yields of aromatization as a function of hydrogen pressure have been reported as early as 1961.[189]. These were also observed for other skeletal reactions in laboratory studies[190] and under industrial conditions.[191] Hydrogen effects on hydrocarbon reactions have been summarized in detail.[78,192,193]

Hydrogen can participate in the active ensemble of a catalytic reaction, especially in those ensembles responsible for the formation of saturated products: isomers, C_5 cyclics, and cyclopentane ring opening.[13,41,43] The idea of reactive chemisorption assumes that dissociative adsorption of hydrocarbons occurs on metal–hydrogen sites, thus including the hydrogen in the adsorption equation. This hypothesis applies a single rate equation over a range of several orders of magnitude of hydrogen pressure assuming that adsorption is the rate-determining step.[82,83] Bond et al. suggested a bimolecular Langmuir–Hinshelwood equation to explain maximal rates of alkane hydrogenolysis over Ru/Al_2O_3[194] and Pt/Al_2O_3[195] catalysts. They proposed a competition of the dissociatively chemisorbed reactive intermediates for hydrogen at lower H_2 pressures, i.e., in the positive hydrogen order range. In the negative hydrogen order range, the reacting molecules have to compete with hydrogen for the empty surface sites. Hence, more hydrogen leads to lower rates. The maximal rates (zero hydrogen order) mean an optimal hydrogen coverage for the reaction. Earlier observations[196] that higher temperatures shift this maximum to larger $p(H_2)$ values could be quantified on this basis.[194b,197]

The *optimal* hydrogen pressure is, as a rule, different for different reactions. Two explanations of these hydrogen effects have been proposed. A *direct hydrogen effect* can be interpreted in terms of competition between hydrogen and hydrocarbon reactant. Displacement of the reactive hydrocarbon species by hydrogen explains adequately the hydrogen response of *overall rates* at and above hydrogen pressures for maximal reaction rates. To interpret these different hydrogen responses, two types of surface hydrocarbon entities may be assumed: one deeply dissociated precursor for benzene and/or coke, and one less dissociated intermediate for saturated products produced *via* C_5 cyclic intermediates,[13,28] as illustrated in Figure 8.

Temperature programmed desorption studies point to the presence of more than one type of surface hydrogen (for reviews, see Refs. 192 and 188). Following H_2 adsorption at 773 K, a Pt/Al_2O_3 catalyst contained almost three times as more high-temperature desorbable hydrogen than after adsorption at 300 K.[198] This 'high-temperature' hydrogen seemed to facilitate benzene formation from 2-hexene,[35] likely because hydrogen had an active part in the *trans* $\rightarrow$ *cis* isomerization step(s) essential for the cyclization of a randomly dehydrogenated pool of surface unsaturated entities (cf. Sec. 2). The temperature-programmed dehydrocyclization of fully deuterated *n*-heptane (C_7D_{16}) gave three peaks of toluene.[198] Apart from small D_2 peaks at lower temperatures (one of which corresponded to desorption of unreacted *n*-heptane), there was only one major desorption peak of D_2, coinciding with that of the highest temperature toluene peak. Thus, hydrogen split off during toluene formation may have joined at lower temperatures the hydrogen pool of the surface (perhaps according to the reactive chemisorption theory[82,83,84]). This peak might correspond to the 'direct' aromatization pathway[29] that has been reported to occur with increasing hydrogen concentration in the gas phase.[199] Aromatization in TPR studies was followed up to 550 K, which is much less than the temperatures of industrial reforming, and the conditions were also different: vacuum instead of constant hydrogen and/or hydrocarbon pressure.

An alternative explanation assumed an *indirect hydrogen effect* and held that 'the surface-held hydrocarbons whose coverage is a function of the partial pressure of hydrogen contribute to the observed hydrogen sensitivity of the product selectivity'.[166,200] In other words, the actual hydrogen concentration governs the rate of transformation of 'Pt–H' to 'Pt–C–H' and 'Pt–C' surfaces as

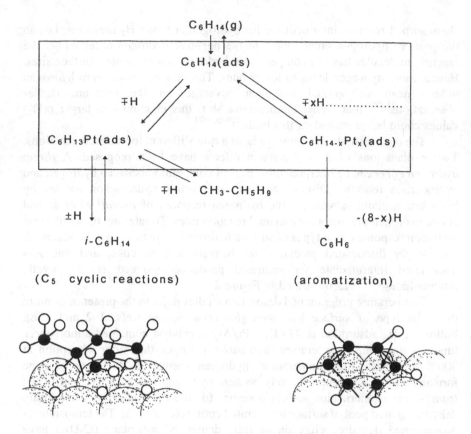

Figure 8 "Direct" hydrogen effect governing C_5 and C_6 cyclization of *n*-hexane on Pt surface, depicted as a fragment of (111) plane. Larger full circles denote C atoms, smaller empty circles hydrogen atoms, both in the reactant and in adsorbed positions. The di-π-chemisorption of the benzene precursor is symbolic; in fact, it may involve Pt–C σ-bond[s].

discussed earlier. The selectivity changes at different hydrogen-to-hexane pressure ratios are in agreement with this hypothesis:[80,168] at lowest H_2 excess, hexenes are the only primary products, as is characteristic of a highly carbonized Pt. The same conclusion has been drawn from transient response curves of the reactions of *n*-hexane on hydrogen-precovered Pt/Al_2O_3 at 793 K.[36] The displacement of hydrogen by hydrocarbon was accompanied by marked changes in relative yields of hydrogenolysis, isomerization, C_5 cyclization, and aromatization. The response curves for 1-hexene formation had a different shape and were hardly affected by the presence or absence of hydrogen precoverage. The curves for all other reactions were essentially different from those of hydrogen-free catalyst.

The two explanations are complementary rather than contradictory. Their essence can be summarized in terms of whether surface hydrogen ultimately determines catalytic selectivity directly or by controlling the amount and/or the state of surface carbon. This latter interpretation could explain why higher hydrogen pressures would increase the selectivity of reactions that require larger clean metallic ensembles. Transient response studies do not exclude either possibility: they indicate that 'the selectivity of n-hexane conversion is controlled by (i) the extent of dehydrogenation and geometry of the primary surface intermediates and (ii) the fraction of sites covered by hydrogen and hydrocarbon'.[36, p. 1290]

Figure 9a shows that maximal rates appear at higher hydrogen pressures with increasing temperatures.[201]. This can give rise to such phenomena as inverse Arrhenius plots as reported for Pt[68] and Ni[81] single crystals and which were explained by shifting from the negative to the zero and then to the positive hydrogen order branch of the bell-shaped curve as the temperature increases at the same hydrogen pressure.[196] This makes it difficult to characterize the host of individual reactions by a single kinetic equation.[195b,197,202] The E_a values measured at zero hydrogen order (at maximum) was 66 kJ mol^{-1} for the overall reaction as well as for the formation of saturated C_6 products from hexane.

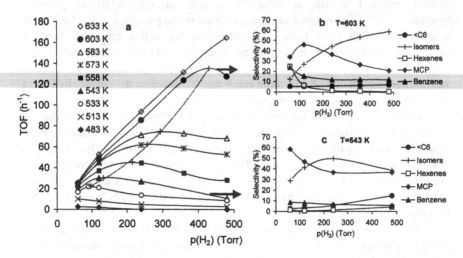

Figure 9 Overall rates of hexane transformation ($p = 10$ Torr) on EUROPT-1 (6% Pt/ SiO$_2$) as a function of the hydrogen pressure (a) and selectivities at two selected temperatures (b and c). (Adapted after Ref. 201.)

Higher values were measured in the negative hydrogen order range, whereas E_a could only be calculated for benzene in the positive hydrogen order branch.[201] Surface hydrogen depletion may be the reason why metal-catalyzed isomerization (which is favored by a high hydrogen content of the catalyst) almost stopped at the low hydrogen pressure and/or at 693 K as opposed to 603 K.[168] At this temperature, reversible carbonaceous deposits have been claimed to determine the catalytic properties, whereas a massive carbonization was assumed to occur above 703 K.[203]

The necessity of optimal surface hydrogen concentration as a function of temperature and pressure is clearly seen when the ratio of C_5 cyclics to aromatics is measured over both single crystals and dispersed Pt catalysts;[58,130,188] cf. Table 2. The metallic character of 8% Pt/NaY with different Pt particle size was more pronounced at higher $p[H_2]$ values.[105] More hydrogen favors the formation of skeletal isomers as opposed to C_5 cyclics from their common surface intermediate,[109,168] as shown in Figure 9b and 9c. Their ratio seems to be the most sensitive indicator of the availability of surface hydrogen on metallic sites, on all types of supports, including neutral,[204] alkaline,[115b,116] and acidic[205] supports. The ratio of fragments to nondegradative products behaves similarly, with lower H_2 pressure favoring hydrogenolysis.[193] This is the situation with MCP ring opening on supported Rh catalysts. In this case, higher temperatures and lower H_2 pressures favor the further splitting of C_6 ring-opening products into fragments, almost exclusively at the expense of 2-methylpentane.[46] The favored position of hydrogenolytic splitting of an alkane chain (expressed by the ω value[54]) is also strongly hydrogen pressure dependent.[58,110,188] Terminal rupture prevailed in the presence of less hydrogen, in agreement with Pt/C being claimed to be the active ensemble for this process.[132] Hydrogen effects seem to be superimposed on structural and particle size effects.[47] Hydrogen should also be considered as one of the factors controlling deactivation of reforming catalysts;[159] see also the chapter by Parera and Figoli in Ref. 1 and Chapter 10 of this book.

In the case of nonacidic supports, hydrogen spillover would also influence the product composition. A graphite nanofiber support has a hydrogen storage capacity. Back spillover from the support can increase the hydrogen concentration on the metallic sites. Thus, the isomer selectivity increased as compared to the values on EUROPT-1 of Figure 9.[206a] On Pt/CeO$_2$ the spillover hydrogen could be consumed by reduction of ceria and more MCP was produced than isomers.[206b]

The maximal yields as a function of hydrogen pressure found at lower absolute pressures in laboratory studies[191] and at higher pressures with industrial catalysts may represent different sections of the same set of curves in an n-dimensional space. Three-dimensional plots have already been published, using $p[H_2]$ and $p[alkane]$ as parameters.[195b] Direct comparison of the same catalysts in a much wider pressure and temperature range would be necessary to elucidate all subtleties of the very complex phenomena occurring during reforming.

6 CONCLUDING REMARKS ON THE POSSIBLE ACTIVE SITES IN METAL CATALYZED REACTIONS

The feature of the catalyst metal considered to be most important in determining its properties was defined either as an electronic factor[125] or as a bond strength requirement.[207] Another important parameter can be catalyst geometry. Balandin[208] attributed catalytic effect to atoms of large, stable, low-index clean areas of metal atoms. Crystal structure and atomic diameter are important (as in Fig. 1) but in the opposite sense: (1) the surface of working catalysts is never clean; (2) the activity is attributed to rough, high Miller index surfaces (cf. Fig. 2), which (3) reconstruct readily during catalysis, as postulated by the 'flexible surface' theory.[67,87]

Statement (1) is supported by surface analysis of Pt-black catalysts[88,172] as well as Pt single crystals[67,68,160] containing high amounts of metallic Pt available for catalytic reactions in spite of their relative large carbon (and some oxygen) content, as postulated in the concept of catalytic system.[89] Statement (2) is in agreement with the importance of ledge structures in skeletal reactions,[74,75] as formulated by Somorjai: 'rough surfaces do chemistry'.[67,87] As for statement 3, the concept of active sites and active centers as entities undergoing constant changes during reaction has been discussed by Burch,[142] who put forward also the concept of *reaction-sensitive structures*.[209] On inhomogeneous surfaces the reactions select their optimal active sites themselves, and additional active sites are created by surface reconstruction induced by the reactant or by other components present, e.g., hydrogen and/or oxygen.[87,88,186,187] The strength of chemical bond between a chemisorbed entity and a metal is close to the lattice energy, and the formation and rupture of such bonds may induce rearrangement in the metal crystal itself ('corrosive chemisorption'[207]). Thus, a catalyst must not be regarded as a constant and immobile entity during reaction. The catalytic reaction itself may still be one of the best means of catalyst characterization.

REFERENCES

1. *Catalytic Naphtha Reforming: Science and Technology*, 1st ed.; Antos, G.J., Aitani, A.M., Parera, J.M., Eds.; Marcel Dekker: New York, 1995.
2. Gates, B.C.; Katzer, J.R.; Schuit, G.C.A. *Chemistry of Catalytic Processes*; McGraw-Hill: New York, 1979; 184–324.

3.   Satterfield, C.N. *Heterogeneous Catalysis in Industrial Practice*; McGraw-Hill: New York, 1991.

4.   Olah, G.A.; Molnár, Á. *Hydrocarbon Chemistry*; Wiley: New York, 1995.

5.   (a) Kazansky, B.A.; Plate, A.F. Ber. Dtsch. Chem. Ges. B **1936**, *69*, 1862; (b) Kazanskii, B.A.; Liberman, A.L.; Batuev, M.I. Dokl. Akad. Nauk SSSR **1948**, *61*, 67.

6.   Kazanskii, B.A.; Liberman, A.L.; Bulanova, T.F. Aleksanyan, V.T. Sterin, Kh.E. Dokl. Akad. Nauk SSSR **1954**, *95*, 77; Kazanskii, B.A.; Liberman, A.L.; Aleksanyan, V.T.; Sterin, Kh.E. Dokl. Akad. Nauk SSSR **1954**, *95*, 281.

7.   Zelinsky, N.D.; Kazansky, B.A.; Plate, A.F. Ber. Dtsch. Chem. Ges. B **1935**, *68*, 1869.

8.   Yuryev, Yu.K.; Pavlov, P.Ya. Zh. Obshch. Khim. **1937**, *7*, 97.

9.   Barron, Y.; Maire, G.; Muller, J.M.; Gault, F.G. J. Catal. **1966**, *5*, 428.

10. (a) Anderson, J.R.; Baker, B.G. Nature, **1960**, *187*, 937; (b) Anderson, J.R.; Avery, N.R. J. Catal. **1966**, *5*, 446.

11. McKervey, M.A.; Rooney, J.J. Sammann, N.G. J. Catal. **1973**, *30*, 330.

12. Gault, F.G. Adv. Catal. **1981**, *30*, 1.

13. Paál, Z. Adv. Catal. **1980**, *29*, 273.

14. de Jongste, H.C.; Ponec, V. Bull. Soc. Chim. Belg. **1979**, *88*, 453.

15. Sinfelt, J.H. Adv. Catal. **1973**, *23*, 91.

16. Paál, Z. Tétényi, P. In *Catalysis Specialists Periodical Reports*; Bond, G.C., Webb, G., Eds.; Royal Soc. Chem.: London, 1982; Vol. 5, 80 pp.

17. van Broekhoven, E.; Ponec, V. Progr. Surf. Sci. **1985**, *19*, 351.

18. Clarke, J.K.A.; Rooncy, J.J. Adv. Catal. **1976**, *25*, 125.

19. Anderson, J.R. Adv. Catal. **1973**, *23*, 1.

20. Ponec, V.; Bond, G.C. *Catalysis by Metals and Alloys*; Elsevier: Amsterdam, 1995.

21. Twigg, G.H. Trans. Faraday Soc. **1939**, *35*, 979.

22. Davis, B.H. Catal. Today **1999**, *53*, 443.

23. Rozengart, M.I.; Mortikov, E.S.; Kazansky, B.A. Dokl. Akad. Nauk SSSR **1966**, *158*, 911; **1966**, *166*, 619.

24. Trunschke, A.; Hoang, D.L.; Radnik, J.; Lieske, H. J. Catal. **2000**, *191*, 456.

25. Hoang, D.L.; Preiss, H.; Parlitz, B.; Krumeich, F.; Lieske, H. Appl. Catal. A **1999**, *182*, 385.

26. Trunschke, A.; Hoang, D.L.; Radnik, J. Brezinka, K.-W.; Brückner, A.; Lieske, H. Appl. Catal. A **2001**, *208*, 371.

27. Paál, Z.; Tétényi, P. Acta Chim. Acad. Sci. Hung. **1967**, *53*, 193; **1968**, *54*, 175; **1968**, *58*, 105.

28. Paál, Z.; Tétényi, P. J. Catal. **1973**, *30*, 350.

29. Dautzenberg, F.M.; Platteeuw, J.C. J. Catal. **1970**, *19*, 41.

30. Paál, Z. Discussion remark. *Proceedings of the 10th International Congress on Catalysis*, Budapest, 1992; Part A, 903 pp.

31. Nogueira, L.; Pines, H. J. Catal. **1981**, *70*, 404.

32. Iglesia, E.; Baumgartner, J.; Price, G.L.; Rose, K.D.; Robins, J.L. J. Catal. **1990**, *125*, 95.

33. (a) Sparks, D.E.; Srinivasan, R.; Davis, B.H. J. Mol. Catal. **1994**, *88*, 325; (b) Srinivasan, R.; Davis, B.H. J. Mol. Catal. **1994**, *88*, 343; (c) Sparks, D.E.; Srinivasan, R.; Davis, B.H. J. Mol. Catal. **1994**, *88*, 359; (d) Davis, B.H. *Proceedings of the 10th International Congress on Catalysis*, Budapest, 1992; Part B, 889 pp.

34. Twigg, G.H. Proc. R. Soc. A **1941**, *178*, 106.
35. Zimmer, H.; Rozanov, V.V.; Sklyarov, A.V.; Paál, Z. Appl. Catal. **1982**, *2*, 51.
36. Margitfalvi, J.L.; Szedlacsek, P.; Hegedüs, M. Tálas, E.; Nagy, F. *Proceedings of the 9th International Congress on Catalysis*, Calgary, 1988; Vol. 3, 1283 pp.
37. Teplyakov, A.V.; Bent, B.E.; J. Phys. Chem. **1997**, *101*, 9052; Teplyakov, A.V.; Gurevich, A.B.; Garland, E.R.; Bent, B.E.; Chen, J.G. Langmuir **1998**, *14*, 1337.
38. Paál, Z.; Thomson, S.J. J. Catal. **1973**, *30*, 96.
39. Maire, G.L.C.; Garin, F.G. In *Catalysis, Science and Technology*; Anderson, J.R., Boudart, M., Eds.; Springer-Verlag: Berlin, 1984; Vol. 6, 161 pp.
40. Shephard, F.E.; Rooney, J.J. J. Catal. **1964**, *3*, 129.
41. Liberman, A.L. Kinet. Katal. **1964**, *4*, 128.
42. Finlayson, O.E.; Clarke, J.K.A. Rooney, J.J. J. Chem. Soc. Faraday Trans. 1 **1980**, *80*, 345.
43. Zimmer, H.; Paál, Z. *Proceedings of the 8th International Congress on Catalysis*, Berlin, 1984; Vol. 3, 417 pp.
44. Zimmer, H.; Paál, Z. J. Mol. Catal. **1989**, *51*, 261.
45. Paál, Z.; Brose, B.; Räth, M.; Gombler, W. J. Mol. Catal. **1992**, *75*, L13.
46. Teschner, D.; Matusek, K.; Paál, Z. J. Catal. **2000**, *192*, 335; Teschner, D.; Paál, Z.; Duprez, D. Catal. Today **2001**, *65*, 185.
47. Bragin, O.V.; Karpinski, Z.; Matusek, K.; Paál, Z.; Tétényi, P. J. Catal. **1979**, *56*, 219.
48. Sachtler, W.M.H. J. Mol. Catal. **1992**, *77*, 99; Alvarez, W.E.; Resasco, D.E. J. Catal. **1996**, *164*, 467.
49. Vogelzang, M.W.; Botman, M.J.P.; Ponec, V. Faraday Disc. Soc. **1981**, *72*, 33.
50. Garin, F.; Girard, P.; Weisang, F.; Maire, G. J. Catal. **1981**, *70*, 215.
51. Maire, G.; Garin, F. J. Mol. Catal. **1988**, *48*, 99.
52. Paál, Z.; Tétényi, P. Nature **1977**, *267*, 234.
53. Paál, Z.; Tétényi, P.; Dobrovolszky, M. React. Kinet. Catal. Lett. **1988**, *37*, 163.
54. Leclercq, G.; Leclercq, L.; Maurel, R. J. Catal. **1977**, *50*, 87.
55. Burch, R.; Paál, Z. Appl. Catal. A **1994**, *114*, 9.
56. Ponec, V.; Sachtler, W.M.H. *Proceedings of the 5th International Congress on Catalysis*, Palm Beach, 1972; Vol. 1, 645 pp.
57. Van Schaik, J.R.H.; Dessing, R.P.; Ponec, V. J. Catal. **1975**, *38*, 273.
58. Paál, Z.; Tétényi, P. React. Kinet. Catal. Lett. **1979**, *12*, 131.
59. Majesté, A.; Balcon, S.; Guérin, M.; Kappenstein, C. Paál, Z. J. Catal. **1999**, *187*, 486.
60. Taylor, J.F.; Clarke, J.K.A.; Z. Phys. Chem. Neue Folge **1976**,*103*, 216.
61. Hammer, B.; Norskov, J.K. Adv. Catal. **2000**, *45*, 71.
62. Bond, G.C. J. Mol. Catal. **2000**, *156*, 1.
63. Biloen, P.; Dautzenberg, F.M.; Sachtler, W.M.H. J. Catal. **1977**, *50*, 77.
64. Biloen, P.; Helle, J.N.; Verbeek, H.; Dautzenberg, F.M.; Sachtler, W.M.H. J. Catal. **1980**, *63*, 112.
65. (a) Anderson, J.B.F.; Burch, R.; Cairns, J.A.; J. Catal. **1987**, *107*, 364; (b) Anderson, J.B.F.; Burch, R.; Cairns, J.A. J. Catal. **1987**, *107*, 351.
66. Bonzel, H. Surf. Sci. **1977**, *68*, 236.
67. Somorjai, G.A. *Chemistry in Two Dimensions*; Cornell University Press: Ithaca, London, 1981; Somorjai, G.A. *Introduction to Surface Chemistry and Catalysis*; Wiley: New York, 1994.

68. Davis, S.M.; Zaera, F.; Somorjai, G.A. J. Catal. **1984**, *85*, 206.
69. Zaera, F.; Godbey, D.; Somorjai, G.A. J. Catal. **1986**, *101*, 73.
70. Nieuwenhuys, B.E.; Somorjai, G.A. J. Catal. **1977**, *46*, 259.
71. Freund, H.-J.; Bäumer, M.; Kuhlenbeck, H. Adv. Catal. **2000**, *45*, 333.
72. Somorjai, G.A.; McCrea, K.R. Adv. Catal. **2000**, *45*, 385.
73. Rupprechter, G.; Somorjai, G.A. J. Phys. Chem. B **1999**, *103*, 1623.
74. Garin, F.; Aeiyach, S.; Légaré, P.; Maire, G. J. Catal. **1982**, *77*, 323.
75. Dauscher, A.; Garin, F.; Maire, G. J. Catal. **1987**, *105*, 233.
76. Blakely, D.W.; Somorjai, G.A. J. Catal. **1976**, *42*, 181.
77. Ehrlich, G. Adv. Catal. **1963**, *14*, 255; a structure is depicted on page 314.
78. Paál, Z. Catal. Today **1992**, *12*, 297.
79. Gnutzmann, V.; Vogel, W. J. Phys. Chem. **1990**, *94*, 4991.
80. Bond, G.C.; Paál, Z. Appl. Catal. **1992**, *86*, 1.
81. Goodman, D.W. Catal. Today **1992**, *12*, 189.
82. Frennet, A. In *Hydrogen Effects in Catalysis*; Paál, Z., Menon, P.G., Eds.; Marcel Dekker: New York, 1988, 399 pp.
83. Parayre, P.; Amir-Ebrahimi, V.; Gault, F.G.; Frennet, A. J. Chem. Soc. Faraday Trans. 1 **1980**, *76*, 1704.
84. Frennet, A.; Liénard, G.; Crucq, A.; Degols, L. Surf. Sci. **1979**, *80*, 412.
85. Shi, B.; Davis, B.H. *Proceedings of the 11th International Congress on Catalysis*, Baltimore, 1996; 1145 pp.
86. (a) Shi, B.; Davis, B.H. J. Catal. **1996**, *162*, 134; (b) Shi, B.; Davis, B.H. J. Catal. **1994**, *147*, 38.
87. Somorjai, G.A. Catal. Lett. **1991**, *9*, 311; Catal. Lett. **1992**, *12*, 17; J. Mol. Catal. A **1996**, *107*, 39.
88. Paál, Z.; Wild, U.; Wootsch, A.; Find, J.; Schlögl, R. Phys. Chem. Chem. Phys. **2001**, *3*, 2148.
89. Tétényi, P.; Guczi, L.; Paál, Z. Acta Chim. Acad. Sci. Hung. **1974**, *83*, 37.
90. Burch, R. In *Hydrogen Effects in Catalysis*; Paál, Z., Menon, P.G., Eds.; Marcel Dekker: New York, 1988; 347 pp.
91. Hayek, K.; Kramer, R.; Paál, Z. Appl. Catal. A **1997**, *162*, 1.
92. Kramer, R.; Zuegg, H. J. Catal. **1983**, *80*, 446; **1984**, *85*, 530.
93. Paál, Z. Catal. Today **1988**, *2*, 595.
94. Rupprechter, G.; Hayek, K.; Hofmeister, H. J. Catal. **1998**, *173*, 409; Rupprechter, G.; Seeber, G.; Goller, H.; Hayek, K. J. Catal. **1999**, *186*, 201.
95. Karpinski, Z. Adv. Catal. **1990**, *37*, 45; Lomot, D.; Juszczyk, W. Karpinski, Z. Appl. Catal. A **1995**, *156*, 77.
96. Mills, G.A.; Heinemann, H.; Milliken, T.H. Oblad, A.G. Ind. Eng. Chem. **1953**, *45*, 134.
97. Silvestri, A.J.; Naro, P.A.; Smith, R.L. J. Catal. **1969**, *14*, 386.
98. Birke, P.; Engels, S.; Becker, K.; Neubauer, H.-D. Chem. Techn. **1979**, *31*, 473.
99. Parera, J.M.; Beltramini, J.N.; Querini, C.A.; Martinelli, E.E.; Churin, E.J.; Aloe, P.E.; Figoli, N.S. J. Catal. **1986**, *99*, 39.
100. Paál, Z. J. Catal. **1987**, *105*, 540.
101. Mériaudeau, P.; Naccache, C. Cat. Rev. Sci. Eng. **1997**, *39*, 5.
102. Paál, Z.; Räth, M.; Zhan, Z.; Gombler, W. J. Catal. **1994**, *127*, 342.

103. Moretti, G.; Sachtler, W.M.H. J. Catal. **1989**, *115*, 205.
104. Sachtler, W.M.H. Adv. Catal. **1993**, *39*, 129.
105. Paál, Z.; Zhan, Z.; Manninger, I.; Sachtler, W.M.H. J. Catal. **1995**, *155*, 43.
106. Zhan, Z.; Manninger, I.; Paál, Z.; Barthomeuf, D. J. Catal. **1994**, *147*, 333.
107. Paál, Z.; Xu, X.L. *Catalysis by Microporous Materials*; Beyer, H.K., Karge, H.G., Kiricsi, I., Nagy, J.B., Eds.; Stud. Surf. Sci. Catal.; Elsevier: Amsterdam, 1995; Vol. 94, 590 pp.
108. (a) Manninger, I.; Paál, Z.; Tesche, B.; Klengler, U. Halász, J.; Kiricsi, I. J. Mol. Catal. **1991**, *64*, 361; (b) Manninger, I.; Zhan, Z.; Xu, X.L. Paál, Z. J. Mol. Catal. **1991**, *65*, 223.
109. Paál, Z.; Groeneweg, H.; Paál-Lukács, J. J. Chem. Soc. Faraday Trans. **1990**, *86*, 3159.
110. Zimmer, H.; Dobrovolszky, M.; Tétényi, P.; Paál, Z. J. Phys. Chem. **1986**, *90*, 4758.
111. Bernard, J.R. In *Proceedings of the 5th International Congress on Zeolites, Naples*, 1980; 686 pp.; Hughes, T.R.; Buss, W.C.; Tamm, P.W.; Jacobson, R.L. *New Developments in Zeolite Science and Technology*; Murakami, Y., Iijima, A., Ward, J.W., Eds.; Kodansha–Elsevier: Tokyo–Amsterdam, 1986; 725 pp.
112. Derouane, E.G.; Vanderveken, D.J. Appl. Catal. **1988**, *45*, L15.
113. Lane, G.S.; Modica, F.S.; Miller, J.T. J. Catal. **1991**, *129*, 145.
114. Mielczarski, E.; Hong, S.B.; Davis, R.J.; Davis, M.E. J. Catal. **1992**, *134*, 359.
115. (a) Becue, T.; Maldonado, F.J.; Antunes, A.P.; Silva, J.M.; Ribeiro, M.F.; Massiani, P.; Kermarec, M. J. Catal. **1999**, *181*, 244; (b) Maldonado, F.J.; Becue, T.; Silva, J.M.; Ribeiro, M.F.; Massiani, P.; Kermarec, M. J. Catal. **2000**, *195*, 342.
116. Menacherry, P.; Haller, G.L. J. Catal. **1998**, *177*, 175.
117. (a) Davis, R.J.; Derouane, E.G. J. Catal. **1991**, *132*, 269; (b) Kazansky, V.B.; Borovkov, V.Yu.; Derouane, E.G. Catal. Lett. **1993**, *19*, 327.
118. Davis, R.J.; Derouane, E.G. J. Catal. **1991**, *132*, 269; Derouane, E.G.; Jullien-Lardot, V.; Davis, R.J.; Blom, N.; Hojlund-Nielsen, P.E. *Proceedings of the 10th International Congress on Catalysis*, Budapest, 1992; Part B, 1031 pp.
119. (a) Mojet, B.L.; Miller, J.T.; Ramaker, D.E.; Koningsberger, D.C. J. Catal. **1999**, *186*, 373; (b) Koningsberger, D.C.; de Graaf, J.; Mojet, B.L.; Ramaker, D.E.; Miller, J.T. Appl. Catal. A **2000**, *191*, 205; (c) Ramaker, D.E.; de Graaf, J.; van veen, J.A.R. Koningsberger, D.C. J. Catal. **2001**, *202*, 7.
120. Iglesia, E. Baumgartner, J.E. *Proceedings of the 10th International Congress on Catalysis*, Budapest, 1992; Part B, 993 pp.
121. Jentoft, R.E.; Tsapatis, M.; Davis, M.E.; Gates, B.C. J. Catal. **1998**, *179*, 565; Jacobs, G.; Ghadiali, F.; Pisanu, A.; Borgna, A.; Alvarez, W.E.; Resasco, D.E. Appl. Catal. A **1999**, *188*, 79.
122. Vaarkamp, M.; Miller, J.T.; Modica, F.S.; Lane, G.S.; Koningsberger, D.C. J. Catal. **1992**, *138*, 675.
123. Paál, Z.; Brose, B.; Räth, M. Gombler, W. J. Mol. Catal. **1992**, *75*, L13.
124. Foger, K.; Anderson, J.R. J. Catal. **1978**, *54*, 318.
125. Garin, F.; Maire, G. Acc. Chem. Res. **1989**, *20*, 100.
126. Paál, Z.; Matusek, K.; Zimmer, H. J. Catal. **1993**, *141*, 648.
127. Paál, Z.; Tétényi, P. J. Catal. **1973**, *29*, 176.
128. Kane, A.F.; Clarke, J.K.A. J. Chem. Soc. Faraday Trans. 1 **1980**, *76*, 1640.

129. Kazansky, B.A.; Liberman, A.L.; Loza, G.V.; Vasina, T.V. Dokl. Akad Nauk SSSR **1959**, *128*, 1188.

130. Leclercq, G.; Leclercq, L.; Maurel, R. Bull. Soc. Chim. Belg. **1979**, *88*, 599.

131. Sachtler, W.M.H.; van Santen, R.A. Adv. Catal. **1977**, *26*, 69.

132. Ponec, V. Adv. Catal. **1983**, *32*, 149.

133. Charron, A.; Kappenstein, C.; Guérin, M.; Paál, Z. Phys. Chem. Chem. Phys. **1999**, *1*, 3817.

134. Ali, A.-G.; Ali, L.I.; Aboul-Fotouh, S.A.; Aboul-Gheit, A.K. Appl. Catal. A **2001**, *215*, 161.

135. Coq, B.; Figueras, F. Coord. Chem. Rev. **1998**, *178–180*, 1753.

136. Palazov, A.; Bonev, Ch.; Shopov, D.; Lietz, G.; Sárkány, A.; Völter, J. J. Catal. **1987**, *103*, 249.

137. Kappenstein, C.; Saouabe, M.; Guérin, M.; Marcot, P.; Uszkurat, I.; Paál, Z. Catal. Lett. **1995**, *31*, 9; Kappenstein, C.; Guérin, M.; Lázár, K.; Matusek, K.; Paál, Z. J. Chem. Soc. Faraday Trans. **1998**, *94*, 2463.

138. Dautzenberg, F.; Helle, J.N.; Biloen, P.; Sachtler, W.M.H. J. Catal. **1980**, *63*, 119.

139. Srinivasan, R.; Davis, B.H. J. Mol. Catal. **1994**, *88*, 343.

140. Llorca, J.; Homs, N.; Garcia-Fierro, J.L.; Sales, J.; Ramirez de la Piscina, P. J. Catal. **1997**, *166*, 44.

141. Sachtler, W.M.H. J. Mol. Catal. **1984**, *25*, 1; Shum, V.K.; Butt, J.B.; Sachtler, W.M.H. J. Catal. **1986**, *99*, 126.

142. Burch, R. Catal. Today **1991**, *10*, 233.

143. Burch, R.; Garla, L.C. J. Catal. **1981**, *71*, 360.

144. Margitfalvi, J.L.; Hegedüs, M.; Göbölös, S.; Kern-Tálas, E.; Szedlacsek, P.; Szabó, S.; Nagy, F. *Proceedings of the 8th International Congress on Catalysis, Berlin*, 1984; Vol. 4, 903 pp.; Margitfalvi, J.L.; Tálas, E.; Göbölös, S. Catal. Today **1989**, *6*, 73; Vértes, Cs.; Tálas, E.; Czakó-Nagy, I.; Ryczkowsky, J.; Göbölös, S.; Vértes, A.; Margitfalvi, J. Appl. Catal. **1991**, *68*, 149.

145. Margitfalvi, J.L.; Borbáth, I.; Tfirst, E.; Tompos, A. Catal. Today **1998**, *43*, 29; Margitfalvi, J.L.; Borbáth, I.; Hegedüs, M.; Göbölös, S.; Lónyi, F.; React. Kin. Cat. Lett. **1999**, *68*, 133.

146. Edreva-Kardjieva, R.M.; Andreev, A.A. J. Catal. **1985**, *94*, 97.

146a. Wootsch, A.; Pirault-Roy, L.; Leverd, J.; Guérin, M.; Paál, Z. J. Catal. **2002**, *208*, 490.

147. Clarke, J.K.A. Chem. Rev. **1975**, *75*, 291.

148. Menon, P.G.; Paál, Z. Ind. Eng. Chem. Res. **1997**, *36*, 3282.

149. Biswas, J.; Bickle, G.M.; Gray, P.G.; Do, D.D. J. Barbier, Catal. Rev. Sci. Eng. **1988**, *30*, 161.

150. Menon, P.G.; Prasad, J. *Proceedings of the 6th International Congress on Catalysis*, London, 1976; Vol. 2, 1061 pp.

151. Pönitzsch, L.; Wilde, M.; Tétényi, P.; Dobrovolszky, M.; Paál, Z. Appl. Catal. **1992**, *86*, 115.

152. Dees, M.J.; den Hartog, A.J.; Ponec, V. Appl. Catal. **1991**, *72*, 343.

153. Ponec, V. Catal. Today **1991**, *10*, 251.

154. Maire, G.; Lindauer, G.; Garin, F.; Légaré, P.; Cheval, M.; Vayer, M. J. Chem. Soc. Faraday Trans. **1990**, *86*, 2719.

155. Kim, Ch.; Somorjai, G.A. J. Catal. **1992**, *134*, 179.
156. (a) Paál, Z.; Matusek, K.; Muhler, M. Appl. Catal. A **1997**, *149*, 113; (b) Paál, Z.; Muhler, M.; Matusek, K.; J. Catal. **1998**, *175*, 245.
157. Barbier, J.; Lamy-Pitara, E.; Marecot, P.; Boitiaux, J.P.; Cosyns, J.; Verna, F. Adv. Catal. **1990**, *37*, 279.
158. Borgna, A.; Garetto, T.F.; Apesteguia, C.R. Appl. Catal. A **2000**, *197*, 11.
159. Parera, J.M.; Figoli, N.S. In *Catalysis Specialists Periodical Reports*; Spivey, J.J., Ed.; Roy. Soc. Chem.: London, 1992; Vol. 9, 65 pp.
160. Somorjai, G.A.; Zaera, F. J. Phys. Chem. **1982**, *86*, 3070.
161. Garin, F.; Maire, G.; Zyade, S.; Zauwen, M.; Frennet, A.; Zielinski, P. J. Mol. Catal. **1990**, *58*, 185.
162. Menon, P.G. J. Mol. Catal. **1990**, *59*, 207.
163. Sárkány, A. J. Chem. Soc. Faraday Trans. 1 **1988**, *84*, 2267.
164. Sárkány, A. Catal. Today **1989**, *5*, 173.
165. Thomson, S.J.; Webb, G. J. Chem. Soc. Chem. Commun. **1976**, 526; G. Webb, Catal. Today **1990**, *7*, 139.
166. Sárkány, A. J. Chem. Soc. Faraday Trans. 1 **1989**, *85*, 1523.
167. Paál, Z.; Zhan, Zh. Langmuir **1997**, *13*, 3752.
168. Paál, Z.; Manninger, I.; Zhan, Zh.; Muhler, M. Appl. Catal. **1990**, *66*, 305.
169. Querini, C.A.; Figoli, N.S.; Parera, J.M. Appl. Catal. **1989**, *53*, 53.
170. Sárkány, A.; Lieske, H.; Szilágyi, T.; Tóth, L. *Proceedings of the 8th International Congress on Catalysis*, Berlin, 1984; Vol. 2, 613 pp.
171. Wolf, M.; Deutschmann, O.; Behrendt, F.; Warnatz, J. Catal. Lett. **1999**, *61*, 15.
172. Paál, Z.; Schlögl, R.; Ertl, G. J. Chem. Soc. Faraday Trans. **1992**, *88*, 1179.
173. Guisnet, M.; Magnoux, P. Appl. Catal. A **2001**, *212*, 83.
174. van Donk, S.; Bitter, J.H.; de Jong, K.P. Appl. Catal. A **2001**, *212*, 97.
175. Barbier, J. In *Catalyst Deactivation 1987*; Delmon, B., Froment, G.F., Eds.; Elsevier: Amsterdam, 1987; 1 pp.
176. Foger, K.; Gruber, H.L. J. Catal. **1990**, *122*, 307.
177. Fási, A.; Kiss, J.T.; Török, B.; Pálinkó, I. Appl. Catal. A **2000**, *200*, 189; Török, B.; Molnár, A.; Pálinkó, I.; Bartók, M. J. Catal. **1994**, *145*, 295.
178. Sárkány, A. *Catalyst Deactivation*; Delmon, B., Froment, G.F., Eds.; Stud. Surf. Sci. Catal.; Elsevier: Amsterdam, 1987; Vol. 34, 125 pp.
179. (a) Matusek, K.; Paál, Z. React. Kinet. Catal. Lett. **1999**, *67*, 246; (b) Matusek, K.; Wootsch, A.; Zimmer, H.; Paál, Z. Appl. Catal. A **2000**, *191*, 141; (c) Matusek, K.; Kappenstein, C.; Guérin, M.; Paál, Z. Catal. Lett. **2000**, *64*, 33.
180. Wootsch, A.; Descorme, C.; Paál, Z.; Duprez, D. J. Catal. **2002**, *208*, 276.
181. Rodriguez, N.M.; Anderson, P.E.; Wootsch, A.; Wild, U.; Schlögl, R.; Paál, Z. J. Catal. **2001**, *197*, 365.
182. Wild, U.; Teschner, D.; Schlögl, R.; Paál, Z. Catal. Lett. **2000**, *67*, 93.
183. Sundararajan, R.; Petö, G.; Koltay, E.; Guczi, L. Appl. Surf. Sci. **1995**, *90*, 165.
184. Wilde, M.; Stolz, R.; Feldhaus, R.; Anders, K. Appl. Catal. **1987**, *31*, 99.
185. Santacesaria, E.; Gelosa, D.; Carra, S. J. Catal. **1975**, *39*, 403.
186. Gao, S.; Schmidt, L.D. J. Catal. **1989**, *115*, 473; Schmidt, L.D.; Krause, K.R. Catal. Today **1992**, *12*, 269, 291–293.

187. Ruckenstein, E.; Sushumna, I. In *Hydrogen Effects in Catalysis*; Paál, Z., Menon, P.G., Eds.; Marcel Dekker: New York, 1988; 259 pp.
188. Paál, Z. In *Hydrogen Effects in Catalysis*; Paál, Z., Menon, P.G., Eds.; Marcel Dekker: New York, 1988; 293 pp.
189. Rohrer, J.C.; Hurvitz, H.; Sinfelt, J.H. J. Phys. Chem. **1961**, *65*, 1458.
190. Paál, Z.; Tétényi, P. Dokl. Akad. Nauk SSSR **1971**, *201*, 1119.
191. Bournonville, J.-P.; Franck, J.-P. In *Hydrogen Effects in Catalysis*; Paál, Z., Menon, P.G., Eds.; Marcel Dekker: New York, 1988; 653 pp.
192. Paál, Z.; Menon, P.G. Cat. Rev. Sci. Eng. **1983**, *25*, 273.
193. Paál, Z. In *Hydrogen Effects in Catalysis*; Paál, Z., Menon, P.G., Eds.; Marcel Dekker: New York, 1988; 449 pp.
194. (a) Bond, G.C.; Slaa, J.C. Catal. Lett. **1994**, *23*, 293; (b) Bond, G.C.; Slaa, J.C. J. Mol. Catal. A **1995**, *98*, 81.
195. (a) Bond, G.C.; Hui, L. J. Catal. **1992**, *137*, 462; (b) Bond, G.C.; Cunningham, R.H. J. Catal. **1997**, *166*, 172.
196. Paál, Z. J. Catal. **1985**, *91*, 181.
197. Bond, G.C. Appl. Catal. A **2000**, *191*, 23.
198. Rozanov, V.V.; Gland, J.; Sklyarov, A.V. Kinet. Katal. **1979**, *20*, 1249.
199. Paál, Z.; Xu, X.L. Appl. Catal. **1988**, *43*, L1.
200. Lankhorst, P.P.; de Jongste, H.C.; Ponec, V. *Catalyst Deactivation*; Delmon, B., Froment, G.F., Eds.; Elsevier: Amsterdam, 1980; 43 pp.
201. Wootsch, A.; Paál, Z. J. Catal. **1999**, *185*, 192; J. Catal. **2002**, *205*, 86.
202. Bond, G.C.; Hooper, A.D.; Slaa, J.C.; Taylor, A.G. J. Catal. **1996**, *163*, 319.
203. Christoffel, E.G.; Paál, Z. J. Catal. **1982**, *73*, 30.
204. Paál, Z. J. Catal. **1995**, *156*, 301.
205. Falco, M.G.; Canavese, S.A.; Comelli, R.A.; Figoli, N.S. Appl. Catal. A **2000**, *201*, 37.
206. (a) Baker, R.T.K.; Laubernds, K.; Wootsch, A.; Paál, Z. J. Catal. **2000**, *193*, 165; (b) Teschner, D.; Wootsch, A.; Röder, T.; Matusek, K.; Paál, Z. Solid State Ionics **2001**, *141–142*, 709.
207. Sachtler, W.M.H. Faraday Disc. Chem. Soc. **1981**, *72*, 7.
208. Balandin, A.A. Z. Phys. Chem. B **1929**, *2*, 289; Adv. Catal. **1969**, *19*, 1.
209. Burch, R. In *Catalysis Specialists Periodical Reports*; Bond, G.C., Webb, G., Eds.; Roy. Soc. Chem.: London, 1985; Vol. 7, 149 pp.

3
Chemistry of Bifunctional Metal–Acid Catalysis

José M. Parera and Nora S. Fígoli
Instituto de Investigaciones en Catálisis y Petroquímica (INCAPE),
Santa Fe, Argentina

1 INTRODUCTION

In Chapter 3 of the first edition of this book[1] the hydrocarbon reactions occurring during naphtha reforming were reviewed. The most important reactions (from most rapid to slowest: dehydrogenation of cyclohexanes, dehydroisomerization of alkylcyclopentanes, dehydrogenation of paraffins, isomerization of paraffins, dehydrocyclization of paraffins, hydrocracking, hydrogenolysis, and coke deposition) were analyzed individually. The thermodynamics, kinetics, and mechanisms of each reaction were studied, and the influence of operational parameters on the activity, selectivity, and stability of the reactions and the reaction models for naphtha reforming were used to conclude which operational conditions are the most convenient. As the most important reactions of naphtha reforming occur through bifunctional catalysis, the chemistry of the bifunctional metal–acid catalyst is emphasized in this chapter.

Many important hydrocarbon reactions, such as paraffin cracking and isomerization, are catalyzed by materials with acid properties. Some reactions require sites of stronger acidity than others, for example, in paraffin cracking, the required strength of the acid sites is stronger than that required for isomerization.[2] When the metallic function is added to the acid function, a bifunctional catalyst is obtained. This catalyst presents in general a more beneficial behavior than the catalyst having only the acid function. The bifunctional metal–acid catalysts are generally porous oxides with acid properties that have a small amount of a metal supported on them. The acid

75

function of the support and the metal function may be tuned to promote the desired reaction selectivity by the addition of promoters.

There are several bifunctional mechanisms proposed in the literature to interpret experimental data. For the main naphtha-reforming reactions (dehydrocyclization, isomerization and hydrocracking of paraffins, and dehydroisomerization of alkylcyclopentanes), a bifunctional metal–acid mechanism was postulated by Mills et al.,[3] Weisz,[4,5] and Sterba and Haensel.[6] This mechanism interpreted the results of the reforming reactions carried out under experimental conditions and with catalysts not very different than those of the commercial process. After nearly half a century since it was proposed, the mechanism is still very useful and is usually named as the "classical" bifunctional reaction mechanism. Under other experimental conditions, other bifunctional mechanisms have been proposed, and these will be discussed as well.

2   THERMODYNAMICS

The chemical thermodynamic equations that can be applied to hydrocarbon reactions are discussed in the following sections.

2.1   Change of Standard Free Energy of Reaction

$$\Delta F_R{}^o = \Sigma \gamma_i \Delta F_{Fi}{}^o \tag{1}$$

where $\Delta F_{Fi}{}^o$ is the standard free energy of formation for each of the i reacting species and γ_i is the stoichiometric coefficient of each in the reaction, which is negative for reactants and positive for products.

2.2   Equilibrium Constant

$$K = \Pi a_i{}^{\gamma i} \tag{2}$$

where Π is the mathematical product of all the values of the equilibrium activity of reactants and products, a_i. K is a comparison of the products activities with the reactants activities, and the reaction is feasible when K is greater than 1. K is related to the change in free energy of reaction by:

$$\Delta F_R{}^o = -RT \ln K \tag{3}$$

$\Delta F_R{}^o$ must be negative to have K values greater than 1. For gaseous reactants, fugacities are used instead of activities. If the system behaves as an ideal gas, fugacities could be substituted by partial pressures, and the equilibrium constant is expressed as

$$K = K_p = \Pi P_i^{\gamma i} \tag{4}$$

Figure 1 shows the free energy of formation per carbon atom of several hydrocarbons as a function of temperature. For the elements, in this case carbon and hydrogen, ΔF_F^o is conventionally taken as zero at any temperature. As shown in Eq. (3), a reaction is thermodynamically feasible when a decrease in free energy is produced, i.e., ΔF_R^o is negative. This means that in Figure 1 the free-energy level of the product must be less than the free energy level of the reactant. The figure shows that the greater the number of carbon atoms for a paraffin, the higher its free energy of formation. Therefore, cracking reactions producing hydrocarbons with a smaller number of carbon atoms are feasible. Methane is the hydrocarbon showing the lowest free energy of formation, and it should be the main component when total cracking reaction equilibrium is reached. This means that total hydrocracking to methane is the most feasible reaction from the thermodynamic point of view and that oligomerization of

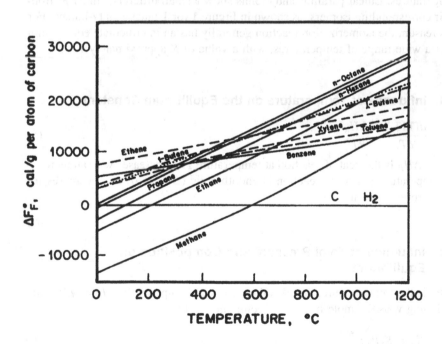

Figure 1 Free energy of formation of hydrocarbons, per carbon atom, as a function of temperature. (From Ref. 7.)

methane is not thermodynamically feasible. To make the last reaction possible, it must be coupled with another reaction that is thermodynamically very feasible, as in the case of oxidative dimerization of methane.

At low temperatures, paraffins have ΔF_F^o values lower than those of the corresponding olefins, rendering the hydrogenation of olefins as feasible, whereas at high temperatures dehydrogenation of paraffins is thermodynamically favored. This can be seen in Figure 1 for ethane–ethene. At temperatures below 820°C, ethene hydrogenation shows a negative change in free energy, so that the reaction is thermodynamically feasible. At 820°C, lines for ethane and ethene intersect, no change of free energy in the reaction is observed, and the equilibrium is at an intermediate position. At higher temperatures, ethane has a higher free energy of formation than ethene, and dehydrogenation is favored. The temperature of inversion of the hydrogenation–dehydrogenation equilibrium is lower when the number of carbon atoms in the molecule is larger. Figure 1 also shows that aromatic hydrocarbons are more stable at high temperatures than the corresponding paraffins. For this reason, paraffin dehydrocyclization is thermodynamically feasible under these conditions. On the other hand, opening of the aromatic ring and hydrogenation to paraffins (hydrogenolysis) are favored at low temperatures. Linear paraffins and olefins show a small difference in ΔF_F^o from their corresponding isomers, as shown in Figure 1 for 1-butene and i-butene. For this reason, the isomerization reaction generally has an intermediate equilibrium over a wide range of temperatures, with a value of K approximately 1.

2.3   Influence of Temperature on the Equilibrium Constant

$$\frac{d(\ln K)}{dT} = \frac{\Delta H_R}{RT^2} \tag{5}$$

where ΔH_R is the heat of reaction at temperature T. K increases with an increase in temperature when the reaction is endothermic ($\Delta H_R > 0$) and decreases for exothermic reactions.

2.4   Influence of Total Pressure and Composition on Equilibrium

If P_T is the total pressure, the partial pressure of component i is $P_i = \gamma_i P_T$, and including y_i as the mole fraction of component i, then

$$K_p = K_y P_T^{\sum \gamma_i} \tag{6}$$

$$K_y = \Pi y_i^{\gamma_i} \tag{7}$$

If the number of moles increases by the reaction, the exponent of P_T is positive and any increase in P_T (at temperature and therefore K_p constants) produces a decrease of K_y or a decrease in the mole fraction of products compared to reactants.

Table 1 shows thermodynamic data for several reactions occurring at 500°C, a normal temperature for the naphtha reforming process, and considering $K = K_p$. Figure 2 shows the change of standard free energy in reactions between hydrocarbons with six carbon atoms as a function of temperature. Considering the data presented in Figure 2 and Table 1, the following observations are made: Reaction A is an isomerization that has low feasibility because ΔF_R^o is positive and K_p has a small value, 8.6×10^{-2}. As the reaction is exothermic, according to Eq. (5), the feasibility decreases with the incremental increase in temperature, but the slope of change of K_p is small due to the low value of ΔH_R. Other hydrocarbon isomerizations have a similar behavior, like the n-hexane isomerization shown as reaction F. Most paraffin isomerizations have K_p values near 1, meaning similar amounts of n- and isoparaffins at equilibrium, and small exothermicity. As the number of moles does not change, total pressure has no influence on the equilibrium composition. Reaction B is a dehydrocyclization with low feasibility ($\Delta F_R^o > 0$, $K_p(500°C) = 1.3 \times 10^{-1}$) and it is endothermic; the cyclohexane produced is easily dehydrogenated to benzene by reaction E ($K_p = 6 \times 10^5$) and the equilibrium is favored with an increase in temperature (endothermic reaction). The equilibrium is disfavored with an increase in hydrogen and in total pressure [Eqs. (6) and (7)]. Increasing the molecular weight of the alkylcyclohexane increases its feasibility to be dehydrogenated to an

Table 1 Thermodynamic Data of Some Reactions That Occur in Naphtha Reforming at 500°C

Reaction			K_p (p in atm)	ΔH_R (cal/mol)
A	Methylcyclopentane	$\leftrightarrow$	Cyclohexane	8.6×10^{-2}	$-3,800$
B	n-Hexane	$\leftrightarrow$	Cyclohexane + H_2	1.3×10^{-1}	$10,800$
C	n-Hexane	$\leftrightarrow$	1-Hexene + H_2	3.7×10^{-2}	$31,000$
E	Cyclohexane	$\leftrightarrow$	Benzene + $3H_2$	6.0×10^5	$52,800$
F	n-Hexane	$\leftrightarrow$	2-Methylpentane	1.1	$-1,400$
G	Methylcyclohexane	$\leftrightarrow$	Toluene + $3H_2$	2.0×10^6	$51,500$
H	n-Hexane	$\leftrightarrow$	Benzene + $4H_2$	7.8×10^4	$63,600$
I	Methylcyclopentane	$\leftrightarrow$	Benzene + $3H_2$	5.2×10^4	$49,000$
J	n-Heptane + H_2	$\leftrightarrow$	Butane + propane	3.1×10^3	$-12,300$
K	n-Heptane + H_2	$\leftrightarrow$	n-Hexane + methane	1.2×10^4	$-14,800$

Source: Calculated from Ref. 8.

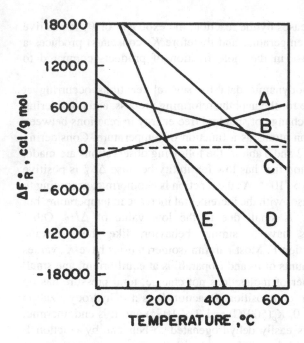

Figure 2 Change in the standard free energy for several reactions of hydrocarbons of six-carbon atoms as a function of temperature. A, Methylcyclopentane → cyclohexane; B, n-hexane → cyclohexane + H_2; C, n-hexane → 1-hexene + H_2; D, n-hexane → 1-butene + ethane; E, cyclohexane → benzene + $3H_2$. (Data from Refs. 7 and 8.)

aromatic, as can be seen comparing reactions E and G. Reactions B, E and H show that, although the dehydrocyclization of n-hexane to cyclohexane has low feasibility, the great feasibility of the dehydrogenation of cyclohexane to benzene makes the dehydrocyclization of n-hexane to benzene very feasible. Reaction I is the dehydroisomerization of an alkylcyclopentane. The five-carbon ring is isomerized to a six-carbon ring (reaction A), which is then dehydrogenated to the aromatic ring (reaction E). Because of the low feasibility of reaction A, the dehydroisomerization of methylcyclopentane (I) is less feasible than the dehydrogenation of cyclohexane. Reaction C is the dehydrogenation of a paraffin, with a low thermodynamic feasibility that increases with an increase in temperature because the reaction is endothermic. As shown in Figure 1 for ethane, at a certain temperature there is an inversion of the paraffin dehydrogenation–hydrogenation equilibrium. In the case of n-hexane, the dehydrogenation is favored at temperatures above 560°C, whereas at lower temperatures the hydrogenation of 1-hexene is more feasible. As the number of moles increases with the production of hydrogen, the conversion of n-hexane at equilibrium decreases with the increase in total and in hydrogen pressures. Reaction D is the cracking of a paraffin producing an olefin and a paraffin; this reaction is endothermic and very feasible at high temperatures. If the produced olefin is hydrogenated, only paraffinic hydrocracking products are obtained. The olefin hydrogenation is very feasible and greatly exothermic; for this reason, hydrocracking is very feasible and exothermic, as shown for n-heptane in reaction J. Hydrogenolysis (C-C rupture with production of methane) has similar thermodynamic behavior to hydrocracking, as shown in reaction K.

3 REACTIONS

Each hydrocarbon has several reaction paths that are thermodynamically feasible. For example, a normal paraffin can be cracked, producing a mixture of paraffins and olefins having a smaller number of carbon atoms. The n-paraffin can also be isomerized to a branched paraffin, dehydrogenated to an olefin with the same number of carbon atoms, or dehydrocyclized producing a naphthenic or an aromatic hydrocarbon. At any given contact time, the product distribution depends on the relative rate of these reactions, on the successive reactions that can occur, and on the thermodynamic equilibrium of the reactions, which are so rapid that they virtually reach equilibrium. Rates of the individual reactions depend on the catalyst selectivity. An important point in hydrocarbon processing is to find a very active catalyst and, above all, one selective enough for the desired reactions.

As noted in Chapter 1, naphthas to be reformed contain paraffins, naphthenes having five or six carbon atom rings, and aromatics. Since one of the main goals of the process is to increase the aromatics concentration, the desired reactions are dehydrocyclization of paraffins, dehydrogenation of cyclohexanes, and dehydroisomerization of alkylcyclopentanes to aromatics. Even the best catalysts are not completely selective for these reactions. Some undesirable reactions that decrease the liquid yield (light hydrocarbon gas formation by hydrocracking) and deactivate the catalyst (formation of carbonaceous deposits) always occur to some extent. Paraffins, mostly linear, are the main components of virgin naphthas. Over the classical bifunctional naphtha-reforming catalyst, the main reactions of paraffins are the following: dehydrogenation to olefins, isomerization, dehydrocyclization to aromatics or naphthenes, and hydrocracking or hydrogenolysis to lighter paraffins. Naphthenic hydrocarbons are alkylcyclopentanes and alkylcyclohexanes, and the possible reactions are dehydroisomerization of alkylcyclopentanes and dehydrogenation of alkylcyclohexanes to produce aromatics, ring isomerization, and ring opening to produce paraffins.

Paraffinic side chains of alkylaromatics can be partially or totally hydrocracked, and the ring can also be hydrocracked to produce a paraffin.

Considering normal heptane as a typical component of a naphtha, a simple scheme for its reactions on the bifunctional reforming catalyst is as follows:

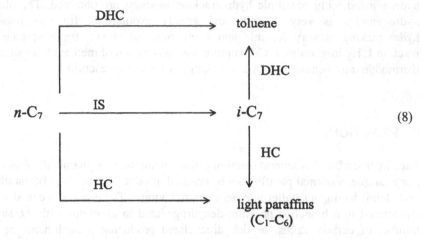

(8)

The desired reaction in this scheme is n-heptane dehydrocyclization to toluene, but i-heptanes and light paraffins are produced simultaneously. i-Heptanes are intermediate products that can be dehydrocyclized or hydrocracked. The catalyst selectivity and the contact time determine the product distribution. At the same time these reactions are taking place, a carbonaceous deposit is slowly deposited on the catalyst surface due to condensation of small amounts of unsaturated hydrocarbons present as reaction intermediates.

The most important reactions of catalytic naphtha reforming are analyzed separately and detailed in the first edition of this book,[1] where the kinetic models for the simultaneous reactions of naphtha reforming were studied. Further discussion of kinetic models may be found in Chapter 15 of this edition. For comparative purposes Table 2, which summarizes the properties of the reforming reactions, is presented in this chapter. The reactions are ordered according to reaction rates, starting with the most rapid, i.e., dehydrogenation of cyclohexanes. All of the reactions produce an increase in octane number. Coking is not included because its product is not present in the reformate.

The most rapid reactions reach thermodynamic equilibrium, and the others are kinetically controlled. High temperature and low pressure favor the thermodynamic feasibility as well as the reaction rate in the three most important reactions: dehydrogenation of cyclohexanes, dehydroisomerization of cyclopentanes, and dehydrocyclization of paraffins. For the other reactions—isomerization, hydrocracking, and hydrogenolysis—temperature and pressure have only a slight influence on equilibrium. Overall, high temperatures and low pressures

Table 2 Thermodynamic and Kinetic Comparison of the Main Reforming Reactions[a]

Reaction	Reaction rate	Thermal effect	Reach thermodynamic equilibrium	Thermodynamic P	T	Reaction P	T	H_2	Vapor pressure	Density	Liquid yield
Cyclohexane dehydrogenation	Very rapid	Very endothermic	Yes	−	+	−	+	Produces	d	i	d
Alkylcyclopentane dehydroisomerization	Rapid	Very endothermic	Yes	−	+	−	+	Produces	d	i	d
Paraffin isomerization	Rapid	Slightly exothermic	Yes	None	sl −	+	+	None	i	sl d	sl i
Paraffin dehydrocyclization	Slow	Very endothermic	No	−	+	−	+	Produces	d	i	d
Hydrocracking, hydrogenolysis	Very slow	Exothermic	No	None	−	++	++	Consumes	gr i	d	gr d

[a] +, an increase in either pressure or temperature produces an increase in equilibrium conversion or reaction rate;
++, produces a great increase;
−, produces a decrease;
i, increase;
d, decrease;
sl, slight;
gr, great.

would seem most desirable, but the same conditions favor deactivation of the catalyst and the reactions of hydrocracking and hydrogenolysis. Thus, the process conditions are a compromise.

The three most important reactions are endothermic and dominate the heat balance, producing a decrease in temperature along the catalyst bed. For this reason, the classical commercial reforming plant (Platforming[TM]) distributes the catalyst in three reactors with intermediate heating. In the first and second reactor, due to the endothermic reactions, temperature decreases; in the last reactor, where the slowest reactions occur, the endothermicity of dehydrocyclization of paraffins is partially neutralized by the exothermicity of hydrocracking and hydrogenolysis. Only these last two molecule rupture reactions consume hydrogen, and as a whole the process is a net hydrogen producer.

4 MECHANISMS OF REACTION

4.1 Classical Metal–Acid Bifunctional Reaction Mechanism

As quoted before, the naphtha-reforming catalyst consists of a metal function supported on a material that provides the acid function. Three of the naphtha-reforming reactions are catalyzed by the metal function (dehydrogenation of cyclohexanes, dehydrogenation of paraffins, and hydrogenolysis) and are practically unaffected by the presence or absence of acid sites. Other reactions (ring enlargement of alkylcyclopentanes, paraffin isomerization, cyclization, and hydrocracking) can be catalyzed by acid sites but require a very strong acidity to occur. The situation is quite different when the metal function is added to the acid function. In this case, the metal function dehydrogenates the saturated hydrocarbons (alkylcyclopentanes or paraffins) producing an unsaturated (olefinic) compound that is more reactive than the original saturated one. The reaction is bifunctional, as it starts on the metal sites and continues on the acid sites. The key of the reaction mechanism proposed by Mills et al.[3] is the formation of very reactive olefins as intermediate compounds. The existence of the olefin intermediate, or its equivalent as a carbenium ion, is reasonable because minute traces are sufficient for the postulation of a mechanism.[9] A typical bifunctional reaction is paraffin isomerization, as shown in scheme 9 for n-hexane isomerization:

$$n\text{-Hexane} \underset{\text{metal}}{\overset{-H_2}{\rightleftarrows}} n\text{-hexene} \xrightarrow[\substack{\text{acid} \\ \text{rate-controlling} \\ \text{step}}]{} i\text{-hexene} \underset{\text{metal}}{\overset{+H_2}{\rightleftarrows}} i\text{-hexane} \qquad (9)$$

The importance of the bifunctional catalyst is shown in the experimental results shown in Table 3. Charging to the reactor catalysts with just the metallic (M) or the acid function (A), only slight conversion is obtained. The mixture of

Table 3 Polystep *n*-Hexane Isomerization on Coarse Catalyst Mixtures[a]

Catalyst charged to reactor	Wt % conversion to *i*-hexanes
10 cm^3 of Pt/SiO$_2$ (M)	0.9
10 cm^3 of SiO$_2$-Al$_2$O$_3$ (A)	0.3
Mixture of 10 cm^3 of M and 10 cm^3 of A	6.8

[a]Reaction conditions: 373°C, 17.2 g *n*-hexane/h, molar ratio H$_2$ *n*-hexane = 5, atmospheric pressure.
Source: Ref. 5.

both functions produces a higher conversion because the dual-site mechanism is now allowed. In this mechanism, hydrogenation–dehydrogenation steps are catalyzed by the metal function and are very rapid; olefin isomerization is the slowest or rate-controlling step. As shown in Table 1, reaction C, the dehydrogenation of *n*-hexane, has a small thermodynamic feasibility, thus giving a very small concentration of *n*-hexene at equilibrium. Although this dehydrogenation is the first step, the reaction proceeds to a great extent because the olefin formed is subsequently isomerized on the acid sites. This allows the continuation of paraffin dehydrogenation, as it is always in thermodynamic equilibrium. The *i*-hexene produced is rapidly hydrogenated. The olefin concentration is very small, nearly undetectable,[5] and it is always in equilibrium with the paraffin. All steps proceed until the *n*-hexane isomerization equilibrium is reached. The thermodynamic data are shown in reaction F, Table 1 for the production of 2-methylpentane. For this bifunctional mechanism the *n*-olefin isomerization is the controlling step and the rate of reaction should be proportional to the concentration of the *n*-olefin adsorbed on the acid sites, which is directly related to the *n*-olefin pressure in the gas phase. This was shown by Sinfelt et al.[1] for the isomerization of *n*-pentane on Pt/Al$_2$O$_3$ at 372°C and 7.7–27.7 atm total pressure. They plotted the rate of *n*-pentane isomerization as a function of the *n*-pentene pressure calculated according to the dehydrogenation thermodynamic equilibrium and compared the results with those obtained using platinum-free alumina (acid function only) and feeding *n*-pentene. Both reaction systems have similar rates, which is a verification that the acid *n*-pentene isomerization is the controlling step. This is the key of the classical bifunctional mechanism. Another important point of the mechanism is the negative effect of the hydrogen pressure, since with increasing hydrogen pressure the equilibrium is disfavored for the olefin. Therefore, hydrogen will have a negative order during the reaction.

Other important bifunctional reaction is the dehydroisomerization of alkylcyclopentanes, as shown in Figure 3 for methylcyclopentane. Table 4 shows that methylcyclopentane does not react on a monofunctional acid catalyst (SiO$_2$-

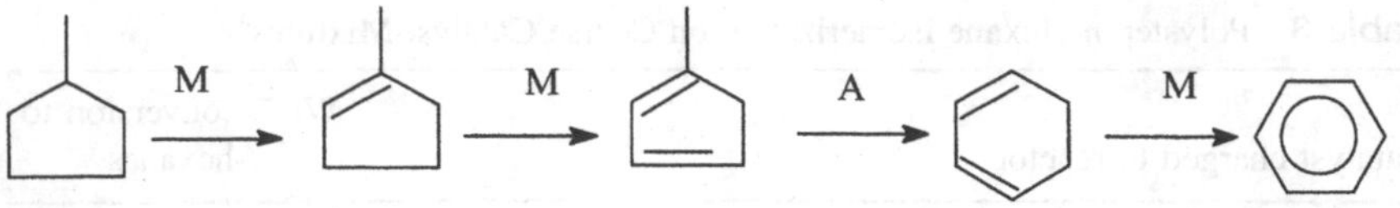

Figure 3 Mechanism of methylcyclopentane dehydroisomerization. M, metal site; A, acid site.

Al$_2$O$_3$) and that on a monofunctional metallic catalyst it can be dehydrogenated up to methylcyclopentadiene. To continue from methylcyclopentadiene to benzene the acid function is necessary to produce the ring enlargement (an isomerization) to cyclohexadiene, which is then dehydrogenated on the metal function to benzene.

The most important reaction in naphtha reforming is the dehydrocyclization of paraffins to aromatics. Figure 4 shows the bifunctional mechanism of dehydrocyclization of n-hexane. Similar to the other cases, isomerizations (cyclization, ring enlargement) occur with the reactive unsaturated hydrocarbons.

Table 5 shows that saturated hydrocarbons do not react over the acid function without Pt. The metal function is necessary because it is the generator of the unsaturated hydrocarbons. The reaction rates are little affected by an increase in Pt content because the acid steps are the controlling ones.

The principles of polystep catalysis were detailed by Weisz[5] and can be applied to the bifunctional metal function supported on acid oxide catalysts used in naphtha reforming, as seen below.

Intermediates, Reaction Sequences, and Conversions

Consider the reaction

$$A \Leftrightarrow B \Leftrightarrow C \tag{10}$$

Table 4 Dehydroisomerization of Methylcyclopentane Catalyzed by Acid, Metal, and Mixed Catalysts (Concentrations in the Liquid Product, mol %)[a]

Catalyst	Methyl cyclopentane →	Methyl cyclopentene →	Methyl cyclopentadiene →	Benzene
10 cm^3 SiO$_2$-Al$_2$O$_3$	98	0	0	0.1
10 cm^3 Pt/SiO$_2$	62	19	18	0.8
SiO$_2$-Al$_2$O$_3$ + Pt/SiO$_2$	65	14	10	10.0

[a]Reaction conditions: 500°C, 0.8 atm hydrogen partial pressure, 0.2 atm methylcyclopentane partial pressure, and 2.5 s residence time. Catalysts: 0.3 wt% Pt on SiO$_2$, SiO$_2$-Al$_2$O$_3$ with Sg = 420 m^2/g.
Source: Ref. 4.

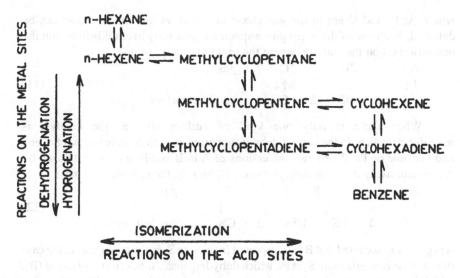

Figure 4 Reaction network for *n*-hexane dehydrocyclization through the bifunctional mechanism.

Table 5 Effect of Pt Content on Reactivities of *n*-Heptane (nC_7) and Methylcyclopentane (MCP) over Pt/Al$_2$O$_3$

	Pt content (wt %)			
Reaction rates[a]	0	0.10	0.30	0.60
Isomerization of nC_7				
744 K	0	0.035	0.035	0.038
800 K	0	0.120	0.130	0.120
Dehydrocyclization of nC_7				
744 K	0	0.0022	0.0027	0.0045
800 K	0	0.0200	0.0250	0.0350
Isomerization– dehydroisomerization of MCP				
744 K	0	—	0.019	0.021
772 K	0	—	0.039	0.043

[a]Moles gram per hour per gram of catalyst at 21 atm and H$_2$/hydrocarbon mole ratio = 5.
Source: Ref. 10.

where A, B, and C are in the gas phase and B is an intermediate that can be detected. Each one of these gas phase species is generally in equilibrium with the one adsorbed on the surface, where the reaction really occurs:

$$
\begin{array}{cccc}
A & B & C & \text{gas} \\
\updownarrow & \updownarrow & \updownarrow & \\
AS & \Delta \quad BS & \Delta \quad CS & \text{catalyst surface}
\end{array} \tag{11}
$$

When there is only one kind of surface site, S, the catalyst is monofunctional, as in the case of dehydrogenation of cyclohexane to cyclohexene and benzene on Pt. When the conversions of A to B and B to C are catalyzed by two qualitatively different catalytic sites, S_1 and S_2, the scheme becomes:

$$
\begin{array}{cccc}
A & B & C & \text{gas} \\
\updownarrow & \nwarrow \searrow & \updownarrow & \\
AS_1 & \Delta \quad BS_1 \quad BS_2 & \Delta \quad CS_2 & \text{catalyst surface}
\end{array} \tag{12}
$$

It is generally accepted that B is transported from S_1 to S_2 in the gas phase. In the case of n-hexane isomerization, S_1 is Pt, which dehydrogenates n-hexane to n-hexene (B), which is isomerized on S_2 (the acid function) to produce i-hexene (C), which after desorption from S_2 must be adsorbed on S_1 to be hydrogenated to i-hexane.

For simplicity, the reaction can be written in similar fashion to Eq. (10), without adsorbed species. In the case of successive irreversible reactions on two different surface sites, the nomenclature is:

$$
\begin{array}{ccc}
S_1 & S_2 \\
A & \rightarrow \quad B & \rightarrow \quad C
\end{array} \tag{13}
$$

If there are two successive reaction zones or steps containing catalysts with individual sites, as shown in Figure 3, and if a high conversion can be achieved for each step under similar operational conditions, then a total conversion of A to C is possible. If the catalysts of the two reaction zones are mixed, a basically similar overall conversion would be expected. This reaction with two irreversible steps is generally described as trivial because the degree of mixing of S_1 and S_2 is not important. The most interesting case is the nontrivial, when the first reaction advances little due to thermodynamic limitations:

$$
A \underset{k_1'}{\overset{k_1}{\rightleftharpoons}} B \overset{k_2}{\rightarrow} C; \quad K_1 = \frac{k_1}{k_1'} \ll 1 \tag{14}
$$

When the catalysts are separated into two zones or reactors (Fig. 3.5a) in this situation, A is converted to B in the first reactor with conversion x, and the rate of reaction will be:

$$
\frac{dx}{dt} = k_1(1 - x) - k_1' x \tag{15}
$$

Integrating between the initial state ($t = 0$, $x = 0$) and t, the conversion of A is:

$$x = \frac{1 - \exp[-t(k_1 + k_1')]}{1 + (k_1'/k_1)} \tag{16}$$

B is converted to C in the second reactor by catalyst S_2 with a conversion y:

$$\frac{dy}{dt} = k_2 x = k_2 \left(\frac{1 - \exp[-t(k_1 + k_1')]}{1 + (k_1'/k_1)} \right) \tag{17}$$

Integrating this equation between $t = 0$ and $y = 0$ and the time t and substituting for $t = \infty$, one obtains:

$$y = \frac{1}{1 + (k_1'/k_1)} = \frac{1}{1 + (1/K_1)} \tag{18}$$

where K_1 is the thermodynamic equilibrium constant of the first step, which has a small value. Therefore, y can never reach the value of 1.

If S_1 and S_2 are separated in several zones, as shown in Figure 5b, A is converted in the first zone to B up to equilibrium, x_{A1}; in the second zone, the produced B is transformed totally or partially into C. Then, A will react again in the third zone to restore the thermodynamic equilibrium and the conversion of A is increased, $x_{A1} + x_{A2}$. Along this sequence the total transformation of A can be higher than the value fixed by the thermodynamic equilibrium A $\rightleftarrows$ B.

If S_1 and S_2 are intimately mixed, as in Figure 5c, the rate of reactions will be expressed as two simultaneous differential equations:

$$\frac{dx}{dt} = k_1(1 - x) - k_1'(x - y)$$

$$\frac{dy}{dt} = k_2(x - y) \qquad \text{initial conditions } t = 0; x = 0; y = 0 \tag{19}$$

Solving this system, an expression of y is obtained; for $t = \infty$ the maximal value of y ($= 1$) is obtained. This is a nontrivial case where all A can be converted to C in spite of the thermodynamic limitations of the first step, and it shows the advantage of the dual site or bifunctional catalyst.

If the second reaction is not irreversible but is limited by thermodynamic equilibrium, the maximal value of y will be that determined by the equilibrium B $\rightleftarrows$ C. If C is transformed into D by an irreversible reaction (the case of olefin hydrogenation during paraffin isomerization), the maximal amount of D will be fixed by the thermodynamics of A $\rightleftarrows$ D (as in the case of n-hexane $\rightleftarrows$ i-hexane).

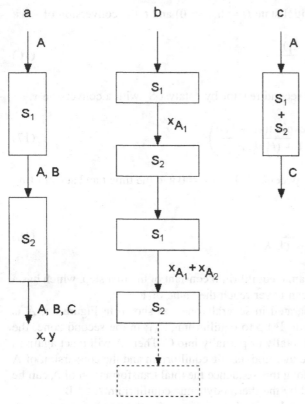

Figure 5 (a) Catalysts separated in two successive reaction zones. (b) Catalysts separated in several small successive reaction zones, as a sandwich of different catalyst layers. (c) Both catalysts uniformly mixed in the reactor.

Thermodynamics

In Sec. 2, thermodynamic equations (1) to (4) state that the change of standard free energy of a reaction must be negative to have a value of the equilibrium constant K higher than 1. In this case, the reaction has thermodynamic feasibility. If the change in free energy is positive, the reaction will not be feasible; for zero change in free energy K is equal to 1 and the equilibrium will have an intermediate position. The changes in free energy for the trivial case of a two-step reaction represented by Eq. (13) are plotted in Figure 6a. Both reactions have thermodynamic feasibility with a decrease in free energy. Figure 6b indicates the changes in a non trivial case represented by Eq. (14), where the free energy of

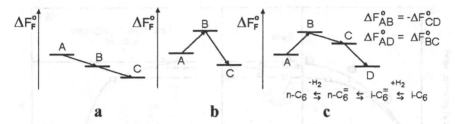

Figure 6 Changes in free energy during the advance of the reaction. (a) Trivial case as indicated in Eq. (13). (b) Nontrivial case as indicated in Eq. (14). (c) Three-step isomerization of *n*-hexane as indicated in Eq. (9).

formation of B is greater than that of A and the total free energy change $\Delta F^0_{A \rightarrow C}$ is negative. Figure 6c shows the case of the bifunctional *n*-hexane isomerization as a three-step reaction. Reaction C → D is very feasible and is the opposite of A → B. The total equilibrium constant K is the product of the three equilibrium constants. As K_3 is practically the reciprocal of K_1, the total K is equal to K_2 (ΔF^0_R of paraffin isomerization is equal to ΔF^0_R of the isomerization of the corresponding olefin). The value for *n*-hexane is around 1 at 500°C (Table 1), representing an intermediate equilibrium.

Mass Transport

The contact between both catalysts or catalytic functions is very important in order to achieve the advantages of the bifunctional mechanism because the intermediate compound B must diffuse from S_1 to S_2. Both active sites or particles are considered to act independently, and the intraparticle diffusion of B can be a limiting factor. Figure 7 shows the results of Weisz and Swegler (11) for *n*-heptane isomerization. The conversion was very small when the catalyst had only the metal function, Pt/SiO$_2$ or Pt/C. When the catalyst had only the acid function, SiO$_2$-Al$_2$O$_3$, the conversion was undetected and not shown in Figure 7. On the other hand, the reaction occurs when a 50:50 by volume mechanical mixture of 1000-μm particles of Pt/SiO$_2$ and SiO$_2$-Al$_2$O$_3$ was placed in the reactor. Decreasing the particles size to 70 μm, the conversion increases due to the decrease in diffusional limitations between particles. In the limit, a mechanical mixture of 5-μm particles produced the same conversion as Pt supported directly on SiO$_2$-Al$_2$O$_3$.

Weisz[5] stated the criterion defining the physical conditions of intimacy between the component systems for no mass transport inhibition. The criterion

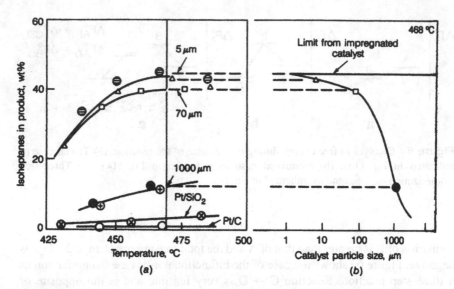

Figure 7 Isomerization of *n*-heptane over a mixed component catalyst, for varying size of the component particles. (a) Conversion vs. temperature. (b) Conversion at 468°C vs. component particle diameter. Reaction conditions: partial pressure of *n*-heptane 2.5 atm, of hydrogen 20 atm, and residence time in catalyst bed 17 s. (From Ref. 11.)

depends on the gas-phase diffusivity of the intermediate and on the distance between catalytic functions. To assure intimacy between the functions in the naphtha-reforming catalyst, the metal particles must be small and uniformly distributed over the acid support.

Selectivity

A monofunctional catalyst (S_1) can produce B as an intermediate species during the reaction A → C. With the introduction of a second catalyst (S_2), B can be intercepted and a new product (D) obtained:

$$
\begin{array}{ccc}
S_1 & & S_1 \\
A \underset{k_1'}{\overset{k_1}{\rightleftarrows}} & B & \overset{k_2}{\rightarrow} C \\
& S_2 \downarrow k_3 & \\
& D &
\end{array}
\tag{20}
$$

If $k_1/k_1' \ll 1$ the amount of B will be very small and this is a nontrivial case. Considering as before first-order reactions, the rate of production of C and D at any moment will have the ratio:

$$\frac{dC/dt}{dD/dt} = \frac{k_2}{k_3} \tag{21}$$

and the amount produced will have the same ratio, representing the selectivity:

$$\frac{C(t)}{D(t)} = \frac{k_2}{k_3} \tag{22}$$

The value of the total conversion of A can be deduced:

$$x_A = 1 - \exp(-k_\varepsilon t); \; k_\varepsilon = \frac{(k_2 + k_3)k_1}{k_1' + k_2 + k_3} \tag{23}$$

From Eqs. (21) and (22) and considering that $A_o x_{(t)} = C_{(t)} + D_{(t)}$ (neglecting the small amount of B produced), it is possible to see the effect of the introduction of catalyst S_2 in the reaction system of catalyst S_1. This was plotted by Weisz[5] for different values of the kinetic constants, showing how the conversions of A to C and D change as a function of the total conversion of A.

An interesting case is the change of selectivity during cyclohexane dehydrogenation to benzene:

$$\text{Cyclohexane} \underset{k_1'}{\overset{k_1}{\rightleftarrows}} \text{cyclohexene} \overset{k_2}{\rightarrow} \text{benzene} \tag{24}$$

The reaction occurs on the metal function (Pt) and, as k_2 is greater than k_1, the intermediate cyclohexene is not detected. The dehydrogenation of cyclohexene is so rapid that it is completed without appreciable desorption of cyclohexene. When Pt is partly deactivated (e.g., by addition of a sulfur compound or by coke deposition), k_2 decreases and a certain amount of cyclohexene can be desorbed. If an acid catalyst is present, the desorbed cyclohexene could be isomerized to methylcyclopentene, which could then be hydrogenated to methylcyclopentane:

$$\text{Cyclohexane} \overset{M}{\Leftrightarrow} \text{cyclohexene} \overset{M}{\underset{k_2}{\rightarrow}} \text{benzene}$$

$$A \uparrow\downarrow k_3 \tag{25}$$

$$\text{methylcyclopentane} \overset{M}{\Leftrightarrow} \text{methylcyclopentene}$$

Table 6 Diversion of Cyclohexane → Benzene Reaction to Methylcyclopentane Formation for Pt/SiO₂ and Pt/SiO₂ + SiO₂-Al₂O₃ Having Different Particle Sizes[a]

	Products per 100 parts cyclohexane charged		
Catalyst	Benzene	Methylcyclopentane	Cyclohexane (unconverted)
Pt/SiO₂	85	1.5	9
1000 μ mixture	84	6	8.5
100 μ mixture	57	20	7.5
5 μ mixture	59	23	8.5
impregnated catalyst	40	40	7

[a]Reaction conditions: 450°C, hydrogen pressure 20 atm and hydrocarbon partial pressure 5 atm.
Source: Ref. 5.

Along this sequence, cyclohexene (an intermediate of the monofunctional dehydrogenation reaction) is intercepted and becomes an intermediate in the dual-function cyclohexane ring isomerization. The formation of methylcyclopentane at the expense of benzene depends on the k_3/k_2 ratio. Table 6 shows results obtained converting cyclohexane on the metal component alone (Pt/SiO₂) and on the mechanical mixtures Pt/SiO₂ + SiO₂-Al₂O₃ (50 : 50) having different particle sizes. The greater contact of cyclohexene with the acid function allows an increase in the production of methylcyclopentane. These concepts are applicable to methylcyclohexane reactions over Pt/SO₄²⁻-ZrO₂ studied by Figueras et al.[12] By the metal only mechanism, methylcyclohexane produces toluene. By the bifunctional mechanism, due to the acidity of SO₄²⁻-ZrO₂, alkylcyclopentanes are produced from the intermediate methylcyclohexene. This intermediate can desorb from Pt before further dehydrogenation occurs because the metallic properties of Pt are decreased due to the strong interaction with the support.

Another interesting reaction is the xylene–ethylbenzene interconversion. The interconversion among xylenes (transfer of methyl groups) occurs on acid catalysts, as mordenite, but the isomerization to ethylbenzene does not occur. Ethylbenzene is produced if a metal function and hydrogen are present and the reaction occurs at a lower temperature in order to favor the aromatic ring hydrogenation to a cyclic olefin, as shown in Figure 8. The mechanism of o-xylene isomerization to ethylbenzene includes acid-catalyzed ring contraction and expansion steps of cycloolefin intermediates. The cycloolefin is produced only when the metal function and H₂ are present and is not produced at a low H₂ pressure.

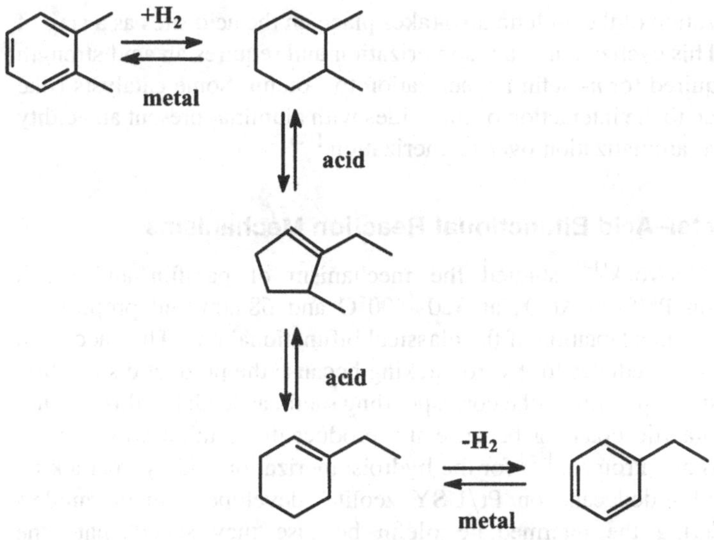

Figure 8 Bifunctional mechanism for the isomerization of o-xylene to ethylbenzene.

An important example in naphtha reforming is the modification or tuning of the selectivity of each catalytic function. Contacting a n-paraffin with a bifunctional catalyst, in addition to isomerization, other reactions including hydrogenolysis on the metal function and cracking on the acid function are possible:

$$n\text{-paraffin} \overset{M}{\rightleftharpoons} n\text{-olefin} \overset{A}{\to} i\text{-olefin} \overset{M}{\to} i\text{-paraffin}$$
$$\downarrow M \qquad\qquad A\searrow \quad \nearrow A \qquad\qquad \downarrow M$$
$$\text{(hydrogenolysis)} \qquad \text{(cracking)} \qquad\quad \text{(hydrogenolysis)}$$
$$\text{lower } n\text{-paraffins} \qquad \text{lower olefins} \qquad\quad \text{lower paraffins} \qquad (26)$$
$$\downarrow M$$
$$\text{(hydrogenation)}$$
$$\text{lower paraffins}$$

For paraffins of seven or more carbon atoms, hydrocracking is more important than hydrogenolysis. Both reactions are harmful because they produce less valuable lower paraffins. To decrease these reactions, the selectivity of each function is modified. As hydrogenolysis requires large ensembles of Pt atoms, this reaction is decreased when the number of these ensembles is reduced by dilution of Pt with a second element (Sn, Ge, ReS). Hydrocracking requires stronger acidity than isomerization, and the acidity is decreased using chlorided alumina as support instead of a stronger acidic material such as SiO_2-Al_2O_3.

The cyclization of the n-olefin also takes place on the acid sites as a step of aromatization. This cyclization is an isomerization and requires an acid strength near the one required for n-olefin isomerization to i-olefin. Some catalysts (like Pt-Sn/Al$_2$O$_3$, due to the interaction of tin oxides with alumina) present an acidity tuned to increase aromatization over isomerization.[12]

4.2   Other Metal–Acid Bifunctional Reaction Mechanisms

Coonradt and Garwook[14] studied the mechanism of paraffin and olefin hydrocracking on Pt/SiO$_2$-Al$_2$O$_3$ at 320–400°C and 68 atm and proposed a mechanism that is an adaptation of the classical bifunctional one. They accepted that olefins are intermediates for hydrocracking because the product distribution was the same using a paraffin or the corresponding olefin as feed; but this was not applicable to catalytic cracking because the product distribution was entirely different. Steijns and Froment,[15] for the hydroisomerization and hydrocracking of n-decane and n-dodecane on Pt/USY zeolite, developed kinetic models without considering the intermediate olefin because they stated that "the occurrence of intermediate olefin is questionable at low temperature (down to 150°C) and hydrogen pressures of 100 bar." Chu et al.[16] found that Pt/H-β zeolite is very active in the skeletal isomerization of n-hexane with high selectivity (99%) and stability, while pure zeolite is only 10% as active as the bifunctional catalyst and only 66% selective. They worked at 250°C and considered that the n-hexene produced by dehydrogenation of n-hexane at the metal sites is not directly involved in the reaction due to the very small equilibrium concentration. The authors proposed a different mechanism: an acid-catalyzed chain reaction involving methyl shifts and hydrogen transfer was made responsible for the reaction. The chain is terminated when an olefin (formed by side reactions) reacts with a C$_6$H$_{13}^+$ species bonded to the zeolite. They stated that the role of Pt is to hydrogenate olefins, so that olefin steady-state concentration remains negligible. Working with Pt/SO$_4^{2-}$-ZrO$_2$, Iglesia et al.[17] pointed out that the equilibrium concentration of olefin intermediates would be extremely small at the low temperature and high hydrogen pressure currently used in n-paraffin isomerizations.

Table 1 shows that the dehydrogenation of n-hexane to n-hexene has a low thermodynamic feasibility; the equilibrium constant at 500°C is $K_p = 0.037$ and, considering the formation of all n-hexene isomers, the olefin concentration will be 0.6%.[5] But at 200°C, $K_p = 10^{-7}$, working at a total pressure of 5.9 bar and molar ratio H$_2$/n-hexane = 7, the pressure of n-hexene in equilibrium with n-hexane is 1.5×10^{-8} bar.[18] If the five possible n-hexenes are produced, the total pressure of n-hexenes will be 7.5×10^{-8} bar. This value is 10^7 times smaller than the value of n-hexane pressure. According to the classical bifunctional mechanism, the rate of isomerization on the acid sites is related, generally proportionally, to the olefin pressure[1] and this rate would therefore be negligible at 200°C. At this low temperature olefins have a high adsorption equilibrium constant over the acid sites. Olefins are not desorbed and instead of being an intermediate in the reaction, they polymerize and deactivate the acid sites.

All of the literature on bifunctional metal–acid catalysis indicates that, in addition to the acid and the metal functions, the presence of hydrogen is necessary. In reactions like *n*-paraffin isomerization, hydrogen is not a reactant; however, in the absence of hydrogen, the catalytic activity is very small. Therefore, hydrogen has an important function that must be taken into account in all the reaction mechanisms. Hattori[19] studied the influence of hydrogen on the skeletal isomerization of *n*-butane at 250°C and atmospheric pressure on a Pt/SO_4^{2-}-ZrO_2 catalyst. He found that the catalytic activity increases when the hydrogen pressure is increased and that this process is reversible. The acid sites were characterized by IR of adsorbed pyridine. Figure 9 shows the amount of Brønsted pyridine (B-Py) and Lewis pyridine (L-Py) as a function of temperature, in the presence of hydrogen and after successive evacuations. The amount of B-Py increases when temperature is increased and, simultaneously, the amount of L-Py decreases. Evacuation produces a decrease in the amount of B-Py, and the amount of L-Py is restored. The formation and elimination of the protonic acid sites are reversible. Hattori[19] interpreted the former results with the mechanism

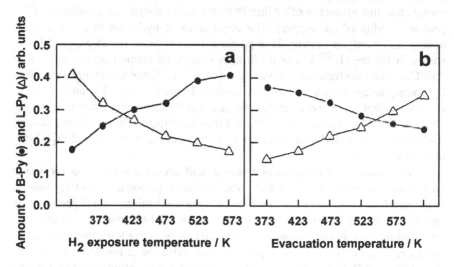

Figure 9   Amount of pyridine adsorbed on protonic sites (●) and Lewis acid sites (△) of Pt/SO_4^{2-}-ZrO_2 at: (**a**) heating in the presence of hydrogen (500 Torr) and (**b**) successive evacuation. Pyridine was adsorbed on the hydrogen-treated sample at 150°C and then evacuated at 400°C for 10 min before introduction of hydrogen. (From Ref. 19.)

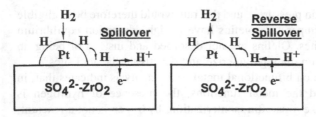

Figure 10   Schematic illustration for the formation and elimination of protonic sites on Pt/SO_4^{2-}-ZrO$_2$. (From Ref. 19.)

presented in Figure 10. The hydrogen molecule is dissociated to form hydrogen atoms on metallic platinum. The hydrogen atoms undergo spillover onto the support (SO_4^{2-}-ZrO$_2$) and migrate to Lewis acid sites. The H atom releases an electron to the Lewis acid site and is converted to a proton. This proton is stabilized by the oxygen atom neighboring the Lewis acid site, acting as an active site for acid-catalyzed reactions. The Lewis acid site weakens its strength by accepting the electron. By hydrogen evacuation, the reverse process proceeds to restore the original Lewis acid sites and to eliminate the protonic acid sites. In this way, the protonic acid sites are generated and eliminated in accordance to the gas-phase hydrogen pressure. In this bifunctional mechanism, olefins are not formed and the advantage of adding Pt to the acid catalyst is the increase in the protonic acidity of the support. The importance of hydrogen in order to have acidity and catalytic activity in *n*-pentane isomerization on Pt/SO_4^{2-}-ZrO$_2$ is shown in Figure 11.[20] Under a nitrogen stream, no isomerization occurred at 250°C or when the temperature was raised to 300°C. When switching nitrogen to hydrogen, isomerization activity appeared. Hattori[19] and Ebitani et al.[20] considered that the presence of hydrogen is beneficial for the catalytic activity. Other authors, like Garin et al.,[21] found that the effect of the partial pressure of hydrogen is negative during isomerization of *n*-butane at 250°C on SO_4^{2-}-ZrO$_2$ and Pt/SO_4^{2-}-ZrO$_2$.

On these hydrogen-generated strong acid sites, carbenium ion formation from paraffins proceeds by hydride abstraction or protonation and H$_2$ release from paraffins. Olefins are not involved in these mechanisms, but if present they are immediately protonated. Hydride abstraction and heterolytic cleavage of H$_2$ are possible only on very strong acid sites that are active at low temperature. For instance, SO_4^{2-}-ZrO$_2$ is active in paraffins isomerization at low temperature and the addition of Pt and hydrogen produces an increase in activity, selectivity, and stability.

The differences between bifunctional catalysts with mild and with strong acidity or classical and non-classical bifunctional mechanism are summarized in

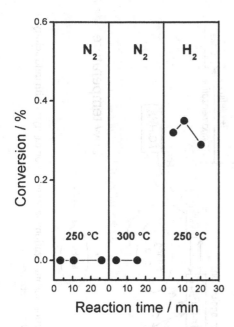

Figure 11 Effect of carrier gas on catalytic activity of Pt/SO$_4^{2-}$-ZrO$_2$. (From Ref. 20.)

Figure 12 and Table 7. In the upper portion of Figure 12 the classical mechanism at high temperature appears, where olefin formation is feasible and the main objective of the metal is to produce the reactive olefins that diffuse in the gas phase and are able to react even on mild acid sites. At the bottom appears the mechanism that occurs at low temperature and with strong acid catalysts. The formation of olefins on the metal sites is not feasible at low temperature and it can be accepted that the main objective of the metal is the dissociation of hydrogen whose products diffuse on the surface and increase the acidity of the support.

The nonclassical bifunctional mechanism cannot be in operation on the catalyst under the operational conditions of naphtha reformers. The two main reasons to discard the nonclassical bifunctional mechanism are (1) the support of the reforming catalyst is Al$_2$O$_3$ or chlorided Al$_2$O$_3$, which has relatively weak Lewis acid sites, and is not able to abstract one electron from the H atom; (2) the reforming temperature is higher than those used with Pt/SO$_4^{2-}$-ZrO$_2$ during isomerization. Therefore, the equilibrium of paraffin dehydrogenation is greater and the amount of the reactive olefins produced during reforming is greater. The olefins can react over weak and strong acid sites but are strongly adsorbed on the strong ones. At high temperatures the amount of hydrogen adsorbed will also decrease. In the case of SO$_4^{2-}$-ZrO$_2$, the catalyst is deactivated at high temperatures due to sulfate elimination.[22,23]

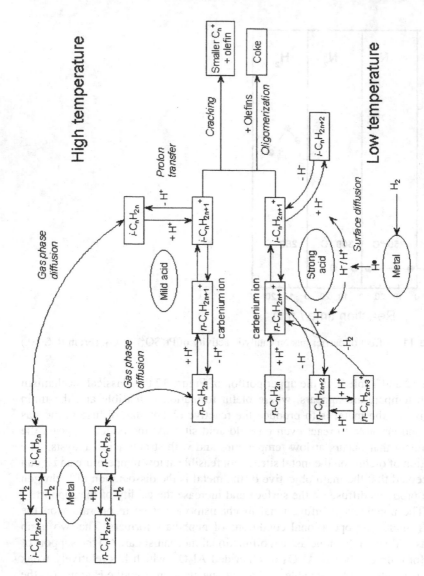

Figure 12 Schematic representation of bifunctional isomerization mechanisms on catalysts with mild acidity at high temperature and with strong acidity at low temperature.

Table 7 Reactions on Mild and Strong Acid Bifunctional Catalysts

Reaction	Mild acid bifunctional catalyst		Strong acid bifunctional catalyst	
	Low temperature	High temperature	Low temperature	High temperature
Isomerization	Negligible activity	Medium activity	Medium activity	Superseded by cracking and coking
Olefin formation	Inhibited thermodynamically (low overall activity)	Favored thermodynamically (increases overall activity)	Inhibited thermodynamically (but not needed)	Favored thermodynamically (increases overall activity)
Carbenium ion formation	By olefin protonation (inhibited by low olefin concentration)	By olefin protonation	By olefin protonation. By hydride abstraction from alkane. By direct alkane protonation	By olefin protonation. By hydride abstraction from alkane. By direct alkane protonation
Cracking	Low activity	Medium activity	Medium activity	Strong activity (dominant)
Oligomerization	Negligible activity	Medium activity (aided by olefins)	Medium activity (aided by high acid strength)	High activity (dominant)
Carbenium ion to gas phase species	—	By olefin deprotonation (weak adsorption)	By hydride transfer from alkane or H. No deprotonation (strong adsorption)	
Acid–metal transfer of intermediates	—	By bulk gas phase diffusion (olefins)	By surface diffusion (activated hydrogen)	
Hydrogen activation	Not needed	Needed for coke removal. Proton formation not possible	Needed for coke removal. Needed for hydride transfer. Needed for proton (acid site) formation.	Not enough for coke removal

The lower part of Figure 12 shows several proposed reaction mechanisms on acid sites. For instance, Iglesia et al.[17] found during n-heptane isomerization that reaction rates show positive hydrogen kinetic orders. This suggests that the reaction proceeds on Pt/SO_4^{2-}-ZrO_2 via chain transfer pathways in which carbenium ions propagate by hydride transfer from neutral species to carbocations, allowing the isomers to desorb. This was verified by the addition of molecules that provide hydride ions during reaction, such as adamantane. The rate-limiting isomer desorption step is faster with the increment in hydride ions transfer. This decrease in the residence time of isomeric carbocations allows its desorption before β-scission reactions (cracking) occur. The importance of this hydride transfer was stressed by Filimonova et al.[24] in the monomolecular pathway of n-pentane isomerization over Pt/WO_x-ZrO_2.

For the isomerization of n-hexane over Pt/WO_3-ZrO_2 it was proposed that the reaction could occur by a nonclassical bifunctional mechanism,[25] like that proposed for the n-heptane isomerization at 200°C over the same catalyst.[26] In this nonclassical mechanism, hydrogen dissociates on platinum spills over the support and is transformed into acid sites by interaction with partially reduced WO_3 species. The acid sites generate the adsorbed ionic isomeric species and desorption is favored by the presence of dissociated hydrogen.[27] Moreover, the rapid desorption of the isomeric ionic species from the surface before β scission avoids cracking and polymerization reactions that decrease selectivity and increase coke deposition.

4.3 Other Types of Bifunctional Catalysts

For the types of reactions studied in this chapter, the most important bifunctional catalysts are the ones with metal and acid functions. But other catalysts with different kinds of active sites have been also investigated. Fung and Wang[28] studied the relation between the acid–base properties of a binary oxide, TiO_2-ZrO_2, and its catalytic activity in hexane reactions. The yield and the selectivity to benzene increased with the relative amount of the acid–base pair. They stated the following mechanism for n-hexane dehydrocyclization: n-hexane is dehydrogenated on an acid–base site, then cyclization occurs on a basic site, and finally dehydrogenation of cyclohexane to benzene occurs on the acid–base pair. Garin et al.[29] used bulk tungsten carbide for paraffin reforming and stated that the reaction occurred by a bifunctional mechanism. In the same laboratories two other catalysts were studied for n-hexane reforming. One was MoO_3-Al_2O_3[30] and a dual-site or bifunctional mechanism was proposed: Mo(V) is responsible for acidic isomerization, and Mo(IV) and Mo(II) species exhibit metallic dehydrogenation properties. The other catalyst for the reforming of hexanes and hexenes was WO_3.[31] WO_3 is inactive but when reduced to $W_{20}O_{58}$ the surface becomes acidic and isomerizes hexenes by a monofunctional acidic mechanism. A further reduction leads to a surface active for isomerization of *n*-hexane and *n*-hexenes, and it was interpreted that the surface is bifunctional.

NOMENCLATURE

a_i	equilibrium activity of reactants and products
ΔF_R^o	standard free energy of reaction
ΔF_{Fi}^o	standard free energy of formation for each of the i reacting species
ΔH_R	heat of reaction
k	forward reaction rate constant
k'	reverse reaction rate constant
K	thermodynamic equilibrium constant
K_p	pressure equilibrium constant
K_y	molar fraction equilibrium constant
P_i	partial pressure of i
P_T	total pressure
R	gas constant
S	surface active site
T	temperature
t	time
x	conversion of reactant A
y	conversion of reactant B
y_i	mole fraction of component i
Π	mathematical product of all the equilibrium values of reactants and products
γ_i	stoichiometric coefficient of i in the reaction, which is negative for reactants and positive for products

ACKNOWLEDGMENT

The authors thank Dr. C. A. Vera for valuable comments and for the elaboration of Figure 12 and Table 7.

REFERENCES

1. Parera, J.M.; Fígoli, N.S. In *Catalytic Naphtha Reforming Science and Technology*; Antos, G.J.; Aitani, A.M.; Parera, J. M., Eds.; Marcel Dekker: New York, 1995; 45 pp.

2. Wojciechowski, B.W.; Corma, A. *Catalytic Cracking Catalysts, Chemistry and Kinetics*; Marcel Dekker: New York, 1986; 34 pp.
3. Mills, G.A.; Heinemann, H.; Milliken, T.H.; Oblad, A.G. Ind. Eng. Chem. **1953**, *45*, 134.
4. Weisz, P.B. *2nd Int. Cong. Catal.*; Technip: Paris, 1961; 937 pp.
5. Weisz, P.B. Adv. Catal. **1962**, *13*, 137.
6. Sterba, M.J.; Haensel, V. Ind. Eng. Chem. Prod. Res. Dev. **1976**, *15*, 2.
7. Parks, G.S.; Huffman, H.M. *Free Energy of Some Organic Compounds*; Reinhold: New York, 1932.
8. Rossini, F.D.; Pitzer, K.S.; Arnett, R.L.; Braum, R.M.; Pimentel, G.C. *Selected Values of Physical and Thermodynamic Properties of Hydrocarbons and Related Compounds*. API Res. Project 44; Carnegie Press: Pittsburgh, 1953.
9. Haensel, V. Ind. Eng. Chem. **1965**, *57*, 18.
10. Sinfelt, J.H.; Hurwitz, H.; Rohrer, J.C. J. Catal. **1962**, *1*, 481.
11. Weisz, P.B.; Swegler, E.W. Science **1957**, *126*, 31.
12. Figueras, F.; Coq, B.; Walter, C.; Carriat, J.-Y. J. Catal. **1997**, *169*, 103.
13. Rangel, M.C.; Carvalho, L.S.; Reyes, P.; Parera, J.M.; Fígoli, N.S. Catal. Lett. **2000**, *64*, 171.
14. Coonradt, H.L.; Garwood, W.E. Ing. Eng. Chem. Proc. Des. Dev. **1964**, *3*, 38.
15. Steijns, M.; Froment, G.F. Ind. Eng. Chem. Prod. Res. Dev. **1981**, *20*, 660.
16. Chu, H.Y.; Rosynek, M.P.; Lundsford, J.H. J. Catal. **1998**, *178*, 352.
17. Iglesia, E.; Soled, D.G.; Kramer, G.M. J. Catal. **1993**, *144*, 238.
18. Falco, M.G.; Canavese, S.A.; Comelli, R.A.; Fígoli, N.S. Appl. Catal. A **2000**, *201*, 37.
19. Hattori, H. Stud. Surf. Sci. Catal. **1993**, *77*, 69.
20. Ebitani, K.; Konishi, J.; Horie, A.; Hattori, H.; Tanabe, K. In *Acid–Base Catalysis*; Tanabe, K.; Hattori, H.; Yamaguchi, T.; Tanaka, T., Eds.; Kodansha Ltd.: Tokyo, 1989; 491 pp.
21. Garin, F.; Andriamasinoro, D.; Abdulsamad, A.; Sommer, J. J. Catal. **1991**, *131*, 199.
22. Ng, F.T.T.; Horvát, H. Appl. Catal. A **1995**, *123*, L 197.
23. Vaudagna, S.R.; Comelli, R.A.; Fígoli, N.S. Catal. Lett **1997**, *47*, 259.
24. Filimonova, S.V.; Nosov, A.V.; Scheithauer, M.; Knözinger, H. J. Catal. **2001**, *198*, 89.
25. Vaudagna, S.R.; Canavese, S.A.; Comelli, R.A.; Fígoli, N.S. Appl. Catal. A **1988**, *168*, 93.
26. Iglesia, E.; Barton, D.G.; Soled, S.L.; Miseo, S.; Baumgartner, J.E.; Gates, W.E.; Fuentes, G.A.; Meitzner, G.D. Stud. Surf. Sci. Catal. **1996**, *101*, 533.
27. Barton, G.; Soled, S.L.; Iglesia, E. Top. Catal. **1998**, *6*, 87.
28. Fung, J.; Wang, J. J. Catal. **1996**, *164*, 166.
29. Garin, F.; Keller, V.; Ducros, R.; Muller, A.; Maire, G. J. Catal. **1997**, *166*, 136.
30. Keller, V.; Barath, F.; Maire, G. J. Catal. **2000**, *189*, 269.
31. Logie, V.; Wehrer, P.; Katrib, A.; Maire, G. J. Catal. **2000**, *189*, 438.

4

Naphtha Hydrotreatment

Syed Ahmed Ali
King Fahd University of Petroleum and Minerals
Dhahran, Saudi Arabia

1 INTRODUCTION

Hydrotreatment is the conventional means for removing of sulfur from petroleum fractions. Removal of sulfur is required to eliminate a major poison in subsequent processing (e.g., by naphtha reforming) or to meet environmental legislation (e.g., in middle distillates). Commercially used catalysts for reformer feedstock hydrotreatment belong to the broad class of molybdenum sulfide catalysts, which are applied for the hydroprocessing of a variety of feedstocks, including middle distillates and gas oils. The major requirements to be met by naphtha hydrotreatment catalysts are high activity and selectivity for hydrogenolysis of C-S bonds, with low selectivity for cracking of C-C bonds. In addition, they should exhibit reduced activity for hydrogenation of aromatic compounds in order to maintain octane number and reduce the hydrogen consumption. The catalyst should also possess long-term stability and low sensitivity to poisons.

Naphtha hydrotreatment is practiced to prepare feedstock for catalytic reforming. All reforming catalysts are gradually poisoned by sulfur compounds in the feed that strongly adsorb on and coordinate with the metal sites. High sulfur levels in the feedstock lead to excessive coking and rapid deactivation.[1] Therefore, in commercial operation, it is required that sulfur content in the feed be extremely low. Basic nitrogen compounds adsorb on acid sites. They inhibit principally the acid function, but to some extent they can alter the metallic properties of the platinum. In the presence of water, they can also result in leaching of chloride. Hence, nitrogen compounds should also be removed from the reformer feedstock.

It is reported that a very small amount of sulfur may be beneficial in mitigating the excessive initial hydrogenolysis activity of reforming catalysts. Presulfidation of the catalyst selectively blocks the hyperactive sites and avoids an increase of reactor temperature due to excessive hydrocracking. When the catalyst is subjected to higher temperature and hydrogen atmosphere, part of the sulfur is eliminated (reversible) and the rest remains as irreversibly adsorbed on the surface. It has been reported that the amounts of reversible and irreversible adsorbed sulfur have equilibrium values dependent on the temperature and partial pressure of hydrogen sulfide.[2]

Despite this beneficial effect, the poisoning effect of sulfur in the feedstock in the initial startup stages of reforming catalysts, the presence of sulfur shortens the operating cycle time. For bimetallic Pt-Re catalysts the accepted guideline for optimal catalyst performance is that the sulfur in the feed should be below 0.25–1 ppm to prevent the reversible sulfur from poisoning the hydrogen transfer capability. As the Re fraction of the metal is increased, sensitivity to poisoning is also increased; therefore, an aggressive sulfur control strategy is needed to provide improved catalyst life.[3]

The cost of removing sulfur from the reformer feedstock is easily justified by the extended life of the reforming catalyst. Sulfur can be removed from hydrocarbon streams by reaction with hydrogen over dedicated catalysts. The hydrodesulfurization (HDS) reaction consumes hydrogen and generates hydrogen sulfide according to the general reaction:

$$R\text{-}S\text{-}R + 2H_2 \rightarrow 2R\text{-}H + H_2S \tag{1}$$

Hydrogen sulfide is removed from the gas stream by washing with an amine solution and finally converted to elemental sulfur in a Claus process unit. This hydrogen sulfide removal process is called *hydrotreatment*, which is the only viable option for removal of the contaminants, such as sulfur and nitrogen, from reformer feedstock. Hydrotreatment of naphtha is accomplished by passing the feedstock, together with the hydrogen or hydrogen-rich gas (usually above 75% hydrogen), over a fixed-bed of catalyst under conditions that depend mainly on the feedstock properties and desired product specifications.[4]

This chapter is devoted to the hydrotreatment of reformer feedstock. Different types of reformer feedstocks are reviewed, with focus on the types of sulfur compounds present in them. The main chemical reactions, along with the thermodynamics and reaction kinetics, are then described. The characteristics and functioning of commercial hydrotreatment catalysts are enumerated together with the features of deactivation, poisoning, and regeneration. Hydrotreatment processes are briefly discussed from the practical point of view. Finally, the future outlook of the reformer feedstock pretreatment is presented in light of the growing importance of cleaner transportation fuels.

2 REFORMER FEEDSTOCKS: GENESIS AND CONSTITUTION

Typical reformer feedstock is a straight-run naphtha distilled directly from crude petroleum. However, sometimes naphthas from fluid catalytic cracking (FCC) and hydrocracker and/or coker units are also processed in the reformer after blending with the straight-run naphtha. Since there are major differences in these naphthas, each is dealt with separately in this section. In addition, the sulfur compounds present in the feedstocks and products of naphtha hydrotreatment are also explored.

1.1 Straight-Run Naphthas

Primary sources of reformer feedstocks are heavy straight-run (HSR) naphtha, with a boiling range of 85–165°C. In practice, light straight-run (LSR) naphtha (C_5 to 85°C) is not a good reformer feedstock because it is largely composed of low molecular weight paraffins that tend to crack to butane and light fractions. On the other hand, hydrocarbons boiling above 180°C are easily hydrocracked resulting in excessive carbon laydown on the reformer catalyst.

The composition and quantity of HSR naphtha, however, vary significantly depending on the origin of crude petroleum and its API gravity. Table 1 gives a summary of selected properties of HSR naphthas from some of the important crude petroleum sources.[4] There is a wide variation in the composition of HSR naphthas according to its genesis. Hence, the preparation of reformer feedstock from different HSR naphthas requires careful control of the hydrotreatment process.

Arabian crudes contain about 10–20% naphtha with sulfur content averaging between 300 and 500 ppm. The paraffin content of these crudes is also higher than that of most other crudes. Olefin content in the HSR naphtha is generally very low (0.1–0.3 wt %) and hence is not reported separately. The naphtha sulfur content varies widely from only 13 ppm in Tapis crude oil of Malaysia to 8000 ppm in Lloyd Minister of Canada (Table 1). The nitrogen content of straight-run naphthas, not often reported, is usually around 1–2 ppm. This is so low that the hydrotreater designs does not include it. Naphthas from some crudes will require more severe treatment to meet the nitrogen requirement. The content of HSR naphtha present in each crude varies from only 5 vol% in Shengli, China to about 28 vol % in Zarzaitina, Algeria. This figure is inversely related to the API gravity of the crude.

1.2 Cracked Naphthas

Products from other processing units in the refinery, which have almost the same boiling range as HSR, are added to HSR naphthas as reformer feedstock.

Table 1 Selected Properties of Typical HSR Naphthas[4]

Origin	API° crude	Naphtha specific gravity (g/ml)	Sulfur (ppm)	Paraffins and olefins (wt %)	Naphthenes (wt %)	Aromatics (wt %)	Vol % of crude
Arabian Extra Light	37	0.746	500	65	20	15	16
Arabian Light	33	0.742	350	65	20	15	15
Arabian Heavy	27	0.737	450	70	20	10	11
Bonny Light, Nigeria	37	0.746	500	65	20	15	16
Brega, Libya	40	0.750	40	65	20	15	23
Brent, North Sea	37	0.769	30	42	34	24	18
Lloyd Minister, Canada	21	—	8000	55	32	13	10
Maya, Mexico	22	—	400	60	27	13	8
Shengli, China	24	—	120	46	40	14	5
Statford, North Sea	38	0.760	37	49	37	14	18
Tapis, Malaysia	44	—	13	40	40	20	22
Urals, Russia	32	0.761	290	38	52	10	12
West Texas Sweet, USA	39	0.754	500	47	13	10	20
Zarzaitina, Algeria	43	0.767	100	45	25	30	28

The important sources of such naphthas are (1) visbreaker unit; (2) coker unit; (3) hydrocracker unit; and (4) FCC unit. Since all these units produce naphtha by cracking of heavier fractions of crude oil (vacuum gas oils or atmospheric residues), they are called *cracked* naphthas. The percentage of *cracked* naphthas that are blended in the HSR naphtha depends on the availibility of these streams and also on the design configuration of the hydrotreater to process the blend naphtha.

Cracked naphthas generally contain more sulfur, nitrogen, and olefins. Moreover, the sulfur and nitrogen compounds present are mostly aromatic, produced during the cracking of very large molecules of heavy oils. In general, these types of sulfur and nitrogen compounds are more difficult to hydrotreat.[5]

Table 2 gives some of the properties of the cracked naphthas obtained from different sources.[6] As can be noted from the table, each heavy oil conversion process produces naphtha with a significantly different composition.

Visbreaker Naphtha

Visbreaker naphtha, specifically that obtained from high-sulfur feedstock (such as Arabian heavy vacuum residue), contains about 15,000 ppm sulfur as well as 500 ppm basic nitrogen. Such high concentrations of heteroatoms require severe hydrotreatment in order to prepare suitable reformer feedstock. Therefore, visbreaker naphtha is typically added to HSR naphtha in small percentages.

Coker Naphtha

Coker naphtha properties do not differ significantly from visbreaker naphtha. However, the process yield is four to five times higher, which means that refineries with coking units have to process more of the coker naphtha into production of reformer feedstock. Different coking processes also result in somewhat different composition of coker naphtha, as shown in Table 2. The research octane numbers of visbreaker and coker naphthas are generally in the range of 65–70.

FCC Naphtha

Fluid catalytic cracking of vacuum gas oil and atmospheric resid produces substantial amounts of naphtha. In fact, the FCC unit is a major source of the gasoline pool, constituting 30–40% of a typical U.S. refinery gasoline pool. As shown in Table 3, the FCC unit contributes only 36% of the gasoline but 98% of the sulfur in the pool. Significant reduction in gasoline sulfur is mandated by regulations in many industrialized countries. Meeting gasoline sulfur specifications of 50 ppm for the pool means that the FCC naphtha sulfur is limited to about

Table 2 Properties of Cracked Naphthas[4,5]

Naphtha type	Bromine number	Diene (g/100 ml)	Sulfur (ppm)	Basic N (ppm)	Paraffins	Olefins	Naphthenes	Aromatics	Wt. % on feed
					(wt %)				
Visbreaker									
High sulfur feedstock	90	2	15,000	500	33	45	11	11	4–5
Low sulfur feedstock	80	2	1,000	200	33	45	11	11	4–5
Coker									
Low sulfur feedstock	80	5	2,000	450	30	38	17	15	20
Flexicoking	122	—	1,100	70	—	—	—	—	15
Delayed coking	90	—	1,100	62	—	—	—	—	18
EUREKA Process	82	—	800	50	—	—	—	—	12
ART Process	80–100	—	400	—	—	—	—	—	17
FCC	50	<0.5	1,200	100	28	25	22	25	10–15
Hydroprocessing									
VGO hydrocracked	0	0	0–10	0–2	30	0	55	15	10–40
Resid hydrocracking	2	0	20–100	50–200	35	0	50	15	5–10
Gasoil HDS and mild hydrocracking	0–2	0	2–50	1–10	40	5	30	25	1–5

Table 3 Typical Gasoline Pool Composition in U.S. Refinery[37]

Gasoline blendstocks	Percent of pool volume	Percent of pool sulfur
FCC naphtha	36	98
Reformate	34	—
Alkylate	12	—
Isomerate	5	—
Butanes	5	—
Light straight-run naphtha	3	1
Hydrocracked naphtha	2	—
MTBE	2	—
Coker naphtha	1	1
Total	100	100

135 ppm. Lower gasoline sulfur specifications or higher FCC naphtha content in the pool requires further reduction in the allowed FCC naphtha sulfur. Refiners are currently looking at a variety of routes of controlling sulfur in FCC naphtha.[7]

There are two broad approaches available to refiners for lowering sulfur in FCC naphtha. The first is to hydrotreat the feed to the FCC unit, and the second is to treat the naphtha produced by the FCC unit. FCC feed hydrotreatment improves naphtha yield as well as quality, and reduces SO_x emissions from the FCC unit. Unfortunately, this is typically a high-pressure, capital-intensive process unit. Furthermore, in an environment where gasoline sulfur content must be below 30 ppm, FCC feed hydrotreatment may not be sufficient to reach the target. For these reasons, many refiners will choose to hydrotreat the naphtha produced by the FCC unit. However, the key to success in hydrotreating FCC naphtha is development of a process, which avoids the large octane losses that occur upon saturation of olefins during conventional HDS of full-range FCC naphtha[8] or the yield loss to cracking that accompanies octane upgrading.

It has long been recognized that sulfur compounds concentrate in the higher boiling portion of FCC naphtha, whereas olefins concentrate in the lower boiling portion as shown in Table 4 and in Figure 1. Fractionation is therefore an attractive first step in many potential process configurations designed to reduce sulfur while maintaining olefins (and thus, octane). Applying conventional hydrotreatment to the heavy FCC naphtha fraction is sufficient in many cases to achieve 100–150 ppm sulfur pool targets, with very little loss of octane. However, where very low gasoline pool sulfur levels are required (in the <50 ppm S range), removal of sulfur from the intermediate and, in some cases, even the light fraction will be necessary as well. If the end point of the light FCC naphtha stream is kept relatively low, a significant portion of the sulfur in this

Table 4 Typical Composition of FCC Naphtha Cuts[4]

Property	IBP–75°C	75–125°C	125–150°C	150°C to FBP	Full range
Wt %	22	30	16	32	(100)
S, ppm	15	20	40	120	50
N, ppm	5	6	20	75	35
Br no., g/100 ml	100	75	45	25	70
PONA, wt %:					
P + N	33	38	30	20	30
O	65	50	30	20	45
A	2	12	40	60	25
RON	95	87	93[a]	93[a]	92
MON	81	77	81[a]	81.5[a]	80

At lower conversion (middle distillate mode) these octane numbers can be even lower than those of the 75–125°C fraction, making heavy FCC naphtha the best choice for reforming; otherwise only the "heart cut 75–125°C."

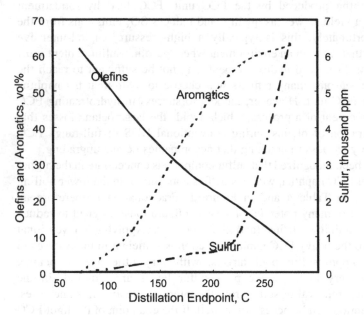

Figure 1 Distribution of aromatics, olefins, and sulfur in FCC naphtha.[37]

stream will be light mercaptans, which can be removed via caustic extraction or other emerging technologies. This avoids hydrotreatment of the light naphtha, which would result in severe octane loss. Desulfurization of the intermediate FCC naphtha without serious octane loss is a much more technically challenging problem.

The incentive to process FCC naphtha in a reforming unit is a net increase in the gasoline pool octane. For most refiners, however, FCC naphtha is not a viable feedstock for the reforming unit since the small gain in octane number for the gasoline pool usually does not justify the costs involved in hydrotreating and reforming the FCC naphtha. Nevertheless, FCC naphtha is processed in several reforming units. It is possible that more FCC naphtha will be reformed in the future as stringent sulfur limits are imposed on gasoline.[9,10]

The heart cut of FCC naphtha (170–300°F) is most suitable for processing in the reformer; however, this material must first be properly hydrotreated to meet reforming unit feed specifications. At typical conditions required to adequately hydrotreat FCC naphtha, any olefin will be hydrogenated into paraffin and approximately 1 vol % of the aromatics will be hydrogenated to naphthenes. FCC naphtha typically contains more nitrogen than does straight-run naphtha; therefore, adequate nitrogen removal would determine required hydrotreatment severity. The impact of processing FCC naphtha on reformer cycle length can be severe, in some cases doubling or more than doubling the coke laydown rate compared to straight-run naphtha.[10] The FCC naphtha has a higher naphthenes and aromatics (N + 2A) content than the virgin feed, so the impact on cycle life is dependent on the relative N + 2A content and the target octane. In addition, FCC naphthas tend to have higher alkylcyclopentane contents at the same N + 2A. In general, the alkylcyclopentanes have poorer reformate yield and higher coking tendency than their alkylcyclohexane counterparts. However, with a middle cut of the FCC naphtha, the differences between the alkylcyclopentane and the alkylcyclohexane yields are fairly small.

Hydrocracked Naphtha

Hydrocracking of heavier fractions of crude oil, such as vacuum gas oil and residues, produces heavy naphtha, which is rich in naphthene fraction and hence a very suitable reformer feedstock. Furthermore, hydrocracked naphtha has very low sulfur content if the feed is vacuum gas oil. When heavy residues are hydrocracked, the naphtha produced has more nitrogen. The severity of hydrotreatment needs to be increased if a large percentage of hydrocracked naphtha is blended into the hydrotreated feedstock. Blending of hydrocracked naphtha with straight-run naphtha in the ratio of 40:60 has been reported in commercial application.[11]

1.3 Sulfur Species in Naphthas

Knowledge of the exact nature of the sulfur compounds in naphtha can help in determining the severity required in HDS. In addition, it can also reveal the nature of sulfur compounds that remain after hydrotreatment. This information provides better understanding of the function of catalysts and aids in the selection of a suitable catalyst on the basis of selectivity for the removal of sulfur compounds. The amounts and types of sulfur compounds present in naphtha are determined by the source of the crude—its genesis. These sulfur compounds include elemental sulfur, hydrogen sulfide, disulfides, sulfides, mercaptans, and thiophenes. Characterization of sulfur compounds in naphtha (particularly in hydrotreated naphtha) is very difficult because of extremely low concentrations. In addition, standard techniques are not available and the quantitative analysis is tedious. The literature cites some methods for the determination of sulfur compounds in petroleum fractions.[12,13]

Anabtawi and coworkers[14] reported that the straight-run naphtha obtained from Arabian light crude contains 897 ppm sulfur, which is present in the form of 55.7 wt % mercaptans, 43.1 wt % thiophenes and sulfides, 1.1 wt % disulfides, and traces of polysulfides, hydrogen sulfide, and elemental sulfur as shown in Table 5. However, the major sulfur type in hydrocracked naphtha was mercaptans (90 wt %). The sulfur species composition in a blend (60% straight run and 40% hydrocracked naphtha) is almost the same as straight-run naphtha since sulfur contribution from hydrocracked naphtha is negligible.

Table 5   Sulfur Species in Naphtha Feedstocks and Product[11]

Sulfur species	Straight-run naphtha	Hydrocracked naphtha	Blend naphtha[a]	Hydrotreated blend naphtha
Total sulfur (ppm)	897	2.2	534	0.30
Mercaptans (ppm)	500	2.0	297	0.29
Disulfides (ppm)	10	tr[b]	5	tr
Sulfides + thiophenes (ppm)	387	Tr	232	tr
Elemental sulfur + H$_2$S	tr	tr	tr	tr
Polysulfides	tr	tr	tr	tr

[a]60% straight run + 40% hydrocracked naphtha.
[b]Present in trace amounts.

A simulated distillation method based on gas chromatography using flame photometric detection (GC–FPD) has been used in determining the boiling range distribution of sulfur-containing compounds in naphtha. The GC-FPD chromatogram of straight-run naphtha, presented in Figure 2, shows 52 sulfur compound peaks. A mixture of 24 sulfur compounds in isooctane was injected to determine their retention times. Seventeen sulfur compounds were identified in straight and blend naphthas by matching the retention times. The DB-1 column used in this study elutes the sulfur compounds in order of increasing boiling point. This characteristic was used to estimate the boiling points of the compounds in the unidentified peaks, which were then matched with the boiling points of known sulfur compounds. Table 6 shows the compounds identified by both methods, with their retention times, matched peak numbers, and boiling points. Of the 42 identified compounds, 15 were mercaptans, 15 thiophenes, 9 sulfides, and 3 disulfides.

It was reported that six sulfur compounds are present in hydrocracked naphtha and four in the hydrotreated product. The major part of the sulfur present in hydrocracked naphtha includes five mercaptans (peaks 2, 4, 6, 7, and 10) and one thiophene (peak 51) in Figure 2. The presence of methylmercaptan, isopropylmercaptan, n-butylmercaptan, and 2,3,4,5-tetramethylthiophene (peak 51) was observed in the hydrotreated product. The presence of 2,3,4,5-tetramethylthiophene was expected because thiophenes, especially with four methyl groups, are difficult to hydrodesulfurize. But mercaptans were expected to

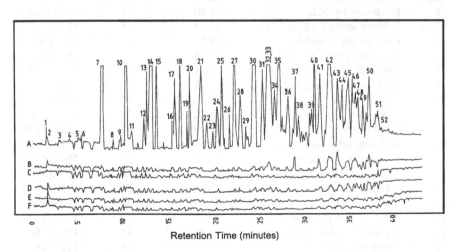

Figure 2 Chromatograms of sulfur compounds in (A) blend naphtha and its hydrotreated products obtained at (B) 250°C, (C) 280°C, (D) 300°C, (E) 320°C, and (F) 350°C.[11]

Table 6 Sulfur Compounds in Straight-Run Naphtha[11]

Serial no.	Compound name	Retention time (min)	Matched peak no.	Boiling point (°C)
Identified by Matching Their Retention Times				
1.	Hydrogen sulfide	1.48	1	—
2.	Ethylmercaptan	3.00	3	36
3.	Isopropylmercaptan	4.07	4	56
4.	*t*-Butylmercaptan	5.04	5	64.2
5.	*n*-Propylmercaptan	5.42	6	67.5
6.	*sec*-Butylmercaptan	7.76	7	84
7.	Diethyl sulfide	9.11	8	92
8.	*n*-Butylmercaptan	10.29	10	100
9.	2-Methylthiophene	13.18	14	112.4
10.	3-Methylthiophene	13.64	15	115.4
11.	*n*-Amylmercaptan	16.88	19	126
12.	2-Ethylthiophene	20.30	24	132.5
13.	2,5-Dimethylthiophene	20.75	25	135.5
14.	Di-*n*-propyl sulfide	22.84	28	142
15.	*n*-Hexylmercaptan	25.25	31	151
16.	*n*-Heptylmercaptan	34.18	44	176
17.	Di-*n*-butyl sulfide	38.18	51	188
Identified by Matching Their Boiling Points				
1.	Methylmercaptan	1.51	2	6 (<10)[a]
2.	Methyl disulfide	12.57	13	109.7 (110)
3.	Diisopropyl sulfide	15.60	17	120.7 (120.5)
4.	*n*-Butyl, methyl sulfide	16.15	18	122.5 (122.5)
5.	Ethyl methyl disulfide	18.51	21	130 (130)
6.	Isopropyl propyl sulfide	19.75	23	132 (132)
7.	2-methyl butyl sulfide	22.11	27	139.5 (139.5)
8.	3,4-Dimethyl thiophene	23.38	29	144 (144)
9.	*t*-Butyl sulfide	24.37	30	149 (148)
10.	2-Isopropyl thiophene	25.25	31	152 (152)
11.	Ethyl disulfide	25.79	32	153 (152.5)
12.	2-Hydroxyethylmercaptan	25.79	33	154 (154.5)
13.	3-Isopropylthiophene	27.01	35	156 (156)
14.	Cyclohexylmercaptan	28.05	36	159 (159)
15.	2 Ethyl 3-methylthiophene	28.85	37	161 (161.5)
16.	3-*t*-butylthiophene	31.59	41	169 (169)
17.	2,3,4-Trimethylthiophene	32.65	42	172.7 (172.5)
18.	1,3-Dithiacyclopentane	33.57	43	175 (175)
19.	2-Benzothiozole thiol	34.85	45	179 (178)

(*Table continues*)

Table 6 Continued

Serial no.	Compound name	Retention time (min)	Matched peak no.	Boiling point (°C)
20.	2-Methyl, 5-propyl thiophene	35.25	46	179.5 (179.5)
21.	2-n-Butylthiophene	35.65	47	181 (180.5)
22.	2,5-Diethylthiophene	35.89	48	181 (181)
23.	3-n-Butylthiophene	36.48	49	183 (183)
24.	3,4-Diethylthiophene	37.18	50	185 (185)
25.	2,3,4,5-Tetramethylthiophene	38.13	51	186 (187)

[a]Values in parentheses are those estimated from peak position.

be removed completely as they are easier to hydrodesulfurize. Their presence indicates the occurrence of recombination reactions between alkenes and hydrogen sulfide resulting in the formation of mercaptans:

$$\text{Olefins} + H_2S \rightleftharpoons \text{mercaptans} \tag{2}$$

The reversible reaction, which shifts to the right at higher hydrogen sulfide partial pressure, is also dependent on temperature, partial pressure of alkenes, residence time, and catalyst type. The recombination of olefins and hydrogen sulfide is discussed later in Sec. 3.

3 HYDROTREATING REACTIONS: REACTIVITIES, THERMODYNAMICS, AND KINETICS

3.1 Hydrotreating Reactions and Reactivities

Catalytic naphtha hydrotreatment can simultaneously accomplish desulfurization, denitrogenation, and olefin saturation. From Sec. 2, it was evident that sulfur compounds present in naphtha are mainly thiols (mercaptans), sulfides, disulfides, and various thiophenes and thiophene derivatives. The basic desulfurization reactions of different types of sulfur compounds are presented in Table 7. These reactions illustrate that desulfurization is accomplished by the cleavage of C-S bonds, which are replaced by C-H and S-H bonds.

The ease of desulfurization is dependent on the types of sulfur compounds. As expected, the lower boiling compounds are desulfurized more easily than the high-boiling ones. The reactivity of sulfur compounds decreases in the order mercaptans > disulfides > sulfides > thiophenes > benzothiophenes > alkylbenzothiophenes. Within a group, the reactivity decreases with increased

Table 7   Representative Hydrotreating Reactions

Hydrodesulfurization Reactions

Mercaptans	$RSH + H_2 \rightarrow RH + H_2S$
Sulfides	$RSR' + 2H_2 \rightarrow RH + R'H + H_2S$
Disulfides	$RSSR' + 3H_2 \rightarrow RH + R'H + 2H_2S$
Cyclic sulfides	
Thiophene	

$$\text{(thiophene)} + 3H_2$$

$$\longrightarrow H_2S + C_4H_8 \quad \text{(mixed isomers)}$$

Benzothiophene

$$\xrightarrow{2H_2} \text{(ethylbenzene } CH_2CH_3) + H_2S$$

Hydrodenitrogenation Reactions

Pyridine

$$\xrightarrow{+H_2} C_5H_{11}NH_2 \xrightarrow{+H_2} C_5H_{12} + NH_3$$

Hydrogenation Reactions

Olefin saturation	$R{=}R' + H_2 \rightarrow HR{-}R'$

---

molecular size and varies depending on whether the alkyl group is an aliphatic or aromatic group. Among thiophene derivatives, the alkylthiophenes are less reactive than thiophene, presumably due to steric effects.

Similar to desulfurization reactions, nitrogen and oxygen are removed in catalytic hydrotreatment by the cleavage of C-N and C-O bonds, forming ammonia and water, respectively. The C-N and C-O bond cleavage is much more difficult to achieve than the C-S bond cleavage. Consequently, denitrogenation and deoxygenation occur to a much lesser extent than does desulfurization. Nitrogen and oxygen are not significant in virgin naphthas, although in cracked stocks or in synthetic naphthas (such as those from coal or shale) concentrations of nitrogen and oxygen can be quite high. In such cases, very severe hydrotreatment conditions are required to reduce the concentrations to levels acceptable for reforming. Finally, olefin saturation is effected simply by the addition of hydrogen to an unsaturated hydrocarbon molecule (a molecule with one or more C-C double bonds) to produce a saturated product. Olefin saturation reaction is very exothermic and proceeds relatively easily. Straight-chain monoolefins are easy to hydrogenate whereas branched, cyclic, and diolefins are somewhat more difficult. Aromatic rings are not hydrogenated except under severe hydrotreatment conditions. As with nitrogen and oxygen compounds, olefins are not found to any great extent in virgin naphthas, but concentrations can get as high as 40 or 50 vol % in cracked or synthetic stocks (see Table 2).

During HDS of reformer feedstock at high temperatures some cracking of alkanes occurs resulting in 0.2–0.4 wt % alkenes. Alkenes are also present in cracked feedstocks. The alkenes and hydrogen sulfide (product of HDS) can recombine to form mercaptans.

3.2 Thermodynamics of Hydrotreating Reactions

Hydrodesulfurization

The HDS of organosulfur compounds is exothermic. The heat of reaction varies significantly from one compound to another. Table 8 lists the standard enthalpies of HDS of representative compounds.[14] The amount of heat release increases with the number of moles of hydrogen required to desulfurize the compound. In addition, there is a marked difference between heat of reactions of different classes of sulfur compounds. For the HDS of mercaptans, the heat of reaction is about 55–70 kJ/mole of hydrogen consumed, while for thiophenes it is 85 kJ/mole of hydrogen consumed. The heat of reaction of HDS can increase the reactor temperature by 10–80°C at typical operating conditions depending on the nature of feedstock. Hence, it must be taken into consideration while designing the HDS reactors.

Estimation methods for standard enthalpies, standard free energies, entropies, and ideal-gas heat capacities involve group contributions based on molecular structure of the reactants and products.[15] Several methods are reported in the literature with each one of them having advantages and disadvantages. For the estimation of equilibrium constants of HDS and heterodenitrogenization (HDN) reactions, *Joback's method* is more suitable as it provides reliable estimates of thermodynamic properties of almost all compounds of interest.[16] Furthermore, this method is much simpler to use and no symmetry or optical isomer corrections are necessary. However, caution should be exercised while applying Joback's method to very complex molecules as this may lead to large errors.

Table 8 Equilibrium Constants and Standard Enthalpies for the Hydrodesulfurization Reactions[14]

Reaction	$\log_{10} K_{eq}$ at					ΔH^{0a}
	25	100	200	300	400°C	
Mercaptans						
$CH_3-SH + H_2 \rightleftharpoons CH_4 + H_2S$	12.97	10.45	8.38	7.06	6.15	−72
$C_2H_5-SH + H_2 \rightleftharpoons C_2H_6 + H_2S$	10.75	8.69	6.99	5.91	5.16	−59
$C_3H_7-SH^b + H_2 \rightleftharpoons C_3H_8 + H_2S$	10.57	8.57	6.92	5.87	5.15	−57
$C_2H_5-SH^b \rightleftharpoons CH_2 = CH_2 + H_2S$	−5.33	−2.57	−0.25	1.26	2.31	78
$C_3H_7-SH^b \rightleftharpoons \Delta CH_2 = CH-CH_3 + H_2S$	−4.54	−2.18	−0.20	1.08	1.97	67
Thiophenes						
$Thiophene + 3H_2 \rightleftharpoons n\text{-}C_4H_{10} + H_2S$	30.84	21.68	14.13	9.33	6.04	−262
$3\text{-Methylthiophene}^b + 3H_2 \rightleftharpoons 2\text{-methylbutane} + H_2S$	30.39	21.35	13.88	9.11	5.82	−258
$2\text{-Methylthiophene}^b + 3H_2 \rightleftharpoons n\text{-pentane} + H_2S$	29.27	20.35	13.33	8.77	5.66	−250
$Thiophene + 2H_2 \rightleftharpoons tetrahydrothiophene^b$	10.51	6.47	3.17	1.12	−0.21	−116
Benzothiophenes						
$Benzothiophene^b + 3H_2 \rightleftharpoons ethylbenzene + H_2S$	29.68	22.56	16.65	12.85	10.20	−203
$Dibenzothiophene^b + 2H_2 \rightleftharpoons biphenyl + H_2S$	24.70	19.52	15.23	12.50	10.61	−148
$Benzothiophene^b + H_2 \rightleftharpoons dihydrobenzothiophene^b$	5.25	3.22	1.55	0.49	−0.23	−58
$Dibenzothiophene^b + 3H_2 \rightleftharpoons hexahydrodibenzothiophene^b$	19.93	11.93	5.47	1.54	−0.98	−230

aStandard enthalpy of reaction in kJ/mole of organosulfur reactant.
bThermodynamic properties of these compounds are calculated by Joback's method.

Equilibrium constants of representative HDS reactions of typical organosulfur compounds present in petroleum light distillates are calculated using the available experimental data or by the Joback group's contribution method and presented in Table 8. The equilibrium constants for the HDS reactions involving conversion of sulfur compounds to saturated hydrocarbons over a wide range of temperatures are all positive. This indicates that the HDS reactions are essentially irreversible and can proceed to completion if hydrogen is present in stoichiometric quantity under the reaction conditions employed industrially (e.g., 250–350°C and 3–10 MPa). Generally, equilibrium constants decrease with an increase in temperature, which is consistent with the exothermicity of the HDS reactions. Values of equilibrium constants much lower than unity are attained only at temperatures (>425°C) considerably higher than those observed in practice.

Comparison of equilibrium constants for HDS of different classes of sulfur compounds indicate that there is a marked difference in the equilibrium constants from one class to another. However, within one class, such as mercaptans, the variation is very small and perhaps within the experimental error. The HDS of sulfur compounds to yield unsaturated hydrocarbons (alkenes) and hydrogen sulfide is not thermodynamically favored as the reactions are endothermic and will not proceed to completion at temperatures below 300°C.

Sulfur removal may occur with or without hydrogenation of the heterocyclic ring as explained in the next section. Reaction pathways involving prior hydrogenation of the heterocyclic ring are affected by the thermodynamics of reversible hydrogenation of sulfur-containing rings or benzenoid rings, which are equilibrium limited at practical HDS temperatures. For example, the equilibrium constant for hydrogenation of thiophene to tetrahydrothiophene is less than unity at temperatures above 350°C. Hydrogenation of benzothiophene to form dihydrobenzothiophene is similar (Table 8). Thus, sulfur removal pathways via hydrogenated organosulfur intermediates may be inhibited at high temperatures and low pressures because of the low equilibrium concentrations of the sulfur species.

Hydrodenitrogenation

Most of the nitrogen components in petroleum fractions are present in the form of five- or six-membered rings, nearly all of which are unsaturated. These compounds can be basic or nonbasic. Pyridines and saturated heterocyclic ring compounds (indoline, hexahydrocarbazole) are generally basic, while pyrroles are not. Nitrogen removal from heterocyclic organonitrogen compounds requires hydrogenation of the ring containing the nitrogen atom before hydrogenolysis of the carbon–nitrogen bond can occur. Hydrogenation of the hetero ring is required to reduce the relatively large energy of the carbon–nitrogen bonds in such rings and thus permit more facile carbon–nitrogen bond scission. Nitrogen is then removed from the resulting amine or aniline as ammonia. The energies of carbon–nitrogen double and single bonds are 615 kJ/mole and 305 kJ/mole, respectively.

The requirement that ring hydrogenation occur before nitrogen removal implies that the position of the equilibrium of the hydrogenation reactions can affect the nitrogen removal rates, if the rates of the hydrogenolysis reactions are significantly lower than the rates of hydrogenation. An unfavorable hydrogenation equilibrium results in low concentrations of hydrogenated nitrogen compounds undergoing hydrogenolysis. Consequently, HDN rates will be lowered. However, high hydrogen pressure can be used to increase the equilibrium concentrations of saturated hetero ring compounds.[17]

Little has been published about possible thermodynamic limitations of HDN reactions, though this could have significant implications. Sonnemans et al.[18] were the first to report the estimated thermodynamic equilibrium constants for pyridine HDN:

$$
\text{pyridine} + 3H_2 \rightleftharpoons \text{piperidine} \rightarrow C_5H_{11}NH_2 \rightarrow C_5H_{12} + NH_3
$$

(3)

They found that the vapor-phase hydrogenation to piperidine is reversible under representative hydrotreatment operating conditions and that the equilibrium favors pyridine at higher temperature. Satterfield and Cocchetto[19] observed a maximum, followed by a decrease, in the extent of pyridine HDN with the increase in temperature in the range of 200–500°C during the experiments at 1.1 MPa. This was associated with the thermodynamic equilibrium limitation. If pyridine hydrogenation in Eq. (3) is rate limiting, the piperidine will undergo hydrogenolysis as fast as it is formed so that the position of the hydrogenation equilibrium does not influence the rate of HDN. If, on the other hand, piperidine hydrogenolysis is rate limiting, the hydrogenation equilibrium can be established. Increased temperature shifts this equilibrium to the left (toward pyridine), decreasing the partial pressure of piperidine in the system. This can lower the rate of cracking and therefore the rate of HDN, resulting in a decrease in pyridine conversion with increasing temperature.

Illustrative values of equilibrium constants for different reactions in HDN networks of some representative compounds are presented in Table 9. The equilibrium constants of all the reactions decrease with temperature, consistent with the fact that all the reactions are exothermic. The hydrogenolysis reactions remain favorable at temperatures as high as 500°C. The overall reactions are highly exothermic but favorable at temperatures up both pyridine and pyrrole, the equilibrium constants for the initial ring saturation steps are favorable ($K > 1$, $\log_{10} K < 0$) above approximately 225°C.

Table 9 Equilibrium Constants and Standard Enthalpies for Selected Hydrogenation, Hydrogenolysis, and Overall HDN Reactions[14]

Reaction	$\log_{10} K_{eq}$ at		ΔH^{0a}
	300	400°C	
Pyridine + 3H$_2$ $\rightleftharpoons$ piperidine	−2.4	−5.1	−199
Piperidine + H$_2$ $\rightleftharpoons$ n-pentylamine	1.3	0.8	−37
n-Pentylamine + H$_2$ $\rightleftharpoons$ n-pentane + NH$_3$	9.9	8.7	−89
Pyridine + 5H$_2$ $\rightleftharpoons$ n-pentane + NH$_3$	8.9	4.4	−362
Pyrrole + 3H$_2$ $\rightleftharpoons$ Pyroldine	−1.3	−2.8	−111
Pyroldine + H$_2$ $\rightleftharpoons$ n-butylamine	2.2	1.3	−66
n-Butylamine + H$_2$ $\rightleftharpoons$ n-butane + NH$_3$	9.3	8.2	−81
Pyrrole + 4H$_2$ $\rightleftharpoons$ n-butane + NH$_3$	10.0	6.1	−288
Indole + H$_2$ $\rightleftharpoons$ Indoline	−2.7	−3.3	−46
Indoline + H$_2$ $\rightleftharpoons$ o-ethylaniline	4.7	3.3	−105
o-Ethylaniline + H$_2$ $\rightleftharpoons$ ethylbenzene + NH$_3$	5.8	5.0	−58
Indole + 3H$_2$ $\rightleftharpoons$ ethylbenzene + NH$_3$	7.8	5.0	−49
Carbazole + H$_2$ $\rightleftharpoons$ o-phenylaniline	1.2	0.3	−66
o-Phenylaniline + H$_2$ $\rightleftharpoons$ biphenyl + NH$_3$	5.8	5.0	−58
Carbazole + 2H$_2$ $\rightleftharpoons$ biphenyl + NH$_3$	6.8	5.1	−126

[a]Standard enthalpy of reaction in kJ/mole of organonitrogen reactant.

Recombination Reaction

The thermodynamic analysis of the alkene–hydrogen sulfide recombination reaction was carried out for five reactions as shown in Table 10. The experimentally determined thermodynamic properties of the compounds involved were reported by Reid et al.[15] The values of standard enthalpies of reactions indicate that these reactions are exothermic. The amount of heat release is between 105–130 kJ/mole when methylmercaptan is the product and it is lower when heavier mercaptans are formed. The equilibrium constants are very high at lower temperatures, indicating the definite likelihood of these reactions in the downstream of reactors such as in coolers, pipes, and heat exchangers. The methylmercaptan-forming reactions are more favorable than other mercaptan-forming reactions. In fact, the formation of heavier mercaptans, such as propylmercaptan, is not thermodynamically favored at temperatures above about 250°C.

Recombination of alkenes and hydrogen sulfide was observed in a study conducted by Ali and Anabtawi[20] while investigating the deep desulfurization of naphtha. The product naphtha obtained at higher temperature contained more sulfur, especially as mercaptan sulfur, than the product obtained at lower

Table 10 Equilibrium Constants and Standard Enthalpies for Some Alkene–Hydrogen Sulfide Recombination Reactions (kJ/mole)

Reaction	$\log_{10} K_{eq}$ at					ΔH^{a*}
	25°C	100°C	200°C	300°C	400°C	
$C_2H_4 + H_2S + H_2 \rightleftharpoons CH_3-SH + CH_4$	16.78	12.22	8.41	5.95	4.25	−130
$C_3H_6 + H_2S + H_2 \rightleftharpoons CH_3-SH + C_2H_6$	12.71	8.91	5.75	3.71	2.31	−108
$1\text{-}C_4H_8 + H_2S + H_2 \rightleftharpoons CH_3-SH + C_3H_8$	12.55	8.81	5.68	3.68	2.29	−107
$C_2H_4 + H_2S \rightleftharpoons C_2H_5-SH$	5.33	2.57	0.25	−1.26	−2.31	−78
$C_3H_6 + H_2S \rightleftharpoons C_3H_7-SH$	4.54	2.18	0.20	−1.08	−1.97	−67

[a] Standard enthalpy of reaction in kJ/mole.
Source: Ref 14.

temperature. It was inferred that higher temperature caused cracking, which resulted in more alkenes available for recombination with the hydrogen sulfide downstream of the reactor to form mercaptans. Analysis of sulfur compounds present in the product naphtha by GC combined with flame ionization detection showed that methylmercaptan is present. Thermodynamic analysis of these reactions also confirms that formation of methylmercaptan is more favored than formation of higher mercaptans.

3.3 Reaction Kinetics

The two main naphtha hydrotreatment reactions, namely HDS and HDN, are generally represented by simple first-order kinetics. Assuming first-order kinetics, at a hydrogen pressure of 300 psig bar and 330°C, the comparative relative first-order rate constants for the various petroleum feedstocks are straight-run naphtha = 100 (by definition); kerosene = 20; light gas oil = 10; and vacuum gas oil = 1. Hence, naphtha HDS is considered a straightforward, catalytically easy reaction. However, there are several difficulties in conducting kinetic studies for naphtha hydrotreatment in the laboratory as described by Lovink.[4]

HDS kinetics of a number of model sulfur compounds has been reported in the literature.[21] Since thiophenic compounds are less reactive sulfur compounds than mercaptans, sulfides, and disulfides, they have been the subject of many kinetic studies. It has been reported that the relative pseudo-first-order constants of typical thiophenic compounds decrease in the following order: thiophene (= 1) > benzothiophene (= 0.59 of thiophene) > dibenzothiophene (0.04 of thiophene).[22] In general, it seems that HDS reactivity depends on the molecular size and the structure of the sulfur-containing compound. However, the reactivities of these compounds change depending on the reaction conditions, and it is difficult to compare the results obtained under different conditions.

Thiophene HDS has been examined by several researchers and recently reviewed by Girgis and Gates.[17] Figure 3 shows the possible reaction pathways for thiophene HDS. Two major reaction pathways have been proposed in the literature:

1. Thiophene is desulfurized prior to hydrogenation to give butadiene which is hydrogenated to butane.
2. Thiophene is hydrogenated prior to desulfurization to give tetrahydrothiophene, which is desulfurized to butene. Small amounts of butadiene are formed, possibly as an intermediate, but this is rapidly hydrogenated to butene. The butenes in turn are more slowly hydrogenated to butane.

Satterfield and Roberts reported the first example of the determination of the kinetics of thiophene HDS.[23] The reaction was performed in a steady-state recirculation flow reactor using Co-Mo/Al_2O_3 catalyst in the absence of mass

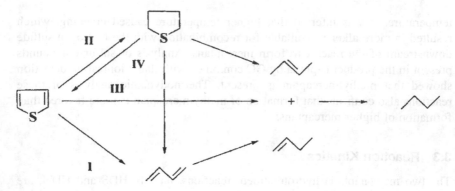

Figure 3 Proposed reaction network of thiophene hydrodesulfurization.[21]

transfer influence (235–265°C). The thiophene disappearance was represented by the following Langmuir–Hinshelwood rate equation:

$$r_{HDS} = \frac{k_{HDS}K_T P_T K_H P_H}{(1 + K_T P_T + K_{H_2S}P_{H_2S})^2} \tag{4}$$

where r_{HDS} is rate of HDS; k_{HDS} is rate constant of HDS; K_T, $K_{H_2}S$ and K_H are adsorption equilibrium constants of thiophene, hydrogen sulfide, and hydrogen, respectively; P_T, $P_{H_2}S$, and P_H are partial pressure of thiophene, hydrogen sulfide, and hydrogen, respectively. The rate expression implies that the surface reaction between thiophene and hydrogen is the rate-determining step and that thiophene and hydrogen sulfide, which retards the HDS reaction, are competitively adsorbed at one type of catalytic site. In a power law expression, the HDS reaction appears to be between half and first order with respect to hydrogen at pressures above atmospheric.

In the case of benzothiophene, substituted or unsubstituted, the thiophene ring is hydrogenated to the thiophane derivative before the sulfur atom is removed, in contrast with the behavior of thiophene. Other steps may also be involved in the reaction network. As with thiophene, methyl substitution reduces reactivity.

4 HYDROTREATING CATALYSTS

4.1 Active Components

A commonly used hydrotreatment catalyst is molybdenum oxide dispersed on γ-Al$_2$O$_3$ support. It is generally promoted by either cobalt or nickel oxides—the so-called CoMo or NiMo catalysts. Typical properties and composition of commercial HDS catalysts are presented in Table 11. Cobalt–molybdenum (Co-Mo/Al_2O_3) catalysts are noted for their desulfurization activity, while nickel-molybdenum (Ni-Mo/Al_2O_3) catalysts are most suited for denitrogenation and olefin saturation. However, this does not mean that Co-Mo/Al_2O_3 catalysts are always a preferred choice for desulfurization. In contrast to the flexibility of Ni-Mo/Al_2O_3 catalysts, Co-Mo/Al_2O_3 catalysts exhibit lower denitrogenation activity. When naphtha must be hydrotreated to remove both sulfur and nitrogen to very low levels, Ni-Mo/Al_2O_3 catalysts are preferred. However, since most straight-run naphthas do not have a high nitrogen content, Co-Mo/Al_2O_3 catalysts are extensively used.[5]

Commercial HDS catalysts, available from many catalyst manufacturers, have compositions in the range of 3–5% CoO and 12–15% MoO_3 on a γ-Al_2O_3 support and have a surface area of about 250 m^2/g. They are usually prepared by impregnation of γ-Al_2O_3 with an aqueous solution of ammonium molybdate and

Table 11 Typical Properties and Composition of Present-Day Naphtha Hydrodesulfurization Catalysts

Catalyst property		Catalyst A	Catalyst B
Physical Properties			
Form		1/16 in, extrudate	1/16 in, extrudate
BET surface area	(m^2/g)	270	250
Total pore volume	(m^3/g)	0.50	0.45
Average pore radius	(Å)	36.7	35.7
Compacted bulk density	(kg/m^3)	742.8	723.1
Chemical Analysis (wt %, dry basis)			
Cobalt oxide	(CoO)	3.6	3.6
Molybdenum oxide	(MoO_3)	18.6	19.0
Aluminum oxide	(Al_2O_3)	72.4	71.3
Silicon oxide	(SiO_2)	4.6	0.4
Phosphorous oxide	(P_2O_5)	0.04	1.84
Iron oxide	(Fe_2O_3)	0.08	0.04
Magnesium oxide	(MgO)	0.10	0.06
Nickel oxide	(NiO)	<0.1	<0.1
Potassium oxide	(K_2O)	0.23	0.12
Sodium oxide	(Na_2O)	0.23	0.19
Mechanical Properties			
Side crushing strength	(N/mm)	33.4	30.3
Attrition resistance	(wt %)	<1.0	6.8

cobalt (or nickel) nitrate. This precursor is dried and oxidized, which converts the molybdenum to MoO_3. The most active form of CoMo or NiMo catalysts is with the metals in the sulfided form. Therefore, the catalyst must be activated by convertion of the metallic oxides to the sulfide forms with a sulfur-containing stream at high temperatures in a pretreatment step. The presulfiding of hydrotreating catalysts is an important step in meeting the requirements for high desulfurization activity for reformer feedstock preparation. The common spiking agents used are carbon disulfide, butanethiol, dimethyl sulfide (DMS), and dimethyl disulfide (DMDS), with DMDS and DMS as preferred choices. It is important to keep the presulfiding temperature under 300°C or to heat very slowly in order to avoid reduction of metal oxides to the metallic state. The method of presulfiding the catalyst is considerably more important in the case of $NiMo/Al_2O_3$ than with $CoMo/\gamma-Al_2O_3$ catalysts.

CoMo and NiMo catalysts will continue to be important in HDS far into the future because of their high activity, selectivity, and stability in the presence of sulfur-containing feedstocks. With the advent of skewed bimetallic reforming catalysts, which require ultralow levels of sulfur in hydrotreated naphtha, highly active catalysts are needed. However, these high-activity HDS catalysts should also possess olefin hydrogenation activity to avoid undesirable olefin-H_2S reactions to form mercaptans.[20] A compound bed of Ni-Mo catalyst (front) and Co-Mo (back) may hydrogenate olefins and offer deeper desulfurization. Catalyst additives, such as phosphorus and fluoride, could also have an important role in deep desulfurization.

4.2 Catalyst Structure

The structure of the sulfided form of HDS catalysts has been the subject of many studies, but there is no complete consensus yet. With the CoMo catalyst, the sulfided form may be represented as MoS_2 and Co_9S_8, but the actual compositions are more complex.[4] It has been proposed that the coordinatively unsaturated sites or exposed Mo ions with sulfur vacancies at the edges and corners of MoS_2 structures are active in hydrogenation and hydrogenolysis reactions. Basal planes are inactive in adsorption of molecules and are probably unimportant in hydrotreatment reactions. For cobalt- or nickel-promoted catalysts several different models have been proposed to explain the role of promoter and its exact location in the catalysts. These include the monolayer model,[24] contact synergy model,[25] intercalation model,[26] Co-Mo-S phase model,[30] and catalytic Co site model.[31] Among these models the Co-Mo-S (or Ni-Mo-S) phase model, proposed by Topsoe and coworkers, has been widely accepted. In this model the promoter atoms are located at the edges of MoS_2-like structures in the plane of Mo cations. The relative amount of Co atoms present as Co-Mo-S phase was shown to correlate linearly with the HDS activity.[29]

Although the catalytically active sites involved in hydrotreatment reactions are generally viewed as sulfur anion vacancies, it is still a matter of debate whether both hydrogenation and hydrogenolysis occur at the same type of site or on different type of sites. Several kinetic studies with model compounds have indicated the existence of two distinct types of catalytic sites, one responsible for hydrogenation and the other responsible for hydrogenolysis of the heteroatom. Some authors have suggested that the hydrogenation site (I) on the catalyst surface can be transformed into the hydrogenolysis site (II) by the adsorption of hydrogen sulfide.[30-33] This means that one type of sulfur anion vacancy present on sulfided catalyst is enough to explain the kinetic data. The distribution of sites I and II would depend essentially on the sulfidation states of the catalyst and the partial pressure of hydrogen sulfide. Interconversion of hydrogenation and hydrogenolysis sites on hydrotreatment catalysts, as proposed by Topsoe, is the most widely accepted explanation of the reported kinetic data.

4.3 Catalyst Deactivation and Poisoning

Hydrotreatment catalysts lose their activity during operation for a variety of reasons. The most common reasons are poisoning by the impurities in the feed, deposition of "coke" on the catalyst surface as a result of side reactions, and loss of dispersion of active components by "sintering". Catalyst deactivation during the course of a process is often unavoidable and represents a technical as well as an economic factor in most refining processes.

Deactivation by coking is caused by high molecular weight carbon compounds, which deposit at the pore entrance thus limiting the diffusion of reactants and products. The rate of catalyst deactivation and ultimate life are dependent on both the feedstock and operating conditions. When treating high concentrations of cracked stocks, organometallic compounds and other heavy contaminants can be deposited on the catalyst. In general, catalyst deactivation due to coke deposition increases with increasing temperature and decreasing hydrogen partial pressure. However, when light straight-run naphtha is treated, catalyst deactivation due to coke deposition occurs very slowly.[5]

For naphtha hydrotreatment catalysts, poisoning is more likely and can cause more damage than coke deposition. Well-known poisons for the naphtha hydrotreatment catalysts are arsenic, sodium, and hydrogen sulfide.[33]

Arsenic

Arsenic is a very severe poison to hydrotreatment catalysts. It is naturally occurring in crude oil, with the concentration highly dependent on the crude source. Arsenic poisoning is primarily observed in distillate and gas oil hydrotreatment but is occasionally observed in lighter feedstocks. Even 0.1 wt %

arsenic on catalyst results in more than 50% reduction in activity as compared to fresh catalyst. Arsenic poisons the catalyst by blocking access to active sites. In general, it is not recommended to regenerate or to use regenerated catalyst containing more than 250 ppm arsenic. However, arsenic does not deposit uniformly in a catalyst bed. If an entire catalyst bed is discharged together, it is possible for the catalyst to have an average arsenic content in excess of 250 ppm while still retaining acceptable activity. In this case, catalyst from the top of the bed will have very low activity but the bulk of the catalyst will have reasonable activity.

To reduce the impact of arsenic on activity during a cycle, it is recommended to minimize the depth of penetration of arsenic into the catalyst bed. Deposition of arsenic on catalyst is controlled by three variables: severity (temperature, pressure), space velocity, and metal capacity. The combination of these variables determines the profile of arsenic penetration into the catalyst bed. Higher severity will increase deposition in the upper portion of the bed. Higher space velocity will drive the arsenic more deeply into the bed.

Sodium

Sodium is a severe poison to hydrotreatment catalyst. In addition, Na can form a crust at the top of the hydrotreatment bed, resulting in pressure drop buildup. It is naturally occurring in crude oil as well as dissolved in water in emulsion with the oil. Desalting is used to control Na content of the feed and mitigate the effects on catalyst and unit performance. However, poor desalting or sodium from another source can result in the contaminant being present in the feed to a hydrotreater, causing short cycles and poor performance.

While Na affects activity within the cycle in which it is deposited, it has a more severe effect during catalyst regeneration. At elevated temperatures of regeneration, Na sinters the catalyst surface, causing destruction of acid sites, reduction in surface area, and reduction of active sites. Sodium also becomes mobile at elevated temperatures, so that a high concentration of sodium on the outer edges of the spent material can move inward. For these reasons, regeneration is not recommended for spent catalysts containing more than 0.25 wt % Na.

Hydrogen Sulfide

Hydrogen sulfide in hydrotreater recycle gas is an activity depressant for HDS reactions. The presence of hydrogen sulfide inhibits the rate of reaction of hydrocarbon molecules with active sites on the catalyst surface. In addition, hydrogen sulfide reduces the hydrogen partial pressure in the reactor. This, in combination with higher operating temperature requirements, can lead to a substantial increase in the deactivation rate of the catalyst. Hydrogen sulfide concentration in the recycle gas builds as the gas moves from reactor inlet to outlet. The recycle gas loop often contains an amine scrubber to remove hydrogen sulfide from the gas stream. In cases where the hydrogen sulfide content is relatively low, a gas purge may be used in place of an amine scrubber to prevent hydrogen sulfide buildup in the system.

Hydrogen sulfide inhibits the activity of hydrotreating catalysts by competitive adsorption on catalytically active sites. This blocks the active sites available for hydrotreatment reactions, resulting in higher temperature requirement to obtain a constant product quality. As recycle gas hydrogen sulfide concentration is raised the number of blocked active sites increases. Hydrodesulfurization, hydrodenitrogenation, and aromatic saturation are all negatively affected by increasing hydrogen sulfide concentration in the recycle gas.

In addition to the inhibiting effect on hydroprocessing reactions, increasing H_2S content in the gas reduces hydrogen partial pressure in the reactor. Combined with a higher temperature requirement for constant product quality, the lower hydrogen partial pressure can cause increased catalyst deactivation rates and ultimately a reduction in cycle length. For this reason, it is recommended that H_2S content in the recycle gas stream be maintained below 2 vol %.

4.4 Catalyst Regeneration

Regeneration of deactivated hydrotreatment catalysts is an attractive process to recover the original activity and it is carried out by oxidation in the presence of air. Earlier, the regeneration process was performed in situ using a mixture of air and steam or oxygen and nitrogen. However, because of the hazards and air pollution involved, this practice has largely been replaced by ex situ regeneration. In addition, ex situ regeneration in theory offers better activity recovery because of better temperature control.[4]

The catalyst regeneration process is started by burning of coke in the presence of oxygen. This is called carbon burn-off regeneration and is used extensively in the industry. In the regeneration procedure, the initial evaporation of oil attached to spent catalysts and the subsequent oxidation of the hydrocarbons occur at lower temperatures (up to 250°C), and finally carbonaceous materials are burned at over 350°C. Metal sulfides are converted to oxides as well. Therefore, it is important to consider the exothermic oxidation and to avoid overheating of the catalysts. The ratio of activity recovery in regeneration depends on the severity of hydrotreatment. When there is no metal deposition and the catalyst is not used under severe conditions, the activity of spent catalysts can be recovered up to the fresh catalyst activity level by regeneration.

During regeneration, physicochemical properties such as surface area, crystallinity, reducibility, and metal distribution change significantly with the regeneration temperature. Increase in the dispersion of promotor species was observed in the catalysts regenerated at low temperatures, and this gave rise to the enhancement of activity in comparison with the fresh in some cases. On the other hand, promoters migrate into the sublayer of alumina support at higher temperature, resulting in the formation of inactive $CoAl_2O_4$ or $NiAl_2O_4$ phases.

Naphtha HDS is important as it protects the sensitive and expensive reforming catalysts. Refiners cannot accept much risk regarding its reuse as it is the only protector of these reforming catalysts. Therefore, due attention should be given to the crush strength, density, and length. If any of these properties has suffered 20–30% loss, it is better to discard the catalyst and start with fresh material. A check of catalyst activity should be carried out in case of doubt; not more than 20% loss relative to fresh catalyst should have occurred. In general, naphtha catalysts are regenerated only once or twice during a total life of 3–6 years.

Metal recovery is an alternative recycle method to regeneration when the activity of regenerated catalysts cannot be recovered to a level appropriate for reuse. Since the metal content in the used hydrotreatment catalysts is higher than that in ore, these catalysts are attractive as mineral resources. However, because of the complex composition and the variety of catalysts, it is not easy to recover the metals economically.

5 PROCESS SCHEME AND OPERATING CONDITIONS

5.1 Process Flow Scheme

Most of the commercial hydrotreating processes have essentially the same flow scheme as shown in Figure 4. Gas- or liquid-phase oil is mixed with recycle hydrogen plus makeup hydrogen and heated in a fired heater to the proper reactor inlet temperature (250–450°C), where, in the case of naphtha, it is totally vaporized. The combined feed flows downward through solid, metal oxide catalyst particles in a fixed-bed reactor under conditions that depend on the feedstock properties and desired product specifications. The hydrogen reacts with the oil to produce hydrogen sulfide, ammonia, and saturated hydrocarbons.

The reactor effluent is cooled and then sent to a separator where most of the hydrogen, methane, ethane, and some of the hydrogen sulfide and ammonia are flashed off as gas. This gas can be scrubbed and either recycled or used for other purposes. The liquid phase from the separator is sent to a stripper (a pressurized distillation column) in order to remove any residual hydrogen sulfide and ammonia that may be dissolved in the liquid. The stripper is also used to separate any C_3 and C_4 from the pretreated naphtha product. In the reformer feedstock

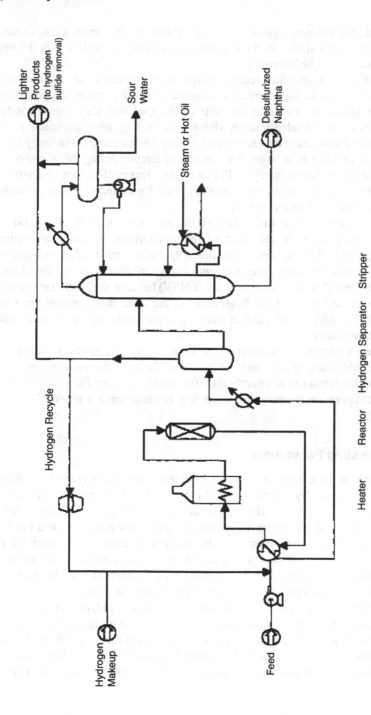

Figure 4 Typical process flow diagram of naphtha hydrotreatment.

application, the efficient operation of the stripper is important since residual dissolved hydrogen sulfide in the naphtha could easily cause less than 0.5 ppm sulfur limitations to be exceeded.

In a fixed-bed downflow reactor design, refiners prefer to use small catalyst particles to (1) obtain high surface-to-volume ratios that minimize diffusional constraints placed on the large reactants in the feed and (2) increase catalyst activities in a dense-loaded reactor. Unfortunately, they generate unacceptably high pressure drops across the reactor bed. Hence, to date the HDS industry limits catalyst size to 0.8 mm or larger. One means of circumventing this limitation is through the use of shaped catalysts. For example, cloverleaf-shaped catalysts can afford higher surface-to-volume ratios than those for cylindrical analogs so that higher effectiveness factors prevail.

Furthermore, typical trickle-bed HDS units suffer from the restriction of end-of-run temperature limits, fixed by hydrometallurgical design parameters. When improved HDS capacity is required, the refiner must either increase the hydrogen partial pressure (pressure and cost limitations) or be faced with reducing the liquid hourly space velocity (LHSV) because the HDS reaction (first and second order) is also time dependent. In the latter case, refiners are often faced with the addition of parallel units to compensate for the loss in space velocity in production.

When a highly unsaturated stock such as coker naphtha is being fed, a separate additional reactor may be installed ahead of the main reactor. The purpose of this reactor is to saturate diolefins under mild conditions to prevent a runaway temperature increase due to the highly exothermic reaction.[34]

5.2 Operating Parameters

By far the two most important factors in determining the operating conditions needed for a given hydrotreater are (1) the feedstock properties and (2) the desired product properties. These two factors set the general severity that is required in the hydrotreatment operation although other considerations (such as catalyst type, heater limitations, or the amount, pressure, and purity of the available hydrogen) will have a bearing on the actual conditions that are used. Typical ranges of conditions used for various types of naphtha hydrotreatment for the product to be charged to a reformer are presented in Table 12.

As the concentrations of sulfur, nitrogen, oxygen, and/or olefins increase, more severe hydrotreatment conditions will be needed to bring the concentrations down to the required levels. As mentioned earlier, the rate of HDS reactions of single sulfur compounds follows first-order kinetics; however, when present in combinations, the compounds taken as a group exhibit apparent reaction orders of up to 1.6.[21]

Table 12 Typical Ranges of Conditions Used for Various Types of Naphtha Hydrotreating[35]

General feedstock type	Straight run	Cracked or blends of HSR and cracked	Synthetic or blends of HSR and synthetic
Severity	Mild	Intermediate	Severe
Feedstock Composition:			
Feedstock sulfur (ppm)	100–700	100–3000	500–8000
Feedstock nitrogen (ppm)	<2	2–30	500–6000
Feedstock oxygen (ppm)	<10	10–100	100–5000
Feedstock olefins (vol %)	<1.0	5–40	<1–45
Operating conditions:			
Space velocity (h^{-1})	7.0–15.0	3.0–6.0	0.8–2.0
Total pressure (psig)	200–400	400–700	800–1500
Hydrogen partial pressure (psig)	60–150	150–400	500–1000
Hydrogen purity in reactor gas (mol %)	70–90	70–90	70–90
Gas rate (m^3/m^3)	36–72	72–180	180–1080
Average temperature (°C)	230–310	280–330	300–350
Cycle life, months	6–24	6–18	3–12

The conditions typically used to hydrotreat straight-run stocks are mild, whereas treating cracked feeds (or blends of cracked and straight-run feeds) requires more severe conditions. Hydrotreatment of synthetic stocks represents the most severe type of operation. Table 12 presents typical ranges of conditions for three hydrotreating cases.[35] Case I is mild hydrotreatment of a straight-run naphtha, case II represents relatively severe treatment of a blend of straight-run naphtha and cracked (coker) naphtha, and case III is the severe pretreatment of shale oil or coal naphtha or a blend of these feeds with petroleum naphthas.

In hydrotreatment, the principal operating variables are temperature, hydrogen partial pressure, and space velocity. In general, an increase in temperature and hydrogen partial pressure increases the reaction rates of sulfur and nitrogen removal, whereas an increase in space velocity has the reverse effect. Specific effects of the three main variables are given below:

Temperature

Higher temperatures increase the rates of HDS and the other desired reactions, which remove contaminants. In the design of hydrotreaters, the initial (start-of-cycle) temperature is set by the design throughput, the operating pressure, and other economic factors. Enough temperature flexibility is built into the unit so that temperatures may be increased throughout the cycle to offset the loss in catalyst activity, which occurs as the catalyst ages. Table 12 shows that hydrotreater temperatures are normally kept between 250°C and 370°C. Above 370°C, hydrocracking reactions and coke deposition become too prominent for economical operation.

Another effect of higher temperature is that it leads to promotion of undesirable hydrogen sulfide and olefin recombination reaction.[11,20] The effect of temperature on product total sulfur and mercaptan sulfur is shown in Figure 5. The total sulfur decreased from 72 ppm at 220°C to a minimum of 0.69 ppm at 300°C. However, upon further increase in the temperature up to 350°C, the total sulfur increased. The mercaptans (which were about 56% of the total sulfur in feedstock) were reduced to about 6% at 220°C and then increased to about 36% at 300°C and to about 57% at 350°C. This indicates the occurrence of a mercaptan-forming reaction at temperatures above 300°C along with removal of other types of sulfur.

GC-FPD results, shown in Figure 2, also indicate that the product obtained at 250°C contained 12 sulfur compounds: 3 mercaptans, 2 sulfides, and 7 thiophenes. At temperatures above 280°C, thiophenes were removed almost completely, but mercaptans were still present suggesting their formation as a result of recombination reactions. Hence, the HDS units should be operated at the lowest temperature possible to minimize mercaptan formation.

Space Velocity

When holding all other operating conditions constant, increasing space velocity will cause product sulfur (and/or other contaminant concentrations) to increase. This effect can in some cases be offset by increasing the reactor temperature and/or the hydrogen partial pressure. As shown in Table 12, in mild operations space velocities range from 7 to 10 or even as high as 15. In more severe operations, space velocities are usually between about 3 and 6, and in the most severe cases space velocities typically vary between 0.8 and 2.

It is reported that, at temperatures above 280°C, higher space velocity inhibits mercaptan-forming reactions.[11,20] The effect of hydrogen gas rate on HDS was insignificant between 67 and 80 Nm^3/m^3. The effect of hydrogen gas rate is more pronounced at lower space velocity where a higher hydrogen gas rate increases desulfurization and suppresses hydrogen sulfide–olefin recombination reactions.

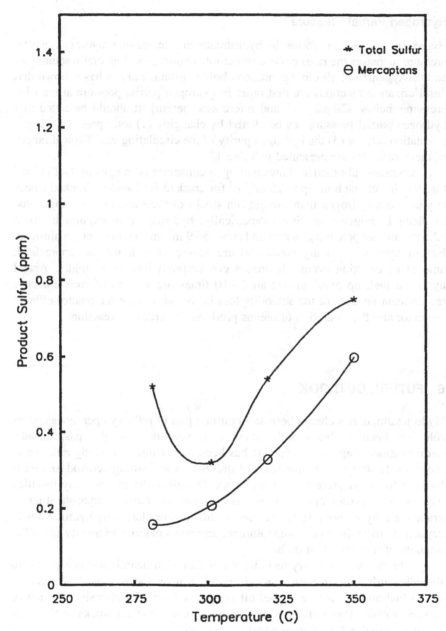

Figure 5 Effect of temperature on product total sulfur and mercaptans.[20]

Hydrogen Partial Pressure

Hydrogen is the vital factor in hydrotreatment. Increasing hydrogen partial pressure increases the rates of desulfurization, denitrogenation, olefin saturation, and deoxygenation; all other conditions hold constant. Studies have shown that desulfurization reactions are first order in hydrogen partial pressure at least for pressures below 420 psig [21 and references therein]. It should be noted that hydrogen partial pressure can be altered by changing (1) total pressure, (2) gas circulation rate, or (3) the hydrogen purity of the circulating gas. Typical ranges of these variables are presented in Table 12.

Commercial practice shows hydrogen consumption ranges of $1-2$ m^3/m^3 for straight-run naphtha up to 25 m^3/m^3 for cracked feedstocks. Cracked stocks require more hydrogen than straight-run stocks of the same boiling ranges due to olefin hydrogenation. Stoichiometrically, hydrogen consumption is about 12.5 m^3/m^3 per percentage sulfur and about 56.9 m^3/m^3 per percentage nitrogen. Furthermore, if operating conditions are severe enough that an appreciable amount of cracking occurs, hydrogen consumption increases rapidly. Actual hydrogen makeup requirements are $2-10$ times the amount of stoichiometric requirement because of the solubility loss in the oil leaving the reactor effluent separator and the saturation of olefins produced by cracking reactions.

6 FUTURE OUTLOOK

Hydrotreatment has always been an important part of refinery operations and its role has been evolving with added responsibilities. In the past decade, hydrotreatment capacity worldwide has been increasing at a steady rate due to increasingly stringent environmental requirements and strong demand for clean-burning fuels. At present there are about 25 million barrels per day installed capacity of hydrotreatment in the world, and the ratio of hydrotreating to crude capacity is about 33%. Refiners worldwide are increasing hydrotreatment capacity to meet the present and future requirements in terms of quality as well as quantity of transportation fuels.

The demand for heavy fuel oil has declined significantly and is expected to dwindle further in the near future. In order to meet the increased demand for diesel fuel and to dispose of fuel oil surplus, refiners are choosing alternative process routes. These changes will produce more cracked stocks, which will become available for reforming and isomerization.

It is expected that significant refining industry investment over the next 5 years will be committed to resolving environmental issues, including production of lower sulfur fuels, replacement of MTBE in gasoline, and the potential reduction of benzene in gasoline. One of the approaches is posttreatment sulfur removal from FCC naphtha and other gasoline blendstocks.[36] Posttreatment HDS processes remove sulfur from gasoline blendstocks, enabling refiners to meet stringent specifications for gasoline sulfur and olefin content. This route may result in greater needs for reforming capacity to produce higher octane gasoline. With reduced demand for MTBE and increased demand for new sources of clean-burning gasoline blendstocks, the need for isomerate will increase. Benzene saturation and isomerization technologies upgrade C_5-C_6 streams to 80–90 RON, depending on the flow scheme and catalyst choice.

Naphtha hydrotreatment will be important in other regards. For example, the future refinery will change, with integration of conventional fuels processes and separation of very specific hydrocarbon compounds. The future refineries will become integrated plants producing high-demand petrochemicals. These will require significant purity improvement using hydrotreatment, as well as catalytic reforming to make the desired molecules. Hydrogen management is another area that will become increasingly important in the future. Hydrogen availability will certainly be an issue for refiners to meet lower sulfur levels in gasoline and diesel fuel. Catalytic reforming will continue to be a key source of refinery hydrogen. Hydrotreatment with minimal hydrogen utilization will be a desirable goal. Finally, fuel cells are becoming very efficient sources of energy, and the ultimate goal is to run fuel cells on light hydrocarbons, such as naphtha with ultralow sulfur content. Research is significant in this area and will clearly have an impact on deep HDS of naphtha.

REFERENCES

1. Biswas, J.; Bickle, G.M.; Gray, P.G.; Do, D.D.; Barbier, J. Catal. Rev. Sci. Engg. **1988**, *30*, 161.
2. Menon, P.G.; Prasad, J. In *Proc. 6th Cong. Catal.*; Bond, G.C., Wells, P.B., Tompkins, F.C., Eds.; Chemical Society: London, 1976; Vol. 2, 1061 pp.
3. Michel, C.G.; Bambrick, W.E.; Ebel, R.H. Fuel Proc. Tech. **1993**, *35*, 159.
4. Lovink, H.J. In *Catalytic Naphtha Reforming*; Antos, G.J., Aitani, A.M., Parera, J.M., Eds.; 1995; 257 pp.
5. Satterfield, C.N. *Heterogeneous Catalysis in Industrial Practice*; McGraw-Hill: New York, 1991.
6. LePage, J.E.; Chatila, S.G.; Davidson, M. *Resid and Heavy Oil Processing*; Editions Technip: Paris, 1990.
7. Miller, R.B.; Macris, A.; Gentry, A.R. Petroleum Technol. Q. **2001**, 69.
8. Halbert, T.R.; Brignac, G.B.; Greeley, J.P.; Demmin, R.A.; Roundtree, M. Hydrocarbon Eng. **2000**, 1.
9. Gentry, J.C.; Lee, F. 2000 NPRA Annual Meeting, Paper AM-00-35, 2000.

10. Furfaro, A. 1985 NPRA Annual Meeting, Paper AM-85-56, 1985.
11. Anabtawi, J.A.; Alam, K.; Ali, M.A.; Ali, S.A.; Siddiqui, M.A.B. Fuel **1995**, *74* (9), 1254.
12. Bradley, C.; Schiller, D.J. Anal. Chem. **1986**, *58*, 3017.
13. Buteyn, J.L.; Kosman, J. J. Chromatogr. Sci. **1990**, *28*, 47.
14. Ali, S.A. *Thermodynamics of Hydrotreating Reactions*, in Proceedings of the Fourth International Conference on Chemistry in Industry, Manama, Bahrain, October 30–November 01, 2000.
15. Ried, R.C.; Prausnitz, J.M.; Poling, B.M. *The Properties of Gases and Liquids*; McGraw-Hill: New York, 1987.
16. Joback, K.G. Sc.D. Thesis, Massachusetts Institute of Technology, June 1984.
17. Girgis, M.J.; Gates, B.C. Ind. Engg. Chem. Res. **1991**, *30* (9), 2021.
18. Sonnemans, J.; Goudrian, F.; Mars, P. *Fifth Int. Congress Catalysis*, Palm Beach, Florida, USA, Paper 76, 1972.
19. Satterfield, C.N.; Cocchetto, J.F. AIChE J. **1975**, *21*, 1107.
20. Ali, S.A.; Anabtawi, J.A. Olefins can limit desulfurization of reformer feedstock. Oil Gas J. **1995**, *48*.
21. Kabe, T.; Ishihara, A.; Qian, W. *Hydrodesulfurization and Hydrodenitrogenation*; Kodansha Ltd.: Tokyo, Japan and Wiley-VCH: Weinham, 1999; 31 pp.
22. Nag, N.K.; Sapre, A.V.; Broderick, D.H.; Gates, B.C. J. Catal. **1979**, *57*, 509.
23. Satterfield, C.N.; Roberts, G.W. AIChE J. **1968**, *14*, 159.
24. Schuit, G.C.A.; Gates, B.C. AIChE J. **1973**, *19*, 417.
25. Delmon, B. In *Proceedings of the 3rd Int. Conference on Chemistry and Uses of Molybdenum*; Barry, H.F., Mitchell, P.C.H. Eds.; Climax Molybdenum Co., 1979; 73 pp.
26. Farragher, A.L.; Cossee, P. *Proceedings of 5th Int. Congress Catalysis*, North-Holland, Amsterdam, 1973; 501 pp.
27. Topsoe, H.; Clausen, B.S.; Candia, R.; Wivel, G.; Morup, S. J. Catal. **1981**, *68*, 433.
28. Duchet, J.C.; Van Oers, E.M.; de Beer, V.H.J.; Prins, R. J. Catal. **1983**, *80*, 386.
29. Topsoe, H.; Clausen, B.S.; Topsoe, N.; Zeuthen, P. Progress in the design of hydrotreating catalysts based on fundamental molecular insight. In *Catalysts in Petroleum Refining 1989*, Trimm, D.L., Akashah, S., Absi Halabi, M., Bishara, A., Eds.; Stud. Surf. Sci. Catal.; Elsevier: Amsterdam, 1990; Vol. 53, 77 pp.
30. Yang, S.H.; Satterfield, C.N. J. Catal. **1983**, *81*, 168.
31. Vivier, L.; Katsztelan, S.; Perot, G. Bull. Soc. Chim. Belg. **1991**, *100*, 801.
32. Perot, G. Catal. Today **1991**, *10*, 447.
33. Akzo-Nobel Hydroprocessing Technical Information Bulletin, 2000.
34. Yui, S. Oil Gas J. 1999.
35. Mcketta, J. In *Encyclopedia of Chemical Processing and Design*; Marcel Dekker: New York, 1982.
36. Penning, R.T. Hydrocarbon Proc. **2001**, *80* (2).
37. Shorey, S.W.; Lomas, D.A.; Keesom, W.H. Hydrocarbon Proc. **1999**, *78*, 11.

5
Preparation of Reforming Catalysts

J. R. Regalbuto
University of Illinois at Chicago, Chicago, Illinois, U.S.A.

George J. Antos
UOP, LLC, Des Plaines, Illinois, U.S.A.

1   INTRODUCTION

Discussion of the preparation of catalysts for naphtha reforming remains predominantly a discussion of the preparation of platinum-containing catalysts. Platinum remains the chief metal component for all commercial reforming catalysts. The catalysts utilized when the first commercial reformers were employed in the late 1940s were monometallic-supported platinum catalysts. Since then, there has been considerable evolution in the reforming catalyst, centering largely on the chemical formulation although support modifications have some importance. Platinum has remained the key component. The bimetallic catalysts introduced in the late 1960s employed a second element, such as Re, Sn, Ge, and Ir, which interacted with platinum to result in catalysts that offered better gasoline selectivity or better overall performance stability. This offered the possibility of operating at lower pressures. Process innovations were developed, and the new direction for catalytic reforming was initiated. This has been the major direction since then, as discussed in Chapter 13 of this book.

Most conventional bimetallic or multimetallic reforming catalysts remain predicated on achieving a well-dispersed platinum function on an alumina support that is promoted with a halogen. This halogen is usually chlorine, and its function is to provide acidity. This bifunctional catalysis is described in Chapters 2 and 3 of this book. For the past 30 years, researchers have looked for methods to improve on this basic recipe for monometallic and bimetallic catalysts. Multimetallic catalysts abound in the patent and scientific literature. Some of these have reportedly been commercialized. All will employ platinum on alumina as a foundation. There was much consideration to employ platinum deposited or exchanged in zeolites, with further ion exchange using alkaline ions such as potassium and barium.[1-4] Catalysts utilizing L-zeolites have reportedly been commercialized by Chevron and UOP. Despite the high selectivity for dehydrocyclization of n-hexane and n-heptane, the number of units in service are not a significant fraction of the total. This may be related to the sensitivity to poisoning by sulfur compounds[5] or a reduced need for aromatics such as benzene. Additional detail on the nature of these catalysts is found in Chapter 6 on characterization. As a comparison, attempts to utilize a zinc/H-MFI catalyst to reform naphtha, and in particular to convert C_6 and C_7 paraffins, are described in Chapter 9. This work is progressing through development scale-up.

This chapter deals with the preparation of platinum-based reforming catalysts. It differs from the chapter in the previous edition of this book[6] in that there is an in-depth exploration of the chemistry involved in impregnating a support with chloroplatinic acid. Many advances in the understanding of the speciation and interaction with the alumina surface have been made since the first edition. This information is useful in the preparation of any platinum-containing catalyst for any process application. In common with the first edition, the first part of this chapter will deal with the alumina support system and the generation of acidity on the alumina surface.

2   REFORMING CATALYST SUPPORTS

Prior to the successful use of platinum in 1949, reforming catalysts consisted of chromium or molybdenum oxides deposited on alumina.[7] Early patents describe the deposition of chloroplatinic acid on an acidic support of fluorided alumina[8] or chlorided alumina.[9] All modern industrial reforming catalysts consist of platinum, frequently combined with one or more metals, deposited on a chlorided alumina. Given its vital importance in the process, alumina is the only support discussed in depth in this chapter. The first part of this section describes the two main forms of alumina used, their synthesis, structural and surface characteristics, and their shaping. The second part describes the steps involved in catalyst preparation, including impregnation, drying, and oxidation of the impregnated alumina support.

2.1   γ- and η-Al$_2$O$_3$

Aluminas constitute a large group including aluminum hydrates, transition aluminas, and α-Al$_2$O$_3$. Aluminum hydrates include amorphous hydroxide, crystallized trihydrates, gibbsite, bayerite, and norstrandite and the monohydrates boehmite and diaspore. Several methods are available for preparing these hydrates:[10] acidification of sodium aluminate, neutralization of an aluminum salt, and hydrolysis of an aluminum alcoholate.

The hydrate conversions to the oxide state are complex and provide several low-temperature transition aluminas: ρ, χ, η, and γ. These can in turn be converted to one or more of the following high-temperature transition aluminas: δ, κ, and θ. The final stage in these transformations is α-Al_2O_3, which is thermodynamically the most stable. Figure 1, based on references 10–23, gives an idea of the complexity of these relationships between the different aluminas.

The two main alumina supports with any use in catalytic reforming are the two transition aluminas η- and, above all, γ-Al_2O_3. γ-Al_2O_3 is usually obtained by the calcination in air of boehmite. Boehmite, which is generally described as an aluminum monohydrate, actually represents a broad continuum of products $Al_2O_3 \cdot nH_2O$ with varying degrees of hydration and organization without any clear borderlines: gelatinous boehmite is the most hydrated with n possibly even exceeding 2, pseudoboehmite or microcrystalline boehmite, crystalline boehmite, and, finally, highly crystallized boehmite in large rhombohedral crystals with n close to 1. Furthermore, depending on the conditions used, boehmite crystals with different shapes can be obtained, e.g., acicular or platelets.[10,21] The transformation of an alumina hydrate is a topotactic reaction in which the morphology and size of the particles are preserved. The properties of γ-Al_2O_3 and the shape of the crystallite formed and its specific surface area depend on the initial boehmite. Pseudoboehmite or microcrystalline boehmite leads to γ-Al_2O_3 with a large area (>350 $m^2 g^{-1}$ at 500°C) that is very poorly organized[28,30]. Calcination of the crystalline boehmite at about 500°C promotes the formation of γ-Al_2O_3 with a surface area close to 250 $m^2 g^{-1}$. Highly crystallized boehmite converts to an alumina similar to γ-Al_2O_3 with a low area (<100 $m^2 g^{-1}$) at 500°C.

These low surface areas cannot generate enough acidity for the reforming reaction catalysis and are therefore not used in commercial reforming catalysts. The precursor selection is very important in the preparation of commercially viable reforming catalysts. Contamination of the boehmites with iron or sodium occurs depending on the raw materials used to make the boehmitic phase. Catalyst manufacturers are very selective regarding the sources of alumina raw materials used in the manufacture of reforming catalyst supports.

η-Al_2O_3 is obtained by calcination above 250–300°C in air or in vacuum of the alumina trihydrate bayerite, or in vacuum of gibbsite. The usual precursor, bayerite, is typically obtained in the form of large crystals ≥ 0.1 μm.[10] However, the liberation of the water caused by calcination apparently generates fine micropores in the large particles obtained. The specific surface area of the η-Al_2O_3 varies widely with the final calcination temperature: from 500 to about 400 $m^2 g^{-1}$ between 250°C and 450°C and from 400 to about 250 $m^2 g^{-1}$ between 450 and 550°C.[11]

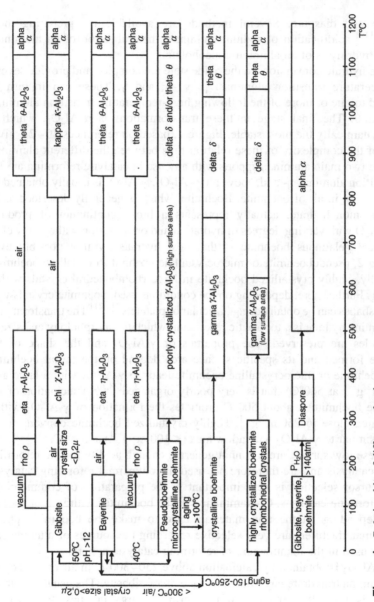

Figure 1 Relationship of different aluminas.

The structures of γ-Al_2O_3 and η-Al_2O_3 are similar.[11,24] They are based on the compact cubic arrangement of the oxygen of the spinel $MgAl_2O_4$ structure,[11,23] but with a slight tetragonal deformation. This deformation, which is a result of a disorder in the stacking of the oxygens, is more pronounced in η-Al_2O_3 than in γ-Al_2O_3. The Al^{3+} cations are distributed over the 32 octahedral sites and the 64 tetrahedral sites. Tetrahedral sites are slightly more occupied in η-Al_2O_3 than in γ-Al_2O_3.[11,24] The nature of the exposed crystal faces at the surface of the two aluminas also appears to be different. Based on spectroscopic data, Knözinger and Ratnasamy[25] concluded that the surface consists of the three faces, (111), (110) and (100) in varying proportions in the two aluminas, with a predominance of the (111) face in η-Al_2O_3 and the (110) face in γ-Al_2O_3. This assumption was confirmed in a further study[21] in which it was observed that the ratio of the proportions of the (110) and (111) faces exposed varies according to the morphology and the size of the crystals of γ-Al_2O_3.

2.2 Surface Acidity

The two main models of the surface acidity–basicity that have been proposed for γ- and η-Al_2O_3 are those of Peri[26] and of Knözinger and Ratnasamy.[25] The latter model considers the environment of the Al^{3+} cations connected to the hydroxyls for the, (111), (110) and (100) faces of the two aluminas. The number of different OH sites is five on the (111) face, three on the (110) face, and one on the (100) face. The authors attributed the catalytic properties of the aluminas to the combination of one of these OH sites with neighboring surface defects that have an energetically less favorable configuration and hence a very low probability of existence. These catalytic sites are formed only above 60% dehydroxylation. The degree of dehydroxylation of the aluminas, which governs the number of defects, appears as an essential parameter for acidity.[25–28] Further information on the characterization of the acidity is to be found in Chapter 6 of this book.

2.3 Influence of Halogen on Alumina Acidity

Although the increase in acidity caused by the fixation of a halogen on the surface of an alumina has been known for many years, the nature of the halogenated surface complexes has still not been fully clarified. Contradictory results have been obtained with fluorided aluminas,[29–37] and the acidity seems to be promoted in different ways according to the halogen used.[29–31] Chloriding of alumina has been studied using CCl_4 or HCl[38–45]. The type of acidity, Brönsted or Lewis, seems to depend closely on the fixed chloride content and the chloride precursor used. The maximal level of chloride fixed on γ-Al_2O_3 using gaseous HCl appears limited to around 2 wt% (i.e., around 2.5 atoms of Cl per nm^2).[40] At such low chloride contents, Gates et al.[23] proposed the following mechanism for the promotion of Brönsted acidity, which will play an important role in some of the numerous reforming reactions:

$$
\begin{array}{ccccccc}
\text{H} & \text{H} & \text{H}^{\delta-} & & \text{H}^{\delta-} & \text{H} & \text{H} \\
| & | & | & & | & | & | \\
\text{O} & \text{O} & \text{O} & \!\!\!\!\text{Cl}\!\!\!\! & \text{O} & \text{O} & \text{O} \\
& \text{Al} & & \text{Al} & & \text{Al} &
\end{array}
$$

The acidity of an OH group is strengthened by the inductive effect exerted by a Cl^- ion adjacent to the OH group. This model agrees with the observations of Tanaka and Ogasawara.[39]

At high chloride contents, 2 wt % < Cl^- < 10 wt %, strong acid sites of the Lewis type are formed.[40–45] They correspond to the formation of a gem-dichlorinated aluminum complex at the surface.[40,43–45] These sites are generated with compounds having at least two chlorine atoms on the same carbon atom: CH_2Cl_2, $CHCl_3$, CCl_4, etc.[45] The acidity corresponding to these high chloride contents is unsuitable for reforming catalysts, for which the usual chloride content is about 1% by weight.

2.4. Forming of Alumina Macroparticles

The use of γ- or η-Al_2O_3 as a reforming catalyst support requires a shaping operation adapted to the type of process: moving bed or fixed bed. For a moving bed, it is necessary to prepare 1- to 4-mm-diameter beads to facilitate circulation and to limit the mechanical abrasion of the catalyst. For the fixed bed, the support can be in the form of either beads or cylindrical extrudates 1–4 mm in diameter. The catalyst support is shaped essentially by three methods: granulation, drop coagulation, and extrusion.[10,46]

In the granulation process, a powder is agglomerated in the form of spherical beads by progressive humidification in a large bowl, called a pan granulator, with granulation seeds and water to which a peptizing agent may be added. As the bead grows, gravity and centrifugal force tend to push its trajectory to the side of the granulator, where it is ultimately ejected upon reaching a certain size.

Drop coagulation, as typified by the "oil drop technique", is a particularly important technique for the formation of alumina beads used as reforming catalyst supports. The first patents issued on this technique date from the 1950s.[47–49] In general, the oil drop technique consists of dropping an alumina hydrosol in a water-immiscible liquid in a vertical column. The alumina hydrosol can be prepared by the hydrolysis of an acid salt of aluminum, such as aluminum chloride,[41] nitrate,[47] or sulfate,[50] or by the digestion of aluminum metal under heat by an aqueous aluminum salt solution.[47–51] The water-immiscible liquid is preferably an oil,[48,50,52] such as a light gas oil[51] or a paraffinic cut[53] with a high interfacial tension with respect to water. The higher the interfacial tension, the greater the sphere forming tendency of the hydrosol in the liquid. The liquid is then heated to 50–105°C for a given time to permit the hydrosol spheres to transform progressively into a hydrogel. To promote gellation of the hydrosol droplets, ammonia precursors such as hexamethylenetetraamine,[47,48,51,54–56] urea,[51,54,55] or a solution of ammonium acetate and ammonium hydroxide[48] may be added to the hydrosol, or the water-immiscible organic liquid can be saturated with ammonia.[44]

In practice, the hydrogel spheres are held, or "aged", in the oil medium for an extended period to harden the spheres so as to avoid deformation of the spheres during subsequent handling. In many cases, a subsequent complete coagulation is achieved by immersion in an aqueous alkaline media, such as an ammonium hydroxide solution, for a further extended period.[44,51] This latter procedure imparts a high mechanical resistance to the spheres. Various modifications of the general oil drop technique have been mentioned in other patents.[52,54,57–60]

The oil drop technique previously described applies easily to the synthesis of γ-Al_2O_3 beads. It can also be used to produce η-Al_2O_3 beads provided that bayerite alumina spheres are formed through a final aging step.[55]

Extrudates are formed using two main steps. In the first mixing step, a peptizing agent is added to a mixture of water and alumina powder (such as pseudoboehmite) and the mixture is thoroughly stirred to form a plastic paste that may typically contain 40% by weight of alumina. In the second extrusion step, the paste is forced through dies of a shape and diameter selected in accordance with the desired end product. Whether beads or extrudates, the supports thus obtained are dried and then calcined in air, generally between 400°C and 600°C. This completes the conversion to the gamma phase.

The shaping operation normally has a negligible effect on the microporosity (pore diameter $d_p < 5$ nm) and hence on the specific surface area of the aluminas because it does not alter the size of the precursor crystallites. It may nevertheless have a significant influence on the macroporosity ($d_p > 50$ nm) and hence on the pore volume of the support beads.[10] This macroporosity stems essentially from the free voids between the more or less large agglomerates of crystallites and is a function of the shaping method. Moreover, it can be adjusted, before shaping, by the addition of a pore-forming substance that is combustible or decomposable, or through the addition of a nondispersive inorganic filler.[10,46,61] The relative importance of micro- and macropore volume and pore radii are explored through modeling and experimentation in Chapter 7 of this book.

3 PLATINUM-CONTAINING CATALYSTS

3.1 Key Features

When preparing the bifunctional monometallic Pt/Al_2O_3-Cl catalyst, the following main features must be achieved to guarantee optimal performance and cost:

Low Pt content (usually <0.5 wt%) with uniform macroscopic distribution across the catalyst particle and maximum accessibility, i.e., maximum atomic dispersion. This high dispersion also reduces metal sintering during high-temperature treatments.

Acid sites in close proximity of the atoms or small particles of Pt.

Adequate mechanical properties ideally undiminished from those of the support.

To deposit platinum on an alumina support, two types of impregnation chemistries can be used: impregnation with or without interaction. In the first case, the platinum precursor forms an electrostatic or chemical bond with the surface of the support. In the second, the precursor displays no affinity for the surface and remains localized in the solution and deposits when the solution evaporates. It has been demonstrated that the impregnation techniques with interaction are substantially superior to the other in terms of metal dispersion and catalyst performance.[62,63] The following discussion of the preparation of a monometallic catalyst is therefore limited to this category.

3.2 Industrial Impregnation

Modern techniques for the preparation of reforming catalysts use solutions of chloroplatinic acid as a platinum precursor with the addition of an acid such as hydrochloric acid. The mechanism and impact of a second acid is described in the next section. These solutions may be placed in contact with the alumina support in several different procedures.

1. The dried alumina is impregnated with a volume of solution corresponding exactly to the quantity required to fill the pore volume V_p ("capillary" or "dry" or "without excess solution" impregnation).
2. The dried alumina is immersed in a volume of solution substantially larger than V_p (impregnation with excess solution).
3. The alumina, which is previously saturated with water or with a solution of hydrochloric acid, is immersed in the aqueous solution containing H_2PtCl_6 (difffusional impregnation).

Strong interaction develops in an acidic medium between the support and H_2PtCl_6, slowing down the diffusion of the latter toward the center of the beads. The rate of diffusion determines the impregnation time. In technique 1, capillary aspiration of the solution by the support enables the rapid penetration (a few dozen seconds to a few minutes) of the solution into the pores.[64-66] The solution that reaches the center of the bead will be depleted of the Pt precursor, owing to the strong interaction of the latter with the support. A homogeneous Pt profile from the outside to the core of support grains is attained in this technique by allowing sufficient time for platinum diffusion in the solution in the pore volume. In technique 3, the H_2PtCl_6 acid must diffuse in the aqueous phase from the external solution to the centers of the beads, and impregnation is purely diffusional. The diffusion is slow and may require many hours to ensure good distribution of the Pt in the bead.[67–69] Technique 2 is intermediate between the other two techniques.

In technique 1, which may also be carried out by spraying the support with the solution, the considerable liberation of heat that occurs may raise the temperature by a few dozen degrees.[64] If uncontrolled, this could lead to insufficient penetration of the metallic precursor. In technique 2, the excess solution favors rapid elimination of liberated heat. If the very dilute solution contains only the quantity of metal to be introduced onto the support in a single operation, a long immersion of a few hours is required to exhaust the metallic precursor from the solution. If, on the other hand, the solution is concentrated to enable the introduction of the desired quantity of platinum by simply filling the pores of the support, the latter must be dipped and then withdrawn very quickly to prevent rapid depletion of the precursor from the solution. In this case, removal of the heat is obviously less effective than in the other methods.

During capillary impregnation carried out in techniques 1 and 2, air bubbles are trapped in the pores, and especially in the micropores, where they are highly compressed.[64] The Young–Laplace law $P = 2\gamma/r$ expresses the overpressure applied to the air bubbles trapped by the solution with a surface tension γ in pores with radius r. If r is very small, very high pressures may be reached (for example, $P = 14$ MPa approximately for $r = 10$ nm), which may be detrimental to the mechanical properties of the beads [64]. Under the effect of such pressures, the air is progressively dissolved and migrates toward the macropores where less compressed and hence larger bubbles tend to appear. Removal of most of the imprisoned air to the exterior of the beads is generally complete after a few dozen minutes.[64] Additional practical information can also be obtained in Ref. 56.

3.3 Impregnation

Fundamental Phenomena of Impregnation

The fundamental phenomena occurring during noble metal impregnation of alumina can be categorized into several areas: chemistry of surface hydroxyl groups, role of alumina dissolution, and nature of the metal complex–oxide surface interaction. In this section, the contrasting adsorption mechanisms that have arisen from these areas will be introduced and examined in light of the most recent literature data.

Surface Chemistry of Hydroxyl Groups. The first consideration is the chemistry of the support surface, which is populated by hydroxyl groups that serve to balance charge at the discontinuity of the solid metal–oxygen framework. It has been known for a great while in the colloid science literature that oxide surfaces can become either positively or negatively charged at different pH values, as these hydroxyl groups become either protonated or deprotonated in response to the solution pH. Different models of hydroxyl chemistry have arisen.[70,71] An elegant method has been developed to identify "proton affinity distributions"—multiple types and chemistries of hydroxyl groups on a single oxide[72]—but most models for adsorption employ only a single site. This site is assumed to act as an amphoteric acid, becoming protonated and positively charged at low pH, and deprotonated and negatively charged at high pH according to the following equations:

$$> SOH_2^+ \leftrightarrow SOH + H^+, \quad K_1 = 10^{-6} (pK_1 = 6)$$

$$> SOH \leftrightarrow SO^- + H^+, \quad K_2 = 10^{-11} (pK_1 = 11)$$

The values of the "surface acidity" constants K_1 and K_2 given above are typical of alumina.[73,74] The pH at which the surface is neutral, termed the point of zero charge (pH$_{PZC}$, or simply PZC), occurs midway between pK_1 and pK_2.[70] Most aluminas have a PZC of about 8.5 $\pm$ 0.5,[75] independent of phase. The difference between pK_2 and pK_1, or ΔpK, can be found experimentally[73,75] and has been predicted theoretically.[76] The surface density of OH groups can also be measured independently and is thought to be about 5 OH/nm^2 for alumina.[77] The surface charging of alumina can then be described by a PZC of 8.5, a ΔpK of 5, and $N_S = 5$ OH/nm^2 which results in the distribution of protonated, neutral, and deprotonated surface species as a function of pH shown in Figure 2. Recent work[73,75,76] has concurred that the value of ΔpK for common oxides such as silica

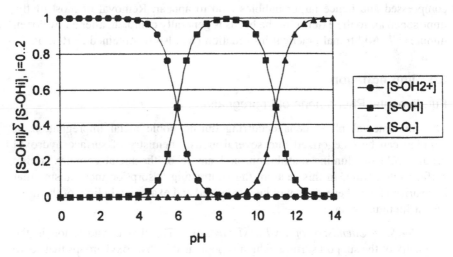

Figure 2 Distribution of species on an alumina surface as a function of pH, using typical PZC and ΔpK parameters (8.5 and 5, respectively).[75]

and alumina is large, such that virtually the entire surface is covered with neutral OH groups at the PZC, and on one side of the PZC or the other, the surface polarizes in only one way.

An issue related to surface charging is the solution-buffering effect of the surface. As protons are taken up or released by the hydroxyl groups at the oxide surface, the bulk pH can be influenced to a tremendous extent. Although this very significant effect has been in the literature for more than a decade, it is still relatively unappreciated by the catalysis community at large. The influence of an oxide surface on bulk pH was first demonstrated in an elegantly simple technique to determine oxide PZC termed "mass titration".[78] In this experiment, small masses of oxide are added progressively to a liquid solution, and with each addition, the solution pH steps up or down toward the oxide pH. A schematic of mass titration taken from[78] is shown in Figure 3.

A microscopic interpretation of mass titration is given in Figure 4 for an oxide placed in a solution at a pH below the PZC of the oxide. With the initial addition of oxide, there exists a driving force for adsorption of protons, and as

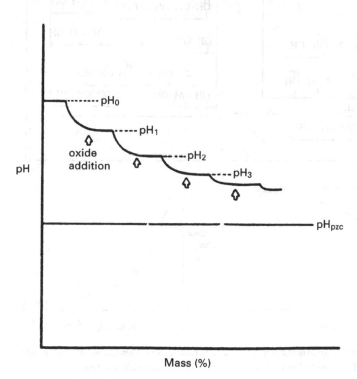

Figure 3 "Mass titration"—the response of liquid pH to the stepwise addition of an oxide.[78]

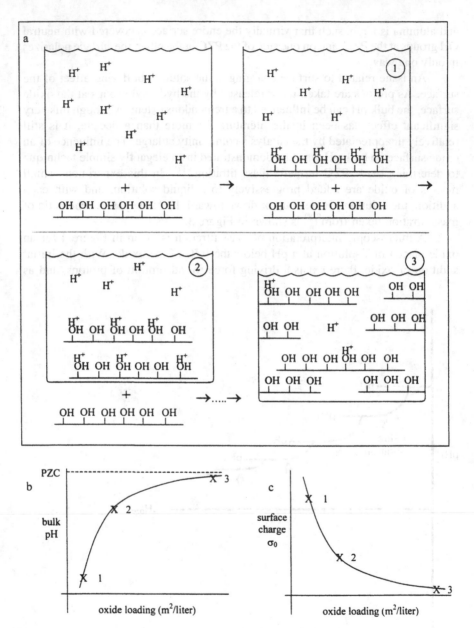

Figure 4 Microscopic interpretation of mass titration. (a) Depiction of the oxide surface. (b) Bulk pH vs. oxide loading. (c) Surface charge vs. oxide loading.

they are lost from the bulk liquid, the solution pH drops (Fig. 4a, b). A surface charge σ_0 is established (Fig. 4c). As more mass is added, and although more protons adsorb (Fig. 4a), causing a further drop in the bulk pH (Fig. 4b), the area density of adsorbed protons actually decreases (Fig. 4a) such that the surface charge σ_0 decreases (Fig. 4c). As more oxide is added, the solution pH increases until the solution pH approaches the PZC, at which point there is no driving force for proton adsorption. At this condition, only a miniscule fraction of the oxide surface is charged (Fig. 4a, c); the amount of oxide in solution or, more appropriately, the number of hydroxyl groups at the oxide surface is very large compared to the number of protons in solution. The surface charge is effectively zero at this condition and the oxide surface acts as a pH buffer.[74]

The first quantification of the oxide buffering effect appeared in 1995.[74] In this model, a proton balance was solved simultaneously with the single-site, amphoteric model of surface OH chemistry shown above, and a surface charge–surface potential relationship assumed from electric double layer theory. Figure 5 is taken from this work and is representative of an alumina surface. Final pH is plotted against initial pH for different values of the critical parameter of oxide surface loading. Oxide loading, the most direct measure of the amount of surface in solution, has units of m^2/liter and is the product of specific surface area (m^2/g)

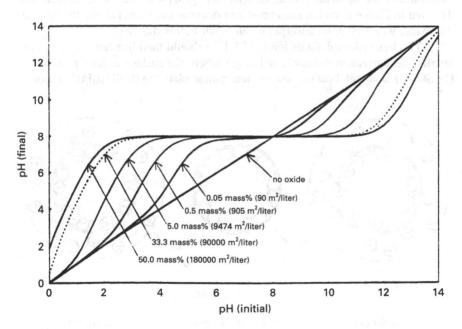

Figure 5 Model results for shifts in pH during impregnation for a representative alumina support.[75]

times mass loading (g/liter). The highest loading, 180,000 m^2/liter, corresponds to pore filling (at 1.1 ml/g) of a 200 m^2/g alumina. At this condition, the model predicts that an impregnating solution initially at a pH as low as 2 will end up near 8, the assumed PZC of the oxide. At this condition, the surface is uncharged and there would be no interaction with an anionic–metal complex. The sets of curves shown in Figure 5 at different surface loadings have been verified experimentally.[74,75] From this model a practical consequence follows for impregnation by pore filling (or incipient wetness). If a strong electrostatic interaction between metal complex and surface is desired, the starting pH must be extremely acidic or basic in order to achieve a significant surface charge.[74,79] Otherwise, no strong interaction will exist and at least the initial dispersion of the metal complex will be poor. For typical γ-aluminas, this would mean starting with a pH of less than 1 if a positive surface charge is desired or more than 13 if a negatively charged surface is desired.

   Electrostatic Adsorption Models.   A charged oxide surface is the basis of many adsorption models. A landmark theory of the late 1970s posited that noble metal adsorption onto common support oxides, such as silica and alumina, was essentially an electrostatic interaction[80] between ionic metal complexes in solution and the oppositely charged hydroxyl groups at the oxide surface. As depicted in Figure 6, cation adsorption can occur at pH values above the PZC of the oxide, whereas anion adsorption can occur below the PZC.

   The hexachloroplatinate ion ($[PtCl_6]^{2-}$) should then interact strongly with an alumina surface in the acidic pH range where the surface is fully protonated ($> SOH_2^+$), and metal cations such as tetraamine platinate ($[(NH_3)_4Pt]^{2+}$) should

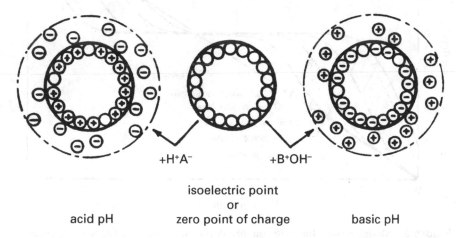

                    +H⁺A⁻   +B⁺OH⁻

                              isoelectric point
                                    or
         acid pH   zero point of charge   basic pH

Figure 6   Qualitative "ion exchange" adsorption theory of Brunelle.[80]

adsorb in the basic pH range where the surface is negative ($>SO^-$). In another seminal paper,[81] just such behavior and a semiquantitative model for it were reported for the palladium complexes $[PdCl_4]^{2-}$ and $[(NH_3)_4Pd]^{2+}$ in the respective pH ranges over alumina.

The first quantitative theory based on electrostatic interactions appeared somewhat earlier in the colloid science literature.[82] In this work the attempt was made to calculate a priori the free energy of adsorption from coulombic and solvation energies, although a "chemical" energy term had to be employed to compensate for an admittedly crude solvation energy calculation. There was a subtle difference between this model and the Brunelle scheme. The maximal adsorbate density was calculated on a steric basis, as a monolayer of close-packed complexes which retained a hydration sheath. In the work of Brunelle[80] and others[83], the process was thought to be more like ion exchange, in which equal equivalents of protons and metal cations or hydroxyls and metal anions exchanged at the surface. Both interpretations, based on a purely physical and mainly coulombic interaction between a charged surface and oppositely charged metal complexes, can be thought of as "physical" adsorption models.

In more recent work the original physical adsorption model of James and Healy[82] has been revised,[79,84,85] initially using a more rigorous calculation of the solvation energy term, which led to a much smaller and less dominant value for this term.[84] In later works it was discovered that the solvation free-energy term could be eliminated entirely[79,85] and the free energy of adsorption, from which the adsorption equilibrium constant is calculated, is based entirely on the coulombic energy.

The three components of the revised physical adsorption (RPA) model are shown in Figure 7. The first is a description of the surface charging, which includes the proton balance illustrated in Figure 5. Second, the adsorption equilibrium constants are calculated a priori based on the coulombic free energy. Third, the speciation of Pt complexes in the liquid phase must be accounted for as a function of pH, as the pH will change greatly upon adsorption. Up to this point, a single Pt species has been employed in the RPA models[79,85,86] based on the assumption that other chlorohydroxo complexes such as $[PtCl_4(OH)_2]^{2-}$ will have the same charge and about the same radius. In a later section on the coordination chemistry of CPA (chloroplatinic acid), this assumption is discussed.

In the model a Langmuir isotherm is assumed:

$$\Gamma = \Gamma_{max} \frac{K_{ads}C}{(1 + K_{ads}C)}$$

where $K_{ads} = \exp[-\Delta G_{coulombic}/RT]$ and C is the equilibrium concentration of the metal species. The coulombic energy $\Delta G_{coulombic}$ is derived from the surface-charging portion of the model discussed above. Earlier experimental

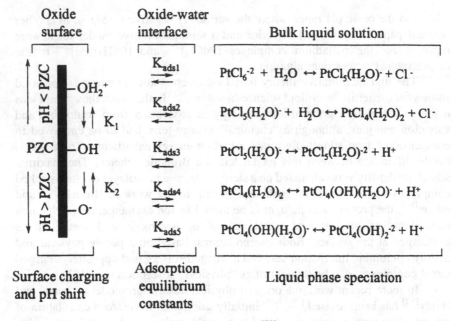

Figure 7 Three regimes of an adsorption model.[79]

work and a survey of the literature[87] suggested that the maximal uptake of anionic Pt and Pd complexes, Γ_{max}, as in James and Healy's model,[82] corresponded to a steric monolayer of complexes that retained one hydration sheath. For CPA this number is about 1.6 μmole/m^2, or about one Pt complex/ nm^2. An illustration of these hydrated complexes, along with a hydroxyl group density of about 5 OH/nm^2, is given in Figure 8. Note that, contrary to an "ion exchange" interpretation of noble metal adsorption, steric limitations prevent electrical saturation of the surface. Cationic complexes such as $[(NH_3)_4Pt]^{2+}$ and $[(NH_3)_4Pd]^{2+}$ have been reported to adsorb at a maximum of 0.8 μmole/nm^2, or one complex per 2 nm^2, and thus appear to retain two hydration sheaths upon adsorption.[87,88]

Representative experimental results and an RPA model simulation are shown in Figure 9. These are measurements of CPA uptake onto alumina vs. pH (Fig. 9a), at a constant initial CPA concentration of 180 ppm and a surface loading of 500 m^2/liter, which implies a great excess of liquid. For a series of aluminas of different phase and specific surface area (m^2/g), the mass of oxide was adjusted to give the same surface loading. When alumina powder is used (as opposed to spheres, pellets, or extrudates), adsorption occurs rapidly; close to 100% of equilibrium coverage was obtained within 1 h, and perhaps 75% within 10 min.[75]

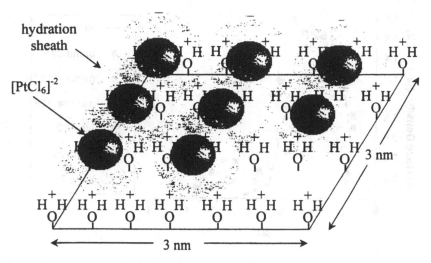

hydration sheath

$[PtCl_6]^{-2}$

3 nm

← 3 nm →

Figure 8 Depiction of the maximal extent of CPA adsorption over an alumina surface.[79]

Since all these aluminas have about the same PZC, near 8.5,[75] they can be modeled with one set of parameters. The shift in pH during adsorption can be calculated from the proton balance portion of the model; plots of final vs. initial pH for these adsorption experiments are shown in Figure 9b. The particulars of the model are given in [79]. In short, at the PZC the surface is uncharged and there is little or no adsorption. As the pH decreases from the PZC the surface charges and uptake increases. In the lowest pH range, Pt uptake is retarded not due to competition from chloride (HCl was used to acidify the solutions) but rather to the electric screening effect;[79,84] high ionic strength effectively reduces the adsorption equilibrium constant. Competitive adsorption of chloride is a commonly cited phenomenon in the impregnation literature[89,90] but does not appear to occur to a significant extent. A careful measurement of Cl^- concentration in the pH range where retardation occurred, in the experiment of Figure 9, showed that Cl^- did not adsorb and is not responsible for the retardation.[75] According to the RPA model, rather than a chloride-filled surface, it is simply empty since all adsorption equilibrium constants are very low at high ionic strength.[75]

The RPA model has been used with one set of independently measured parameters to simulate all extent sets of CPA/alumina adsorption data to a reasonable degree.[79] Many groups have recorded data as isotherms, in which uptake was measured as CPA concentration was increased.[91-94] To apply the RPA model to these datasets, the final pH of the solution had to be estimated using an initial estimate of pH from the concentration of the CPA, and then

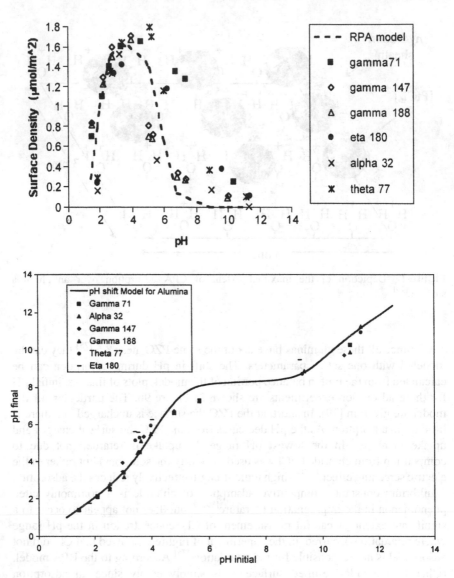

Figure 9 (a) Measured Pt uptake over a series of aluminas vs. pH, and the parameter free RPA model simulation of the data.[75] Surface area and phase of alumina is given in the figure legend, (b) measured and simulated pH shifts.

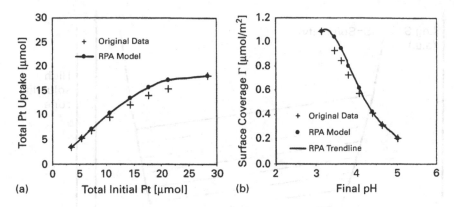

Figure 10 RPA model simulation of literature data. (a) Uptake vs. CPA concentration. (b) Uptake vs. final pH.[79]

employing the proton balance portion of the model. A representative simulation of literature data[91] is shown in Figure 10, which can be plotted as uptake vs. CPA concentration (Fig. 10a), or uptake vs. final pH (Fig. 10b). The same set of model parameters (PZC = 8.5, $\Delta pK = 5$, $N_S = 5$ OH/nm^2) was used for the simulation of Figure 10 as employed for Figures 2, 5, and 9.

Effect of Dissolved Aluminum: "Coordination" Adsorption Models. Other models have arisen from the complicating fact that aluminum dissolution often occurs during impregnation, especially in the acidic pH range. Figure 11 from Ref. 80 below indicates the solubility of alumina in the acidic and basic pH ranges. Superimposed onto this figure are the pH ranges of anion and cation adsorption. It is not surprising that several adsorption theories have predicted an impact of dissolved alumina and in entirely different ways. One group, noting that the adsorption of Pt onto alumina diminished at low pH as aluminum dissolved, compared to higher pH, suggested that the decrease in Pt uptake was due to the loss of adsorption sites from the alumina surface.[95]

In a completely contrary hypothesis, an adsorption mechanism has been proposed in which platinum species stemming from CPA are deposited onto the surface only after aluminum has dissolved from the alumina surface and formed a hetero complex with platinum.[94] From a kinetic analysis, the adsorption step was postulated as shown in Figure 12. In a later work,[96] this mechanism was inferred from a series of alumina supports over which the extent of Pt adsorption (from CPA) directly correlated with the degree of Al dissolution. In both of these papers the adsorption of Pt-containing species is postulated to occur only after the dissolution of aluminum and the formation of a heterogeneous coordination complex. This sort of mechanism might then be thought of as a "coordinative" mechanism.

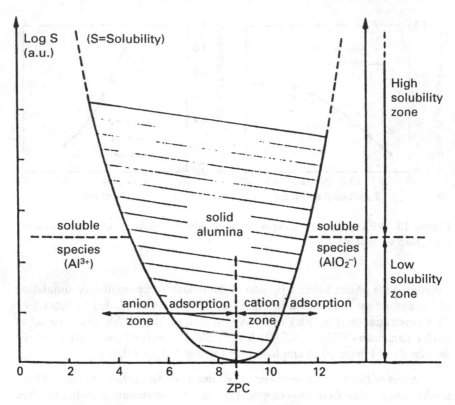

Figure 11 Aluminum solubility as a function of pH.[80]

Evidence against a direct role of dissolved aluminum has accrued since these works[94,95] were published. In the first place, similar adsorption capacities have been demonstrated for aluminas that differ widely in solubility. This is seen in the Pt uptake results of Figure 9, from Ref. 75, where all phases of alumina behave similarly. The alumina solubility for this series of samples is shown below in Figure 13, taken from Ref. 75. α-Alumina is sparing soluble, η and θ are

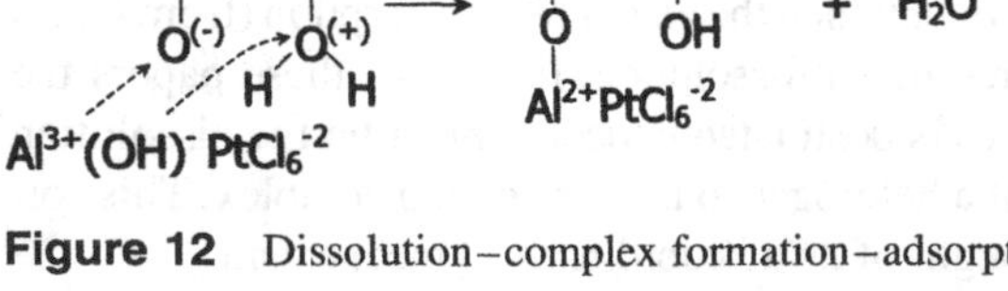

Figure 12 Dissolution–complex formation–adsorption mechanism.[96]

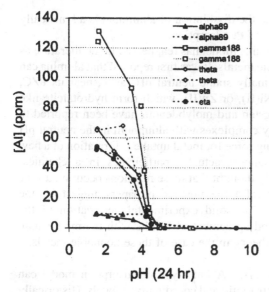

Figure 13 Dissolved alumina vs. pH in the presence and absence of Pt.[75]

moderately so, and γ-alumina is the most soluble. Since the same surface loading was employed for all samples, the trend seen here should truly reflect the difference in intrinsic dissolution rates. In the earlier work in which Pt uptake was correlated to Al dissolution,[97] the same mass of alumina was used for aluminas with different specific surface areas. Not surprisingly, the samples with the highest surface area exhibited the highest degrees of both aluminum dissolution and Pt uptake. It was seen[87] that if the uptake results from that study are normalized for surface area, in fact all samples adsorbed Pt in an amount close to the monolayer capacity of 1.6 μmole/m^2.

The influence of dissolved aluminum is manifested by upward shifts in pH, as the overall dissolution equation

$$Al_2O_3 + 6H^+ \rightarrow 2Al(OH)_3$$

dictates that three protons are consumed for each aluminum dissolved. These upward shifts have been observed in addition to proton adsorption (see figures 4 and 7 of Ref. 75) and were more severe for the more soluble aluminas. Above a pH of about 3.5, dissolution is negligible.[75,80]

Finally, in Figure 13 the solid lines are for dissolved aluminum measured during the adsorption experiment, while the dashed lines were recorded for a control experiment, in the absence of Pt, in aqueous HCl in the same pH range. In essentially all cases, the amount of dissolved aluminum is independent of the presence of Pt. Thus, it appears that the dissolution of aluminum is a kinetically limited process that is independent of Pt adsorption.

As a quick aside, a version of this model has reappeared in the literature as a "geochemical" model for nonnoble metals. It was first reported that alumina can be induced to dissolve in the normally stable neutral pH range, near its PZC, when in the presence of Co(II), Ni(II), or Zn(II), which form hydrotalcite-like complexes.[97] Most recently, tungsten and molybdenum have been reported to form stable Keggin-like heteropoly complexes with aluminum in the neutral pH range.[98,99] In this model the driving force for metal uptake is formation of a new solid compound. The alumina support actually participates in a chemical reaction, and while it was originally thought that these reactions occur on a very large time scale (hence, "geochemical"), the most recent papers suggest that the reactions occur in minutes.[98,99] The solid experimental verification of the heteropoly complexes and the good agreement of results with thermodynamic theory[98,99] strongly support this theory in the case of these nonnoble metals.

"Chemical" Adsorption Models.   A final type of adsorption model can be thought of as a blend of the electrostatic and coordinative models. Historically the electrostatic model of James and Healy was abandoned, perhaps due to the magnitude of the "chemical" or adjustable free-energy term. The size of this term usually swamped the coulombic and solvation terms;[84] the model was too often a one-parameter fit of an adsorption uptake curve. In its place arose the "triple layer" or "site binding" model[100] in which metal complexes adsorb at an outer layer, and small electrolytes such as Cl^- and Na^+ were postulated to adsorb, via some unspecified "chemical" interaction, at the inner layer. This is the extent of the chemical interaction; there is no dissolution and reaction of the aluminum in the support with the impregnated metal complexes as occurs in the previous model. The presence of the inner layer ions can be used to attenuate the outer layer surface potential and so adjust the uptake of metal complexes.

A triple-layer model was employed relatively recently for CPA adsorption over alumina.[89] The data that were modeled are presented in Figure 14. This is a relatively complete set of data, since not only was the uptake of Pt recorded, but also shifts in pH and the amount of dissolved aluminum at a particular pH were monitored. The trend in Pt uptake vs. pH has the familiar volcano shape. Besides the retardation of Pt adsorption at low pH, the shifts in pH must also be simulated.

The triple-layer model equations used to simulate this set of data are shown in Figure 15. First, there are the hydroxyl protonation–deprotonation reactions as seen in Figure 7. The surface ionization constants K_1 and K_2 have the same meaning as before. In addition, the inner layer adsorption of Cl^- and Na^+ is modeled with four additional equilibrium constants. The retardation was simulated by fitting the data using large values of the equilibrium constants of equations (3) and (5) of Figure 15, which give rise to large values of "chemically"

Number	Initial pH	Final pH	Pt$_{ads}$ 10^{-3} (mol/l)	Al$_{sol}^{3+}$ %[a]
1 (+HCl)	1.04	1.13	0.88	2.20
2 -	2.07	4.02	4.1	0.35
3 (+NaOH)	2.45	5.33	2.7	0.03
4 (+NaOH)	2.83	6.40	1.6	<0.01
5 (+NaOH)	3.00	6.60	1.2	<0.01
6 (+NaOH)	4.32	7.3	0.54	0.01
7 (+NaOH)	5.50	8.20	0.28	0.01
8 (+NaOH)	6.44	8.11	0.08	0.01

[a] Al^{3+} in the filtrate/g Al$_2$O$_3$.

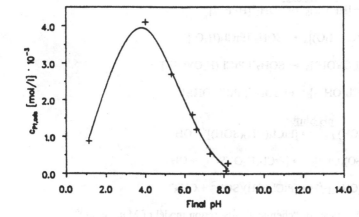

Figure 14 pH shift, Pt uptake, and Al dissolution data from Mang et al.[89]

adsorbed Cl$^-$. Metal adsorption is described by either four or seven additional adsorption equilibria, the last three of which (11–13) of Figure 15 are surface–ligand exchange reactions that release OH$^-$ and so can account for the observed upward shifts in solution pH.

The fit of the chemical adsorption model for this set of Pt uptake data is shown in Figure 16a. With the large number of adjustable parameters it contains, the fit of the chemical adsorption parameter is good, although the number of adjustable parameters this sort of model contains is a major drawback. However, the fit of the RPA model using the unadjusted set of parameters as in the earlier figures is even better, as seen in Figure 16b. In addition, the pH shifts are well predicted by the proton balance of the RPA model; these data are shown in Figure 16c. The pH shifts are calculated independently from Pt adsorption. Again, the

Surface charging:
(1) $SOH_2^+ \leftrightarrow SOH + H_S^+$; K_1

(2) $SOH \leftrightarrow SO^- + H_S^+$; K_2

Inner layer adsorption/metal adsorption retardation:
(3) $SOH_2^+ + Cl_S^- \leftrightarrow SOH_2^+Cl^-$; K_{an}

(4) $SO^- + Na_S^+ \leftrightarrow SO^-Na^+$; K_{cat}

(5) $SOH + H_S^+ + Cl^- \leftrightarrow SOH_2^+Cl^-$; K^*_{an}

(6) $SOH + Na^+ \leftrightarrow SO^-Na^+ + H_S^+$; K^*_{cat}

Adsorption:
(7) $SOH_2^+ + [HPtCl_6]_S^- \leftrightarrow SOH_2^+[HPtCl_6]^-$

(8) $SOH_2^+ + [PtCl_5(H_2O)]_S^- \leftrightarrow SOH_2^+[PtCl_5(H_2O)]^-$

(9) $SOH_2^+ + [PtCl_4(H_2O)(OH)]_S^- \leftrightarrow SOH_2^+[PtCl_4(H_2O)(OH)]^-$

(10) $SOH_2^+ + [PtCl_4(OH)_2]_S^{2-} \leftrightarrow SOH_2^+[PtCl_4(OH)_2]^{2-}$

pH Shifts:
(11) $SOH + [PtCl_4(OH)_2]_S^{2-} \leftrightarrow [PtCl_4(OH)(SOH)]^- + OH_S^-$

(12) $SO^- + [PtCl_4(H_2O)(OH)]_S^- \leftrightarrow [PtCl_4(H_2O)(SO)]^- + OH_S^-$

(13) $SO^- + [PtCl_4(OH)_2]_S^{2-} \leftrightarrow [PtCl_4(OH)(SO)]^{2-} + OH_S^-$

Figure 15 The triple layer or "chemical" adsorption model of Mang et al.[89]

most recent experimental evidence also seems to support the RPA model; chloride adsorption has been shown not to occur in the low-pH regime of Pt adsorption retardation,[75] and pH shifts (Fig. 16c) are observed to occur in the absence of Pt adsorption.[74,75]

The most recent appearance of a chemical adsorption mechanism stems from a molecular level characterization of Pt coordination complexes.[101–103] Extended X-ray absorption for fine structure (EXAFS) and [195]Pt nuclear magnetic resonance (NMR) spectroscopy have been used to study the coordination chemistry of adsorbed Pt complexes. This work will be discussed in more detail in the following section. Here it will be pointed out that based on this characterization the adsorption process is thought to occur in large part as an electrostatic mechanism, but also with some "specifically adsorbed" and some "grafted" complexes.[103] A molecular depiction of the specifically adsorbed Pt complexes is shown in Figure 17a. In this picture, the protonated hydroxyl groups

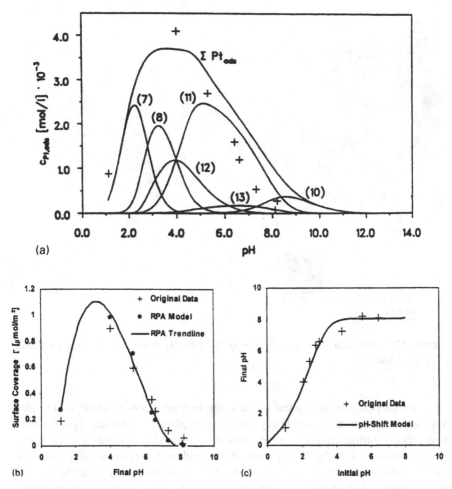

Figure 16 Models of Pt adsorption. (a) "Chemical" adsorption and (b) RPA simulations of the Mang et al. uptake data.[79] (c) RPA proton balance for final vs. initial pH.

at the alumina surface have replaced one or two water molecules in the hydration shell surrounding the Pt complex. This hydrogen bonding (and thus, "specific") interaction was inferred from small shifts in the NMR signal. In more basic pH solutions, the NMR signal for the penta- and tetrachloride species disappears altogether, which is taken as a signal of a "grafting" reaction in which the chloride ligands in the Pt complex are replaced by OH groups from the alumina surface (species B1 in Fig. 17b) or OH groups at the alumina surface are replaced by Cl ligands from the Pt complex (species B2 in Fig. 17b). An alternative explanation for the disappearance of the NMR signal will be given in the next section.

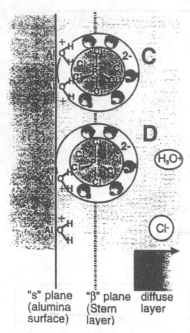

"s" plane   "β" plane   diffuse
(alumina   (Stern   layer
surface)   layer)

Figure 17 Molecular depiction[103] of (a) "specifically adsorbed" Pt complexes and (b) grafted species.

    Of all possible adsorption models, the revised physical adsorption model appears to capture most completely and most simply the fundamental phenomena that occur during catalyst impregnation of powdered catalysts. The key phenomena appear to be the charging of the oxide surface, the influence of bulk pH by proton transfer to the surface, and electrostatic adsorption. The charging of the surface and electrostatic adsorption appear to be independent phenomena, coupled by the bulk liquid pH. The dissolution of aluminum and Pt adsorption also appear to be independent phenomena, weakly coupled by solution pH and ionic strength. Recent advances in the molecular characterization of dissolved and adsorbed Pt complexes will enable the adsorption mechanism of CPA onto alumina to be further refined. At present, this is the cutting edge of CPA/alumina impregnation research and is the subject of the following section.

Coordination Chemistry of Dissolved and Adsorbed Pt Complexes from CPA

In recent years, sensitive in situ spectroscopic characterization of the coordination chemistry of Pt complexes has been directed to a molecular level description of the mechanism of Pt adsorption and a more precise determination of the composition of adsorbed and adsorbing Pt complexes.

Chloroplatinic Acid in Solution. As a first consideration, contradictory sets of formation constants on CPA speciation are available in the literature. Sillen and Martell[104] give the most comprehensive and most cited set of equilibrium constants. Their speciation pathway, shown in Figure 7, presumes that up to two chlorides can exchange successively for water ligands:

$$[PtCl_6]^{2-} + H_2O \leftrightarrow [PtCl_5(H_2O)]^- + Cl^- \tag{1}$$

$$[PtCl_5(H_2O)]^{1-} + H_2O \leftrightarrow [PtCl_4(H_2O)_2]^0 + Cl^- \tag{2}$$

and at basic pH the chloroaquo complexes behave as weak acids and deprotonate as follows:

$$[PtCl_5(H_2O)]^- \leftrightarrow [PtCl_5(OH)]^{2-} + H^+ \tag{3}$$

$$[PtCl_4(H_2O)_2]^0 \leftrightarrow [PtCl_4(OH)(H_2O)]^- + H^+ \tag{4}$$

$$[PtCl_4(OH)(H_2O)]^- \leftrightarrow [PtCl_4(OH)_2]^{2-} + H^+ \tag{5}$$

Dissociation of the chloroplatinic acid is presumed complete and up to two chloride ligands may be exchanged. The Pt species predicted at different pH values according to the Sillen and Martell data are shown in Figure 18. Since the equilibrium constants were determined at low pH and high Pt concentrations, it is

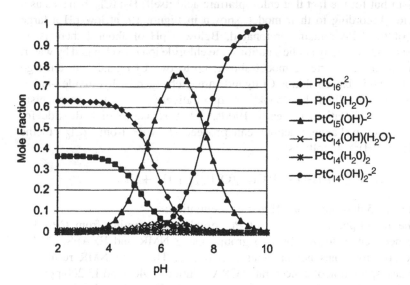

Figure 18 Speciation of CPA according to the mechanism and formation constants reported by Sillen and Martell.[104]

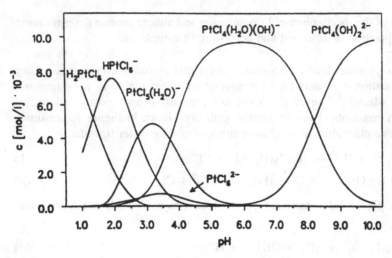

Figure 19   Speciation of CPA reported by Mang et al.[89]

likely that the equilibrium constants only accurately predict the Pt species in acidic-to-neutral aqueous solutions. They acknowledge that at higher pH more extensive hydrolysis is possible,[104] perhaps leading to $Pt(OH)_6^{2-}$ in strongly basic solution.

Based on the data of Cox and Peters[105] and Davidson and Jameson,[106] Mang et al.[89] showed a speciation of CPA similar to that of the Sillen and Martell mechanism but for the fact that chloroplatinic acid itself, H_2PtCl_6, behaves as a weak acid. According to their model, shown in Figure 19, at low pH a large fraction of the CPA remains protonated. Below a pH of about 2 there is no hydrolysis, and in weakly acidic solutions one chloride ion is exchanged by water. As with Sillen and Martell's model above, up to two Cl ligands can undergo hydrolysis, and in basic solution Cl ligands are exchanged by hydroxide ion.

Finally, Gmelin's handbook[107] qualitatively gives the mechanism established by Miolati in the early 1900s.[108] A sequence of hydroxide for chloride ligand exchange reactions can proceed stepwise from $PtCl_6^{2-}$ to the insoluble $Pt(OH)_6^{2-}$ as follows:

$$PtCl_{6-n}(OH)_n^{2-} + H_2O \leftrightarrow PtCl_{5-n}(OH)_{n+1}^{2-} + H^+ + Cl^-$$

where $n = 0-5$ depending on pH and concentration.

The liquid-phase speciation of Pt complexes stemming from CPA has recently been characterized by two groups using NMR and EXAFS,[101–103] sensitive spectroscopies not previously available. The ^{195}Pt NMR results on liquid-phase speciation of concentrated CPA solutions (2400 and 13,200 ppm Pt) obtained by Lambert and coworkers[10] are shown in Figure 20 and appears much

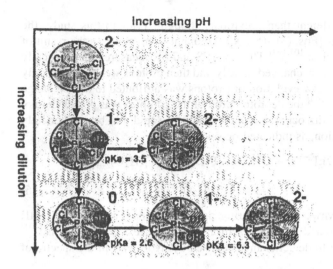

Figure 20 Speciation of concentrated CPA solutions reported by Lambert et al.[99]

in line with the Sillen and Martell scheme. The water–chloride ligand exchanges are thought to occur slowly, and the chloroaquo complexes are thought, as weak acids, to dissociate quickly.

Even more recent EXAFS results obtained for dilute solutions at the highly sensitive Advanced Photon Source at Argonne National Laboratory reveal that the speciation pathway is more complicated in the low-concentration regime[109] and calls into question the mechanistic interpretation of rapid deprotonation.[101,104]. Figure 21 shows that the Pt-Cl coordination number decreases approximately linearly with decreasing Pt concentration below 2400 ppm. The Pt-Cl coordination of the 2400 ppm CPA solution determined by the Pt NMR study[101] is included in this figure and is in good agreement with these EXAFS determinations. But many more than two chlorides are exchanged at low Pt concentration. The calculated coordination for a 200 ppm sample at a pH of about 2.7 is an average of 2.7 Cl^- and 3.3 OH^- or H_2O ligands, and with aging the chloride coordination decreases even further to a value of about 2.[109] Figure 22 illustrates the effect of pH and aging on the speciation of 200 ppm CPA. (It is important to study the speciation in the basic pH range, to account for the pH buffering effect of alumina that will pull the liquid pH into this range.) In the more basic pH range with aging, up to five chlorides are exchanged. It is clear from these EXAFS results that dilute CPA solutions clearly undergo more extensive hydrolysis than predicted by the Sillen and Martell model.

The slow, large shifts in pH after 24 h aging, noted with arrows in Figure 22, give rise to a mechanism for the formation of chlorohydroxoaquo complexes that is significantly different than previously thought. Prior to this study, the deprotonation reactions [Eqs. (3)–(5)] were thought to be rapid, as the result of dissociation of weak acids.[101,104] The EXAFS and pH measurements basically showed that the chlorides exchanged rapidly and then protons were released only slowly thereafter. This slow evolution of protons was attributed to the exchange of hydroxide for water;[109] the deprotonation of the complex was postulated not to occur via weak acid dissociation, which would be rapid, but by a kinetically slow substitution reaction, as indicated below:

$$[PtCl_x(OH)_y(H_2O)_z]^{4-x-y} + OH^- \leftrightarrow [PtCl_x(OH)_{y+1}(H_2O)_{z-1}]^{4-x-y}$$

$$+ H_2O \tag{6}$$

While these large downward shifts during aging were also noted in the NMR study, they were attributed in some fashion to the process of hydrolysis.[101] However, it is difficult to detect changes caused by hydrolysis in the spectra of fresh and aged samples.

While the actual mechanism of the formation of chloroaquo and chlorohydroxo species is still under debate, an overall speciation pathway independent of mechanism can be formulated from the EXAFS study. This overall scheme is shown in Figure 23. This comprehensive pathway subsumes

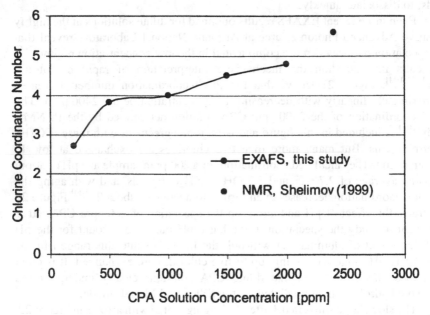

Figure 21   Decrease in chloride coordination with decreasing CPA concentration.[109]

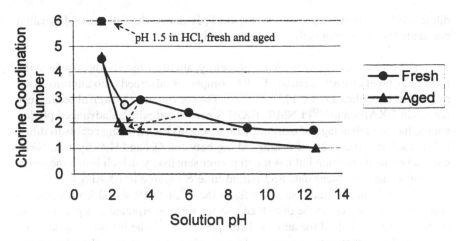

Figure 22 Changes in the chloride coordination of 200 ppm CPA with respect to pH and aging. (Adapted from Ref. 109.)

earlier versions: the Gmelin pathway is seen as the bottom diagonal of the figure, and the Sillen and Martell pathway is seen in the first three columns and is represented by the horizontal and vertical pathways.

Since aquo species definitely form, the pathway of Gmelin is incorrect. The chloride coordination was also seen to be a function of pH at constant excess chloride concentration.[109] With the Sillen and Martell pathway composed of the horizontal water–chloride exchanges [Eqs. (1) and (2)] and the vertical hydroxyl–water exchanges [Eqs. (3)–(5)] in Figure 23, this functionality is impossible to obtain. Thus, an additional pathway must be added in which hydroxyls directly exchange with chloride ligands;[40] these new pathways are represented in Figure 23 as the diagonal arrows, and the new path is indicated in Eq. (7).

$$[PtCl_x(OH)_y(H_2O)_z]^{4-x-y} + OH^- \leftrightarrow [PtCl_{x-1}(OH)_{y+1}(H_2O)_z]^{4-x-y}$$
$$+ Cl^- \tag{7}$$

Careful pH measurements were combined with the EXAFS analysis to determine the most prevalent species in this scheme.[109] These measurements suggest a valence of 0 for the 1000 ppm CPA solution, i.e., $[PtCl_4(H_2O)_2]^0$ would appear to be the dominant species at low pH. Furthermore, at dilute concentrations and at low pH, it appears that zero valent species are often preferred. The $[PtCl_6]^{2-}$ complex is only seen at high CPA concentration and low pH, or at low CPA concentration in a great excess of chloride. With further characterization of these dilute CPA solutions, especially in the mid-pH range, the full set of formation constants for the overall pathway might soon be obtained.

   CPA Adsorbed onto Alumina.   Recently, attention has also been paid to the molecular level characterization of CPA complexes adsorbed onto alumina, such as used to formulate Figure 17.[102,103,110] Two of the most powerful techniques are again EXAFS and [195]Pt NMR. EXAFS performed with an advanced photon source has the advantage of extremely high sensitivity, allowing access to dilute solutions, but is incapable of distinguishing between OH and H_2O ligands. NMR can make this distinction but has much poorer sensitivity, which limits the range of dilution and lengthens data acquisition time compared to EXAFS.

   In the Lambert study, small shifts in the [195]Pt NMR signal from adsorbed $[PtCl_6]^{2-}$ with respect to the dissolved species were interpreted as a perturbation of the hydration shell of the adsorbed complex.[102,103] The interactions depicted in Figure 17 are thought to be partly electrostatic and partly "specific adsorption" owing to a hydrogen bond between a proton from a protonated surface hydroxyl group and a chloride ligand from the complex. NMR analysis also revealed a conversion of $[PtCl_6]^{2-}$ to $[PtCl_5(OH)]^{2-}$ during the aging of a sample impregnated with an initial pH of 1.15, and that the integrated signal of the hexa-, penta-, and tetrachloro species decreased greatly over time when the impregnation was conducted with a solution initially at pH 2.8.[102] The loss of NMR signal was taken as evidence of a grafting reaction, illustrated in Figure 24, in which hydroxide ligands from the support exchange with chloride.

   EXAFS analysis of this latter sample after drying revealed a chloride coordination of only 3, which is somewhat difficult to explain by grafting. In fact, an alternate explanation for the decrease in chloride coordination upon aging in the one adsorbed sample and the loss of the NMR signal in the other can be proposed from just-completed EXAFS work of the Regalbuto group.[110] In that work the adsorption of $[PtCl_6]^{2-}$ was compared to that of Pt complexes with about half the chloride exchanged for hydroxide or water. The experiments performed and the results are summarized in Table 1. In this series of experiments, the target Pt loading was varied by changing the oxide loading. The highest oxide loading, samples 1 and 2, yielded the lowest Pt loading and the highest upward pH shifts, as expected in view of the proton balance of Figure 25. The $[PtCl_6]^{2-}$ complex was made by adding excess chloride to half the samples; the increased ionic strength in these samples partially inhibited Pt uptake, also as expected in view of Figure 9. A perhaps unexpected result is that the adsorbed complex in either series, whether initiating as $[PtCl_6]^{2-}$ or "$PtCl_3O_3$" (possibly $[PtCl_3(OH)(H_2O)_2]^0$, from Figure 23) ended up as "$PtCl_2O_4$".

   The interpretation of these data relied on the earlier liquid-phase speciation study[109] and on the RPA model. The electrochemical equilibrium at the oxide interface results in a Boltzman distribution of protons such that below the PZC of an

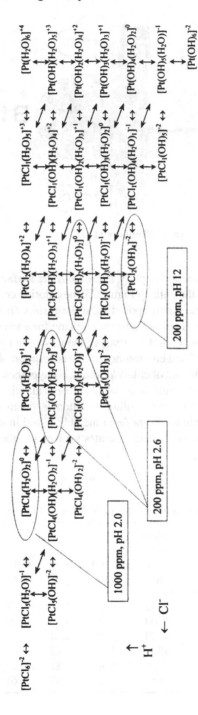

Figure 23 A comprehensive pathway of CPA speciation.[109]

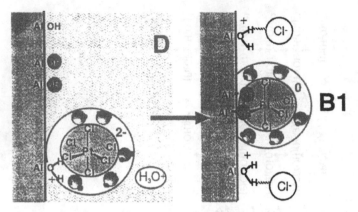

Figure 24 Grafting mechanism of Shelimov et al.[103]

oxide, the pH at the surface, and slightly away from the surface at the adsorption plane, is higher than that of the bulk. In Figure 25, the chloride coordination numbers of the adsorbed species are superimposed onto the results from the liquid-phase study, using the respective pH of the adsorption plane for each solid sample. Both sets of solid samples fall between the curves of the fresh and aged samples. It was concluded not only that the local environment at the adsorption plane is at higher pH than the bulk but that it is devoid of chloride ions. This hypothesis is consistent with the earlier observation that chloride does not adsorb onto alumina.[75] An illustration of the adsorption of $[PtCl_6]^{2-}$ from dilute solution is given in Figure 26. The hydrated hexachloroplatinate ion is stable in the bulk liquid in excess chloride and low pH, but at the adsorption plane it speciates as if in a higher pH, low-chloride solution.

Table 1 Comparison of Adsorptive Behavior of $PtCl_6$ and $PtCl_3O_3$

Sample	Target wt % Pt	NaCl (mol/ liter)	Surface load (m^2/ liter)	Initial pH	Final pH	CN Cl liquid	CN Cl solid	Actual wt % Pt
1	1	—	5000	2.63	4.34	2.7	1.6	1.0
2	1	0.1	5000	2.60	5.59	6.0	1.5	0.7
3	4.8	—	1000	2.59	2.84	2.7	2.1	3.7
4	4.8	0.1	1000	2.54	3.43	6.0	1.9	1.9
5	7.2	—	650	2.50	2.81	2.7	2.1	4.1
6	7.2	0.01	650	2.55	2.87	(6)	2.2	3.2

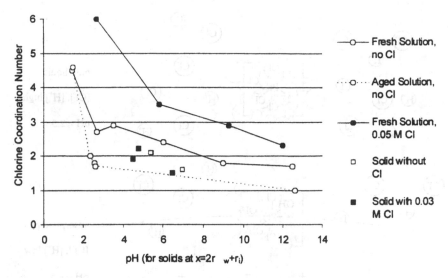

Figure 25 Chloride coordination numbers of adsorbed species (squares) and liquid phase species (circles).[110]

There is still work to be done to confirm this interpretation. The adsorbed species shown in Figure 26 is the dianion $[PtCl_2(OH)_4]^{2-}$, which is consistent with an electrostatic model. Liquid-phase speciation in the pH range above 2.6 and below 12 has not been characterized, however, and the valence of this Pt complex is only assumed. At pH 2.6, as indicated in Figure 23, the Pt complexes are zero valent. The assumption made in drawing Figure 26 is that in the mildly acidic range above pH 3–4, the aquo ligands will exchange with (or deprotonate to) hydroxo ligands. Additional EXAFS studies will be conducted to confirm or disprove this model.

The purely electrostatic interpretation can be extrapolated to higher Pt concentrations to arrive at an alternative explanation to grafting as proposed by the Lambert group. The transformation of adsorbed $[PtCl_6]^{2-}$ to $[PtCl_5(OH)]^{2-}$ during the aging of an impregnation with a solution initially at pH 1.15 can be explained by the complex being subject to increased pH upon adsorption and aging. This would come about in a number of ways. First, from Figure 5 it can be appreciated that the consumption of protons by the surface (the oxide buffering effect) at the surface loading employed, about 150,000 m^2/liter, would raise the pH significantly from 1.15. The pH at the adsorption plane would be even higher, and dissolution of aluminum might slowly increase the pH even further. While the transformation of adsorbed $[PtCl_6]^{2-}$ to $PtCl_2O_4$ was seen to be rapid in dilute solutions,[110] in concentrated solutions it might be slower. The disappearance of

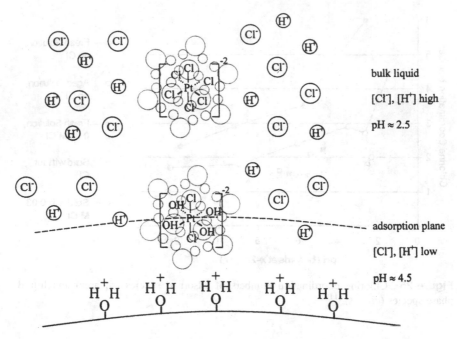

Figure 26 Microscopic interpretation of the speciation of adsorbed Pt complexes.

the integrated NMR signal from hexa-, penta-, and tetrachloroplatinates for the sample impregnated at an initial pH of 2.8[102] can be explained by the same reasoning. At this relatively high initial pH, the shift in pH upon impregnation would be very large and likely approach the oxide PZC. At this high pH, the loss of chloride ligands might be expected in solution for even a concentrated solution.

Independent of the grafting or the "localized electrostatic" speciation theories, the fact is that adsorbed Pt complexes lose chloride upon adsorption except at high chloride concentrations. Further characterization of this system should shed more light on the coordination chemistry of the dissolved and adsorbed species and the adsorption mechanism in general.

Pt uptake in Zeolites and MAPSO and SAPO Zeotypes

As opposed to an electrostatic adsorption mechanism, the uptake of Pt onto zeolites is generally considered to occur via ion exchange at the charge-unbalanced aluminum sites in the zeolite framework. In a classic and comprehensive review of zeolites and their industrial use, ion exchange isotherms for many zeolites and ion pairs are given.[111] This implies, of course, that only cationic complexes of Pt can be easily incorporated into the zeolite and this indeed is the standard recipe [112 and references within]. Since the exchange occurs at a very localized site, another implication of an ion exchange mechanism is that the uptake should be independent of pH.

While a full review of Pt uptake onto zeolites is beyond the scope of this chapter, a limited number of recent data will be presented that are pertinent to the fundamental considerations being presented. A recent study has been conducted[113] to confirm the ion exchange mechanism over zeolites and to contrast zeolite uptake behavior to that of several common oxides and "zeotype"[114] aluminum phosphate materials. In this study, the uptake of anionic and cationic Pt complexes over zeolites and zeotypes with different PZCs was monitored as a function of pH and was directly compared to alumina and silica materials with similar PZCs.

Figure 27 shows the platinum adsorption results over the zeolites/zeotypes and silica and alumina from 120 ppm CPA solutions. First, alumina and silica both show the expected characteristics of electrostatic adsorption, i.e., uptake

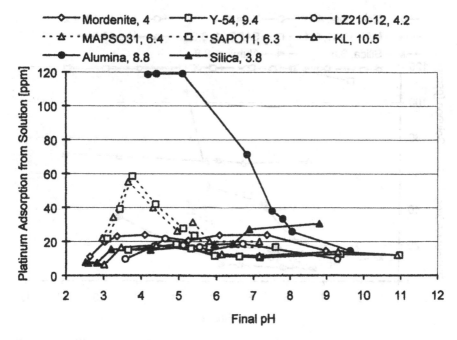

Figure 27 Platinum adsorption from 120 ppm CPA solutions over zeolites, zeotypes, alumina, and silica as a function of the solution pH.[113]

over alumina (filled circles) below its PZC and negligible uptake over silica (filled triangles). The PZC values of the materials are given in the legend. All tested zeolites show generally very little uptake over the entire pH range. The characteristics of Y-54 are especially noteworthy, as it has a PZC similar to that of alumina, yet it does not show the typical electrostatic uptake behavior. The two zeotypes, MAPSO-31 and SAPO-11, seem to exhibit a limited adsorption with the characteristics of the electrostatic mechanism with uptake below their PZCs (6.4 and 6.3) and drop off toward the low-pH region.

In the same fashion, Figure 28 shows the platinum uptake from cationic platinum tetraamine (TAPC) solutions. Again, alumina and silica display the expected electrostatic adsorption, with the silica surface accruing a much stronger negative charge as pH is increased due to its lower PZC. The zeolites generally show a pH-independent, PZC-independent, complete uptake of cationic platinum, which is likely due to ion exchange with the equilibrium far on the side of Pt^{2+} in the zeolite. Head-to-head comparisons between alumina and Y-54, with similar high PZC values, and mordenite and silica, with similar low PZCs, again show that platinum adsorption over zeolites cannot be explained with an

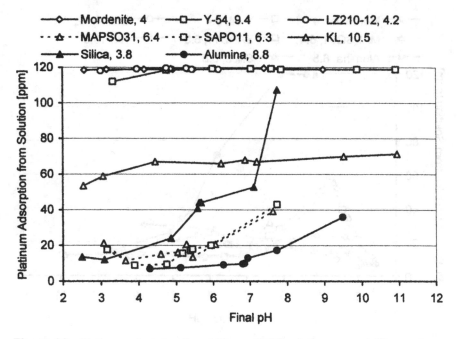

Figure 28 Platinum adsorption from 120 ppm TAPC solutions over zeolites, zeotypes, alumina, and silica as a function of the solution pH.[113]

electrostatic surface adsorption mechanism. The largely pH-independent but lower uptake of the K-L zeolite is most probably due to its lower exchange capacity. The maximal ion exchange observed here corresponds to only roughly 40% of the theoretically determined values based on aluminum content. It is thought that potassium in L zeolites only partially be exchanged, as some negatively charged framework sites seem to be inaccessible to larger ions.[115] The zeotypes again seem to follow the physical surface adsorption mechanism applicable to silica and alumina, as the uptake of the cationic platinum is negligible below their PZC values of 6.3 and 6.4.

3.4 Drying, Oxidation, and Reduction of the Impregnated Support

Drying is intended to eliminate most of the aqueous solution from the pores. It is well known that this operation can cause substantial movements of the solution and hence a significant redistribution of the precursors still present in the dissolved state in the particles of the support. For reforming catalyst in which the uniformly impregnated precursor in the presence of HCl acid is virtually entirely fixed, such a redistribution can be ignored. However, drying may modify the form of the adsorbed platinum complex. In particular, the ligand balance between chloride and hydroxide may change.

Oxidation in air causes further progressive replacement of the chloride ligand by oxygen, with the formation of a complex close to "$[Pt^{IV}O_xCl_y]$" between 500°C and 600°C, such as the four-ligand complex $[PtO_2Cl_2]^{2-}$ according to Lieske et al.[116] or the six-ligand complex "$[PtO_{4(5}Cl_{1(5)}]$" according to Berdala et al.[117] This complex $[Pt^{IV}O_xCl_y]$ could preferentially occupy the high-energy sites of the alumina,[38] first saturating the kink sites, then the step sites, and finally occupying the terrace sites.

Simultaneously, oxidation in air causes a decrease in the chloride content of the alumina, whether it is η- or γ-Al_2O_3.[118] The higher the chloride content after drying, the higher is the decrease. This decrease is also accentuated by the presence of small amounts of water in the air.[118] Under identical conditions, chloride retention depends on the alumina employed; thus, after oxidation at 550°C, a γ-Al_2O_3 with a surface area of $190 \, m^2 \, g^{-1}$ containing 0.9 wt % of chloride loses 40% of the chlorine, whereas a η-Al_2O_3 with $200 \, m^2 \, g^{-1}$ containing 1 wt % of chloride loses only about 17%.[118]

The reduction step, typically conducted between 500°C and 550°C, converts the above platinum species to highly dispersed platinum metal.[119] EXAFS indicates the formation of small metal particles in which the metal atoms have four to six immediately neighboring platinum atoms and also indicates the presence of some Pt-O bonds.[117]

4 BIMETALLIC CATALYSTS

The deposition of chloride and platinum on alumina occurs through complex reactions as discussed in Sec. 3. Both elements must be uniformly distributed in the grains of the support, and platinum must be well dispersed at the end of the impregnation step. The good dispersion, obtained as a result of a strong metallic precursor–support interaction, and a uniform macroscopic distribution, obtained in spite of this strong interaction, must be preserved in the subsequent steps of drying, oxidation, and reduction in order to attain excellent catalytic performance and resistance to sintering. These considerations are still valid for multimetallic catalysts, although every additional metal brings its own unique chemistry, which may impose adaptation and optimization of the means of introducing the various metals. In this section the main preparation methods for Pt-Re, Pt-Sn, and Pt-Ir reforming catalysts are described.

4.1 Platinum–Rhenium

Formulations based on Pt and Re (typically containing between 0.1 and 0.8 wt % of each metal) are among the most important industrial reforming catalysts. Such combinations, initially developed by Chevron in 1969,[120,121] increase the stability of catalytic performance, allowing much longer operating periods of the unit between regenerations. This improved stability is explained by a double effect of rhenium: higher resistance to deactivation by coking and stability of the metallic phase on the support.[7,23,122] However the presence of rhenium induces a high degree of hydrogenolysis, which necessitates a sulfurization step during the initial startup of the catalyst.

It is generally accepted that the optimal Pt-Re phase consists of Pt and Re completely reduced to the metal state[123–126] and modified with trace sulfur. A fraction of rhenium is alloyed with platinum as bimetallic clusters, and this fraction may vary according to the conditions of preparation. Industrial catalysts have utilized mole ratios of Re/Pt in all proportions from >1 to <1. Equal molar is another well-represented ratio. The reduction of rhenium oxides is catalyzed by the presence of platinum. Moreover, water increases the mobility of Re oxide on the surface of alumina; hence, its movement toward neighboring platinum. This favors its reduction and the alloy formation.[127,128]

The widely used platinum precursor for the bimetallic catalyst is the $PtCl_6^{2-}$ ion. Two types of rhenium precursors can be distinguished: mineral compounds and organometallic compounds.

Mineral Compounds

The main mineral compounds reported in the literature are the heptaoxide, perrhenates (especially NH_4ReO_4), and some halides ($ReCl_3$). The heptaoxide, which is very soluble in an aqueous acidic solution, easily gives ReO_4^- anions, which can be fixed on the alumina surface.[129,130] Impregnation chemistry has not been as well studied as chloroplatinate.

Usually, alumina is impregnated with an aqueous solution containing H_2PtCl_6, $HReO_4$, and an acid such as HCl. A mathematical model based on diffusion and adsorption phenomena and successfully predicting the radial profiles of Pt and Re has been proposed by Ardiles et al.[131] Since both precursors show different affinities for the support, it seems rather difficult to obtain similar distribution profiles of metals even with adjusted conditions.[131–133] This is probably why successive impregnations of the two metals are claimed in some patents.[134,135]

After impregnation of the precursors, the impregnated supports are thermally treated. The reactions occurring during the thermal treatment are complicated and poorly understood. Besides the removal of impregnation solvent from the pores of the support, precursors are transformed into oxidized species and then reduced to a highly dispersed bimetallic phase. During these treatments, it is important to control the mobility of Re species on the support surface.

An important feature of bulk rhenium precursors or oxides is their high volatility: the volatilization, which begins at a lower temperature (near 100°C), is complete above 400–450°C. The volatility is decreased considerably by interaction with the alumina support. Therefore, the oxidation temperature of the impregnated catalyst can be adjusted so as to decompose the precursor while preventing the volatilization of rhenium.[136] Thermal oxidation enhances the dehydration of the alumina surface, leading to more potential interaction between Re_2O_7 and the support.[127] A more stable Re precursor requires a higher temperature for its volatilization and/or decomposition, and this leads to an increased dehydroxylation level of the alumina with greater interaction between rhenium and alumina. According to Reyes et al.,[137] ammonium perrhenate, which exhibits the highest stability in oxygen, seems to be the best-suited precursor. A strong interaction between Re_2O_7 and the support limits the volatization of rhenium oxides and the mobility of rhenium species on the support. As mentioned above, such mobility is needed to favor the platinum–rhenium interaction.[128] Moisture is thus expected to be an important parameter. The oxidation step for these rhenium precursors has only a slight effect on the macroscopic distribution, which is determined during impregnation, but has a strong effect on the "bimetallic nature" of the resulting sites.

Organometallic Compounds

The use of $Re_2(CO)_{10}$ dirhenium decacarbonyl as an organometallic precursor has been reported in numerous papers and patents.[131,138] Techniques using such a precursor require either impregnation with an organic solvent or sublimation in an inert gas in the 100–200°C range. This compound can interact either with the alumina support or with the supported platinum. The interaction with alumina is accompanied by a partial decarbonylation, and ligands exchange with the surface of the support.[138,139] However, the affinity of rhenium carbonyl is higher for reduced platinum. This limits the fixation of the precursor on the support sites and favors the formation of an alloy.[140,141] The total decomposition step is performed in an inert gas in order to avoid platinum oxidation.[140] The decomposition can be complicated by the presence of hydroxyl groups, which can lead to oxidized surface complexes.[142] Other types of Re organometallic compounds have been decomposed on reduced platinum. From an industrial point of view, considering the problem of handling and maintaining reduced catalysts and of using nonaqueous solvents, this kind of precursor is less attractive than water-soluble mineral precursors.

4.2 Platinum–Tin

The appearance of platinum–tin catalysts in reforming began in the late 1960s with the first patent claiming the use of this type of catalyst for dehydrocyclization reactions.[143] Compared with monometallic systems, tin increases the selectivity and stability of the catalyst. Moreover, resistance to agglomeration of the bimetallic catalysts during coke combustion has increased compared with the corresponding monometallic systems.[144]

From an industrial point of view, the platinum and tin salts used most frequently as precursors are likely to be the chloro derivatives such as H_2PtCl_6, $SnCl_2$, and $SnCl_4$. The tin and platinum contents of industrial catalytic systems are always less than 0.8 wt %.

After reduction at temperatures higher than 400°C, it is generally accepted that platinum is in a metallic state. The oxidation state of tin is still subject to discussion. Generally, tin is found as Sn(II) together with a more or less important fraction of metallic tin. The ratio of these two tin species depends on the catalyst preparation conditions, analysis techniques used, and treatments undergone by the sample before analysis. This parameter is therefore not treated in this chapter. Further information can be found in Chapter 6 on characterization.

In contrast to rhenium oxide, the high interaction between tin oxide and alumina does not favor mobility of tin on the surface at temperatures achieved during oxidation treatments. This is why the interaction between platinum and tin in platinum–tin bimetallic catalysts must take place during the impregnation step.

Alumina-Supported Tin Oxide Preparation

Sol-Gel Techniques. Techniques using the sol-gel transition have been used to prepare reforming catalyst supports containing tin. These techniques consist of introducing a tin salt directly in the alumina sol. In a general way this preparation is performed in an aqueous environment by introducing stannic chloride in the sol obtained by hydrocloric acid attack of aluminum.[145] After calcination, tin is present only in its oxidized state (stannic oxide-like compound). No tin chloride species can be detected.[146] This technique has been used with organometallic complexes (tetrabutyltin and tributoxide aluminum) in an alcoholic medium.[147] With this technique, tin is uniformly distributed on the support, and tin oxide particles larger than 10 nm cannot be detected.[148] Such a distribution of tin seems important to ensure good performance of the final catalyst.[145,148]

Impregnation of Tin Salts. Although impregnation of tin salts (most often $SnCl_2$ or $SnCl_4$) is often the first step in preparing a reforming catalyst,[149] the interaction between tin salts and an alumina surface has not been well studied in the literature. The surface complexes formed during the impregnation of stannic chloride on alumina have been studied by Li et al.[150] They showed that whatever the impregnation medium of tin may be (water, alcohols, ketones), the surface species formed are the same. Spectroscopic data can be interpreted in terms of tin complexation by a surface hydroxyl.

However, for stannous chloride impregnation on nonchlorided alumina, the mechanism proposed by Kuznetsov et al.,[151] and confirmed by Homs et al. on silica,[152] occurs via the substitution of a chloride ligand by the oxygen of an OH group, much like the ligand substitution observed with chloroplatinate. On chlorided alumina, stannous chloride is only weakly bonded on the support surface probably through interaction with surface chloride. The tin–support interaction thus obtained should be weaker than the one obtained on nonchlorided alumina in which a chloride ligand is lost by tin. During the drying that follows impregnation, oxidation of tin(II) to tin(IV) is only partial.[146]

In all the cases described, calcination at a high temperature (500°C) results in the formation of supported tin(IV) as SnO_2-like compounds, and partially decomposed SnO_xCl_{4-x}. Contrary to the systems obtained with a sol-gel preparation, tin can keep chloride ligands even after calcination.[146] Moreover, SnO_2 crystallites cannot be detected after calcination.

Reduction at around 500°C produces supported tin(II). The nature of this tin species has not yet been clearly described in the literature (SnO-like compounds, surface tin aluminate, etc.). Except for high tin contents ($>$10 wt %), the reduction of alumina-supported tin oxide never gives metallic tin.

Impregnation of Platinum Derivatives on Alumina-Supported
Tin Oxide

When tin on alumina is obtained by a first impregnation of tin chloride without
subsequent oxidation, part of the tin, which is weakly linked on the support, can
be dissolved during the subsequent impregnation in the chloroplatinic acid
solution. In this case, the impregnation of H_2PtCl_6 can be described as an
impregnation of a mixture of tin and platinum salts.

If the impregnation of platinum salts is performed on an oxidized tin on
alumina, two kinds of reaction can occur. Chloroplatinic acid can react with
alumina as described in Sec. 3.3 or with supported tin oxide. The reaction of
chloroplatinic acid with the tin oxide surface has been described.[153] As far as
Sn-Pt interaction is concerned, hydroxyl groups in the coordination sphere of the
hydrated platinum salt would be displaced by an oxygen from a surface Sn-OH
group:

$$\rightarrow SnOH + Pt(OH)_nCl_{6-n}^{2-} \rightarrow \quad \rightarrow SnOPt(OH)_{n-1}Cl_{6-n}^{2-} + H_2O$$

The surface complexes can then be decomposed by hydroxylation and loss of
chloride. It is obvious that in order to increase the interaction between tin and
platinum during impregnation, the anchoring of the platinum complex must be
promoted on tin oxide sites rather than on alumina.

Coimpregnation

The coimpregnation technique offers the advantage of a reduced number of steps
in comparison with successive impregnation techniques.[157] During the
dissolution of stannous chloride and chloroplatinic acid, the formation of
bimetallic complexes has been observed.[55] Baronettí et al.[159] have shown that
the first step of the reaction is a reduction of platinum(IV) by tin(II):

$$PtCl_6^{2-} + SnCl_3^- + Cl^- \rightarrow PtCl_4^{2-} + SnCl_6^{2-}$$

For tin/platinum ratios greater than 1, the first reaction is followed by the
formation of bimetallic complexes:[159–162]

$$PtCl_4^{2-} + SnCl_3^- \rightarrow PtCl_3(SnCl_3)^{2-} + Cl^-$$

$$PtCl_3(SnCl_3)^{2-} + SnCl_3^- \rightarrow PtCl_2(SnCl_3)_2^{2-} + Cl^-$$

These complexes are generally stable in an acidic environment in an inert
atmosphere. During impregnation of such complexes on nonchlorided alumina,
precipitation of platinum can occur. This precipitation is due to adsorption of
chloride on the support, leading to destruction of the bimetallic complex.
Ensuring the stability of the bimetallic complex during all the impregnation steps

is important in preserving the platinum–tin interaction. For example, increasing the chloride concentration of the impregnation solution or using a prechlorided alumina can prevent platinum precipitation. After drying and oxidation, this preparation method leads to catalytic systems with a better platinum–tin interaction compared with successive impregnations of tin and platinum salts.[144,163–166]

Impregnation of Tin Derivatives on Alumina-Supported Platinum

Hydrolysis of tin(IV) organometallic complexes on supported metals has been studied.[154] This method leads to different kinds of bimetallic catalysts and especially alloy formation.[155] Impregnation of tetraethyltin, in argon, on a reduced alumina-supported platinum precursor gives a selective deposit of tin on the support.[156] On the other hand, if impregnation of the tin complex is performed in hydrogen, tetraethyltin reacts preferentially on platinum particles leading to a metal surface complex:[154,156]

$$Pt - H + Sn(C_2H_5)_4 \rightarrow PtSn(C_2H_5)_3 + C_2H_6$$

At high temperature the surface complex is decomposed, resulting in a supported platinum–tin alloy.[154,156] These catalyst preparation techniques are not practiced commercially.

4.3 Platinum–Iridium

A bimetallic catalyst of some industrial importance in the past, although less frequently utilized now, is platinum–iridium on chlorided alumina, which appeared in the early 1970s.[167] It can be manufactured from chloroplatinic and chloroiridic acids, either by two successive impregnations,[167] where the platinum is the first metal introduced, or by coimpregnation. The respective behaviors of platinum and iridium in reducing and oxidizing atmospheres are so different that there is doubt about easily maintaining intimate contact between both metals once it has been obtained. Foger and Jaeger[168] concluded that obtaining a single-phase Pt-Ir alloy is possible only if the concentrations of Pt and Ir are nearly equal. After oxidation below 300°C, iridium alone is transformed to oxide, and above this temperature iridium oxide crystals segregate. Above 550°C, IrO_2 is transported through the gas phase. Such results are completely in accordance with those obtained by Garten and Sinfelt.[169] Highly dispersed bimetallic clusters are obtained through coimpregnation of chloroiridic and chloroplatinic acids if the exposure to air is maintained below 375°C. Around 600°C, large crystallites of iridium oxide are formed, and after reduction the catalyst consists of highly dispersed platinum or platinum-rich clusters and of large iridium or iridium-rich crystallites. Sequential impregnation of platinum and then iridium gives exactly the same results. These considerations emphasize the importance of the physical treatment subsequent to the catalyst impregnation.

Huang et al.[170] concluded that incorporation of platinum into iridium clusters retards the oxidative agglomeration of iridium. When bimetallic catalysts were oxidized at 320°C, the majority of the surface species were bimetallic Pt-Ir oxychlorides; no significant IrO_2 agglomerates were observed. A few patents claim improved results with the addition to the Pt-Ir couple of a third element capable of promoting a significant change in the metal–support interaction: chromium oxide[171,172] and silicon, calcium, magnesium, barium, or strontium.[173] Other techniques and chemistries have been employed. Further details may be found in Chapter 4 of the first edition of this book.[6]

4.4 Platinum–Germanium

In general, germanium compounds have chemistries very similar to those of tin. Similar methods of preparation as employed for platinum–tin bimetallic catalysts would be anticipated. The cost of germanium compounds is however substantially higher than that of tin. The preceding consideration on the preparation of bimetallic platinum–tin catalysts with organotin compounds are applicable to germanium. Among the organogermanium compounds that may be employed are tetrabutyl or tetramethylgermane. Some commercial catalysts are cited to contain platinum and germanium, but no precise information is obtainable from the suppliers on content or methods of preparation.

A patent issued to UOP cites coimpregnation of alumina with a mixed solution of chloroplatinic acid and germanium dioxide.[174] Obtaining a homogeneous solution of both compounds requires selection of the hexagonal germanium dioxide, the only soluble form. To improve the solubility, the solution is prepared with hot water immediately before impregnation.

Other multimetallic catalysts containing platinum and germanium have been claimed to have substantially higher selectivities than platinum and platinum–rhenium catalysts. For example, Antos[173] has shown a drastic increase in the hydrogen purity and the gasoline yield with a catalyst containing platinum, rhenium, and germanium. The source of rhenium and germanium is a mixed carbonyl species: $[ClGeRe(CO)_5]_3$. A drawback of this preparation is that it requires the use of anhydrous organic solvents, with problems as noted in the section on using carbonyls to prepare bimetallic Pt/Re catalysts.

4.5 Other Bimetallic and Multimetallic Catalysts

Some of the patent literature on bimetallic reforming catalysts relates extensive examples of metallic formulations, but, to our knowledge, none have really been utilized commercially. The same may be said for the multimetallic combinations of elements. Chapter 4 in the first edition of this book[6] describes some of the earlier experimental formulations. It is to be understood that very high precursor costs and unusual (i.e., nonaqueous) impregnation techniques pose high barriers in terms of capital investment in catalyst production equipment and high catalyst cost. Any increased value of products must be able to justify the higher catalyst prices that would result.

With this stated, two observations may be added. First, there are certain of the multimetallic combinations that would not be expected to be significantly higher in cost. There are those that employ the already-used components of the bimetallic catalysts—platinum, rhenium, tin, and germanium. In certain applications, or given certain conditions, an advantage will be present for tri- and even tetrametallic combinations. Since the catalyst costs are somewhat contained—known precursors, known chemistries—various refiners will be able to increase profitability through the use of these multimetallic combinations. The second observation is that multimetallics will be one of the advances for the future of naphtha reforming. Bimetallic combinations usually offer improved performance in only one aspect, such as gasoline yield, or activity. It is easy to conceive that with multiple components, new catalyst formulations will result that offer improvements in both selectivity and activity. Further discussions of new catalyst advances may be found in Chapter 8 of this book.

5 METAL PROFILES IN CATALYST PELLETS

Another consideration into which the fundamental focus of this chapter penetrates is the genesis of metal profiles in formed catalyst particles such as spheres or extrudates. Again, while a comprehensive review of the subject cannot be accomplished, some intriguing results consistent with the recent electrostatic interpretation of adsorption will be presented.

The macroscopic distribution of Pt in formed catalysts has been in the catalysis literature since at least the 1950s, with the work of Maatman and Prater.[175,176] A classic work is that of Lee and Aris,[177] who modeled many cases of this diffusion–adsorption process. A comprehensive semitheoretical and experimental survey of this arena was accomplished by the group of Schwarz.[95,178–181] The group of Varma has been particularly central in controlling metal profiles within pellets to achieve optimal selectivity and activity; their work involves both experimental and theoretical components (see their review article[182] as well as references within and thereafter). A recent article by Khinast and coworkers pertains to the impact of drying on metal movement and also contains a good review of the literature.[183]

A central theme in the diffusion–adsorption literature, dating from the work of Maatman,[90,175] is the role of competitor species in obtaining Pt profiles that are uniform throughout the pellet (as opposed to surface-deposited "eggshell", annulus-deposited "egg white", or center-deposited "egg yolk" distributions). It is widely recognized that uniform distributions require the addition of an excess acid such as hydrochloric or citric acid to the CPA. Maatman proposed that the role of the acid is to provide competitor ions such as Cl^-, which forces the chloroplatinate ions to move deeper into the pellet in order to adsorb.[90] This explanation has received widespread support with several exceptions. Heise and Schwarz identified high ionic strength as a key cause of adsorption retardation.[178] Ruckenstein and Karpe[184] also demonstrated that electrostatic effects contribute significantly to the evolution of metal profiles in pellets.

Further indications have appeared that support the electrostatic interpretation and not physical competition. First, as mentioned several times earlier in this chapter, the uptake of Cl^- in the regime where Pt adsorption is retarded has been measured directly;[175] these results are shown in Figure 29. This pH range corresponds to the low end of Figure 9; virtually no uptake of Cl^- is seen as the Pt adsorption drops to 40% of monolayer coverage. The electrostatic interpretation is that high ionic strength effectively diminishes the adsorption equilibrium constant;[74,79,178] rather than chloride-filled, as assumed by most "chemical" adsorption models, the surface is empty.

The relevance of these considerations to industrial catalyst preparation was suggested in Regalbuto's comprehensive adsorption model, which is shown in Figure 30.[79] In this figure the uptake of CPA is plotted as function of CPA

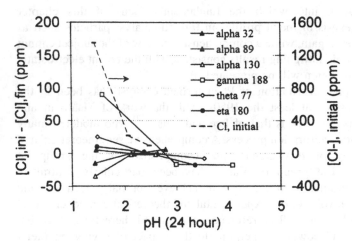

Figure 29   Chloride uptake as a function of pH in the regime of Pt adsorption.[75]

concentration and either equilibrium (Fig. 30a) or final (Fig. 30c) pH. The final pH is plotted as a function of initial pH and metal concentration (Fig. 30b), and the uptake efficiency, or fraction of Pt that is adsorbed, is plotted in Figure 30d. At the high ratio of solid to solution typically employed by industry, and the low weight loadings of Pt employed, surface coverage of Pt is always quite low. The most telling features of this calculation is that when no additional acid is added to the impregnating solution, the amount of alumina surface is sufficiently in excess of protons in solution that the surface is never significantly charged. This is manifested by a final pH value (Fig. 30b) near the alumina PZC. The uptake efficiency for the CPA-only case is low (Fig. 30d). In order to achieve high uptake efficiency through high surface charge (and a corresponding pH appreciably below the PZC; Fig. 30b), an acid such as HCl must be added. It is the proton, however, and not the anion that is the important species in maximizing uptake.

To investigate the effects of diffusion during the impregnation of formed catalysts, a comparison of proton and Pt uptake has been recently made between alumina catalyst spheres and powder made from crushed spheres.[185] The uptake-pH results, in a format similar to Figure 9, are shown in Figure 31a for the spheres and Figure 31b for the crushed spheres. The crushed material behaves as the powdered materials seen in Figure 9. Uptake is essentially complete after 1 h in the mid-to-low pH range. For the spheres, however, practically no Pt uptake was observed in this time frame. The metal profiles at 1 and 48 h were obtained by analytical electron microscopy for the samples impregnated at initial pHs of 1.4 (excess HCl) and 2.5 (CPA only); these results are shown in Figure 32. The depth of 800 μm is the center of the sphere. At 1 h, it is the sample prepared in excess HCl that shows a more homogeneous distribution (diamond symbols, solid line), and this changes relatively little in time (diamonds, dashed line). On the other hand, the CPA-only solution begins with a sharp edge distribution at 1 h, but over time distributes more evenly and attains a much higher loading than does the pH 1.4 sample. The equilibrium metal loading is in fact in agreement with the powder uptake–pH plots seen in Figures 9 and 31. The difference in this behavior can once again be explained on the basis of different proton concentrations; at pH 1.4 the proton concentration was sufficiently high to quickly charge the interior of the particle, whereas at the higher initial pH the interior of the sphere took much longer to charge.

6 CONCLUSION

The maturity of the catalytic reforming process and the catalysts utilized in the process is a factor in the information presented in this chapter. Platinum continues to be the workhorse of reforming catalysts. The primary promoters

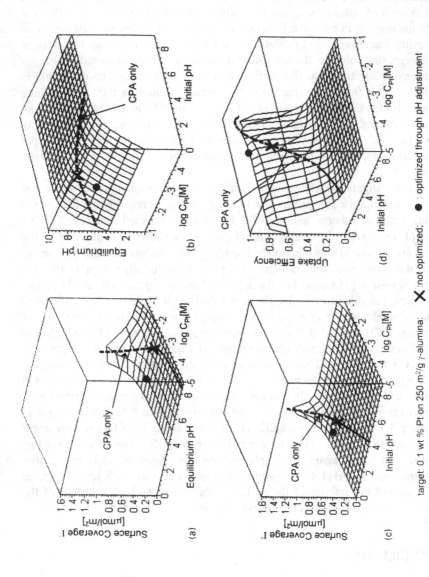

Figure 30 Comprehensive model of CPA uptake onto alumina at high surface loadings. [79]

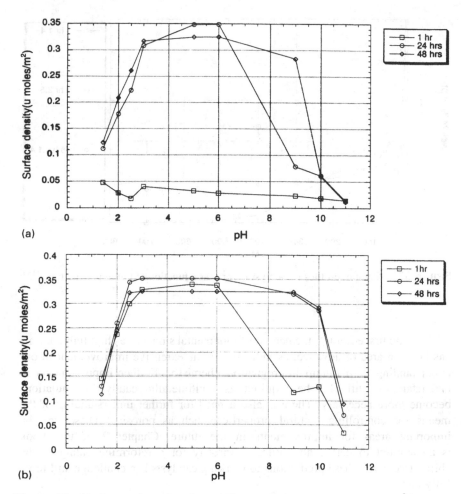

Figure 31 Platinum adsorption from 180 ppm CPA solutions over 500 m²/liter γ-alumina as a function of the solution pH and time, for (a) alumina spheres and (b) crushed alumina spheres.[185]

continue to be tin, rhenium, and germanium. Therefore, the preparation chemistries remain similar to those exemplified in the first edition. Alumina remains the support of choice. Catalyst advances have been made in the intervening time. These have encompassed modifications in the physical and textural aspects of the catalyst support. There have been some changes in metal combinations that have been employed. These changes have probably not required changes in impregnation technology.

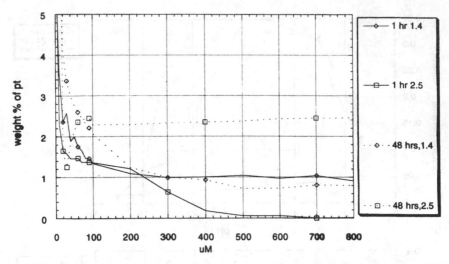

Figure 32   Platinum profiles in catalyst spheres as a function of initial pH and time.[185]

In the first edition[6] the need for fundamental studies was highlighted. This has been an area of advance, as indicated in the extensive improvement in our understanding of platinum impregnation chemistry. Further improvements in understanding will also be required as multimetallic catalyst formulations become more necessary. There is also a need for further understanding of the means for controlling textural properties, such as porosity. These will be important areas for improvements in the future. Chapter 7 of this book is an attempt to derive an optimum porosity for a reforming catalyst. The ability to convert these predictions to working catalysts is a challenge still to be fully met.

REFERENCES

1.   Bernard, J.R. *Proc. 5th Int. Zeolite Conf.*, Heyden, London, 1980; 686–695.
2.   Bezuhanova. C.; Guidot, J.; Barthomeuf, D.; Breysse, M.; Bernard, J.R. J. Chem. Soc. Faraday Trans. 1 **1980**, *77*, 1595–1604.
3.   Hughes, T.R.; Buss, W.C.; Tamm, P.W.; Jacobson, R.L. Stud. Surf. Sci. Catal. In *New Developments in Zeolite Science and Technology*; Murakami, Y., Lijima, A., Ward, J.W., Eds.; 1986; Vol. 28, 725–732.
4.   Tamm, P.W.; Mohr, D.H.; Wilson, C.R. Stud. Surf. Sci. Catal. In *Catalysis 38*; Ward, J.W., Ed.; 1988; 335–353.

5. Kao, J.L.; McVicker, G.B.; Treacy, M.M.J.; Rice, S.B.; Robbins, J.L.; Gates, W.E.; Ziemiak, J.J.; Cross, V.R.; Vanderspurt, T.H. *Proc. 10th Int. Congr. Catal.*, Budapest, 1993; 1019–1028.

6. Bortiaux, J.P. et al. In *Catalytic Naphtha Reforming, Science & Technology*; Antos, G.J., Aitani, A.M., Parera, J.M., Eds.; Marcel Dekker: New York, 1995; 79–111.

7. Little, D.M. *Catalytic Reforming*; Pennwell Publishing Co.: Oklahoma, 1985.

8. Haensel, V. U.S. Patent, 2,611,736, UOP, 1952.

9. Haensel, V. U.S. Patent, 2,623,860, UOP. 1952.

10. Poisson, R.; Brunelle, J.P.; Nortier, P. *Catalyst Supports and Supported Catalysts*; Stiles, A.B., Ed.; Butterworth: Boston, 1987; 11–55.

11. Lippens, B.C. University of Delft: Netherlands, 1961; Ph.D. Thesis.

12. Lippens, B.C.; Steggerda, J.J. In *Physical and Chemical Aspects of Adsorbents and Catalysis*; Linsen, B.G., Ed.; Academic Press: New York, 1970; Chap. 4.

13. Lippens, B.C. Chem. Week B1 **1966**, *6*, 336.

14. Cocke, D.C.; Johnson, E.D.; Merrill, R.P. Catal. Rev. Sci. Eng. **1984**, *26*, 163–231.

15. Lostaglio, V.J.; Carruthers, J.D. Chem. Eng. Prog. **1986**, *82*, 46–51.

16. Rhône-Poulenc Documentations: Spheralite, Catalyst Carriers from Rhône-Poulenc and Activated Alumina.

17. Montarnal, R. I.F.P. Internal Report 21074; 1973.

18. Brett, N.H.; Mackenzie, D.K.J.; Sharp, J.H. Q. Rev. **1970**, *24*, 185–207.

19. Rosinski, E.J.; Stein, T.R.; Fischer, R.H. U.S. Patent 3,876,523, Mobil, 1975.

20. Vishnyakova, G.P.; Dzis'ko, V.A.; Kefeli, L.M.; Kokotko, L.F.; Olen'kova, I.P.; Plyasova, L.M.; Ryzhak, I.A.; Tikhova, A.S. Kinet. Katal. **1970**, *11*, 1287–1292.

21. Nortier, P.; Fourre, P.; Mohammed Saad, A.B.; Saur, O.; Lavalley, J.C. Appl. Catal. **1990**, *61*, 141–160.

22. Leonard, A.J.; Van Cuwelaert, F ; Fripiat. J.J. J. Phys. Chem. **1967**, *71*, 695–708.

23. Gates, B.C.; Katzer, J.R.; Schuit, A.G.C. *Chemistry of Catalytic Processes*; McGraw-Hill: New York, 1979.

24. Papee, D.; Tertian, R. J. Chim. Phys. **1958**, *341*.

25. Knözinger, H.; Ratnasamy, P. Catal. Rev. Sci. Eng. **1978**, *17*, 31–70.

26. Peri. J.B. J. Phys. Chem. **1965**, *69*, 211–219; J. Phys. Chem. **1965**, *69*, 220–230; J. Phys. **1965**, *69*, 231–239.

27. Knözinger, H. Adv. Catal. **1976**, *25*, 184–271.

28. Vit, Z.; Vala, J.; Malek, J. J. Appl. Catal. **1983**, *7*, 159–168.

29. Webb, A.N. I.E.C. **1957**, *49*, 261–263.

30. Boehm, H.P. Adv. Catal. **1966**, *16*, 179–274.

31. Berteau, P.; Delmon, B. Catal. Today **1989**, *5*, 121–137.

32. Chapman, D.; Hair, M.L. J. Catal. **1963**, *2*, 145–148.

33. Antipina, T.V.; Bulgarov, O.V.; Uvarov, A.V. *Proc. 4th Int. Congr. Catal.*, Moscow, 1968; 376–387.

34. Ballou, E.V.; Barth, R.T.; Flint, R.T. J. Phys. Chem. **1961**, *65*, 1639–1641.

35. Mirschler, A.E. J. Catal. **1963**, *2*, 428–439.

36. Chernov, V.A. and Antipina. T.V., *Kinet. Katal. 7*, 651–653 1966.

37. Holm, V.C.F.; Clark, A. Ind. Eng. Chem. **1963**, *2*, 38–39.

38. He, J.; Ai, J.; Wu, K.; Luo, X. *5th Natl Conf. Petrol. Petrochem.*, Shandong, China, 1989.
39. Tanaka, N.; Ogasawara, S. J. Catal. **1970**, *16*, 157–163; J. Catal. **1970**, *16*, 164–172.
40. Basset, J.M. University of Lyon, 1969; Ph.D. Thesis.
41. Garbowski, E.; Candy, J.P.; Primet, M. J. Chem. Soc. Faraday Trans. I **1983**, *79*, 835–844.
42. Melchor, A.; Garbowski, E.; Mathieu, M.; Primet, M. J. Chem. Soc. Faraday Trans I **1986**, *82*, 1893–1901.
43. Roumegous, A. University of Paris VI, 1978; Ph.D. Thesis.
44. McClung, R.G.; Sopko, J.S.; Kramer, R.; Casey, D.G. NPRA Annual Meeting, San Antonio, 1990.
45. Goble, A.; Lawrence, P.A. Proc. 3rd Int. Congr. Catal., Amsterdam, 1964; 320–324.
46. Brunelle, J.P.; Poisson, R. *Matériaux de l'Avenir*; Rhône-Poulenc, 1991.
47. Hoekstra, J. U.S. Patent 2,620,314, UOP, 1952.
48. Hoekstra, J. U.S. Patent, 2,666,749, UOP, 1954.
49. Wankat, C. U.S. Patent 2,672,453, UOP, 1954.
50. Hoekstra, J. U.S. Patent, 2,774,743, UOP, 1956.
51. Schoonover, M.W. U.S. Patent 4,318,896, UOP, 1982.
52. Wankat, C. U.S. Patent 2,733,220, UOP, 1956.
53. Michalko, E. U.S. Patent 3,027,232, UOP, 1962.
54. Hayes, J.C. U.S. Patent 3,887,492, UOP, 1975.
55. Hayes, J.C. U.S. Patent 3,887,493, UOP, 1975.
56. LePage, J.F. *Applied Heterogeneous Catalysis*, Technip Ed.; IFP Publications: Paris, 1987.
57. Moehl, R.W., U.S. Patent 2,759,898, UOP, 1956.
58. Murray, M.J. U.S. Patent 2,736,713, UOP, 1956.
59. Vesely, K.D. U.S. Patent 3,496,115, UOP, 1970.
60. Michalko, E. U.S. Patent 4,216,122, UOP, 1980.
61. Trimm, D.L.; Stanislaus, A. Appl. Catal. **1986**, *21*, 215–238.
62. Brunelle, J.P.; Sugier, A. Compt. Rend. Acad. Sci. Serie C **1973**, *276*, 1545–1548.
63. Dorling, T.A.; Lynch, J.B.W.; Moss, R.L. J. Catal. **1971**, *20*, 190.
64. Marcilly, C.; Franck, J.P. Rev. Inst. Fr. Petr. **1984**, *39*, 337–364.
65. Maatman, R.W.; Prater, C.D. Ind. Eng. Chem. **1957**, *49*, 253–257.
66. Neimark, A.V., Kheifez, L.I.; Fenelonov, V.B. Ind. Eng. Chem. Process Des. Dev. **1981**, *20*, 439–450.
67. Weisz, P.B. Trans. Faraday Soc. **1967**, *63*, 1801–1806.
68. Weisz, P.B.; Hicks, J.S. Trans. Faraday Soc. **1967**, *63*, 1807–1814.
69. Weisz, P.B.; Zollinger, H. Trans. Faraday Soc. **1967**, *63*, 1815–1823.
70. Healy, T.W.; White, L.R. Adv. Coll. Interf. Sci. **1978**, *9*, 303.
71. James, R.O.; Parks, G.A. Surf. Coll. Sci. **1982**, *12* 119.
72. Jagiello, Contescu and Schwarz.
73. Parks, G.A. Chem. Rev. **1965**, *65*, 177.
74. Park, J.; Regalbuto, J.R. J. Colloid Interf. Sci. **1995**, *175* 239–252.
75. Regalbuto, J.R.; Navada, A.; Shadid, S.; Bricker, M.L.; Chen, Q. J. Catal. **1999**, *184*, 335.

76. Sverjensky, D.A.; Sahai, N. Geochimica et Cosmochimica Acta **1996**, *60*, 3773.
77. Boehm, H.P.; Knozinger, H. In *Nature and Estimate of Functional Groups on Solid Surfaces*; 1982; Chap. 2.
78. Noh, J.S.; Schwarz, J.A. J. Colloid. Interf. Sci. **1989**, 130, 157.
79. Spieker, W.A.; Regalbuto, J.R. Chem. Eng. Sci. **2001**, *56*, 3491.
80. Brunelle, J.P. Pure Appl. Chem. **1978**, *50*, 1211.
81. Contescu, C.; Vass, M.I. Appl. Catal. **1987**, *33*, 259.
82. James, R.O.; Healy, T.W. J. Coll. Interf. Sci. **1972**, *40*, 65.
83. Benesi, H.A.; Curtis, R.M.; Studer, H.P. J. Catal. **1968**, *10*, 328.
84. Agashe, K.B.; Regalbuto, J.R. J. Coll. Interf. Sci. **1997**, *185* 174.
85. Hao, X.; Spieker, W.A.; Regalbuto, J.R. J. Colloid. Interf. Sci., **2003**, *267*, 259.
86. Regalbuto, J.R. et al. Stud. Surf. Sci. Catal. **1998**, *118*, 147.
87. Santhanam, N.; Conforti, T.A.; Spieker, W.A.; Regalbuto, J.R. Catal. Today **1994**, *21*, 141–156.
88. Schreier, M.; Regalbuto, J.R. Manuscript in preparation.
89. Mang, T.; Breitscheidel, B.; Polanek, P.; Knözinger, H. Appl. Catal. A Gen. **1993**, *106*, 239–258.
90. Maatman, R.W. Ind. Eng. Chem. **1959**, *51*, 913.
91. Shyr, Y.; Ernst, W. J. Catal. **1980**, *63*, 425.
92. Papageorgiou, P. et al. J. Catal. **1996**, *158*, 439.
93. Jianguo, W.; Jiayu, Z.; Li, P. *Catalysts III*; Poncelet, K G., Grange, P., Jacobs, P.A., Eds.; Elsevier, Amsterdam, 1983; 57 pp.
94. Santacessaria, E.; Carra, S.; Adami, I. Ind. Eng. Chem. Prod. Res. Dev. **1977**, *16*, 41.
95. Heise, F.J.; Schwarz, J.A. J. Coll. Interf. Sci. **1990**, *135*, 51.
96. Xidong, W.; Yongnian, Y.; Jiayu, Z. Appl. Catal. **1988**, *40*, 291.
97. Paulhiac, J.L.; Clause, O. J. Am. Chem. Soc. **1993**, *115*, 11602.
98. Carrier, X.; Lambert, J.F.; Che, M. J. Am. Chem. Soc. **1997**, *119*, 10137.
99. Carrier, X.; Lambert, J.F.; Che, M. Stud. Surf. Sci. Catal. **2000**, *130*, 1049.
100. Davis, J.A.; James, R.O.; Lechkie, J.O. J. Coll. Surf. Sci. **1978**, *63*, 480.
101. Shelimov, B. et al. J. Am. Chem. Soc. **1999**, *121*, 545.
102. Shelimov, B. et al. J. Catal. **1999**, *185*, 462.
103. Shelimov, B. et al. J. Mol. Catal. **2000**, *158*, 91.
104. Sillen, L.G.; Martell, A.E. *The Stability Constants of Metal Ion Complexes*; Special Publication No. 25; The Chemical Society: Burlington House, London, 1971; Suppl. No. 1.
105. Cox, L.E.; Peters, D.E. Inorg. Chem. **1970**, *9*, 1927.
106. Davidson, C.M.; Jameson, R.F. Trans. Farad. Soc. **1965**, *61*, 2462.
107. Gmelin, L.; Meyer, R.J. Gmelins Handbuch der anorganishen Chemie, Platin, Teil C Lieferung 1, 8. Auflage, Verlag Chemie: Weinheim, 1939.
108. Miolati, A.; Pendini, U. Z. Anorg. Chem. **1903**, *33*, 251.
109. Spieker, W.A. et al. Appl. Catal. A: Gen **2002**, *232*, 219.
110. Spieker, W.A. et al. Appl. Catal. A: Gen **2003**, *243*, 253.
111. Breck, D.W. *Zeolite Molecular Sieves. Structure, Chemistry and Use*; Wiley, New York, 1974.

112. Ertl., G.; Koningsberger, D.; Weitkamp, J. *Preparation of Solid Catalysts*; Wiley-VCH: Weinheim, Germany, 1999.
113. Spieker, W.A. University of Illinois at Chicago, 2001; Ph.D. Dissertation.
114. Dyer, A. *An Introduction to Zeolite Molecular Sievers*; Wiley: Chichester, 1988.
115. McNicol, D. J. Catal. **1977**, *46*, 438.
116. Lieske, H.; Lietz, G.; Spindler, H.; Völter, J. J. Catal. **1983**, *81*, 8–16.
117. Berdala, J.; Freund, E.; Lynch, J. J. Phys. **1986**, *47*, 269–272.
118. Sivasanker, S.; Ramaswamy, A.V.; Ratnasamy, P. Stud. Surf. Sci Catal. *Preparation of Catalysts II*; Delmon, B., Grange, P., Jacobs, P.A., Poncelet, G., Eds.; 1979; Vol. 3, 185–196.
119. Lietz, G.; Lieske, H.; Spindler, H.; Hanke, W.; Völter, J. J. Catal. **1983**, *81*, 17–25.
120. Jacobson, R.L.; Kluksdahl, H.E.; McCoy, C.S.; Davis, R.W. Proc. Amer. Petrol. Inst. Div. Refining **1969**, *49*, 504.
121. Jacobson, R.L.; Kluksdahl, H.E.; Spurlock, B. U.S. Patent 3,434,960, Chevron, 1969.
122. Bertolacini, R.J.; Pellet. R.J. Stud. Surf. Sci. Catal. In *Catalyst Deactivation*; Delmon, B., Froment, G.F., Eds.; 1980; Vol. 6, 73–77.
123. Johnson, M.F.; Leroy, V.M. J. Catal. **1974**, *35*, 434–440.
124. McNicol, D. J. Catal. **1977**, *46*, 438.
125. Charcosset. French-Venezuelian Congress, Caracas, 1983.
126. Betizeau, C.; Bolivar, H.; Charcosset, R.; Frety, G.; Leclercq, R.; Maurel; Tournayan, L. Stud. Surf. Sci. Catal. In *Preparation of Catalysts I*; Delmon, B., Jacobs, P.A., Poncelet, G., Eds.; 1976; 525–536.
127. Ziemecki, S.B.; Jones, G.A.; Michel, J.B. J. Catal. **1986**, *99*, 207–217.
128. Wagstaff, N.; Prins, R. J. Catal. **1979**, *59*, 434–445.
129. Acres, K.B.J.; Bird, A.J.; Jenkins, J.W.; King, F. The design and preparation of supported catalysts. Spec. Period. Rep. Catal. **1981**, *4*, 1–30.
130. Pascal, P. *Chimie Minérale*; Masson: Paris, 1978; Vol. 10.
131. Ardiles, D.R.; De Miguel, S.R.; Castro, A.A.; Scelza, O.A. Appl. Catal. **1986**, *24*, 175–186.
132. Hegedus, L.L.; Chou, T.S.; Summers, J.C.; Poter, N.M. Stud. Surf. Sci. Catal. In *Preparation of Catalysts II*; Delmon, B., Grange, P., Jacobs, P.A., Poncelet, G., Eds.;1979; Vol. 3: 171–183.
133. De Miguel, S.R.; Scelza, O.A.; Castro, A.A.; Baronetti, G.T.; Ardiles, D.R.; Parera, J. Appl. Catal. **1984**, *9*, 309–315.
134. Hayes, J.C. U.S. Patent 3,775,301, UOP, 1973.
135. Kluksdahl, E. U.S. Patent 3,558,477, Chevron, 1968.
136. Bolivar, C.; Charcosset, H.; Frety, R.; Primet, M.; Tournayan, L.; Betizeau, C.; Leclercq, G.; Maurel, R. J. Catal. **1975**, *39*, 249–259.
137. Reyes, J.; Pecchi, G.; Reyes, P. J. Chem. Res. **1983**, 318–319.
138. Smith, A.K.; Theolier, A.; Basset, J.M.; Ugo, R.; Commereuc, D.; Chauvin, Y. J. Am. Chem. Soc. **1978**, *100*, 2590–2591.
139. Danilyuk, A.F.; Kuznetsov, V.L.; Shepelin, A.P.; Zhdan, P.A.; Maksimov, N.G.; Magomedov, G.I.; Ermakov, Y.I. Kinet. Katal. **1983**, *24*, 919–925.

140. Antos, G.J. U.S. Patent 4,136,017 and 4,159,939, UOP, 1979.
141. Bernard, J.R.; Breysse, M. French Patent 2479707, Elf, 1980.
142. Brenner, A.; Hucul, D.A. J. Catal. **1980**, *61*, 216–222.
143. Compagnie Francaise de Raffinage, French Patent 2031984, 1969.
144. Yining, F.; Jingling, Z.; Liwu, L. J. Catal. (Cuihua. Xuebao) **1989**, *10*, 111–117.
145. Rausch, R.E. U.S. Patent 3,745,112, UOP, 1973.
146. Li, Y.X.; Klabunde, K.J.; Davis, B.H. J. Catal. **1991**, *128*, 1–12.
147. Gomez, R.; Bertin, V.; Ramirez, M.; Zamudio, T.; Bosch, P.; Schifter, I.; Lopez, T. J. Non. Cryst. Solids **1992**, *147*, 748.
148. Chee, T.; Targes, W.M.; Moser, M.D. U.S. Patent 4,964,975, UOP, 1990.
149. Engelhard, P.; Szabo, G.; Weisang, J.E. U.S. Patent 4,039,477, CFR, 1977.
150. Li, Y.X.; Zhang, Y.F.; Klabunde, K.J. Langmuir **1988**, *4*, 385–391.
151. Kuznetsov, V.I.; Belyi, A.S.; Yurchenko, E.N., Smolikov, M.D.; Protasova, M.T.; Zatolokina, E.V.; Duplyakin, V.K. J. Catal. **1986**, *99*, 159–170.
152. Homs, N.; Clos, N.; Muller, G.; Sales, J.; Ramirez de la Piscina, P. J. Mol. Catal. **1992**, *74*, 401–408.
153. Cox, D.F.; Hoflund, G.B.; Laitinen, H.A. Langmuir **1985**, *1*, 269–273.
154. Travers, C.; Bournonville, J.P.; Martino, G. Proc. 6th Int. Congr. Catal., Berlin, 1984; 891–902.
155. Vértes, C.; Talas, E.; Czako-Nagy, I.; Ryczkowski, J.; Göbölös, S.; Vertes, A.; Margitfalvi, J. Appl. Catal. **1991**, *68*, 149–159.
156. Stytsenko, V.D.; Kovalenko, O.V.; Rozovski, A.Y. Kinet. Katal. **1989**, *32*, 163–169.
157. Kluksdahl, H.E.; Jacobson, R.L. French Patent 2076937, Chevron, 1971.
158. Young, J.F.; Gillard, R.D.; Wilkinson, G. J Chem. Soc. **1964**, 5176–5189.
159. Baronetti, G.; De Miguel, S.; Scelza, O.; Fritzler, A.; Castro, A. Appl. Catal. **1985**, *19*, 77–85.
160. Berndt, V.H.; Mehner, H.; Völter, J.; Meisel, W. Z. Anorg. Allg. Chem. **1978**, *429*, 47–58.
161. Yurchenko, E.N.; Kuznetsov, V.I.; Melnikova, V.P.; Sartsev, A.N. React. Kinet. Catal. Lett. **1983**, *23*, 113–117.
162. Jin. Appl. Catal. **1991**, *72*, 33–38.
163. Baronetti, G.; De Miguel, S.; Castro, A.; Scelza, O.; Castro, A. Appl. Catal. **1988**, *45*, 61–69.
164. Davis, B.H. Proc. 10th Int. Congr. Catal., Budapest, 1992; 889–897.
165. Sachdev, A.; Schwank, J. Proc. 9th Int. Congr. Catal., Calgary, 1988; 1275–1283.
166. Baronetti, G.; De Miguel, S.; Scelza, O.; Castro, A. Appl. Catal. **1986**, *24*, 109–116.
167. Buss, W.C. U.S. Patent 3,554,902, Chevron, 1971.
168. Foger, K.; Jaeger, H. J. Catal. **1981**, *70*, 53–71.
169. Garten, R.L.; Sinfelt, J.H. J. Catal. **1980**, *62*, 127–239.
170. Huang, Y.J.; Fung, S.C.; Gates, W.E.; Vicker, G.B.M. J. Catal. **1989**, *118*, 192–202.

171. Anders, K.; Becker, K.; Birke, P.; Engels, S.; Feldhaus, R.; Hager, W.; Lausch, H.; Mahlow, P.; Neubauer, H.D.; Sager, D.; Wilder, M.; Vieweg, H.G. DDR Patent 212 192, Leuna Werke, 1982.
172. Kresge, C.T.; Chester, A.W.; Oleck, S.M. Appl. Catal. **1992**, *81*, 215–226.
173. Antos, G.J. U.S. Patent 4,312,788, UOP, 1992.
174. McCallister, K.R.; O'Neal, T.P. French Patent 2078056, UOP, 1971.
175. Maatman, R.W. and Prater, C.D. Ind. Eng. Chem. **1957**, *49*, 253.
176. Maatman, R.W., Ind. Eng. Chem. **1959**, *51*, 913.
177. Lee, S.-Y.; Aris, R. Catal. Rev. Sci. Eng. **1985**, *27*, 207.
178. Heise, F.J.; Schwarz, J.A. J. Coll. Interf. Sci. **1985**, *107*, 237.
179. Heise, F.J.; Schwarz, J.A. J. Coll. Interf. Sci. **1986**, *113*, 55.
180. Heise, F.J.; Schwarz, J.A. J. Coll. Interf. Sci. **1988**, *123*, 51.
181. Heise, F.J.; Schwarz, J.A. J. Coll. Interf. Sci. **1990**, *135*, 461.
182. Gavriilidis, A.; Varma, A.; Morbidelli, M. Catal. Rev. Sci. Eng. **1993**, 399.
183. Azzeddine, L.; Glasser, B.J.; Khinast, J.G. Chem. Eng. Sci. **2001**, *56*, 4473.
184. Ruckenstein, E.; Karpe, P. Langmuir **1989**, *5*, 1393.
185. Regalbuto, J.R. Manuscript in preparation.

6
Characterization of Naphtha-Reforming Catalysts

Burtron H. Davis
University of Kentucky
Lexington, Kentucky, U.S.A.

George J. Antos
UOP, LLC
Des Plaines, Illinois, U.S.A.

1 INTRODUCTION

Naphtha reforming involves heterogeneous catalysis so that the catalyst constitutes a separate phase.[1] Furthermore, naphtha reforming occurs by bifunctional catalysis.[2] This means that for a Pt-Al$_2$O$_3$-reforming catalyst, some of the processes occur at the surface of platinum or other metal(s) and others at the acidic sites on the alumina or other support. For optimal performance, these two or more types of sites are intermixed on the same primary particles. Characterization of naphtha-reforming catalysts therefore presents many obstacles. One must be aware of the assumptions that enable one to convert the experimental data to conclusions that define the catalyst structure. Unfortunately, all too often the assumptions are overlooked in developing models of reforming catalysts.

At the most elementary level, characterization of a reforming catalyst involves only two topics: (1) a measure of the amount together with the strength and distribution of the acid function and (2) a measure of the amount and activity of the metallic function. Needless to say, topic 2 becomes more difficult when the catalyst contains two or more metallic components.

Two levels of characterization data may be distinguished. One type of data is needed to address engineering applications. Here one is concerned with a characterization of those features that will (1) permit the catalyst manufacturer to prepare repetitively a catalyst with the same properties and (2) provide the

199

process operator with the ability (a) to bring the catalyst on stream with the required activity and selectivity, (b) to monitor catalyst performance, and (c) to adjust the state of the working catalyst to maintain performance specifications for a long period of operation. The engineering approach requires only that the properties that are characterized be related to the performance of the catalyst; it does not have to provide an accurate measure of the absolute value of a particular feature of the catalyst. For example, if it is found that 20 ppm of chlorine in the exit gas provides the optimal activity and selectivity for naphtha reforming with a particular catalyst, the engineering method requires only a measure of the chlorine in the exit gas. It is not necessary to know the amount of chloride incorporated or its chemical state in the working catalyst. Characterizations for engineering purposes are essential for the successful application of catalysts in commercial naphtha operations.

However, the discovery of new catalyst formulations or the improvement of existing formulations normally results from application of scientific models of the catalyst. For this purpose, the engineering characterizations seldom have value. What is needed for new catalyst design may be considered to be standard characterization procedures that allow for an accurate measure of the absolute value and the chemical nature of a specific catalyst feature or catalytic site. Here one must define a specific feature of the catalyst and devise a method to make an accurate and exact measure of the feature. Of necessity, this involves definitions and terminology.[1] At first glance, this requirement appears to be easy to meet; however, in practice it proves frequently to be an extremely demanding task.

It would be desirable to describe in detail the experimental techniques that are appropriate for the characterization of naphtha-reforming catalysts, along with results from studies and their interpretation. However, a large volume would be required to do this. Thus, the present chapter will emphasize results obtained using many experimental techniques rather than an in-depth investigation of the techniques. Interpretations of these studies for a number of reforming catalysts are presented. The reader may see the diversity of experimental approaches and interpretations that are in the literature and may use this chapter as a guide for further investigations.

2 ALUMINA SUPPORTS

2.1 Surface Area and Porosity

The physical characteristics of naphtha-reforming catalysts are determined primarily by the material that serves as a support for the metal or bimetallic function. Alumina is the support for nearly all reforming catalysts. The strength of the macroscopic catalyst particle is an important property. However, measurement techniques are not straightforward. For most fixed-bed operations, if a catalyst can survive the handling during manufacturing and loading, it has adequate strength. Moving-bed operations have their own strength requirements. The reader is best referred to catalyst suppliers or the American Society for Testing and Materials (ASTM) for further information. Some discussion of strength issues may also be found in Chapters 8 and 11 of this edition.

The surface area is one of the most important physical properties of the catalyst. One of the first techniques introduced for the characterization of catalysts was the measurement of the surface area by the use of gas adsorption and the application of the Brunauer–Emmett–Teller BET equation.[3] This method grew from a desire to learn whether it was the specific nature or the extent of the surface that controlled catalytic activity and selectivity.[4] The Brunauer–Emmett–Teller (BET) method was introduced more than 50 years ago and is still the most widely utilized catalyst characterization technique. Detailed descriptions of this method are plentiful.[5] Today the gas adsorption is accomplished by automated instruments that permit measurements to be simultaneously made on multiple samples. These instruments can convert the data to provide measurements of both the surface area and the porosity.

The validity of the BET technique as an absolute method is still debated. The BET equation is based on a simple model, and the validity of some of the assumptions made for its derivation is questionable.[6] It is a very adequate technique for measuring the total surface area and as an engineering method the technique is quite acceptable.[7] In spite of numerous attempts to place the BET equation on a firmer basis or even to supplant it, such as with the introduction of fractal theory,[8] a realistic assessment would lead one to conclude that these attempts have led to more complex, and not more accurate, equations. For the current naphtha-reforming catalysts, the BET equation therefore provides both the engineering and the absolute characterization technique for measuring the total surface area.

A complete assessment of porosity is usually obtained by a combination of gas adsorption and mercury penetration measurements. Gas adsorption is applicable for those pore sizes falling within the range of about 0.5–40 nm; however, some of the newer instruments are claimed to be capable of making measurements that permit the upper limit to be extended to 100 nm. For pores in the range of about 10–5000 nm diameter, mercury porosimetry is applicable. For most materials, a direct comparison of the results from the two measurements can be made.[9] Early calculations of the porosity followed the approach of Barrett et al.[10] and then the one formulated by Wheeler[11] in his classic treatment of the role of diffusion in catalysis. With the introduction of the automated instrumentation, the isotherm of Dollomire and Heal[12] was utilized frequently. Other more complex approaches have been utilized for the calculation.[13–23]

Nearly all physisorption isotherms may be grouped into the six types shown in Figure 1a.[24,25] From the type of isotherm, a general idea of the structure of the

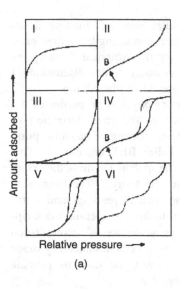

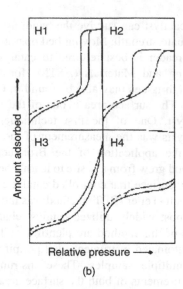

Figure 1 (a) Types of physisorption isotherms. (b) Types of hysteresis loops (from Ref.[24]).

material may be deduced. Four types of hysteresis are illustrated in Figure 1b. In most cases, the naphtha-reforming catalyst will produce either a type II or IV isotherm with H1, H2, or H3 type of hysteresis. In these cases an analysis of the pore size can be obtained from either the adsorption or desorption isotherm, with the desorption isotherm being utilized more frequently.[26–31] For comparative purposes, consistency of use of model and standard isotherm is probably more important than the actual choice of model.

A method with potential to characterize naphtha-reforming catalysts has recently been described.[32] The authors use an apparatus similar to the standard flow instruments but with a modification to permit in situ surface area measurements at various points in the catalyst characterization. For example, the surface area of the catalyst could be measured, without removal from the reactor, at various stages of chloriding or fluoriding. To compensate for the gas flow changes that occur as the sample is cycled from the temperature of the pretreatment procedure to that for the surface area measurement, a standard gas flow bypasses the catalyst chamber. Without correcting for the change in gas flow, errors as great as 50% can be introduced for the surface area measurement.

An approach utilizing the pore size distribution to provide an improved catalyst has been disclosed.[33] A Pt-alumina catalyst using a support of primarily η-alumina that possessed a narrow pore size and acidity exhibited an isomerization activity that was related to the pore–acidity index. The index is defined as

[(100)(pore diameter; d)(acidity; mmoles TMP (trimethylphosphine)/g)]/(surface area (m^2/g)). It is believed that a pore–acidity index of at least 7.0 results in a balance of mass transfer and reaction kinetics to provide both high activity and superior isomerization selectivity.

The theory of mercury penetration used for measuring porosity was developed by Washburn in 1922, but the first measurements were not made for more than 20 years when Ritter and Drake[34] introduced an experimental approach that was eventually developed into a commercial instrument. Today commercial, computer-controlled instruments are available. For materials with pore sizes that permit a direct comparison between the results of the two techniques—gas desorption isotherm and mercury penetration—the two methods provide reasonable agreement. There are small differences in the distributions that are calculated using different models[26–31] with nitrogen adsorption or desorption isotherm data, and there is reasonable agreement with distributions calculated from mercury penetration data. The difference between the two methods becomes less than 20% when various correction procedures are used. The presence of metals on a support may cause a significant alteration of the contact angle needed to provide agreement of the two methods, and the angle needed may depend on the metal loading.[35]

In spite of its maturity, the adsorption technique remains of interest. One reason is that materials with well-defined porosity continue to become available. Zeolites were the first, but these have now been extended to 20 nm with the introduction of the MCM-41-type materials. These materials have allowed one to test theoretical equations against model catalysts to improve the confidence in the theoretical equations. The other reason for this is that more advanced theories of diffusion and kinetics require better definitions of the nature of the porosity of the materials. Thus, during the past 10 years, Chemical Abstracts has received more than 2000 entries for surface area and porosity measurements.

Leofanti et al.[36] have recently reviewed approaches to the measurement of surface area and pore texture of catalysts and provide their choices. An authoritative summary has been provided in a recent volume by Rouquerol et al.[37] In addition to a characterization technique, the porosity may be used as a guide to catalyst preparation. For example, Sharma et al.[38] have studied the role of mesopores of the alumina support on the Pt dispersion in a Pt-Sn-alumina catalyst. They indicate that mesopores in the 2- to 10-nm-diameter range are of paramount importance in Pt-Sn-alumina used in a continuous catalyst regeneration process and at least 30% pore volume should lie in this range to attain a Pt dispersion greater than 70%. A compromise has to be made between pore volume, pore size distribution, and metal dispersion. However, much more work is needed, based on open literature, before this compromise can be defined.

Other techniques are available for porosity measurements but they are not ordinarily utilized with reforming catalysts. For example, Ritter utilized the data from small-angle X-ray scattering measurements to calculate a pore size distribution. While there have been significant advances in the theory needed to calculate a pore size distribution from small-angle X-ray scattering, currently it is not frequently used to characterize reforming catalysts. As noted below, this will probably change.

2.2 Acidity

The other important feature of the support is a measure of its acidity. Measurements of acidity have assumed an increasingly important role in catalyst characterization. Benesi and Winquist[39] provided a concise and precise description of what is involved in terms of electron pair acceptors for Lewis and Brønsted acid sites on surfaces of metal oxides. A characterization of acidity should provide a measure of at least three quantities: (1) the acid type (Brønsted or Lewis); (2) the acid site density; and (3) the acid strength distribution. A characterization of acidity therefore involves a definition of all three quantities. A number of approaches have been utilized for the characterization of these, and they include:

1. *Hammett indicators*. These are compounds that combine with the acid to form a color that differs from the uncombined molecule.
2. *Probe molecules*. These molecules are strong enough bases that they appear to react irreversibly with the acid site.
3. *Probe reactions*. These involve a simple reaction whose rate or selectivity depends on the acidity of the catalyst. Since reactions are covered in other chapters, this topic will not be discussed here.

Hammett acidity measurements involve the adsorption of indicators from suitable nonaqueous solvents that do not interact with the catalyst acid sites. Walling[40] defined the acid strength of a solid as its proton-donating ability, H_0. Benesi[41] utilized a number of indicators that could be related through their color change to the composition of a sulfuric acid solution that gave an equivalent color. Thus, the use of a series of indicators allows for a bracketing of the acidity of the catalyst between a high and low range.

Hammett indicators have several disadvantages when they are employed to measure the acidity of solids: (1) visually it is difficult to detect color changes; (2) many of the indicators are too large to penetrate any microporosity that is present; (3) the measurements are nearly always made far from the reaction conditions; and (4) they may not distinguish Lewis and Brønsted acid sites. Alternatively, arylmethanols react with strong protonic acids with the resulting conjugate acid being a colored carbenium ion. These arylmethanol–

carbenium ion equilibria have been used to define an alternate acidity function,[42-44] which has been previously designated as C_o, J_o, and H_R. The adsorption of these bases may provide a measure of the number and strength of the acid sites, but they reveal very little information about the structure of the catalyst site.[45-47]

Alumina is not the ideal material to characterize using Hammett indicators. Following activation at moderately high temperatures (400–700°C) in either vacuum, air, or oxygen, alumina loses many, but not all, of the hydroxyls. Those that remain do not exhibit strong acidity[48] and exist in a variety of coordination states.[49] One of the types of hydroxyls is considered to be basic because it will react with CO_2 to form the bicarbonate ion.[50-52] The dominant portion of the activated alumina surface is composed of several types of oxide ions; many are a result of the elimination of water from two hydroxyl groups during the activation process. The surface also contains coordinatively unsaturated sites (cus): aluminum ions that impart Lewis acidity to the activated alumina.[48,53] Hammett indicators cannot begin to adequately characterize the complete surface of aluminas.

Infrared Spectroscopy

In general, infrared (IR) is not very well suited to the direct examination of the reforming catalyst. Thus, the characterization nearly always involves the adsorption of one or more molecules on the reforming catalyst using IR to distinguish the interaction of the probe molecule with a feature of the catalyst. Eischens and Pliskin showed the utility of IR spectroscopy for the measurement of acidity and metal–adsorbate interactions.[54] These authors showed that the adsorption of a base such as ammonia could provide a quantitative measure of Brønsted and Lewis acid sites. Brønsted sites were identified with bands characteristic of the formation of the ammonium ion and the Lewis sites by bands characteristic of covalent bonding. Since then there have been tremendous advances in the instrumentation with computer-controlled instruments providing subtraction of background absorption and allowing detection of adsorbed species at much lower levels than was previously possible.[55-58]

The spectra of hydroxyl groups of alumina are shown in Figure 2.[59] Three bands with maxima at about 3800 (I), 3745 (II), and 3700 (III) cm^{-1} are reported in most papers.[60-63] Depending on the extent of dehydration, additional absorption bands at 3780, 3760, and 3733 cm^{-1} may be obtained. The bands do not appear to depend on alumina crystal phase to a significant extent.[59] Based on extensive study, Peri[49] proposed a detailed model of an alumina surface (Fig. 3). In this model a surface hydroxyl group may have 0, 1, 2, 3, or 4 oxygen ions in their nearest environment, and this classification of the hydroxyl types made it

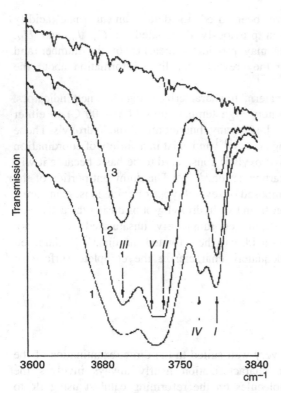

Figure 2   Spectrum of hydroxyl groups of alumina: (1) after evacuation at 700°C; (2) the same, at 800°C; (3) the same, at 900°C; (4) emission background. The Roman numerals correspond to the interpretation of the bands given in Fig. 5 (from Ref.[59]).

possible to explain the position and relative intensities of the bands shown in the spectra in Figure 2 after different heat treatments. The concentration of a completely hydroxylated γ-Al_2O_3 was determined by deuterium exchange to be 1.3×10^{15}/cm^2. The concentration of a similar γ-Al_2O_3 after evacuation at 500°C was 3.6×10^{14} per cm^2.[64]

The adsorption of ammonia was followed by combined gravimetric measurements and IR spectrometry.[65] It was found that ammonia bonded with all five types of hydroxyl groups, and in addition to molecular adsorption of ammonia, surface reactions may occur at higher temperatures. The spectra of pyridine following adsorption on alumina and subsequent evacuation at increasing temperatures (Fig. 4)[48] shows molecularly adsorbed pyridine, which is removed by evacuation at 150°C, and also bands at 1632 and 1459 cm^{-1} which are not removed even at 565°C. These bands are similar to those that result

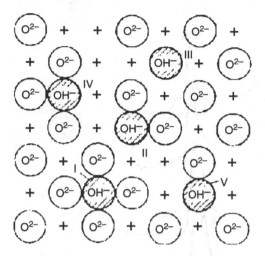

Figure 3 Five types of hydroxyl ions (designated by Roman numerals) on the surface of partially dehydroxylated alumina. The sign ' + ' designates the Al^{3+} ion in a deeper layer (from Ref.[59]).

from the complex formed between pyridine and gas-phase Lewis acids; therefore, it was concluded that these bands are characteristic of bonding due to Lewis acids. The spectrum of adsorbed pyridine on alumina did not show the absorption band at 1540 cm^{-1} characteristic of the pyridinium ion formed by interaction with a Brønsted site.

The acid strength of the Lewis sites is significant and shows a broad distribution. The acid strength of the Lewis site depends on the degree of unsaturation of the Al^{3+} ion and the tetragonal Al^{3+} ion exposed in a vacancy is a stronger site than an octahedral Al^{3+} site.[66] Heats of chemisorption for pyridine on γ-alumina after heating at 770 K range from 90 to over 120 kJ/mole, and chemisorbed pyridine cannot be quantitatively desorbed at temperatures below 750 K where it begins to decompose.[66] Steric hindrance is critical with bulky probe molecules for IR studies. For this reason, work has been typically carried out with smaller molecules to probe acid sites.

Busca[67] indicates that 'in spite of its toxicity, bad smell, low volatility and solubility in greases and rubber, giving rise to pollution of the vapor-manipulation pumps, pyridine is the most largely used basic probe molecule for surface acidity characterization' (Fig. 5). However, as it continues to be used, the specific assignment of the bands to specific sites on the catalyst becomes a more demanding task, and theoretical approaches are being used to assist in this task.[68]

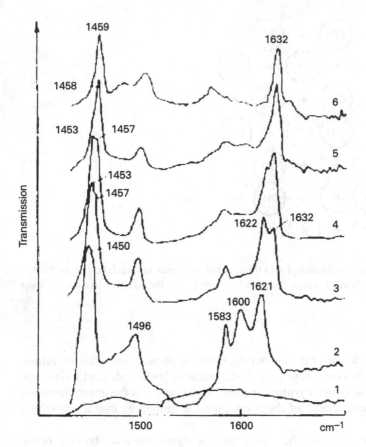

Figure 4 Spectra of pyridine adsorbed on alumina: (1) spectrum of γ-Al$_2$O$_3$ after evacuation at 450°C for 3 h; (2) after adsorption of pyridine at 25°C; (3) after evacuation for 3 h at 150°C; (4) the same, at 230°C; (5) the same, at 325°C; (6) the same, at 565°C. (From Ref.[48].)

Pyridine may also react with the surface to produce yet another complication in interpretation. Lavalley[69] reports data obtained by Bovet[70] for adsorption of pyridine on ZnO. Bovet postulated that dissociative adsorption of pyridine led to cleavage of the C—H bond at the ortho position to provide a band at 1546 cm^{-1}. Using pyridine to characterize sulfated zirconia, it was found that a band at about 1640 cm^{-1} was formed.[71] This band was attributed to oxidation of pyridine in the ortho position with the formation of a 'pyridone-like' structure. The pyridone-like band was similar to one reported to be formed as pyridine adsorbed on alumina was heated.[72]

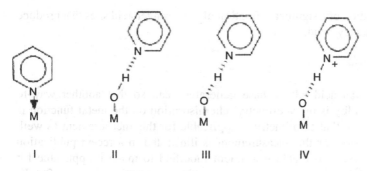

Figure 5 Scheme 2 (from Ref.[67]).

The variety of molecules used in IR studies to define acidity continues to expand and has been extended to a definition of the basic sites.[69,73] In addition, the use of Raman and IR spectroscopies in combination provides additional information and increases the reliability of the conclusions.[74–76]

A potential complication is that the coadsorption of pyridine and hydrogen on a platinum surface leads to the formation of pyridinium ions.[77] However, as the measurement is usually conducted with reforming catalysts, this should not be a significant factor.

Results obtained by Kazansky et al.[78] show that the low-temperature adsorption of dihydrogen is a promising approach for the characterization of Lewis acid sites. A frequency shift of $180 \, \text{cm}^{-1}$ toward lower values relative to the H-H stretching frequency of the free molecule was observed for η-alumina pretreated at 870 K. This shift in frequency is taken to be a measure of the polarizing power of the Al^{3+} (cus).

Carbon monoxide has also been utilized to probe the acidity of alumina.[79] The spectra recorded for adsorption of CO on γ-alumina with increasing partial pressure at 77 K resulted in bands ascribed to CO σ-bonded to strong cationic Lewis acid sites ($2238 \, \text{cm}^{-1}$), CO σ-bonded to bulk tetrahedral Al^{3+} ions on the surface (2210–$2190 \, \text{cm}^{-1}$), CO σ-bonded to octahedral Al^{3+} ions on the surface ($2165 \, \text{cm}^{-1}$), and physically adsorbed CO (3135–$2140 \, \text{cm}^{-1}$). For adsorption of CO at room temperature, the peaks are not well resolved and depend on the alumina sample. Thus, for well-crystallized and pure-phase η-alumina two relatively well-resolved bands are found, whereas for microcrystalline specimens ex-boehmite (γ- and δ-alumina) there are probably more than two bands that are not well resolved.[80] For the γ-alumina dehydrated at a low temperature, the two CO bands are centered at about 2230 and $2200 \, \text{cm}^{-1}$; dehydration at 1023 K causes the whole 2250–$2190 \, \text{cm}^{-1}$ spectral range to be occupied by a broad, asymmetrical, and unresolved band. For aluminas calcined at a higher temperature to produce mixed δ, θ-alumina phases, the band becomes sharper with the elimination of a significant fraction of the stronger acid sites that produce the higher frequency band.

Calorimetry

The reaction of an acid with a base generates heat, so that another way to determine the acidity is by calorimetry; chemisorption on the metal function is also exothermic, so that calorimetry is applicable for this measurement as well. The equipment used for this measurement is illustrated in a recent publication using a commercially available instrument modified to make it applicable for catalyst studies.[81] The instrument was capable of operation down to 200 K. In addition, a number of changes were made in the gas-handling system and the calorimeter that enhanced the sensitivity and accuracy by minimizing baseline perturbations after switching from the purge gas to a stream containing the adsorbate.

The investigation of the acidity and basicity of 20 metal oxides was conducted by Gervasini and Auroux[82] with microcalorimetry. Alumina was among the group of oxides showing amphoteric character by adsorbing both ammonia and carbon dioxide. Cardona-Martinez and Dumesic[83,84] included alumina in their study of the differential heat of pyridine adsorption. They found three regions of nearly constant heats of adsorption with increasing amine coverage; this implies that there are three sets of acid sites of different strength. It is surprising that more use has not been made of calorimetry in the characterization of naphtha-reforming catalysts.

The application of microcalorimetric techniques for characterization of acid–base catalysts has been reviewed recently.[85] The usefulness of the technique has been demonstrated for the Pt-L-zeolite catalyst.[86] The results indicated that the adsorption of CO on fresh Pt-SiO$_2$ and Pt-L-zeolite catalysts evolved similar quantities of heat and Pt adsorption strength; however, as carbon was deposited the adsorption heat decreased on the Pt-silica catalyst but remained constant with the Pt-L-zeolite catalyst. In this example, the heat may be generated by adsorption on both the metal and acid site.

2.3 Chlorided Catalysts

The regulation of the Cl concentration on the surface of the bifunctional catalyst is a key factor in the optimization of the reforming process. An optimal chloride concentration allows enhancement of the acidic function of the catalyst[87] and an improvement in the self-regeneration capability of the 'coked' Pt/γ-Al$_2$O$_3$ system, presumably because of more effective H$_2$ spillover.[88,89] In addition, oxidative treatment ($400 \leq T \leq 550°C$) in the presence of chlorine compounds (i.e., CCl$_4$, CHCl$_3$, HCl, etc.) is a procedure that is claimed to provide redispersion of sintered $Pt/\gamma-Al_2O_3$-reforming catalysts.[90] Several studies have been devoted therefore to rationalize the factors controlling the retention and the leaching of chloride either during activation treatments or under reaction conditions.[91-95] The effects of chloride adsorption on the physicochemical properties and reactivity of $\gamma-Al_2O_3$ surfaces have also received attention.[94-97]

Arena et al.[97] followed the loss of chloride upon heating in dry or wet conditions. They found that the initial rate of chloride loss is proportional to the initial chloride content of the sample and loss could be expressed by a first-order expression involving the actual surface Cl^- concentration. The activation energy for loss was 6.2 kcal/mole, which is similar to the value of 6.0 ± 0.1 kcal/mole reported by Bishara et al.[98] for chloride loss from a $Pt-Cl-Al_2O_3$ catalyst during air calcination. It was found that steam increased the rate of chloride removal. The activation energy obtained for wet and dry conditions is the same; the increased rate results from a considerably higher pre-exponential factor for the wet conditions. Ayame et al.[99] characterized the form of dehydrated alumina halogenated with chlorine at 773–1273 K. They reported that the higher temperature chlorine treatments resulted in the formation of adjacent strong Lewis acid sites, which were induced by the chloride ions bonding to aluminum cations. Treatment with HCl at 773 K produced materials with strong Brønsted acidity.[100] It was shown by Garbowski and Primet[101] that aluminas chlorided by CCl_4 or HCl at 573 K strongly adsorbed benzene to form coke precursors and hexadienal cation, respectively. Arena et al.[97] compared the change in the zero point charge with chloride addition to the catalytic activity for the isomerization of cyclohexene to methylcyclohexene and found a linear relationship between the two.

Alumina is being extensively studied as a catalyst and/or catalyst support for the conversion of environmentally detrimental halogenated compounds. As such, much attention is devoted to the subject of halogen–alumina interaction. An ultrafine alumina ($550\ m^2/g$) was prepared and then reacted with CCl_4 as outlined in Figure 6.[102] The probable mechanism was by dissociative adsorption and heterolytic decomposition of the C-Cl bonds. Overchloriding may result in the loss of alumina as $AlCl_3$.

Since the conversion was accomplished in the absence of oxygen, the CO_2 must be formed from oxygen from the alumina. The mechanism also allows for the formation of the minor by-products. Vigué et al.[103] compared the chlorination of two γ-aluminas with chloroform (Degussa C alumina and Condea alumina). They found distinct differences with the Condea alumina reacting more slowly. Chlorination was described as occurring on the monocoordinated OH groups linked to tetrahedral or octahedral alumina atoms.

While not directly concerned with naphtha reforming, the results of Gracia et al.[104] are of interest. These authors used Fourier transform infrared (FTIR) and controlled-atmosphere extended X-ray absorption fine-structure analysis techniques to study the effect of chlorine on Pt-supported catalysts during

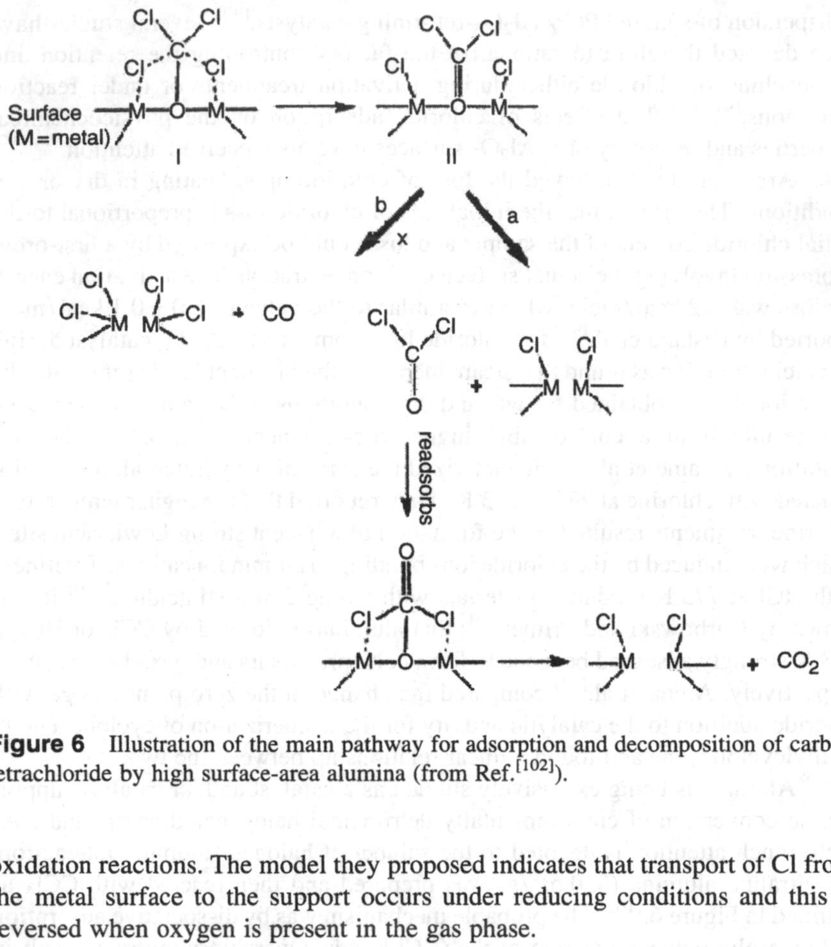

Figure 6 Illustration of the main pathway for adsorption and decomposition of carbon tetrachloride by high surface-area alumina (from Ref.[102]).

oxidation reactions. The model they proposed indicates that transport of Cl from the metal surface to the support occurs under reducing conditions and this is reversed when oxygen is present in the gas phase.

2.4 Fluorided Alumina

Several authors have claimed that strong Brønsted acid sites are formed by fluorination of alumina.[105–110] The extent of acidity, determined from ammonia adsorption, was shown to increase and then decrease as the concentration of fluoride increased (Fig. 7);[111] IR was used to follow the reaction of sterically hindered nitrogen bases[112,113] with fluorided aluminas. A model of the surface modifications has been proposed.[60]

Hirschler[114] obtained a measure of the acidity of an Alcoa F-10 alumina as well as a sample of the fluorided alumina (both calcined at 500°C). The acidity titrations showed that the treatment with HF greatly increased the acid strength

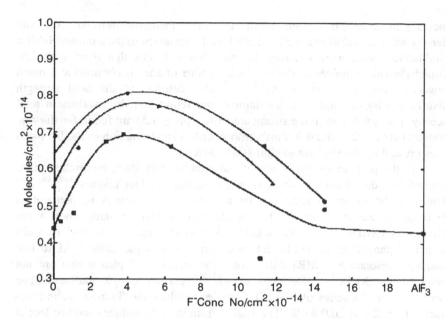

Figure 7 Effect of fluoride concentration on the adsorption of ammonia (from Ref.[111]).

using H_R indicators. Webb,[105] on the basis of the effect of temperature of ammonia chemisorption, concluded that HF treatment of alumina did not increase the number of acid sites but did considerably increase their strength. Weber[115] showed that fluorided alumina impregnated with platinum salts has a high hydrocracking activity, whereas on a chlorided alumina the hydrocracking activity is low. Holm and Clark[116,117] reported that fluoriding alumina considerably increased its activity for n-octane cracking, o-xylene isomerization, and propylene polymerization. The heats of ammonia adsorption and the observations with arylmethanol indicators could be reconciled if it was assumed that the acid centers on fluorided alumina are of a different type (e.g., protonic) from those on alumina. The fact that fluorided alumina converted 1,1-diphenylethylene to the carbenium ion and that it showed no increase in acid strength with the Hammett indicators supported this assumption, from which it followed that protonic acids rather than Lewis acids convert arylmethanols to their corresponding cations.

Typical IR spectra of pyridine adsorbed on samples with increasing F show bands for Lewis sites (1455, 1496, 1580, and 1620 cm^{-1}).[105,112] Brønsted acid sites (1545, 1562, and 1640 cm^{-1}) are not detected on alumina but are present in the fluorided alumina samples. The number of Brønsted sites were found to increase to about 0.6 site/nm^2 as the F content increased; however, this site density was attained only after 10–20% F had been added to the alumina. Similar qualitative results were obtained for the Brønsted sites that react with 2,6-dimethylpyridine; however, the maximal number of sites is obtained at a much lower F content. Corma et al.[105,112] also determined the acid strength distribution by titration with butylamine and found that the maxima in total acidity (pK $\leq$ 6.8) and in the strong acid sites (pK $\leq$ 1.5) are found for fluorine content between 2% and 4%. Furthermore, only a small fraction of the Brønsted sites created by fluorination exhibit strong acidity.

In the past, to resolve some of the contradictory data, researchers have studied fluorided aluminas using a variety of instrumental techniques.[109,118–122] Some of the inconsistencies can be attributed to variations in preparation methods; however, most of the inconsistencies involve data from X-ray diffraction (XRD). It is very possible that some preparations lead to well-dispersed phases that cannot be detected using XRD. In particular, ^{27}Al nuclear magnetic resonance (NMR) data reveal the presence of phases that are not detected by XRD.[123] DeCanio et al.[123] carried out a multitechnique investigation of a series of F/Al$_2$O$_3$ samples in which the fluorine loading was varied from 2.0 to 20.0 wt %. The results from these techniques showed that at low levels fluoride served to block Lewis acid sites but at higher levels its predominant role was to increase the Brønsted acidity of the alumina surface, and that fluoride strengthens the remaining Lewis acid sites.

Two-dimensional magic angle spinning (MAS) NMR is being applied to elucidate the nature of the sites formed by fluoridation; three different Al-F sites have been identified.[124] At levels above that needed for full fluoridation of the surface, AlF$_3$ was formed. Earlier Davis et al.[125] had reached similar conclusions based on their studies using ultrasoft X-ray absorption spectroscopy. In other studies, radiotracers of F exchange with Cl are being used to elucidate the nature of the catalytic site.[126]

3 PLATINUM–ALUMINA CATALYSTS

For both platinum and the bimetallic systems there have been long-term conflicting views of the chemical state of the metal under typical reforming conditions. The early view that platinum was present as the metal was shattered by the report by McHenry et al.[127] that a significant fraction of the platinum in a reforming catalyst could be extracted with dilute HF or with acetylacetone. Acetylacetone extraction had been developed as a method to recover the metallic platinum from reforming catalysts so that XRD could be utilized to obtain an average crystal size in the absence of the interfering peaks of the alumina support.[128] Results[129–131] supporting the view of soluble platinum appeared following this observation. Others questioned the presence of soluble platinum, indicating it was present only when the reduced catalyst had been exposed to air following extraction.[132-135] The latter view is probably most widely accepted today. However, the absence of platinum ions in the reduced reforming catalyst is not universally accepted.[136,137]

The experimental observation that the best metallic function would be one of the expensive group VIII metals dictated that the metallic function of the naphtha-reforming catalyst be optimized. The cost of the noble metal required that the dispersion of the metal, platinum, be maximized. In the following we consider first the definition and a measurement of dispersion of a single function, and then of the function dispersed on a high surface area support.

3.1 Dispersion

Dispersion is easy to define but almost impossible to precisely measure. The dispersion may be defined as the number of atoms in the exposed surface N_s divided by the total number of atoms present in the catalyst (N_{total}). While N_{total} can be determined precisely, N_s will depend on the definition of surface as well as the experimental approach used for the measurement and the model used in the calculations. For example, some of the crystal faces of a metal are more densely packed than others; the openness of the outermost layer of metal atoms will determine the extent that the second layer will be exposed to the gaseous phase.

Except for the chemisorption techniques described below, dispersion is obtained from calculations based on the particle size. Thus, to make comparisons of the results from various techniques one needs to consider the relationship between dispersion and the particle size. The most common shape of particle is a sphere, or a hemisphere, especially for platinum. However, as the dispersion approaches unity, two-dimensional plates ('rafts') may be encountered. Thus, the dispersion is related to the crystal size. The relationship is not simple. The supported particles seldom, if ever, have a uniform size (monodispersed) but will have a distribution of particle sizes (polydispersed). The 'average size' may depend on the experimental technique used to make the measurement. Lemaitre et al.[138] consider these problems in some detail, so that only a brief outline is given.

3.2 X-ray Diffraction

The XRD and line broadening (XLBA) techniques are based on the fact that the breadths of the X-ray reflections, apart from an instrumental contribution, are related to the dimensions of the crystals giving rise to the reflections. The metal reflection must be intense enough to give a signal measurable above the background of the support. This requirement is easily met in the case of the Pt-SiO_2 catalyst but not with the Pt-Al_2O_3 catalyst. The two most intense peaks for

Pt metal fall at angles of 2θ values where intense peaks from the crystalline alumina support mask the Pt peaks. Thus, XLBA has limited utility for the characterization of Pt-Al$_2$O$_3$ reforming catalysts. This was the reason that Adams et al.[139] utilized a Pt-SiO$_2$ catalyst for a comparison of the metal sizes obtained by microscopy, XLBA, and chemisorption techniques.

The crystallite size is calculated from XLBA.[140] The more common approach is to measure the linewidth at half-maximum (LWHM). Instrumental line broadening is usually taken into account by measuring the LWHM for a sample with very large crystallites and using this to correct the experimental LWHM for the sample. The XLBA technique returns 'crystallite' size rather than 'particle' size. For small particles, such as Pt in the naphtha-reforming catalyst, the two are the same. However, as the particles become larger, they may be composed of two or more crystals; in this case the crystallite size can be considerably smaller than the particle size.[141] Ascarelli et al.[142] provide calculations to use XLBA to distinguish between two different Pt growth mechanisms: the coalescence and Ostwald ripening processes. They apply the method to Pt-carbon samples and conclude that the coalescence process applies to their data.

3.3 Transmission Electron Microscopy

Adams et al.[139] measured the distribution of Pt particle sizes for a particular Pt-SiO$_2$ catalyst and found the number average diameter for the size distribution to be 2.85 nm; the surface average diameter, 3.05 nm; and the volume average diameter, 3.15 nm. These authors considered the probable error in these average diameters to be about 10%; this was based on the variation of the values obtained among eight observations.

Rhodes et al.[143] prepared a series of Pt-Al$_2$O$_3$ catalysts by sintering a 46% dispersion sample to produce lower dispersions of 26% and 15%. From the transmission electron microscopy (TEM) size distribution, the dispersions of the samples were estimated. The results of the calculation are largely independent of the shape of the crystal used.[144] Evaluating the dispersions, they obtained 16%, 22%, and 61%, in reasonable agreement with those measured by chemisorption. They attributed the poor agreement for the sample with the highest dispersion to a breakdown of the model for small particles. Thus, the model would not be applicable for naphtha-reforming catalysts that are of the most interest, i.e., the highly dispersed materials. The measurement of the particle diameter will be subject to more error as the particles become smaller. This will result from the inability to measure particles below some size that depends on the resolution of the electron microscope used for the measurement.

Overlapping contrast from the support, especially one that is crystalline, may impact the ability to observe metal particles. White et al.[145] found this to be the case for a high-resolution electron micrograph of a Pt-Al$_2$O$_3$-reforming catalyst. The contrast from the support tended to obscure the metal particles, making them harder to detect. The effects on resolution and contrast from the support on the determination of the sizes and shapes of small metal particles have been considered in theoretical papers.[146] Based on image calculations, a 1.2-nm cubo-octahedron could be undetected when viewed through an amorphous support of 1.9-nm thickness in a microscope having a 0.2-nm point resolution. Particles larger than about 1 nm in diameter can usually be readily detected by brightfield and darkfield microscopy.[147] For smaller particles, high-angle annular darkfield imaging is useful because the electrons scattered at high angles are more sensitive to atomic number; hence, the sensitivity for Pt with respect to the support increases at high angles. This technique, coupled with digital image processing, produced images of particles containing as few as three Pt atoms on alumina support[147] and three atom clusters of Os on γ-alumina.[148–150] Datye and Smith[147] contend that the difficulty in detecting the presence of highly dispersed metallic species may be caused as much by the mobility of the species as by the problem of obtaining sufficient contrast. Thus, it is not surprising that Huang et al.[151] found that Pt-Sn-γ-Al$_2$O$_3$ catalysts contained a significant number of TEM-invisible platinum particles less than 1 nm in diameter in addition to larger TEM-visible particles.

3.4 Chemisorption Techniques

The chemisorption of hydrogen by metals had been related to the number of exposed metal atoms of platinum and other metals.[152,153] However, it was the work of Emmett and Brunauer that showed the utility of the technique for the characterization of more complex catalytic materials that are composed of two or more components.[154–156] Chemisorption measurements are easily applied to Pt-alumina reforming catalysts. Several gases, including CO, H$_2$, O$_2$, and NO, have been used for this purpose. Hydrogen adsorption isotherms on silica-supported platinum catalysts are typical; adsorption corresponding to chemisorption is completed at pressures of 0.1 mm Hg or less, and adsorption at saturation is greater at lower temperatures. The amount of adsorption at saturation will depend on the evacuation temperature following reduction and cooling in hydrogen. Adams et al.[139] found that chemisorption volume increased with increasing temperature of evacuation up to 250°C, remained constant up to 800°C, and then decreased at 900°C, presumably due to sintering of the platinum and/or support.

Although there was much background information available, Adams et al.[139] appear to be among the first to have made a detailed comparison of the dispersion calculated from the results of several techniques using hydrogen adsorption at −78°C or 0°C. To calculate the dispersion, it is first necessary to define an approach to obtain the volume of hydrogen or other chemisorbed gas

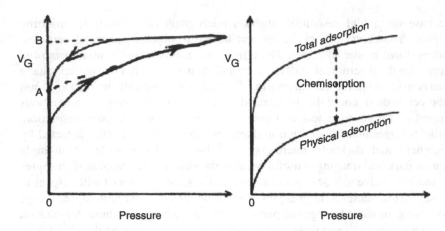

Figure 8 Typical measurement of chemisorption: (a) intrapolation of Langmuir-type isotherm to zero pressure; (b) total adsorption at 90 K and physical adsorption at 90 K after evacuation at 195 K. The difference between the two gives the chemisorbed amount.

that corresponds to complete coverage of the metal. One can utilize two approaches to obtain the volume of gas corresponding to a monolayer of coverage. First, one can extrapolate the adsorption isotherm to zero pressure (Fig. 8) and take this as the appropriate volume of gas. The second approach is to make a first measurement of the isotherm so as to obtain 4–10 data points up to pressures of about 40 cm Hg, then to evacuate the sample at the adsorption temperature at a vacuum of 10^{-3} mm Hg or better, and then to measure a second adsorption isotherm. At any pressure the difference between the first and second isotherms should correspond to the amount of gas that is chemisorbed. For most Pt-Al$_2$O$_3$ catalysts, the two methods of obtaining the amount of hydrogen chemisorbed provide similar volumes.

To calculate the available platinum surface, one must know the area occupied by a chemisorbed hydrogen. The surface area for a platinum black sample calculated from the chemisorption data should be equal to that of the BET surface area calculated from the volume of nitrogen adsorbed. The area occupied by a hydrogen atom on platinum black is 1.12 nm^2. The particle size can then be calculated from the chemisorption data assuming the same density as bulk platinum.

Via et al.[157] have measured hydrogen adsorption at room temperature on a Pt-Al$_2$O$_3$ catalyst and on the support (Fig. 9). The difference between the curves A (adsorption on Pt-Al$_2$O$_3$) and B (adsorption on the alumina support) remains constant over the hydrogen pressures used for the measurement (5–15 cm Hg). However, extrapolation of the hydrogen adsorption on the support alone to zero

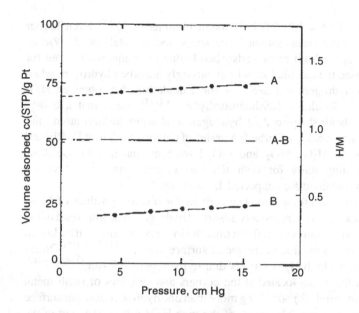

Figure 9 Typical chemisorption isotherms at room temperature. The isotherms are for hydrogen chemisorption on the platinum on alumina catalyst. Isotherm A is the original isotherm, while isotherm B is a second isotherm determined after evacuation of the adsorption cell for 10 min subsequent to the completion of isotherm A. The difference isotherm A-B is obtained by subtracting isotherm B from isotherm A (from Ref.[137]).

pressure does not give the expected value of zero adsorption. For this particular catalyst, the difference between the two curves (A − B) returns a value of $H/Pt = 0.9$. In contrast, if the adsorption isotherm for the $Pt-Al_2O_3$ catalyst is extrapolated to zero hydrogen pressure (curve A), a $H/Pt = 1.2$ value is obtained. The 'correct' procedure for obtaining a measure of H/M has been widely debated at ASTM Committee D32 meetings and the ASTM Committee E42 on surface analysis. The E42 committee has approved surface analysis techniques that are now published in volume 3.06 of the *Annual Book of ASTM Standards*; several subcommittees are active in this area.[158]

It is still not certain which H/Pt stoichiometry is correct. For most widely used platinum metal catalysts an H/M stoichiometry of unity has been used and this assumption has been tested using XRD and TEM data.[159–161] Surface science measurements also show that a maximum of one hydrogen atom per metal atom could be chemisorbed on the (111) faces of FCC metal single crystals.[162] It is generally assumed that metal particles larger than 1–2 nm consist for the most part of (111) faces, and the use of $H/Pt = 1$ seems reasonable. However, reports of stoichiometries greater than one date back over 30 years. For Pt/Al_2O_3

catalysts values of H/Pt = 1.2–2.5 have been obtained.[163–169] Even higher values of near 3.0 have been obtained for supported Ir catalysts. McVicker et al.[170] found an upper limit of two adsorbed hydrogen atoms per Ir atom for Ir/Al$_2$O$_3$ if they based the calculation only on strongly adsorbed hydrogen; when the total adsorbed hydrogen was used for the calculation they obtained H/Ir values exceeding 2. Similarly, Krishnamurthy et al.[171] found that a 0.48% Ir/Al$_2$O$_3$ catalyst adsorbed up to 2.72 hydrogen atoms per iridium atom, and about H/Ir = 0.28 was weakly adsorbed. A series of catalyst studies of Pt, Rh, and Ir metals supported on Al$_2$O$_3$, SiO$_2$, and TiO$_2$ have been made, and these show H/M values exceeding unity for both Rh and Pt catalysts;[172–174] values exceeding 2.0 were obtained for supported Ir catalysts.[175,176]

A number of explanations have been given for obtaining values of H/M greater than 1. In most catalysts, reversibly adsorbed hydrogen makes up a part of the total adsorbed hydrogen; many contend that only the irreversibly adsorbed hydrogen should be used for determination of the metal surface area.[170,171,177–180] Others have explained that the high H/M ratio is due to hydrogen spillover.[169,181–183] Some contend that the atoms located at the corners and/or edges of small metal particles may be responsible for adsorbing more than one hydrogen atom per surface metal.[170,171,184] Another possible source of the high H/M ratio is that part of the hydrogen can be bonded to atoms under the outermost surface layer.[185–187] Kip et al.[176] offered the explanation of adsorption beneath the metal surface or multiple adsorption on parts of the metal surface. In some instances, such as Pt/TiO$_2$, the high value of H/M may be due to a partial reduction of the support itself.

3.5 Titration Methods

A hydrogen–oxygen titration method was introduced by Benson and Boudart in 1965.[188] This technique was expected to reduce or eliminate errors introduced by hydrogen spillover and to provide an increase in the sensitivity over that of the adsorption of hydrogen. The titration should be the result of simple stoichiometries as shown by the following reactions:

$$Pt + 1/2H_2 = Pt\text{-}H \text{ (H chemisorption, HC)}$$

$$Pt + 1/2O_2 = Pt\text{-}O \text{ (O chemisorption, OC)}$$

$$Pt\text{-}O + 3/2H_2 = Pt\text{-}H + H_2O \text{ (H titration, HT)}$$

$$2Pt\text{-}H + 3/2O_2 = Pt\text{-}O + H_2O \text{ (O titration, OT)}$$

Menon[189] summarized the conflicting stoichiometries for HC:OC:HT as found in the literature. Prasad et al.[190] reported that much of this controversy was perhaps due to the fact that every research group used the very first hydrogen chemisorption on a fresh catalyst as the basis for all calculations, and if the surface was given an 'annealing' treatment by a few H_2-O_2 cycles at ambient temperature, it can behave normally in subsequent titrations. Furthermore, if the H_2 titer value was used as the basis for calculations after the H_2-O_2 cycles, the stoichiometry was always found to be 1:1:3, independent of Pt-crystallite size and independent of the pretreatment of the catalyst. Whether the freshly reduced catalyst or one that has been subjected to repeated H_2-O_2 cycles is representative of the working reforming catalyst is not defined at this time. Isaacs and Petersen[191] also found that the dispersion from hydrogen chemisorption was greater than obtained for either oxygen chemisorption or by the titration technique. The authors point out that the ratio (dispersion from hydrogen titration/dispersion from hydrogen chemisorption) of 0.82 is in very good agreement with the value of 0.81 obtained by Kobayishi et al.[192] and by Freel.[193]

O'Rear et al.[194] recently addressed the stoichiometry of the titration technique using a clean surface platinum powder. They point out that several other authors have studied the chemisorption of oxygen on various types of platinum surfaces using ultrahigh vacuum techniques; these are reviewed by Gland.[195] O'Rear et al. imply that the observed low values are due to inefficient clean-off of the gas used for the reduction or that the oxygen pressure was not sufficiently high to cause reconstruction of some Pt single-crystal faces to a 'complex' phase where high values of oxygen chemisorption are obtained. The authors[194] also compiled literature data to make comparisons of the platinum area average particle size calculated from hydrogen chemisorption and the number average particle size measured by TEM or the volume average crystallite size measured by XLBA. There was good agreement between the particle size measured by hydrogen chemisorption and the independent techniques (Fig. 10).[141,163,196–198]

Measurements for a commercial reforming catalyst (CK-306; the same formulation but not the same sample) obtained in three labs are summarized in Table 1, and they show similar values even when different experimental procedures are utilized.[191,199,200]

3.6 Small-Angle X-ray Scattering

X-rays experience scattering at the interface of materials with sufficient differences in density. An alumina support provides a two-phase system—the alumina particles and the porosity represented by the void space. When the alumina contains platinum there are now three interfaces. Small-angle X-ray scattering (SAXS) can be used to obtain interphase surface areas of a system such as alumina-supported platinum catalysts. Recent developments in the experimental techniques and the theory for interpreting the data promise to make this a useful technique for catalyst characterization. Summaries of the general scattering

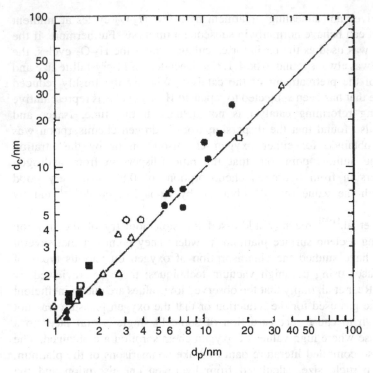

Figure 10 Supported platinum particle size as measured by hydrogen chemisorption (d_c) assuming a stoichiometric coefficient $y = 1$ and measured by transmission electron microscopy, X-ray line broadening, and X-ray small-angle scattering (d_p). Symbols denote the sources of the data (from Ref.[194]).

Table 1 Comparison of Pt Dispersion Values from Different Methods

Method[a]	Gas adsorbed (ml STP/g)	Dispersion (%)
A	0.282	82
B	—	73
C	0.278	81
D	0.271	79

[a]A, Volumetric chemisorption; B, hydrogen titration of Pt-O surface volumetrically; C, hydrogen titration of Pt-O surface gas chromatographically; D, oxygen titration of Pt-H surface gas chromatographically.
Source: Ref.[200].

principles for such systems[201–206] and the experimental techniques[207] have been published.

One approach to obtain information about the platinum particles is to reduce the three-phase system to a two-phase system by filling the pore structure with a liquid of the same electron density as the alumina support. Compounds such as CH_2I_2 are suitable for this purpose. It is difficult, however, to completely fill the pores because of wetting problems and the inaccessibility of some pores. Whyte et al.[208] utilized the masking technique to investigate the Pt-Al$_2$O$_3$ system. Most results were in reasonable agreement with their chemisorption data. Cocco et al.[209] made a detailed study of supported Pd and Pt materials. They excluded scattering from inaccessible voids by subtracting the scattering of the masked, metal-free support and used experimental techniques to make intensity measurements on an absolute basis. Calculated size distribution functions were found to be bimodal. Somorjai et al.[210] used extremely high pressure to reduce the pores present in the alumina to such a small size that they would not contribute significantly to the scattering. It is necessary to assume that the compaction at these very high pressures does not modify the platinum particles.

Brumberger and coworkers have made several experimental studies of catalysts using the SAXS technique.[202,211–215] These workers used the 'support-subtraction' method, assuming that addition of metal does not change the support morphology to an appreciable extent. This technique requires that both support and metal–support receive the same preparative and pretreatment procedures and that both respond exactly the same to these treatments. They generally obtained good agreement between BET and SAXS surface areas.[213] Brumberger et al.[215] utilized SAXS to follow the sintering of Pt-Al$_2$O$_3$ catalysts with high Pt loadings (up to about 11 wt %) in air. The data indicate that there is an initial redispersion at 400–500°C and that the surface then decreases as the temperature is raised further, in agreement with chemisorption data. The SAXS data show that the sintering response of the catalyst to temperature changes is rapid. The authors emphasize that the SAXS method is capable of following changes in surface areas accurately, nondestructively, and continuously *in situ*, for a wide variety of temperature, pressure, and ambient atmosphere conditions.

Small-angle neutron scattering (SANS) can also be utilized. Hall and Williams[206] have recently made a comparison of the surface area and porosity of a range of silicas, aluminas, and carbons obtained from SANS and BET surface areas. They reported that the SANS areas of the nonporous and mesoporous aluminas and silicas were of the correct order of magnitude compared with the BET areas. In summary, SAXS and SANS measurements still have a long way to go before they become a reliable approach for naphtha-reforming catalyst characterization. However, rapid advances are now being made in both the experimental and theoretical approaches so that the methods must be considered to show promise.

Lemaitre et al.[138] contrasted the results obtained using the above methods and showed schematically how the dispersion is approached through the various experimental methods. They concluded that usually the dispersion values are most easily compared by expressing them all in terms of an average particle size.

3.7 X-ray Absorption Fine-Structure Analysis

X-ray absorption fine-structure analysis (XAFS) is an experimental technique which is atom specific and can give structural information about supported metal catalysts. XAFS gives information on the atomic level, but very limited and indirect information on the morphology of the catalyst particles (e.g., size, shape, crystalline imperfections, etc.). XAFS is, however, unique in providing a method to determine coordination numbers, interatomic distances and from this the 'average' nearest-neighbor atoms, and the vibrational motions of the metal atoms. While the phenomenon was first observed more than 60 years ago,[216] it is only recently that XAFS has become a useful analytical tool. One reason for this is that the use of synchrotron radiation increased the available flux by $10^5 - 10^6$, allowing faster, more accurate XAFS experiments (for experimental details, see[217,218]).

Extended XAFS

Extended XAFS (EXAFS) spectra generally refer to the region $40 - 1000$ eV above the absorption edge. Excellent reviews on EXAFS in general[219,220] and its application in catalysis[221-228] are available. The pre-edge region contains valuable bonding information such as the energetics of virtual orbitals, the electronic configuration, and the site symmetry. The edge position also contains information about the charge on the absorber. In between the pre-edge and the EXAFS regions is the X-ray absorption near-edge structure (XANES).

Transmission is just one of several modes of EXAFS measurements. The fluorescence technique involves the measurement of the fluorescence radiation (over some solid angle) at right angle to the incident beam. Other more specialized methods include (1) surface EXAFS (SEXAFS) studies, which involve measurements of either the Auger electrons or the inelastically scattered electrons (partial or total electron yield) produced during the relaxation of an atom following photoionization; and (2) electron energy loss (inelastic electron scattering) spectroscopy (EELS). These latter methods, which require high vacuum, are useful for light-atom EXAFS with edge energies up to a few kilo electron volts. The accuracy of the structural parameters depends on many factors. Typical accuracies for the determination of parameters obtained in an

Table 2 Comparison of Chemisorption and XAFS Data

Catalyst[a]	H/M[b]	CO/M[b]	N[c]	R (Å)[d]
Os-SiO$_2$	1.2	1.0	8.3	2.702 (2.705)
Ir-SiO$_2$	1.5	0.8	9.9	2.712 (2.714)
Ir-Al$_2$O$_3$	1.3	0.9	9.9	2.704 (2.714)
Pt-SiO$_2$	0.7		8.0	2.774 (2.775)
Pt-Al$_2$O$_3$	0.9	0.9	7.2	2.758 (2.775)

[a]Catalysts contained 1 wt % metal.
[b]The number of H or CO chemisorbed per metal atom.
[c]Average coordination number (nearest metal atom neighbors).
[d]Interatomic distance (nearest metal atom neighbor); data in parentheses are for bulk metal.
Source: Ref.[157].

unknown system have been quoted as 1% for interatomic distances, 15% for the coordination numbers, and 20% for the thermal mean-square displacements.[229]

In the pioneering work in the application of EXAFS, Sinfelt, Lytle, and their coworkers[157] compared EXAFS and hydrogen chemisorption data for Os, Ir, and Pt catalysts. The hydrogen chemisorption data indicated a high dispersion with H/M near 1 (Table 2). The data for the interatomic distances are the same as for the bulk metal, within experimental error, except possibly for Pt-Al$_2$O$_3$. This suggests that Pt interacts more strongly with alumina as the coordination number (7.2) for Pt-Al$_2$O$_3$ was the lowest of the three catalysts.

The metal dispersion, as represented by the coordination number, is one of the more important EXAFS parameters for catalyst characterization. While the experimental value of 7.2 is lower than for the bulk metal (12), it still appears high when compared to the H/M value of 0.9. Zhao and Montano[230] point out the small coordination numbers obtained by Sinfelt and coworkers are consistent with very small clusters (dimers, trimers, etc.) but that the interatomic distances correspond to those of the bulk metal. Based on detailed analysis, Zhao[231] contended that the metal particles in the 1% Pt catalyst contain more than 50 atoms and that the real first-shell coordination number could be as large as 10.

Lytle et al.[232,233] reported additional data for *in situ* studies of a 1% Pt on Cab-O-Sil in the presence of He, H$_2$, or benzene. In this instance the hydrogen-reduced sample has Pt present as a mixture of 10- to 15-Å disks and polyhedra. The authors included the phase shift in the transformation of the Pt EXAFS data that was shown by Marques et al.[234] to be very useful in sharpening and simplifying the peaks of the Fourier transform of elements such as Pt which have very nonlinear phase shifts. These latter results showed a clear demonstration of Pt-O bonds to the support at low temperatures. At temperatures above 600 K the Pt-O bonds break and the raft-like clusters curl up to be more sphere-like.

Concurrently with the bond breaking, electrons flow to the Pt d-band to provide a d-electron surplus in the Pt clusters relative to bulk Pt.

Although *in situ* high-temperature[235] and high-pressure[236] cells for EXAFS were known, Guyot-Sionnest et al.[208] claim the first *in situ* cell used to examine the activation of a Pt-alumina-reforming catalyst. Dexpert[237] reported that on heating a catalyst prepared using chloroplatinic acid from room temperature to 200°C in air, there was a destabilization of the Pt-Cl bond with formation of a Pt-O bond. With heating at 200°C in hydrogen the complex rapidly decomposed with a decrease in the Cl/Pt nearest-neighbor and an increase in Pt-Pt nearest-neighbor bonding. The value of the coordination number of Pt-Pt bonds was consistent with a raft structure for Pt.

Guyot-Sionnest et al.[238] found that a sample yielded a well-formed oxide specie after oxidation, with three coordination shells of the oxide visible (Fig. 11). Following reduction at 460°C Pt-Pt bonds having an average interactive distance of 2.67–2.68 Å became evident. This is a much shorter bond than in bulk Pt (2.75 Å); it was believed that the low Pt-Pt coordination numbers (4–5) could cause each Pt atom to share more electrons with neighbors, thereby making a shorter bond. Since the coordination of Pt was lower in the chlorided catalyst, it was concluded that Cl aids in maintaining Pt particle dispersion. Following reduction at 460°C, the pressure was increased to 5 atm and *n*-heptane passed over the catalyst for 4 h. A peak at the appropriate Pt-C bond distance was

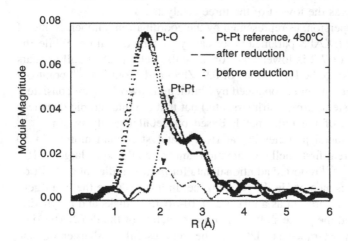

Figure 11 Evolution of the Fourier transform modules for chlorinated 1.0 wt % Pt/ Al$_2$O$_3$ before reduction under flowing H$_2$ ($T = 25°C$) and after reduction ($T = 260°C$), p(H$_2$ total) = 1 atm. Disappearance of Pt-O coordination and formation of Pt-Pt coordination is shown as temperature is raised (from Ref.[238]).

observed. The authors conclude that, since the number of Pt-C bonds detected by EXAFS remained unchanged while dehydrocyclization activity decreased, deactivation of the catalyst was not due to changes in the number of Pt-C bonds.

X-ray Absorption Near-Edge Structure

At an absorption edge, the X-ray absorption coefficient increases abruptly with increasing energy. In an absorption spectrum for a given element, absorption edges are observed at certain energies characteristic of the element. The abrupt rise in absorption occurs when the energy of the X-ray photons are equal to that required to excite electrons from an inner level of the absorbing atom to the first empty electronic states. Edges are identified by the letters K, L, M, etc., to indicate the particular electronic shell from which the electrons are excited by the X-ray photons. By comparing with measurements on well-characterized systems, one can use $L_{II,III}$ edge studies to determine d-band occupancy in transition metal compounds and alloys.[239]

Lytle[240] reported results for the L_{III} absorption edges for supported catalysts. Meitzner et al.[226,241] more recently presented similar data for metals more frequently associated with naphtha-reforming catalysts. The L_{III} absorption edges of platinum, present in a Pt/Al_2O_3 catalyst with a dispersion approaching unity and with a metal dispersion of 0.2 corresponding to large crystallites, are shown in Figure 12. A careful inspection of the curves shows that the resonance may be slightly more intense for the small metal clusters but the differences

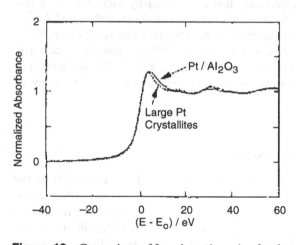

Figure 12 Comparison of L_{III} absorption edge for the platinum clusters in a Pt/Al_2O_3 catalyst (metal dispersion = 1.0) with that for large platinum crystallites (dispersion = 0.2) (from Ref.[241]).

are near to that of the experimental uncertainty. Similar results were obtained for supported Ir and Os catalysts. The earlier reports[242,243] showing larger effects attributed to metal dispersion were apparently confounded by sample thickness effects. Thus, it appears that the metal atom in a 10-Å crystal (a size where the ratio of surface to total atoms approaches unity) of Pt, Os, or Ir, supported on either alumina or silica, exhibits electronic properties that are not very different from an atom present in a metal crystal representative of the bulk metal. Similar electronic effects apparently are experienced in metal crystals that are 10 Å and larger; likewise, the extent of electronic interaction of the metal clusters with alumina or silica support is minimal. This conclusion appears to conflict with the EXAFS data described above.[238]

The chemisorption of a monolayer of oxygen on highly dispersed clusters of Pt, Ir, or Os results in a substantial increase of the intensity of the L_{III} edge. Data for an Ir/Al_2O_3 catalyst containing 1 wt % Ir, with clusters of Ir on the order of 10 Å, indicate that the number of unoccupied d states of the metal increases as a result of interaction with the chemisorbed oxygen. While the metal becomes more electron deficient, chemisorption of oxygen at room temperature does not lead to a bulk oxide.

Whereas most of the above studies focused on one feature of XAS, today this is not the case. For reforming catalysts, as well as for catalysis in general, the use of XAS characterization expands at a rapid rate. The increased availability of instrumentation and the availability of improved theoretical approaches have contributed to this development. Even so, it is easy to be misled and investigators new to the area should benefit from the overview of common artifacts and problems with data interpretation that are offered by Meitzner.[244] Some recent papers offer insight to new opportunities for the technique.[245–251] Lagarde[252] provides a short overview of the principles and some examples of the use of surface EXAFS. A major review of XANES analysis of catalytic systems under reaction conditions has recently appeared.[253] The application of the technique ranges from studies of catalyst preparation[254,255] to study of chemisorbed species.[256]

3.8 Nuclear Magnetic Resonance

Two types of NMR measurements are possible: a study of the nuclei of the catalyst itself and a study of nuclei of molecules adsorbed on the surface of the catalyst. The ^{195}Pt nuclei have sufficient natural isotopic abundance, a reasonably strong gyromagnetic ratio, and a spin of $I = 1/2$, which eliminates the complexity due to electric quadrupole effects. Furthermore, ^{195}Pt has one of the largest Knight shifts of any metal (-3.37%) and therefore offers the possibility of resolving the NMR peak of surface layers of atoms from that of the bulk.

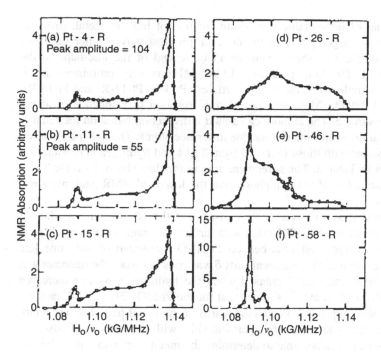

Figure 13 NMR absorption line shapes for six samples at 77 K and $v_o = 74$ MHz (from Ref.[143]).

Yu et al.[257] obtained ^{195}Pt NMR spectra for small unsupported Pt particles that corresponded to the bulk metal. Rhodes et al.[143] obtained spectra for a Pt supported on alumina. The dispersion of these catalysts were measured by chemisorption and by TEM. The line shapes for six samples scaled to the same area are shown in Figure 13. They are broad and extend the full range of Knight

Table 3 Dispersion (in %) of Pt Catalyst Samples Measured by Three Methods

Sample	Chemisorption	Microscopy	NMR
Pt-4-R	4	—	5
Pt-11-R	11	—	8
Pt-15-R	15	16	10
Pt-26-R	26	22	19
Pt-46-R	46	61	40
Pt-58-R	58	—	79

Source: Ref.[143].

shifts from that of nonmetallic compounds to that of bulk Pt metal. In going from samples of larger particles to those of progressively smaller particles, the intensity of the line shifts from the metallic end of the lineshape to the nonmetallic end. The bulk peak at 1.13 kG/MHz is very prominent in the lineshapes of samples with the larger particles (Pt-4-R, Pt-11-R, and Pt-15-R). The peak at 1.089 kG/MHz was taken to correspond to resonance of atoms present in the surface. The authors calculated a dispersion based on the relative area of the total lineshape and the one due to surface atoms. These dispersions are presented together with those obtained from TEM and hydrogen chemisorption measurements in Table 3. The agreement among the data shown in Table 3 was taken as confirmation of the hypothesis that the low-field NMR peak measures surface atoms.

de Ménorval and Fraissard[258] showed that the NMR chemical shift of hydrogen adsorbed on Pt-Al$_2$O$_3$ varied with surface coverage and Pt particle size. The spectrum of hydrogen adsorbed on Pt-Al$_2$O$_3$ consisted of only one line whatever the coverage. The chemical shift δ was constant when the diameter was greater than 70 Å and then decreased when the hydrogen coverage increased beyond 0.5. Rouabah et al.[259] showed that for the Pt-SiO$_2$ EUROPT-1 catalyst, the chemical shift depended on coverage, size of the metal particles, and temperature. From the variation of the chemical shift with the number of adsorbed hydrogen atoms, it was possible to determine the metal dispersion. It is claimed that the NMR technique does not require the stoichiometry of the chemisorption on the metal to be known. The average particle size determined by ^{1}H NMR was smaller than the dispersion based on chemisorption or electron microscopy.

^{129}Xe NMR has been used to estimate the average number of Pt atoms per cluster for samples of Pt contained in a NaY zeolite (Pt/NaY).[260] Thus, an average of four to eight Pt atoms per cluster was estimated when the progressive chemisorption of H$_2$ at room temperature in a bed of Pt/NaY powder was followed by Xe NMR. These results agreed with an average value of six Pt atoms per cluster which had been inferred earlier from the rapid exchange with D$_2$ of OH groups in a Pt/CaY zeolite as followed by IR spectroscopy.[261]

The determination of the number of Pt atoms per cluster attracted attention as it was the first time that the number of atoms in a supported metallic cluster smaller than about 1 nm had been reported.[262] However, the study of similar or identical samples of Pt/Y zeolites by TEM,[263] SAXS,[198] wide-angle X-ray scattering (WAXS),[264] and EXAFS[265] indicates Pt clusters of about 1 nm in size in the supercages of the Y-zeolite (1.3 nm in diameter), with about 16–40 Pt atoms per cluster.[143] In particular, a combined investigation of Pt/Y samples by TEM, SAXS, WAXS, and EXAFS suggested that the Pt clusters were substantially larger than those probed by IR and Xe NMR. Detailed analysis of the experiments has reconciled both types of observations to the higher value.[259,260,266–270]

Boudart et al.[271] utilized Pt supported on SiO_2, γ-Al_2O_3, or Y-zeolite for their NMR studies. The percent metal exposed for each sample was measured by titration of prechemisorbed O by H_2,[163] by irreversible H chemisorption, and by irreversible O chemisorption at room temperature in a standard volumetric system. The average size for the Pt clusters was obtained from the percent metal exposed, assuming a spherical shape and an average Pt surface number density of 1.10×10^{19} m^{-2}.[271]

The Xe NMR spectra were then taken at various values of nominal surface coverage, H/Pt$_s$. All spectra show two peaks: one with $\delta = 0.5$ ppm, assigned to Xe in the gas phase,[269] and another single peak with $\delta = 85-285$ ppm, corresponding to the average interaction of adsorbed Xe with the Pt surface and the support.[267] Boudart et al.[271] reported a puzzling observation in the discrepancy between the value of the chemical shift δ measured on prereduced Pt samples without any chemisorbed hydrogen. With all details of the reduction and evacuation procedures being the same,[267,271] straight lines for δ vs. H coverage extrapolate to the value measured for zero H coverage for all samples except Pt/γAl$_2$O$_3$. For this sample, the measured value of δ on the sample without H is considerably lower than that extrapolated from the δ vs. H coverage line. For as little as 10% coverage by H, the measured value of δ is exactly on the δ vs. H coverage line.

Boudart et al.[271] rationalized the anomaly by the following speculation. In vacuo, the shape of Pt clusters on SiO_2 or in Y-zeolite is approximately spherical, and the interface between metal and support is minimal. But with a γ-Al_2O_3 support, clusters of platinum about 1 or 2 nm in size probably have the shape of a pillbox or raft in vacuo as a result of the high interfacial energy. Indeed, high-resolution TEM reveals epitaxial growth of 2 nm clusters of Pd on $\gamma - Al_2O_3$[270] and with comparable lattice parameters, both metals should behave similarly on γ-Al_2O_3. In the raft-like shape, xenon interacts with the free Pt surface but senses the surface of the cluster in contact with the Al_2O_3 platelets. At this contact, there are weak bonds between Pt and oxygen ions of the Al_2O_3. Hence, the chemical shift of Xe is smaller than that corresponding to bare Pt, just as δ of Xe decreases as oxygen is added to bare Pt clusters.[267] When hydrogen is introduced to the sample, the weak bonding between the metal and the support disappears as a result of chemisorption of hydrogen.

NMR studies of CO adsorbed on L-zeolite and silica-supported small Pt particles show evidence of changes in the metallic nature of these particles with size.[272] Large Pt crystals on silica or on the exterior of the zeolite have conduction-band electrons that cause a Knight shift for adsorbed CO. On the other hand, small particles in zeolite cavities are diamagnetic and yield spectra that are similar to those of transition-metal cluster compounds. Thus, the NMR technique was advanced as a means of defining the metal dispersion due to particles in the zeolite cavities and on the exterior of the framework.

Han et al.[273] obtained dramatically different results for the platinum particles even though the static NMR and dispersion by chemisorption were similar. Unlike Sharma et al., Han et al. did not find the Pt particles to be diamagnetic; they emphasized the need to use multiple techniques to characterize catalysts.

^{13}C and ^{2}H NMR resonance data were obtained for carbon species (ethene or propene) adsorbed on Pt-alumina catalysts.[274,275] While not directly applicable to high-temperature naphtha reforming, the results indicate that progress toward that goal is underway.

3.9 Temperature-Programmed Desorption or Reduction

Temperature-programmed reduction (TPR) is one of the chemical methods for catalyst characterization; however, it does not provide a direct measure of chemical structure or chemical state. It provides a quantitative measure of the extent of reduction but does not provide direct information about what is being reduced. An analogous technique is temperature-programmed desorption (TPD). TPR suffers from a lack of information regarding the species undergoing reduction and whether the weight loss is due to reduction or desorption. It is possible to eliminate some of the uncertainty about the nature of the weight loss by monitoring the exit gas stream with a mass spectrometer. The TPR provides a 'fingerprint' of the reducibility of the platinum group metal catalysts.[276] For example, the reduction of Pt-, Ir-, and Pt-Ir-Al$_2$O$_3$ catalysts has been studied.[277] The reduction of platinum in the Pt-Al$_2$O$_3$ catalyst occurs over a broad temperature range that peaks at about 400°C while iridium oxide is reduced over a narrow but higher temperature range that is centered at about 550°C. In contrast to the two single metals, the reduction of Pt-Ir-Al$_2$O$_3$ is not a composite of the two curves representing the reduction of each metal on the support; rather, the reduction is effected at a lower temperature. This data are consistent with the formation of a Pt-Ir alloy on the support; however, the data do not provide direct evidence for the formation of an alloy.

Temperature-programmed techniques are being applied frequently to characterize coke on used catalysts. Querini and Fung[278] have utilized the technique to show (1) the morphology of the coke and (2) the gasification of coke on the alumina support. Jess et al.[279] showed that the coke burnoff is dominated by the coke deposited on the acid sites on the support.

3.10 X-ray or Ultraviolet Photoelectron Spectroscopy; Electron Spectroscopy for Chemical Analysis (ESCA)

Photoelectron spectroscopy is based on the photoelectron effect where ionization occurs and an electron is expelled from the sample with a kinetic energy that

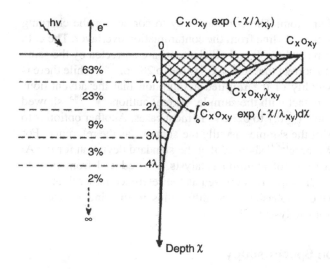

Figure 14 Contribution of successive layers of thickness 8 to the total XPS intensity for $2 = 90°$ (from Ref.[280]).

depends on the energy of the incident photon, the element that is ionized, and the valence of that element in the sample. Depending on the energy of the incident photon, emission of the photoelectrons will be from the valence band only (ultraviolet photoelectron spectroscopy, UPS) or from both the valence and core levels (X-ray photon spectroscopy, XPS). XPS provides a measure of the concentration and chemical valence of only those elements in the surface of the sample. However, a knowledge of the mean free paths of electrons in solids is not completely defined even today. A schematic of the escape probabilities for an electron through surface layers to the vacuum of the instrument is illustrated in Figure 14.

For the characterization of naphtha-reforming catalysts, it is usually necessary to pretreat the sample prior to analysis by UPS or XPS. This is commonly accomplished in a sample cell, attached to the chamber used for the measurement. Following the pretreatment the sample is transferred without exposure to the atmosphere to the high-vacuum chamber where the actual measurements are made. In nearly all of the instruments it is necessary to form the sample into a pellet so that a reasonably smooth surface is provided to the photon beam and to maintain sample integrity during the pretreatment operations.

Semiconductor or insulator samples will undergo charging since, in general, ions are formed more rapidly than the charge can be dissipated through conduction to the instrument. Thus, the measured binding energy must be corrected to account for the charging effect if the data are to have value for defining the valence of an atom. The common way to correct for the charging effect is to reference it to the C 1s line from the contamination overlayer. The C 1s line is normally taken to be 284.6 eV, and all elements are corrected by the same amount that is required to correct that of the C 1s line to 284.6.[280] While there is some question of the validity of the implied assumption that the adventitious carbon layer is in good contact with the sample, several authors[281–283] showed that it is sufficiently reliable ($\pm$ 0.1 to 0.2 eV) in most cases. Another option is to use one of the elements of the sample, usually the support, as the reference. For example, Ogilvie and Wolberg[284] showed that the standard deviation for the Al 2p binding energy in a series of alumina catalysts was reduced from 0.49 to 0.14 eV when O 1s from the support is chosen as the reference rather than C 1s. Round-robin studies have emphasized the difficulties in arriving at accurate binding energy values for catalysts.[285]

3.11 Auger Electron Spectroscopy

Auger electrons are ejected when an atom containing core level vacancy, created as, for example, in the XPS measurements where an X-ray photon interacted so as to eject an electron, de-excites by emitting an electron from an upper orbital. The energy needed for ejecting this electron comes from the nearly simultaneous decay of a third electron from another upper level into the inner core vacancy.

Because XPS and Auger electron spectroscopy (AES) are similar, spectrometers are frequently built that allow one to obtain either type of spectrum. AES and XPS have approximately the same intensity toward the surface. AES is usually more sensitive with the ability to detect surface species at lower concentrations than XPS. Consequently, AES is utilized more frequently in depth profiling than is XPS. To effect depth profiling, successive layers of the material are sputtered from the sample using a beam of ions and recording the surface concentration of the elements of interest between the periods of sputtering. One application of depth profiling would be to determine whether there was an enrichment of an element on the surface, e.g., tin in a Pt-Sn-Al_2O_3 catalyst.

Recent advances suggest that AES may find wider applications in catalyst characterization. Angle-resolved AES (ARAES) provides a method whereby the primary beam incidence angle and the collection angle can be varied independently from 0 to 80° off sample normal.[286] A surprisingly large increase in surface sensitivity is obtained by using grazing incidence and collection angles. Techniques have been developed for collecting high-energy-resolution AES (HRAES) spectra rapidly, making it useful for obtaining chemical state information from AES. It is possible to combine ARAES and HRAES to obtain depth-sensitive, chemical-state information as illustrated in the study by Asbury and Hoflund.[287]

3.12 Ion Scattering Spectroscopy

Ion scattering spectroscopy (ISS) is extremely sensitive to the outermost layers of a surface and this makes it a particularly effective tool for studying the adsorption of a species onto a surface. Sensitivity factors have been used for quantitations of the surface of binary alloys, but little has been done on real catalyst surfaces. A linear relationship exists between the ISS signal and the coverage at the beginning of the adsorption. The scattered ion yield is directly related to the number of surface atoms. Attenuation of the ion yield may result from a number of factors, including matrix effects, preferential sputtering, surface roughness, and a changing background. A determination of neutralization probabilities is difficult and for this reason most catalyst ISS data are given in terms of ratios instead of absolute intensities. A number of factors prevent a calculation of neutralization probabilities from first principles.

Dwyer et al.[288,289] used ISS and TPD to show that the amount of chemisorption of hydrogen and CO was directly dependent on the number of exposed platinum atoms (Fig. 15). The authors interpreted the data to show that the deposited TiO_2 suppressed CO or H_2 chemisorption by blocking the adsorption sites of the platinum. The chemisorptive capacity of the system decreased upon reduction but was recovered upon reoxidation.

It is also possible to study the dispersion of a supported oxide or metal using this technique. Ewertowski et al.[290,291] impregnated γ-alumina with chloroplatinic acid to various extents and then calcined and reduced the materials. They observed that with increasing coverage by Pt metal a decrease in intensity of the narrow component of the momentum distribution curve was obtained. It was verified that the change was due to Pt and not the Cl added during catalyst preparation.

3.13 Infrared Spectroscopy for Metallic Function

The adsorption of hydrogen onto platinum supported on γ-alumina produces bands at 2105 and 2055 cm^{-1}.[292] The high-frequency band is due to more weakly absorbed H. These IR bands gave the shift expected for the isotope effect. The structure associated with the more weakly bonded hydrogen was believed to be due to two hydrogen atoms adsorbed per platinum and the more strongly bound hydrogen to a single adsorbed hydrogen atom per metal atom. Adsorption of CO on the metallic function likewise produces two bands.[293] This has been interpreted as CO bonded to one Pt and to multiple Pt atoms through direct and bridging bonding, respectively. Bridging-type bonding becomes more pronounced at higher surface coverage. In some instances, especially at higher coverage, there may be multiple CO bonding to a single metal atom. This finding

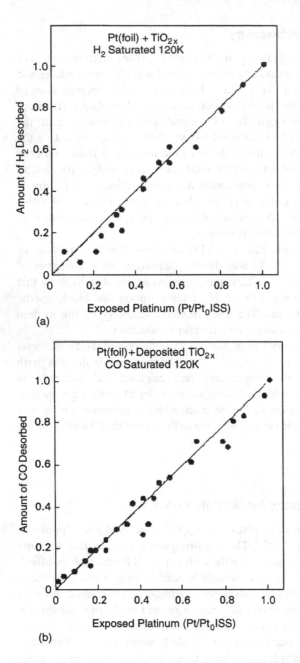

Figure 15 Fractional coverage of (a) H_2 and (b) CO that can be chemisorbed at 120 K vs. the fraction of exposed Pt as measured by ISS (from Ref.[288]).

made investigators aware of the uncertainty associated with chemisorption measurements that utilize $CO/Pt = 1$ to calculate dispersion.

3.14 Calorimetry

Lantz and Gonzalez[294] constructed a calorimeter and measured the heats of adsorption of hydrogen on platinum as a function of Pt crystal size. They found that the initial heat of adsorption decreased only slightly with increasing dispersion. A maximum in the heat of adsorption was obtained for all samples at a surface coverage, θ, of 0.4. The heat of adsorption should not increase with coverage and several possibilities to account for this, including diffusion into the powder, were discussed. Sen and Vannice[295] later reported results that were similar.

4 CHARACTERIZATION OF PLATINUM–RHENIUM CATALYST

TPR data have been used more extensively for the Pt-Re system than for many of the other reforming catalyst formulations. Johnson and LeRoy[296] obtained evidence that the reduction stopped with rhenium in the oxidized state of Re^{4+}. Webb[297] obtained data using this technique that showed both platinum and rhenium were reduced to the metallic state. Johnson[298] pointed out that there was not necessarily any contradiction between these data since Webb used very dry conditions whereas Johnson and LeRoy employed a water partial pressure

Table 4 Isothermal and Dynamic TPR of Pt-Re-Alumina Catalysts[a]

Temp. (°C)	Isothermal[a]			Dynamic TPA[c]		
	Extent of reduction (%)[b]			Extent of reduction (%)		
	15 min	30 min	60 min	Temp. range (°C)	Pt	Re
200	11	20	34	25–300	100	0
300	57	65	72	25–500	100	76
500	70	75	81	25–700	100	100
550	84	89	95			

[a]Catalyst containing 0.375 wt % Pt/0.2 wt % Re/alumina calcined at 525°C.
[b]Assuming that Pt and Re are in the +4 and +7 valency states, respectively, in the oxidized material.
[c]Heating rate of 1°C/min.
Source: Ref.[299].

typical of reforming conditions. Water presumably would serve to maintain rhenium in an oxidized state.

Subsequent TPR studies indicate that rhenium is reduced to the metal even under rather mild conditions. McNicol[299] carried out the reduction in TPD experiments as well as in fixed-bed reactors (Table 4). McNicol did not observe an influence of Pt on the reduction as had been reported in an earlier study by Bolivar et al.[300] and concluded that the difference between the observations was because the studies started with calcined and uncalcined samples, respectively.

Yao and Shelef[301] utilized chemisorption, TPR, and electron paramagnetic resonance spectroscopy (EPR) techniques to investigate the reduction of Re/alumina catalysts. They reported that Re interacts strongly with γ-alumina and can be reduced to the metal in hydrogen only at temperatures greater than 500°C. Even when reduced to the metal, Re did not chemisorb hydrogen at room temperature. Reoxidation, even at 500°C, produces Re^{4+}, and the limiting coverage of alumina surface by this species is about 10%. This limitation in oxidation explains why Re is not volatilized as Re_2O_7 during catalyst regeneration. At high loadings, the three-dimensional phase can be reduced by hydrogen to Re^0 at 350°C. The interaction of oxygen with the reduced dispersed phase at 25°C produces species identified by EPR as surface nondissociated oxygen molecule-ions (O_2^{2-}) and Re^{2+} ions.

Further studies[302–304] confirmed the reducibility of rhenium to Re^0 when present as Re/alumina or Pt-Re/alumina. Moreover, these studies showed that the ease and extent of reduction is dependent on the sample history. Studies indicated that the reduction temperature decreases with increasing loading of Re and with increasing oxidation temperature;[304] that for a freshly oxidized sample about 90% reduction of Re^{7+} to Re^0 took place up to 600°C;[302] that after reoxidation the associated TPR spectra exhibited a shift to higher temperatures with increasing oxidation temperature; and that rhenium could easily reoxidize to the Re^{7+} state. The Re(II) state was considered to reflect oxygen chemisorption. Wagstaff and Prins[302] found that adsorption of oxygen at temperatures up to 100°C leaves the bimetallic clusters largely intact, but subsequent high-temperature treatment in the absence of extra oxygen leads to segregation of Pt and Re species. These authors therefore take the view that 'bimetallic clusters' are formed and that these, in the presence of oxygen, are thermodynamically unstable, but that under mild conditions the rate of segregation is slow.

The results shown in Figure 16 are typical of those obtained by most investigators for Pt-Re/alumina catalysts since they show that at least a portion of the Re is reduced at a lower temperature in the bimetallic catalyst than in a Re/alumina catalyst. Increasing oxidation temperature of the freshly prepared Pt-Re/alumina catalyst causes an increase in the fraction of Re that is not reduced at the lower temperature (Fig. 17); however, for these bimetallic catalysts never does the TPR of the Pt-Re/alumina catalyst become equal to one that is the sum of a

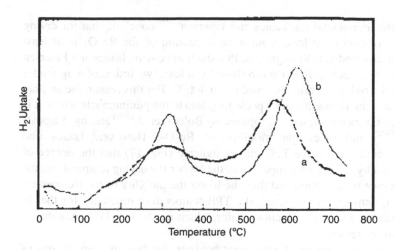

Figure 16 TPR of Pt-Re vs. superimposed Pt + Re. (a) 0.35 wt % Pt-0.35 wt % Re/Al$_2$O$_3$. (b) 0.4 wt % Al$_2$O$_3$ and 0.35 wt % Re/Al$_2$O$_3$. Preoxidized at 500°C (from Ref.[305]).

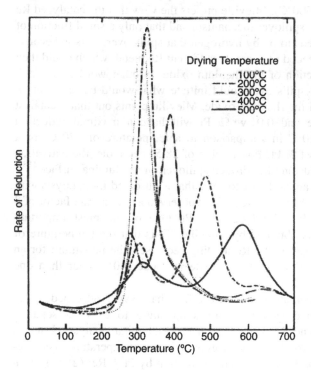

Figure 17 TPR of Pt/Re/Al$_2$O$_3$ (from Ref.[76]).

mixture of the monometallics. Isaacs and Petersen[303] conclude that for drying temperatures of 500°C or less, a substantial fraction of the Re_2O_7 is at least partially hydrated and able to migrate to Pt reduction centers. Isaacs and Petersen conclude that hydrogen spillover is too slow (by at least two orders of magnitude) to allow for the reduction rate observed up to 400°C. For this reason the authors conclude that some rhenium oxide specie migrates to the platinum site where it is reduced; the presence of water, as proposed by Bolivar et al.[300] and by Wagstaff and Prins,[302] influences the mobility of Re_2O_7. However, Isaacs and Petersen[303] contend that the TPR curves indicate (Fig. 17) that the degree of hydration is set by the drying temperature: the higher the drying temperature, the lower the degree of hydration and thus the lower the mobility of the Re species. As the drying temperature is raised, the TPR temperature necessary for the Re species to migrate to the Pt reduction center, which results in a TPR peak due to Re reduction, increases.

There are many points of agreement between the data and conclusions of Mieville[304] and Isaacs and Petersen;[303] however, there are some major differences. Mieville found that the superimposed Pt + Re TPD curve was essentially the same as that of a Pt-Re-alumina catalyst that had been preoxidized at the same temperature (500°C). Mieville prefers the view that the catalyzed Re reduction may be due to a spillover mechanism and that only a small fraction of the total Re reduction needs to be by hydrogen that spills over. This is because only a small fraction is needed to generate sufficient Re metal, which could then catalyze the further reduction of the rhenium oxide. In other words, the small amount of hydrogen that spills over could initiate what would be in effect an autocatalytic reduction of the rhenium oxide. Mieville points out that a catalyst containing 0.35 wt % Re and 0.01 wt % Pt will have a maximal reduction temperature of Re of 410°C in comparison to a temperature of 370°C for a catalyst containing 0.35 wt % Pt. For a value of 28.5 kcal/mole, the activation energy for the rate of diffusion of hydrogen spillover on an alumina surface,[181] the rates should differ by about the factor of 4 that is observed for catalysts with the above Pt loadings. Ambs and Mitchell[305] observed a similar rate factor of 3 for catalysts containing 0.5 and 0.02 wt % Pt. One of the most important implications of the results obtained by Mieville[304] is that the temperature of maximal TPR rate increases as the Re loading decreases; the maximum for an Re/alumina catalyst containing 0.35 wt % Re is about 150° higher than one containing 1.1 wt % Re.

Menon et al.[306] took advantage of the fact that oxygen adsorbed on Re cannot be titrated by hydrogen at room temperature to effect a separate determination of the Pt and Re dispersion in Pt-Re-alumina catalysts. Using a pulse technique, they carried out the following room temperature measurements: (1) oxygen pulsing to measure chemisorption by Pt + Re; (2) reduction in hydrogen to convert Pt-O to Pt-H; (3) renewed oxygen pulsing to yield chemisorption by Pt only; (4) the difference between (1) and (3) for the chemisorption by Re only. These authors found that the dispersion of Pt is little altered in a composite Pt-Re catalyst compared to that in a Pt catalyst of the same metal content; however, the dispersion of Re seemed to be considerably enhanced in the composite catalyst. Eskinazi[307] compared hydrogen and oxygen chemisorption and hydrogen titration data for a series of catalysts with different Pt/(Pt + Re) ratios. The maximal metal surface areas were obtained with 55 mol% rhenium, alumina supports with surface areas greater than $150 \, m^2/g$, and a catalyst pretreatment involving oxidation at 500°C prior to reduction.

Kirlin et al.[308] used XPS to follow the reduction of two Pt-Re-alumina catalysts as well as an Re-alumina catalyst. They reported that reduction of the bimetallic catalysts with hydrogen at 500°C and 1 atm gave predominantly Re^{4+}. However, the addition of small amounts of water to the hydrogen feed effected the reduction to Re^0 and the complete removal of chloride from the catalyst. In this case water assists in reduction whereas Johnson[298] agreed that water maintained a higher oxidation state. Reduction of the catalyst in butane + hydrogen at 500°C resulted in the reduction to Re^0 with retention of the chloride. Under identical conditions, the Re in an Re-alumina catalyst was reduced only to Re^{4+}, indicating that Pt is necessary to effect the reduction to Re^0. The XPS data are therefore consistent with the metals being present in the zero valent state in the bimetallic-reforming catalyst under reaction conditions. Adkins and Davis[309] confirmed the results of Kirlin in their XPS study.

EPR data indicated that the majority of Re was present as Re^0 in a Pt-Re-alumina catalyst.[310] The surface of the reduced bimetallic Pt-Re-alumina catalyst contains both Re^0 and Re^{4+}, although the Re^{4+} could account for only about 10% or less of total Re at the loadings investigated (0.2–0.3 wt %). The results showed that the Re^{4+} surface sites do not chemisorb CO, whereas Re^0, either separately dispersed or in contact with Pt, strongly chemisorbs CO. The authors also concluded that Re^0 was the valence state of Re that is decisive for the catalytic function of the Pt-Re-alumina-reforming catalyst.

Onuferko et al.[311] utilized XPS, EXAFS, and XANES in their study of Pt-Re-alumina catalysts. Following *in situ* reduction, all of their results show that Pt^0 and Re^{4+} are the dominant species. The Pt in the reduced catalyst was different from the bulk metal or the monometallic catalyst. The data also suggested that Re was not significantly associated with Pt.[311,312] Kelley[313] utilized ISS and energy dispersive X-ray analysis (EDX) in TEM for analysis of the Pt-Re-alumina catalyst and found that Re was not significantly associated with Pt but was widely dispersed on the support surface. Sulfiding essentially covered the Re but left most of the Pt exposed.

Meitzner et al.[314] obtained a different structure for reduced 1:1 atomic ratio Pd-Re- and Pt-Re-alumina catalysts using EXAFS. In agreement with most of the TPR data, essentially all of the rhenium had an oxidation state of zero after reduction in hydrogen at 775 K. The EXAFS data also provided evidence for significant coordination of rhenium to platinum in bimetallic clusters. These bimetallic clusters were not characterized by a single interatomic distance. This finding could be an indication of intercluster or intracluster variation of composition. This latter possibility is commonly associated with one of the components concentrating at the boundary or surface of a cluster as a consequence of differences in the surface energies of the components. If this were so, the average number of the nearest-neighbor metal atoms about one component will be larger than the number about the other. This was not observed, suggesting intercluster variation. For both catalysts, the interatomic distance for the unlike pair of atoms was significantly shorter than that for either pair of like atoms in the clusters or the pure metal. This indicates that the Pt-Re bonds in the cluster are stronger than would be anticipated.

Meitzner et al.[314] also examined a reduced Pt-Re-alumina catalyst after exposure to H_2S. The rhenium EXAFS results indicated that the sulfur had little disruptive influence on the clusters. There was no evidence for the formation of ReS_2. The Pt-Re and Re-Re bond distances do not change significantly when sulfur is adsorbed, and the extent of coordination of rhenium to platinum was essentially unchanged. It was, therefore, concluded that the catalyst contained Pt-Re bimetallic clusters with sulfur strongly chemisorbed on the rhenium.

Sinfelt and Meitzner[226] report XANES studies of a Pt-Re-alumina (1% Pt, 1% Re). The data for the L_{III} edges of rhenium show that the Re in the catalyst does not differ significantly from that in bulk rhenium and is much less intense than the edge for rhenium in the $+7$ oxidation state in ReO_4^-. Similar measurements were made for the reduced catalyst following sulfiding. These data show that the edge of rhenium after exposure to hydrogen sulfide does not differ significantly from that before exposure to sulfur but is much less intense than for ReS_2. These edge absorption data are consistent with the EXAFS data in showing that the Pt-Re clusters are not disrupted by sulfur. The sulfur appeared to be in the form of sulfur atoms chemisorbed on exposed rhenium atoms in the Pt-Re cluster.

Bazin et al.[315] obtained somewhat different results in their study of Pt-Re-alumina (240 m^2/g alumina). Both the white line and the EXAFS data for platinum indicated that it was present in the zero valent state. However, whereas an average environment of one Pt atom in a monometallic catalyst was composed of 6 metal atoms at 2.75 Å and 0.7 oxygen atom at 2.04 Å, in the bimetallic Pt-Re catalyst the analysis leads to 3 (reduction at 300°C) and 5 (reduction at 500°C) metal atoms around one Pt with hardly a detectable amount of oxygen. This shows that Pt is present in a more highly dispersed state in the bimetallic catalyst than in the monometallic one. The white line of the rhenium L_{III} edge does not resemble the metallic state and the EXAFS indicates that one rhenium atom is surrounded by 6 to 8 metal atoms and 2 oxygens, and hence in an oxidized state. In their model, a rhenium oxide interface is between the platinum and the alumina, which they contend is in line with other experimental and theoretical results.[314]

Hilbrig et al.[316] utilized coupled XANES and TPR techniques to study the reduction of Pt-Re-alumina catalysts. They measured the chemical shift of the XANES for the two metals in the catalyst relative to metallic Pt and Re. The authors then compare the derivative of the change in the chemical shift with that for hydrogen uptake (both as a function of temperature of reduction) on the reference monometallic catalysts, and find that the two methods of following the reduction are quite similar in the case of each metal. Pt is completely reduced

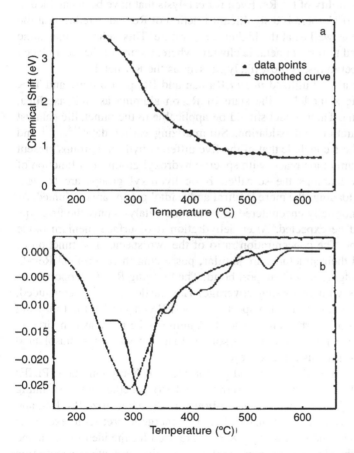

Figure 18 TPR of 0.3 wt % Pt-0.3 wt % Re/Al$_2$O$_3$ pretreated at 250°C. (a) The chemical shift (relative to metallic Re) as a function of temperature of reduction. (b) The derivative of the chemical shift (–) and H$_2$ consumption (– –) as a function of temperature of reduction (from Ref.[316]).

when the temperature has reached 380°C as indicated by a chemical shift of 0.0 eV. A sharp peak at 460°C for Re corresponds to reduction to Re^{4+}; further reduction occurs so that the average Re oxidation state is $+1.5$ at 650°C. There is a considerable difference in the reduction profile for the Pt-Re-alumina catalyst compared to the monometallic catalysts. For bimetallic catalyst with higher Re loading, the Re^{7+} to Re^{4+} reduction occurs around 275°C, and for lower Re loading at 310°C. Pt in both catalysts shows a reduction profile that is similar to the Pt monometallic catalyst. In both samples, however, the chemical shift is not zero even at 450°C; this is interpreted to be due to Pt-Re interaction rather than incomplete reduction of Pt (Fig. 18). Pretreatment at higher temperatures decreases the reducibility of the Re. Even for catalysts that have been calcined at either 400°C or 580°C, there is a significant fraction of Re that is reduced at the lower temperature as well as at the higher temperature. This indicates that some of the Re is reduced to form bimetallic clusters whereas some of the Re, reduced at the higher temperatures, is most likely present as the Re metal.

Vuurman et al.[317] utilized in situ Raman and IR spectroscopy and TPR data in developing a model of the state of Re on alumina as well as silica, zirconia, and titania. Their model should be applicable to the bimetallic catalyst for catalyst preparation and oxidation. Summarizing earlier data[318-320] and their data, the authors conclude that of the five different hydroxyl groups present on alumina, rhenium oxide reacts with specific hydroxyl groups as a function of coverage. At low loadings the so-called basic hydroxyl groups are titrated, whereas at higher loadings the more 'neutral and acidic' groups are consumed. At the low loadings normally encountered in reforming catalysts only the first type of bonding would be expected. After dehydration two surface rhenium oxide species are present. The concentration ratio of the two species is a function of the coverage, and their structures are similar, possessing three terminal Re=O bonds and one bridging Re−O-support bond. The bridging Re−O-support bond strength decreases with increasing coverage. Hardcastle et al.[321] concluded, based on XANES and laser Raman spectroscopy (LRS) data, that the following reversible changes in the rhenium species on alumina take place as dehydration proceeds: perrhenate ion in solution → solvated surface species → unsolvated surface species → gaseous dimeric Re_2O_7.

Jothimurugesan et al.[322] utilized proton-induced X-ray emission (PIXE) and Rutherford backscattering spectrometry (RBS) together with scanning electron microscopy (SEM) and chemisorption data to characterize Pt-, Re-, and Pt-Re-alumina catalysts. PIXE utilizes a beam of energetic protons to excite the inner shell electrons, and the resulting X-rays are used for the identification and quantification of the elements of interest. PIXE can give in-depth concentration profiles of a metal. RBS is based on the elastic backscattering of ions from the catalyst sample. This technique has the possibility of high sensitivity, and surface concentrations on the order of a monolayer of heavy atoms present on a lighter atom support can easily be detected. RBS data present surprising results. For the monometallic Pt catalyst, the depth profile indicated that Pt was uniformly distributed from the surface to the interior of the pellet for samples representing the oxidized, reduced, or used catalyst. For the monometallic Re catalyst, the concentration increased slightly from the surface to the interior for the oxidized, reduced, and used catalyst. For the bimetallic catalyst, the Pt-Re depth profile for the oxidized catalyst resembles that of the Re monometallic Re catalyst; however, there is about a five-fold increase in the metal concentration near the surface of the support pellet for the reduced and used catalyst samples.

Kelly et al.[313,323] used ISS to study Pt-Re-Al$_2$O$_3$. They found that Re was highly dispersed on the surface. Upon sulfiding of the catalyst, they found that rhenium was covered and platinum was exposed. The increase of the Pt + Re peak was strongly associated with a decrease in either the sulfur or chlorine peak.

The bimetallic nature of the catalyst continues to be debated. A study using scanning transmission electron microscopy (STEM) and energy-dispersive X-ray spectrometry (EDX) produced data to show Pt-Re bimetallic particle formation in the reduced catalyst.[324] The highest degree of interaction was obtained if the catalyst was not dehydrated prior to reduction. Later studies by this group utilized EXAFS to extend the results, showing the importance of the pretreatment temperature and atmosphere, and support the STEM/EDX data.[325]

The role of sulfidation continues to be an area of interest. EXAFS data[326] provided support for the general model proposed by Biloen et al.[327] and amplified by Sachtler[328] where the mixed metal cluster contains 'chains' of Pt-Pt and Re-Re, which are connected by Pt-Re bonds (Fig. 19). Bensaddik et al. found that after sulfidation both Re and Pt atoms are coordinated to about two sulfur atoms. Following a second reduction, the Re environment is essentially unchanged but the Pt atoms are coordinated by about one sulfur. Chlorine was shown to have no effect on the stability of the metal particle during sulfidation and reduction.

5   CHARACTERIZATION OF PLATINUM–TIN–ALUMINA CATALYSTS

TPR studies of Pt-Sn-Al$_2$O$_3$ catalysts suggest that Sn is not reduced to the zero valent state.[329,330] Lieske and Völter[331] reported, based on TPR studies, that a

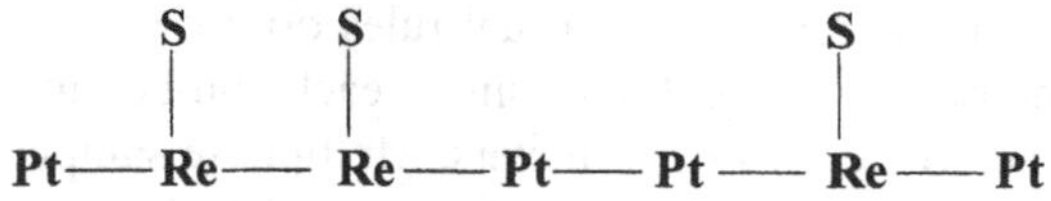

Figure 19   Biloen's model (from Ref.[327]).

minor part of the tin is reduced to the metal, which combined with Pt to form 'alloy clusters,' and the major portion of the tin is reduced to the Sn(II) state. The amount of alloyed tin increased with increasing tin content. This disagrees with TPR data reported by Burch and Garla[332,333] who conclude that the tin is reduced only to Sn(II) and not to elemental tin. They believe that the altered Pt properties are due to an interaction with Sn(II) species, which is considered by Burch and Mitchell[334] to be the primary influence on the reforming properties. This agrees with the work of Müller et al.[335] who used hydrogen and oxygen adsorption measurements to show that neither metallic tin nor a Pt/Sn alloy was present in their catalysts.

In the following, data are reported from methods that make a more direct measure of the chemical or physical state of the elements present in Pt-Sn-alumina catalysts. XPS studies permit one to determine the chemical state of an element but the data do not permit one to define, for example, whether Sn(0), if present, is in the form of a Pt-Sn alloy. Furthermore, the major Pt XPS peak is masked by a large peak of the alumina support. Thus, XPS can only provide data to show that an alloy is possible; it cannot be used to prove the presence of a Pt-Sn alloy.

The early XPS studies revealed that tin is present only in an oxidized state.[330,336,337] Although some observations were due to the use of oxygen-containing pump oils to maintain the high vacuum. Li et al.[338,339] later reported that a portion of the tin in Pt-Sn-alumina catalysts was present in the zero valent state; furthermore, the apparent composition of the Pt-Sn alloy contains increasing PtSn ratios as the Pt/Sn ratio present on the catalyst is increased. However, it must be kept in mind that XPS provides a measure of the surface, and not the bulk, composition; Bouwman and Biloen[340] showed that tin is concentrated in the surface of reduced unsupported Pt/Sn alloys, in general agreement with others.[341–347]

^{119}Sn Mössbauer data have been utilized in a number of studies.[348–363] Direct evidence for PtSn alloy formation was obtained from Mössbauer studies;[348,358–361] however, many of these studies were at high metal loadings and even then such a complex spectrum was obtained that there was some uncertainty in assigning Sn(0) to the exclusion of tin oxide phases. Mössbauer results clearly show changes upon reduction in hydrogen, but the width of the peaks for the reduced sample prevents a specific assignment for some states of tin. This was exemplified by the results of Kuznetsov et al.[363]

There are many complicated relations between the electronic state of the tin ion and the associated Mössbauer spectrum. The usual rules do not always apply. However, this is normally not a problem since each tin-bearing phase usually has a distinct characteristic set of parameters. Multiphase sample spectra can be resolved using standard curve fitting and statistical techniques so that the contributions of each phase may be separated. The relative amount

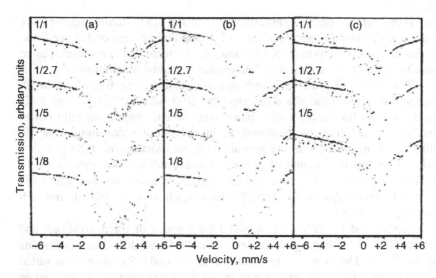

Figure 20 ^{119}Sn Mössbauer spectra for reduced Pt-Sn catalysts: (a) Pt-Sn/Al$_2$O$_3$ (250 m^2/g), (b) Pt-Sn/Al$_2$O$_3$ (110 m^2/g), and (c) Pt-Sn/SiO$_2$ (700 m^2/g). (Pt:Sn mole ratios are indicated by the numbers shown in the figure) (from Ref.[362]).

of tin in each phase can be determined from the peak area associated with that phase.

Li et al.[362] utilized a series of 1% Pt/variable-% Sn catalysts for Mössbauer studies. Tin was observed to be present in forms whose isomer shifts were similar or the same as SnO$_2$, SnO, SnCl$_4$, SnCl$_2$, Sn(0), and PtSn alloy when alumina was the support. Representative Mössbauer spectra are shown in Figure 20. If it is assumed that the Pt/Sn alloy corresponds only to a 1:1 alloy, one finds that for lower SnPt ratios (5 or less), little difference is observed in the extent of alloy formation and the distribution of the oxidized species for a low and high surface area alumina support. The fraction of Pt present in an alloy phase increases with increasing tin composition and only approaches complete alloy formation at SnPt > about 5.

For Pt supported on a particular coprecipitated tin oxide–alumina catalyst, a smaller extent of alloy formation was found than a material prepared by impregnation with the chloride complex.[360,362] In another study, a catalyst was prepared by impregnating a nonporous alumina with a surface area of 110 m^2/g with an acetone solution of Pt$_3$(SnCl$_3$)$_{20}$$^{2-}$.[364] A 5 wt % Pt catalyst was reduced *in situ* in the chamber of an XRD instrument;[365] the diffraction pattern matches very well the pattern reported for PtSn alloy. In the pattern it can be seen that with the 5 wt % Pt catalyst, a small fraction of the Pt is present as crystalline Pt. Similar results for the PtSn alloy are obtained for a catalyst that contains only

0.6 wt % Pt, with the same Sn/Pt ratio. These *in situ* XRD studies support alloy formation with a stoichiometry of PtSn = 1:1. The Sn in excess of that needed to form this alloy is present in an X-ray 'amorphous' form and is postulated to be present in a shell layer with a structure similar to that of tin aluminate.

XRD studies of other variations indicated that, irrespective of the Sn/Pt ratio, the only crystalline phase detected by XRD was SnPt (1:1). The XRD intensity of lines for the SnPt alloy phase increase with increasing SnPt ratios, indicating the presence of unalloyed Pt in the samples containing low tin loadings. The integrated intensity measured for the most intense PtSn peak[123] clearly shows that the amount of Pt present as crystalline alloy increases with increasing Sn/Pt ratios. Likewise, the crystallite size calculated from the XRD data show that the radius of the crystallite increases from about 10 to 16 nm with increasing tin content.

XANES and EXAFS were obtained for a series of dried, oxidized, and reduced (773 K, 1 bar hydrogen) preparations of Pt/Sn-loaded silica and alumina supports.[366,367] The Pt was maintained at 1 wt % and the Sn content was varied from 0.39% to 3.4%. In general, it was found that increasing the Sn loading on either support decreased the d-band vacancy for Sn. In contrast, alloying Pt with Sn leads to an increase in the d-band vacancies, in general agreement with Meitzner et al.[368] The Sn K-edge spectra indicate that the arrangement of atoms about the tin species was essentially the same for the high-area alumina and high-area silica supports. Sn-O atom pairs dominate even after reduction, but their contribution decreases somewhat with increasing Sn loading. The two supports led to profoundly different configurations of atoms about the platinum.

An electron microdiffraction technique was employed to identify crystal structures developed in two Pt-Sn-alumina catalysts. One catalyst was prepared by coprecipitating Sn and Al oxides and then impregnating the calcined material with H_2PtCl_6 to give a PtSn = 1:3 atomic ratio. The second catalyst was prepared by coimpregnating Degussa alumina with an acetone solution of chloroplatinic acid and stannic chloride to provide PtSn = 1:3. PtSn alloy was not detected by X-ray diffraction for the coprecipitated catalyst although evidence for PtSn alloy was found for the coimpregnated catalyst.[369] Electron microdiffraction studies clearly showed evidence for PtSn = 1:1 alloy phase in both catalysts. Evidence for minor amounts of the PtSn = 1:2 phase was also found for the coprecipitated catalyst.

The state of Pt and Sn remains an area of great interest and one that is unresolved. The connection of the state of the metals to catalyst performance is of even greater interest. The importance of characterization data can be inferred from the title of a recent patent application: 'Bimetallic supported platinum–tin naphtha reforming catalysts with strong Mössbauer interaction of tin with platinum'.[370] Pt-Sn was included in a group of multimetallic-reforming catalysts that were characterized by TEM and EDX to define the influence of alloy formation on catalyst activity.[371] Sn and Ge were claimed to modify the properties of the catalyst mainly by a geometrical effect. More importantly, it was claimed that the formation of bulk Pt-Sn and Pt-Ge alloys contributed to the overall rate of deactivation of these catalysts. The activity was based on n-octane conversion at low pressures and presumably with an acid-supported catalyst. It has been shown that for Pt-Sn-alumina catalysts the activity and product selectivity has a pressure dependence; furthermore, the bifunctional conversion pathway was shown to be more rapid that the metal monofunctional pathway.[372–374]

6 CHARACTERIZATION OF OTHER BIMETALLIC CATALYSTS

Following the successful introduction of the Pt-Re-Al$_2$O$_3$ bimetallic catalyst, a rash of metal combinations appeared in the open and patent literature. These catalyst combinations can be conveniently grouped into three general classifications of alumina-supported catalysts: (1) a group VIII metal, almost always Pt, together with a nonreducible metal; (2) a group VIII metal, usually Pt, together with a metal at least partially reducible to the metal; and (3) a combination of group VIII metals or a combination of a group VIII and group IB metal. In view of the large number of catalysts reported (e.g., more than 500 papers concerning Pt and Ir catalysts appeared during the past 10 years), space permits only a brief outline of the results for these catalysts.

6.1 Group VIII or Group VIII/Group IB Bimetallics

Of these catalyst types, the Pt-Ir combination appears to have been the most widely studied for naphtha reforming and is reported to have been used in commercial operations.[375] Much of the early characterization work on Pt-Ir-Al$_2$O$_3$ catalysts has been reviewed by Sinfelt.[375] Chan et al.[376] followed the steps in the preparation of Pt-Ir-alumina catalysts using LRS: impregnation with the appropriate chloroplatinic acid and chloroiridic acid solutions, and heating in air at varying temperatures. The spectra indicated significant differences in the nature of the surface species remaining when Pt/Al$_2$O$_3$ and Ir/Al$_2$O$_3$ preparations were subjected to such thermal treatments. After drying at 110°C, the Pt/Al$_2$O$_3$ preparation exhibits Raman bands assigned for the PtCl$_6$$^{-2}$ ion, whereas Ir/Al$_2$O$_3$ yields a spectrum significantly different from that reported for the IrCl$_6^{2-}$ ion. With increasing temperature, both the platinum and the iridium species decompose with loss of chloride ligands. After being heated in air at 500°C, the Ir/Al$_2$O$_3$ sample exhibits a Raman spectrum characteristic of crystalline IrO$_2$. The presence of platinum in the bimetallic Pt-Ir/Al$_2$O$_3$ sample inhibits the formation of crystalline IrO$_2$ to some degree.

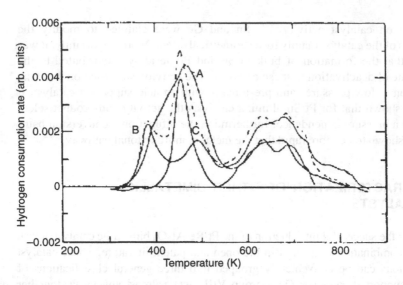

Figure 21 Temperature-programmed reduction profile for Pt-Ir/Al₂O₃, Pt/Al₂O₃, and Ir/Al₂O₃ precursors (from Ref.[377]).

Subramanian and Schwarz[377] utilized TPR and thermogravimetry to show that PtCl₄, IrCl₃, and PtCl₄-IrCl₃ exist as supported entities when the catalyst, prepared from H₂PtCl₆ and H₂IrCl₆, are dried in air at 150°C. These authors reported that bimetal formation does not occur when the dried precursors are directly reduced. The TPR results indicate that there is no interaction between platinum and iridium (prior to reduction) when the precursors are dried at 423 K in air. Although they could not rule out bimetallic formation during the reduction process, the spectrum for the bimetallic precursor (Fig. 21) resembled the profile generated by adding those observed for the corresponding monometallic precursors (dotted line). However, Huang et al.[378] have reported that for samples dried at 383 K, the TPR spectrum of the bimetallic catalyst is not the superposition of the spectra of the individual components. Subramanian and Schwarz explained this apparent discrepancy on the basis of the methods used to prepare the catalysts. A wet impregnation procedure (50 cm³ of impregnation solution/20 g of alumina) was used by Huang et al., rather than the incipient wetness technique using an amount of impregnant corresponding to the pore filling volume of the support (0.5 cm³/g). Recent studies[379] have shown that, at constant metal weight loading, a change in the amount of solution used to prepare the precursors results in a significant change in their reducibility when supported on porous alumina.

Huang et al.[378] found evidence for platinum and iridium in the hydrogen-reduced catalyst to be in their metallic state forming highly dispersed bimetallic Pt-Ir clusters. Huang and Fung[380] reported that the formation of IrO_2 during oxidation, and subsequent reduction to Ir metal, was inhibited by the presence of chloride. Furthermore, the presence of water vapor during the reduction removed chloride with increased formation of IrO_2. The degree of IrO_2 agglomeration of the 693 K oxidized Ir/Al_2O_3 catalyst was much larger than that of $Pt-Ir/Al_2O_3$ catalysts, and indicated that the Pt-Ir interaction reduced the degree of IrO_2 agglomeration. Chloride promotes the interaction between Pt and Ir to form mixed oxychlorides and oxides in an oxidizing environment. The effect of Cl on IrO_2 agglomeration is, therefore, more significant for the $Pt-Ir/Al_2O_3$ catalyst than the Ir/Al_2O_3 catalyst. McVicker and Ziemiak[381] employed XRD and H_2 and CO chemisorption in their studies of the agglomeration behavior of a commercial Pt-Ir catalyst (0.3% Pt–0.3% Ir and 0.67% Cl) and showed that 4-h air oxidation at 773 K resulted in 58% IrO_2 agglomeration. The Cl/M ratio in this case is 1.1. A substantial increase in the Cl/M ratio is required to suppress IrO_2 agglomeration when the oxidation temperature is greater than 733 K. Thus, a subtle difference in water concentration during air oxidation could result in a large change in catalytic properties.

^{193}Ir Mössbauer studies have been conducted for Pt-Ir catalysts. von Brandis et al.[382] reported that silica-supported materials prepared by impregnation with H_3IrCl_6 and H_2PtCl_6 contained Ir in a trivalent state, presumably $(IrCl_6)^{3-}$, following drying. Oxidation at 400°C converted the Ir to IrO_2. Following reduction at 200°C of the oxidized material, there was a strong tendency toward segregation of Ir and Pt. Similar results were obtained for coimpregnation using a solution containing $Ir(NH_3)_5Cl_3$ and $Pt(NH_3)_4Cl_2$.

Sinfelt et al.[383] utilized EXAFS to investigate their alumina or silica supported Pt, Pt-Ir, and Ir catalysts. These studies extended their earlier investigations of the 5–20 wt % Pt + Ir catalysts with XRD, which indicated the presence of 2.5- to 5.0-nm bimetallic clusters. The EXAFS data for the lower loading of these metals showed that these catalysts with highly dispersed metals consisted of bimetallic clusters even as the ratio of surface to total atoms approached unity. The data obtained to date are typical of much of the work reported on catalysts that are utilized commercially. Exxon workers appear to be able to prepare catalysts that consist of bimetallic clusters even at high dispersion. These workers point out the sensitivity of the Pt-Ir catalyst formulations to the drying, oxidizing, and activating procedures. Most of the other workers obtain data that suggest little interaction of the supported Pt and Ir following reduction. It therefore appears that small differences leading to and/or during reduction determine whether bimetallic clusters are obtained. Presumably most of the investigators have not been conducting their experiments under conditions that reproduce the Exxon catalyst preparation and activation conditions.

6.2 Platinum and Partially Reduced Metal

The Pt-Sn catalyst system has been the most widely studied member of this class of catalysts. However, considerable data exist for two other combinations of Pt–group IVA metals: Pt-Ge and Pt-Pb. These systems will be covered briefly in this section.

Romero et al.[384,385] studied the hydrogenolysis of n-butane and the hydrogenation of benzene over fluorided platinum–germanium catalysts and characterized them using IR spectroscopy, electron microscopy, and H_2-O_2 adsorption. They concluded that germanium enhanced the selectivity for the isomerization of n-butane while hydrogenation of benzene went through a maximum at about 0.3% Ge. Electronic effects were used to account for these results. A marked decrease in the dispersion of the platinum was found, as was an increase in the CO stretching frequency when germanium was added to the catalyst. Electron microscopy showed a bimodal distribution of particle size for the bimetallic catalyst. The authors suggested that the decrease in the dispersion of the metal as the amount of germanium increased was related to competitive exchange during preparation.

Goldwasser et al.[386] prepared two series of Pt-Ge-alumina catalysts. In one series the Ge-containing alumina was calcined at 873 K prior to the addition of Pt (series B); both series of catalysts were oxidized at 673 K after both metals had been added. For the series A catalysts (Ge added, then dried at 383, and then Pt added) there was a slight decline in platinum dispersion, calculated from the amount of hydrogen adsorbed, as the germanium content increased. For series B, the dispersion did not depend on Ge content. For higher contents of Ge in series A, it appeared that Ge as well as Pt was reduced, in contrast to the lower loading. Thus, the catalysts containing 0.3% Ge and no Ge gave similar TPR curves, whereas the one containing 1.0% Ge showed a second reduction peak when the temperature was increased above about 1000 K. Since a reforming catalyst would normally not be reduced or operated at 1000 K or higher, the data suggest that Ge would not be reduced in a reforming catalyst.

Bouwman and Biloen[387] showed by XPS that a Pt-Ge catalyst contained Ge^{2+} and Ge^{4+} after reduction at 823 K whereas after reduction at 923 K it contained Ge^{2+} and Ge^0 species; the latter was found to be alloyed with platinum. de Miguel[388] reported that Ge reduction to Ge(0) at 450°C ranged from 15% to 26%; however, these Pt-Ge catalysts were prepared using GeO_2. It is therefore quite likely that discrete GeO_2 particles remained intact during preparation and oxidation and may be responsible for the extent of reduction. However, a later report[389] utilized $GeCl_4$ in the preparation procedure, and again Ge was reduced. From the hydrogen consumption, the authors calculated that 42% of the germanium was Ge^{2+} and 58% was Ge^0, and for samples reduced to 1123 K, 80–100% of the Ge was present as Ge^0.

The major difference between the series B catalysts utilized by Goldwasser et al.[386] and de Miguel et al.[388,389] was that the former utilized an oxidation temperature of 673 K whereas the latter utilized a temperature of 773 K. It is likely, just as was the case with Pt-Ir catalysts, that the formation of GeO_2 particles during oxidation leads to a different extent of reduction of Ge.

Lead-containing Pt-Al$_2$O$_3$ catalysts have been characterized by a variety of techniques. Völter et al.[390] found that with increasing amounts of Sn or Pb in a Pt-Al$_2$O$_3$ catalyst, the weak adsorption of hydrogen decreased and the adsorption of oxygen increased. There was a correlation between the amount of weakly adsorbed hydrogen and the activity of the catalyst for dehydrocyclization. It was found that the addition of Pb to Pt-Al$_2$O$_3$ catalysts decreased the adsorbed CO stretching frequency,[391] consistent with the transfer of s-electrons from Pb to Pt. Their IR and catalytic results indicated that both ligand and ensemble effects were important for this catalyst system. Lieske and Völter[392] concluded, on the basis of TPR results, that the major part of Pb is not reduced but exists as carrier-stabilized Pb^{2+} species; a minor amount of the lead exists as Pb^0 in an alloy with Pt. The literature on multimetal and multielement support is increasing very rapidly, and space does not permit an extensive coverage of this area.

7 CHARACTERIZATION OF ALKALINE L-ZEOLITES—Pt

The availability of high-resolution TEM permitted the extension of this technique to the characterization of small metal clusters in zeolites. Gallezot and coworkers[264] were among the early workers to utilize this technique for this purpose. For example, they reported on the states of Pt dispersed in a Y-zeolite: isolated atoms in sodalite cages, 10-Å-diameter agglomerates fitting into the supercages, and 15- to 20-Å crystallites occluded in the zeolite crystals. In at least some cases it appeared that the metal crystallites had ruptured the zeolite structure. This mode of characterization was of interest for naphtha-reforming catalysis particularly following the observation by Bernard[393] that Pt-L exchanged with alkaline cations had exceptional activity and selectivity for n-hexane dehydrocyclization to benzene.

Besoukhanova et al.[394] utilized a combination of XRD, electron micrography, and IR studies of CO adsorption to characterize Pt-L zeolite catalysts containing alkali cations. They concluded that there were four types of Pt particles: large 100- to 600-Å particles outside the channels, crystals 10- to 25 Å in diameter and small metallic cylinders inside and outside the channels, and very small particles in the cavities of the zeolite structure. They related the catalytic activity of their catalysts to the 10- to 25-Å crystals and the small cylinders of Pt.

From the IR band at 2060–2065 cm^{-1} they concluded that the Pt particles have an excess of electrons and/or atypical faces, corners, or edges.

Hughes et al.[395] determined the Brønsted acid site density of alkaline earth–exchanged L-zeolite from the IR spectrum of adsorbed pyridine. Increasing the oxidation temperature decreased the acidity, and a BaK-L sample oxidized at 866 K had insufficient acidity accessible to pyridine for it to be determined. Shifts in frequency during oxidation permitted them to show that oxidation at 866 K caused the alkaline earth ions to migrate to the locked position in the small cage (inaccessible to pyridine) and the potassium ions to occupy predominantly the open sites in the large channels of zeolite L.[396] Newell and Rees[397] reported a similar movement of ions in zeolite L during oxidation. It was contended that any Brønsted acidity present in the BaK-L zeolite was located in the locked sites and could not participate in acidic reactions.

Hughes et al. also reported the apparent dispersion of Pt (0.8 wt %)/BaKL, assuming that the average metal particle diameter is related to the ratio of surface to total metal atoms by the equation suggested by Anderson[398] (Fig. 22). The particles detected by TEM were about 0.8–1 nm in diameter. The dispersion data presented in Figure 22 are not in agreement with the TEM data for higher temperatures of reduction. The authors explain this by assuming that the particles are located in the large intercrystalline zeolite channels and, as the particles grow either with increasing reduction time or temperature, the majority of the surface becomes inaccessible to the CO used to measure the dispersion. Thus, the decrease in dispersion shown in Figure 22 is considered to be due to this restricted

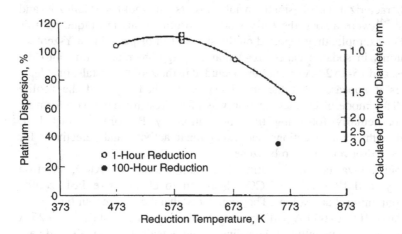

Figure 22 Effects of reduction temperature and time on Pt dispersion in 0.8 Pt/BaKL-866 (from Ref.[397]).

accessibility caused by the Pt particle packing within the large channels, and not due to Pt particle growth beyond about 1 nm diameter.

Rice et al.[399] considered the merits of high-resolution brightfield imaging in the TEM and high-angle annular darkfield imaging in the STEM for the detection and measurement of small noble-metal clusters in zeolites, using Pt-K-L zeolite. They confirmed that high-resolution brightfield imaging is better suited for resolving the zeolite framework. However, even with contrast enhancement, brightfield images are ineffective for detecting clusters containing fewer than about 20 Pt atoms in supports thicker than about 10 nm. This was attributed mainly to phase contrast patterns associated with beam-damaged regions of the zeolite framework. STEM measurements were shown to be capable of detecting single Pt atoms against about 20-nm-thick zeolite support, but the ability to establish the position of the atom relative to the unit cell is limited by beam damage–induced distortion of the zeolite framework.

Miller et al.[400] utilized TEM as one means of characterizing their Pt-BaKL zeolite catalyst. EDX spectroscopy was used to analyze several areas of the catalyst to provide evidence that the platinum is present in the interior of the imaged crystallites. Since the platinum particles were not imaged by the electron microscope, the authors inferred that the platinum particles were very small and most likely well dispersed in the zeolite crystallites. Hydrogen chemisorption data indicated that extrapolation to zero pressure showed no adsorption for the KL zeolite and that the hydrogen adsorption on the Pt/BaKL zeolite sample

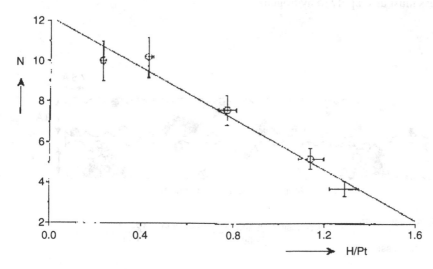

Figure 23 Correlation of hydrogen chemisorption and first-shell Pt-Pt EXAFS coordination number data. Data represented by triangles from Kip et al. (from Ref.[176]).

paralleled the higher pressure isotherm for the zeolite. The data gave an H/Pt value of 1.3. EXAFS data provide a Pt-Pt coordination number of 3.7 that, for a bulk-like FCC packing of the Pt atoms in a cluster, corresponds to a cluster of five or six atoms. The absence of higher order Pt-Pt coordination shells are consistent with this nearly unique cluster size. These data showed that extremely small metal particles covered with chemisorbed hydrogen had metal–metal distances equal to the bulk metal; this was consistent with earlier data.[174,401] These authors compared their data for H/Pt and coordination number determined by EXAFS (Fig. 23) with earlier data reported by Kip et al.[176] for a series of Pt supported on γ-Al$_2$O$_3$. Another feature of the EXAFS data was evidence for a Pt-Ba interaction; this is surprising in light of the observation by Hughes et al.[395] that K preferentially locates in the channel. However, the sample used by Miller et al.[400] was oxidized only at 400 K and then reduced at 773 K, both lower temperatures than used by Hughes et al.[395] (886 K).

Larsen and Haller[402] reported a ratio of CO/H of about 0.7 for the adsorption of these two gases on Pt in KL catalysts and attribute the difference to some degree of reversibility of the hydrogen uptakes at 298 K. These authors preferred the CO/Pt ratio as a measure of the dispersion of platinum for these catalysts. These authors also obtained a linear relationship between the EXAFS coordination number and the H/Pt ratio; however, for a particular value of H/Pt Larsen and Haller obtain a larger coordination number than Miller et al.[400] Larsen and Haller obtained a dispersion, defined as H/Pt, vs. oxidation temperature that was very similar to the one obtained by Hughes et al.,[395] suggesting that restricted access must also apply to hydrogen.

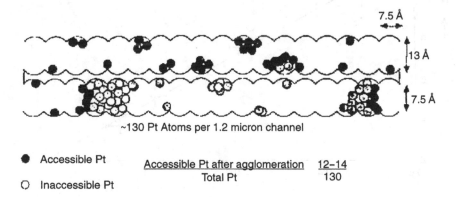

~130 Pt Atoms per 1.2 micron channel

● Accessible Pt

○ Inaccessible Pt

$$\frac{\text{Accessible Pt after agglomeration}}{\text{Total Pt}} \quad \frac{12\text{--}14}{130}$$

Figure 24 Proposed model for pore mouth blockage of Pt/KL catalysts (from Ref.[403]).

McVicker[403] reported on the characterization of both fresh and sulfur poisoned 0.6% Pt-KL zeolite catalysts. A catalyst oxidized at 623 K and then reduced at 773 K yielded H(irreversible)/Pt ratios in the range 0.9:1.1. Brightfield TEM images showed few Pt clusters greater than 1.0 nm in diameter within the channels of the fresh catalyst, and this was confirmed by in-depth Z-contrast STEM studies. IR spectra of adsorbed CO were consistent with deactivation by sulfur occurring by the systematic loss of accessible Pt sites by a plugging of channels with Pt agglomerates sufficiently large to 'double plug' a channel since only the intensity differed for the two catalyst samples. Furthermore, the benzene selectivity from the conversion of *n*-hexane was the same for the sulfur-free and sulfur-poisoned catalysts. These authors therefore consider pore mouth blockage as the reason for the extreme sensitivity of this catalyst to sulfur poisoning and depict this as illustrated in Figure 24. They offered a histogram, obtained using STEM, of Pt particle size for catalyst samples aged in a sulfur-containing feed that showed particle growth. They obtained brightfield TEM micrographs that showed Pt agglomerates clustered predomi-

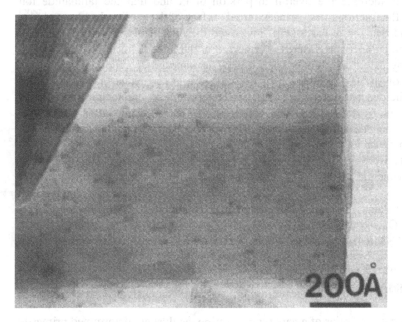

Figure 25 Bright-field TEM micrograph of an unbound 0.6% Pt/KL catalyst showing the presence of Pt clusters elongated along the KL channel. The catalyst was recovered from an aromatization experiment after 100 hr on a feed containing 50 wppm sulfur. KL-zeolite channel spacings of 15.9 Å are visible and provide an internal standard for Pt cluster size measurement (from Ref.[403]).

nantly at the channel entrance (Fig. 25). Iglesia and Baumgartner[404] report that Pt crystallites located outside the zeolite channels are large (about 100 Å) and are coated with graphitic carbon during n-hexane conversion at 783 K; those particles located in the channel retain their activity for n-hexane conversion. Vaarkamp et al.[405] found that H/Pt was 1.4 for a fresh catalyst with a Pt-Pt coordination number of 3.7 as determined from EXAFS data, suggesting an average cluster size of five to six atoms. Following sulfur poisoning, the first-shell Pt-Pt coordination number increased to 5.5, indicating a growth of the average platinum cluster size to 13 atoms and a decrease in H/Pt to 1.0. They conclude that growth of the platinum particle was sufficient to block the pore.

IR spectra of adsorbed CO showed that the low-frequency C-O stretching band was closely related to catalytic performance of a series of Pt/KL catalysts.[406] These authors also reported that halogen anions profoundly affect the electronic state of Pt metal as well as its dispersion. KF and KCl were effective in establishing high electron donating ability of Pt, and this results in an excellent catalyst for hexane aromatization.

FTIR data combined with H_2-O_2 titration and TPR data showed that lanthanides increase the overall dispersion of Pt and that the lanthanide ion increases the fraction of Pt on the external surface of the zeolite. Jacobs et al.[407] also found that a rare earth (Tm) improved the dispersion but only for certain catalyst preparation procedures. Furthermore, the added Tm improved catalyst lifetime by acting as a getter for sulfur. The same group reported that, in the presence of sulfur, the addition of Ce and, to a lesser extent, Yb significantly inhibits the deactivation of the catalyst. This occurs by the anchoring of the Pt clusters and by the added ions capturing sulfur.[408] This latter paper also provides a concise summary of results for recent characterization studies of Pt-L zeolite catalysts.

CO adsorption on Pt clusters on L-zeolite has been studied using theoretical and experimental methods.[409] The distribution of heats of adsorption with CO coverage on Pt measured by microcalorimetry were found to agree with calculated values for a 10 Pt cluster with a combination of linear and bridge-bonded CO.

^{13}C NMR has been used to characterize Pt-K-L zeolite catalysts.[410]

9 SUMMARY

The physical properties of a naphtha-reforming catalyst are determined primarily by the alumina support. Thus, the surface area and pore structure result from the support. Likewise, the crystal structure and the physical shape of the alumina particles determine the physical features of the catalyst. Platelet-like alumina may therefore provide steps that can provide a sink for trapping Pt or other metal atoms formed during catalyst reduction and use. General features of the morphology of the catalyst can be obtained from the shape of the nitrogen adsorption/desorption isotherm and the hysteresis between the two. The surface area is best obtained from the nitrogen adsorption isotherm using the BET equation. The pore size distribution is a very demanding assignment that produces at best an engineering, and not absolute, data. Factors that introduce the major uncertainties in the pore size distribution are the technique used to correct for the thickness of the adsorbed layer, whether the adsorption or desorption isotherm is used, and the assumed shape of the pore. From the engineering point of view, one is usually safe to use any model provided data are obtained for comparative purposes. For most samples, the method of calculation by Broekoff and de Boer[30,31] is most likely to produce a pore size distribution that produces a surface area that closely agrees with the one using the BET equation.

An absolute measure of the acidity of a reforming catalyst is probably not possible. Adsorption data indicate that alumina adsorbs nitrogen bases from either the gas or liquid phase at room temperature to such an extent that it cannot be reliably used to estimate catalytic activity. One potential approach to measure the acidity responsible for the bifunctional character of the reforming catalyst is to utilize the chemisorption of a sterically hindered base, e.g., 2,6-dimethylpyridine, at high temperatures (300–400°C). With chloride introducing additional or enhanced acidity, the use of chemisorption to measure acidity becomes much more difficult. For the engineering approach, chemisorption of a base is most likely the technique that will provide data to aid in evaluating the potential of a catalyst formulation or for following changes in catalyst performance. It must be emphasized that the technique is not absolute and will therefore provide reliable performance prediction or postmortem examination only after reliable correlations of the plant performance with chemisorption data has been developed. Furthermore, the ability to extrapolate correlations from one catalyst formulation to one that is significantly different, e.g., Pt-alumina and fluorided Pt-alumina, is risky.

Two instrumental techniques that should provide additional understanding of the data from the chemisorption of base are IR spectroscopy and calorimetry. Calorimetry offers the potential to measure the distribution of the acidity, at least on a relative basis. Calorimetry may be utilized to determine acidity distributions at elevated temperatures, closer to those used in commercial reformers. The IR measurement of chemisorbed molecules provides a measure of the distribution of the acid sites between those of a Lewis or Brønsted type. It therefore would appear that a preferred method of acidity measurement is to conduct a high-temperature adsorption of a sterically hindered base for quantitation and to augment the primary adsorption data by characterization of the acid site strength distribution with calorimetric measurements as well as acid site–type measurements with *in situ* measurements using IR techniques.

In a commercial operation, the pretreatment and operation procedures appear to provide a catalyst that has a high dispersion of platinum. The combination of high dispersion and low metal loading causes most of the experimental techniques to have little or limited value. For the engineering analysis it appears that chemisorption measurements are the most appropriate. As in the case of the determination of acidity by adsorption, the ability to convert the experimental data to an exact measure of dispersion is limited. This is primarily the result of the lack of knowledge of the 'correct' ratio of adsorbate to metal to use in making the calculations as well as the geometry of the 'cluster' of platinum atoms. For the engineering approach, a comparison of samples of a given catalyst formulation can be made from a comparison of the value of the ratio of adsorbate to metal, e.g., the H/Pt ratio for the chemisorption of hydrogen. For a catalyst suitable for commercial operation, the ratio should closely approach or exceed 1.

High-resolution electron microscopy is applicable, under suitable conditions, for dispersions that approach the atomic level. Even under conditions where the contrast between the crystalline support and dispersed metal is not ideal, it should be possible to provide an upper limit for the size of the metal particles. TEM is most suited for the examination of materials where the supported material is of a somewhat uniform size distribution. Great care must be exerted by spectroscopists to prevent a focus on structures that are of unusual shape or large size as representative the state of the metal because it is not, in general, possible to define quantitatively the fraction of metal represented by the large particles. The tendency will be to consider the larger particles as being representative of the metal function when in fact they may represent only a minor fraction of the whole sample.

The definition of the state of two or more metals in the polyfunctional reforming catalyst becomes a more demanding task. Consider the case of the Pt-Sn-alumina catalyst. The data show that the Pt is reduced to the metal but that only a fraction of the Sn is reduced from the Sn^{4+} state. This means that one can only obtain an 'average' valence of the Sn. For catalyst formulations prepared by impregnation of both tin and platinum compounds, it has been demonstrated that some of the tin is reduced to the metallic state. Furthermore, the extent of tin reduction depends on the Sn/Pt ratio; the higher the ratio, the greater the reduction of tin to the zero valence state. For most catalyst compositions, the major Pt-Sn alloy is PtSn. For very small Sn/Pt ratios, one can observe some Pt-rich alloys; likewise for high Sn/Pt ratios, higher tin alloys may be present. Considering the data from the characterization studies using a variety of techniques (Mössbauer, XRD, TEM, microdiffraction, XPS) for a series of these catalysts, a consistent model can be constructed for this catalyst system. A fraction of the tin is present in an oxidized state (Sn^{2+}, Sn^{4+}); some of the tin is found to be present as a material approximating a tin aluminate. Depending on the Sn/Pt ratio, some or all of the Pt will be present as an alloy with the dominant species being PtSn. As the surface area of the support increases, the fraction of tin present in an oxidized form increases. Thus, for the same Sn/Pt ratio, less alloy will be present when the support has a high surface area than when it has a low one. The extent of tin reduction also depends on the support; it is much easier to reduce tin to the zero valence state when the support is silica than when it is alumina. A variety of species was proposed by Srinivasan and Davis.[364] A fraction of the surface is covered with a 'monolayer' structure that resembles tin aluminate; presumably this tin species retards agglomeration of the Pt species present in the catalyst. Pt is distributed between metallic Pt and PtSn alloy forms; the fraction of each form of Pt depends on the Sn/Pt ratio and the surface area of the support.

In some commercial catalyst formulations, platinum may be supported on a tin/aluminum oxide prepared by coprecipitation. Catalysts prepared this way may have a very different structure than those prepared by coimpregnation. In the case of one type of coprecipitated catalyst, essentially all of the Pt is present in an unalloyed form even for an Sn/Pt ratio of 3. It appears that for coprecipitated supports, one key role of tin is to provide surface traps to retard Pt diffusion and agglomeration. Alloy formation may be less of a factor.

In other cases, Pt-Re or Pt-Ir, the model described above is not appropriate. In these catalysts, both Pt and the second metal appear to be mostly or completely reduced to the zero valence state but alloy formation does not occur to a significant extent. It does appear, however, that there is significant interaction of the two metals in the case of the Pt-Re catalyst. Some authors view the Re to form monolayer rafts on the alumina support, and these rafts serve as an interface between the alumina support and the Pt metal cluster, thereby altering the properties of the supported Pt and preventing agglomeration. While many authors indicate that some portion of the additional metals utilized together with platinum are not completely reduced, the role of the ions of the second component is not clear. One of the roles most frequently encountered in the literature is that of inhibiting the agglomeration of platinum.

In summary, it appears that many of the instrumental techniques for catalyst metal function characterization can be applied to the commercial naphtha-reforming catalyst only with great difficulty and with limited success. Those techniques that depend on the presence of metallic crystals, e.g., XRD, are applicable only for postmortems to identify the reason for catalyst failure or aging. For the very active commercial catalyst, the characterization technique must be applicable to those situations where the dispersion is at or near the atomic level. For these characterizations, it appears that chemisorption offers the 'best' characterization capabilities. Some of the newer instrumental techniques, e.g., EXAFS, offer the potential to provide a characterization at these high dispersions, but further advances in the ability to derive structures from the experimental data must be attained.

The current and future trends in naphtha catalyst characterization are driven by the demand placed on the techniques due to the low concentrations of the metallic and acidic functions of the commercial catalysts. The future characterizations will be driven by in situ techniques.[411–413] Furthermore, the in situ studies will be increasingly directed to operating under conditions that attain, or at least approach, commercial conditions.[414,415] In spite of this there will continue to be studies that involve model reactants and/or conditions that are far removed from commercial conditions.[416,417]

REFERENCES

1. Burwell, R.L., Jr. Manual of symbols and terminology of physicochemical quantities and units, Appendix II. Part II: Heterogeneous catalysis. Adv. Catal. **1977**, *26*, 351.
2. Haensel, V. In *Chemistry of Petroleum Hydrocarbons*; Brooks, B.T., Boord, C.E., Kurtz, S.S., Jr., Schmerling, L., Eds.; Reinhold: New York, 1955; Vol. 2, p. 189.
3. Brunauer, S.; Emmett, P.H.; Teller, E. J. Am. Chem. Soc. **1938**, *60*, 309.
4. Emmett, P.H. Citation classic. Commentary on J. Am. Chem. Soc. **1938**, *60*, 309. Current Contents **1977**, *35* (9), August 29.
5. Adamson, A.W. *Physical Chemistry of Surfaces*, 5th Ed.; Wiley Interscience: New York, 1990.
6. Davis, B.H. Chemtech **1991**, *21*, 18.
7. Halsey, G. J. Phys. Chem. **1948**, *16*, 931.
8. Avnir, D. J. Am. Chem. Soc. **1987**, *109*, 2931.
9. Davis, B.H. Appl. Catal. **1984**, *10*, 185.
10. Barrett, E.; Joyner, L.; Halenda, P. J. Am. Chem. Soc. **1951**, *73*, 373.
11. Wheeler, A. Adv. Catal. **1951**, *3*, 249; **1955**, *2*, 105.
12. Dollomire, D.; Heal, G. J. Appl. Chem. **1964**, *14*, 109.
13. Broekhoff, J.C.P. *Adsorption and Capillarity*; Ph.D. Thesis, University of Delft, 1969.
14. Broekhoff, J.C.P.; Linsen, B.C. In *Physical and Chemical Aspects of Adsorbents and Catalysts*; Linsen, B.G., Ed.; Academic Press: New York, 1970.
15. Adkins, B.A.; Davis, B.H. Langmuir **1987**, *3*, 722.
16. Innes, W.B. In *Experimental Methods in Catalytic Research*; Anderson, R.B., Ed.; Academic Press: New York, 1968; Vol. 1, 45–99.
17. Innes, W.B. Anal. Chem. **1957**, *29*, 1069.
18. Anderson, R.B. J. Catal. **1964**, *3*, 50 and 301.
19. Wayne, L.G. J. Am. Chem. Soc. **1951**, 73, 5498.
20. Lippens, B.C.; Linsen, B.G.; de Boer, J.H. J. Catal. **1964**, *3*, 32.
21. Schull, C.; Elkin, P.; Roess, L. J. Am. Chem. Soc. **1948**, *70*, 1410.
22. Pierce, C. J. Phys. Chem. **1953**, *57*, 149.
23. Cranston, R.W.; Inkley, F.A. Adv. Catal. **1957**, *9*, 143.
24. Pure Appl. Chem. **1985**, *57*, 603–619.
25. Brunauer, S.; Deming, L.S.; Deming, W.S.; Teller, E. J. Am. Chem. Soc. **1990**, *62*, 1723.
26. Adkins, B.D.; Davis, B.H. Ads. Sci. Technol. **1988**, *5*, 168.

27. Adkins, B.D.; Davis, B.H. Ads. Sci. Technol. **1988**, *5*, 76.
28. Foster, A.G. Trans. Faraday Soc. **1932**, *L8*, 645.
29. Cohan, L.H. J. Am. Chem. Soc. **1938**, *60*, 433.
30. Broekhoff, J.C.P.; de Boer, J.H. J. Catal. **1967**, *9*, 15.
31. Broekhoff, J.C.P.; de Boer, J.H. J. Catal. **1968**, *10*, 368.
32. Frennet, A.; Chitry, V.; Kruse, N. Appl. Catal. A: Gen. **2002**, *229*, 273.
33. Gillespie, R.D. U.S. 6,214,764 B1, April 10, 2001.
34. Ritter, H.L.; Erich, L.C. Ind. Eng. Chem. Anal. Ed. **1948**, *20*, 665.
35. Quantachrome Corp. Effect of Coating of Catalyst Supports in Mercury Porosimetry. Powder Technology Note 16.
36. Leofanti, G.; Padovan, M.; Tozzola, G.; Venturelli, B. Catal. Today **1998**, *41*, 207–219.
37. Rouquerol, F.; Rouquerol, J.; Sing, K. *Adsorption by Powders and Porous Solids: Principles, Methodology and Applications*; Academic Press: New York, 1999.
38. Sharma, I.D.; Kumar, M.; Saxena, A.K.; Chand, M.; Gupta, J.K. J. Mol. Catal. A: Chem. **2002**, *185*, 135.
39. Benesi; Wenquist. Adv. Catal. **1978**, *27*, 97.
40. Walling, C. J. Am. Chem. Soc. **1950**, *72*, 1164.
41. Benesi, H.A. J. Catal. **1973**, *28*, 176.
42. Gold, V.; Hawes, B.W. J. Chem. Soc. **1951**, 2102.
43. Deno, N.C.; Jaruzelski, J.J.; Schriesheim, A. J. Am. Chem. Soc. **1955**, *77*, 3044.
44. Deno, N.C.; Berkheimer, H.E.; Evans, W.L.; Peterson, H.J. J. Am. Chem. Soc. **1959**, *81*, 2344.
45. Tanabe, K. *Solid Acids and Bases*; Academic Press: New York, 1970.
46. Tanabe, K. In *Catalysis by Acids and Bases*; Imelik, B., Naccache, C., Coudurien, G., Ben Taaret, Y., Vedrin, J.C., Eds.; Elsevier: Amsterdam, 1985; Vol. 20, p. 1.
47. Tanabe, K.; Misono, M.; Ono, Y.; Hattori, H. *New Solid Acids and Bases*; Elsevier: Amsterdam, 1989; Vol. 51.
48. Parry, E.P. J. Catal. **1963**, *2*, 371.
49. Peri, J.B. J. Phys. Chem. **1965**, *69*, 220.
50. Parkyns, N.D. J. Phys. Chem. **1971**, *75*, 526.
51. Fink, P. Z. Chem. **1967**, *7*, 324.
52. Hightower, J.W.; Hall, W.K. Trans. Faraday Soc. **1970**, *66*, 477.
53. Peri, J.B., *2nd Int. Congr. Catal.* (Paris 1960), **1961**, 1333.
54. Eischens, R.P.; Pliskin, W.A. Adv. Catal. **1958**, *10*, 2–95.
55. Knözinger, H. NATO ASE Ser., B **1991**, *265*, 167.
56. Bell, A.T. Springer Ser. Chem. Phys. **1984**, *35* (Chem. Phys. Solid Surf. 5), 23–38.
57. Paukstis, E.; Yurchenko, E.N. Usp. Khim. **1983**, *52*, 426.
58. Van Woerkom, P.C.M.; De Grott, R.L. Appl. Opt. **1982**, *21*, 3114.
59. Kiselev, A.V.; Lygin, V.I. *Infrared Spectra of Surface Compounds*; John Wiley and Sons: New York, 1975.
60. Peri, J.B. J. Phys. Chem. **1965**, *69*, 211.
61. Peri, J.B.; Hannan, R.B. J. Phys. Chem. **1960**, *64*, 1526.

62. Carter, J.L.; Lucchesi, P.J.; Cornell, P.; Yates, D.J.C.; Sinfelt, J.H. J. Phys. Chem. **1965**, *69*, 3070.
63. Dunken, H.; Fink, P. Z. Chem. **1966**, *6*, 194.
64. Hall, W.K.; Leftin, H.P.; Cheselsee, P.J.; O'Reilly, D.E. J. Catal. **1963**, *2*, 506.
65. Peri, J.B. J. Phys. Chem. **1965**, *69*, 231.
66. Knözinger, K. In *Catalysis by Acids and Bases*; Imelik, B., et al., Eds.; Elsevier: Amsterdam, 1985; 111–125.
67. Busca, G. Catal. Today **1998**, *41*, 191–206.
68. Ferwerda, R.; van der Maas, J.H.; van Duijneveldt, F.B. J. Mol. Catal. A: Chem. **1996**, *104*, 319.
69. Lavalley, J.C. Catal. Today **1996**, *27*, 377.
70. Bovet, C. Thesis, University of Caen, 1981.
71. Davis, B.H.; Koegh, R.A.; Alerasool, S.; Zalewski, D.J.; Day, D.E.; Doolin, P.K. J. Catal. **1999**, *183*, 45–52.
72. Knözinger, H. Adv. Catal. Rel. Subj. **1976**, *25*, 184.
73. Larrubia, M.A.; Gutièrrez-Alejandre, A.; Ramìrez, J.; Busca, G. Appl. Catal. A: Gen. **2002**, *224*, 167.
74. Wachs, I.E. Catal. Today **1996**, 437–455.
75. Wachs, I.E. Top. Catal. **1999**, *8*, 57–63.
76. Wachs, I.E. (I.R. Lewis and H.G.M. Edwards, eds.) *Handbook of Raman Spectroscopy*; 2001; Vol. 28, 799–833.
77. Lee, I.C.; Masel, R.I. J. Phys. Chem. B **2002**, *106*, 368–373.
78. Kazansky, V.B.; Borovkov, V.Yu.; Kustov, L.M. Proc. 8th Int. Congr. Catal., Berlin, 1984; Vol. III, 3–14.
79. Escalona Platero, E.; Otero Aren, C. J. Catal. **1987**, *107*, 244–247.
80. Morterra, C.; Magnacca, G.; Filippi, F.; Giachello, A. J. Catal. **1992**, *137*, 346–356.
81. Vannice, M.A.; Hasselbring, L.C.; Sen, B. J. Catal. **1985**, *95*, 57.
82. Gervasini, A.; Auroux, A. J. Thermal Anal. **1991**, *37*, 1737.
83. Cardona-Martinez, N.; Dumesic, J.A. J. Catal. **1991**, *128*, 23.
84. Cardona-Martinez, N.; Dumesic, J.A. J. Catal. **1991**, *127*, 706.
85. Solinas, V.; Ferino, I. Catal. Today, **1998**, *41*, 179–189.
86. Sharma, S.B.; Ouraipryvan, P.; Nair, H.A.; Balaraman, P.; Root, T.W.; Dumesic, J.A.; J. Catal. **1994**, *150*, 234.
87. Gates, B.C.; Katzer, J.R.; Schuit, G.C.A. In *Chemistry of the Catalytic Processes*; McGraw-Hill: New York, 1979; p. 289.
88. Musso, J.C.; Parera, J.M. Appl. Catal. **1987**, *30*, 81.
89. Parmaliana, A.; Frusteri, F.; Mezzapica, A.; Giordano, N. J. Catal. **1988**, *111*, 235.
90. Lieske, H.; Lietz, G.; Spindler, H.; Völter, J. J. Catal. **1983**, *81*, 8.
91. Sivasanker, S.; Ramaswamy, A.V.; Ratnasamy, P. *Preparation of Catalysts II*; Delmon, B., Grange, P., Jacobs, P.A., Poncelet, G., Eds.; Elsevier: Amsterdam, 1979; p. 185.
92. Castro, A.A.; Scelza, O.A.; Benvenuto, E.R.; Baronetti, G.T.; Parera, J.M. J. Catal. **1981**, *69*, 222.

93. Castro, A.A.; Scelza, O.A.; Baronetti, G.T.; Fritzler, M.A.; Parera, J.M. Appl. Catal. **1983**, *6*, 347.

94. Bishara, A.; Murad, K.M.; Stanislaus, A.; Ismail, M.; Hussain, S.S. Appl. Catal. **1983**, *7*, 337.

95. Grau, J.M.; Jablonski, E.L.; Pieck, C.L.; Verderone, R.J.; Perera, J.M. Appl. Catal. **1988**, *36*, 109.

96. Mardilovich, P.P.; Lysenko, G.N.; Kupchenko, G.G.; Kurman, P.V.; Titova, L.I.; Trokhimets, A.I. React. Kinet. Catal. Lett. **1988**, *36*, 357.

97. Arena, F.; Frusteri, F.; Mondello, N.; Giordano, N.; Parmaliana, A. J. Chem. Soc. Faraday Trans. **1992**, *88* (22), 3353.

98. Bishara, A.; Murad, K.M.; Stanislaus, A.; Ismail, M.; Hussain, S.S. Appl. Catal. **1983**, *6*, 347.

99. Ayame, A.; Sawada, G.; Sato, H.; Zhang, G.; Ohta, T.; Izumizawa, T. Appl. Catal. **1989**, *48*, 25.

100. Tanaka, M.; Ogasawara, S. J. Catal. **1970**, *16*, 157.

101. Garbowski, E.G.; Primet, M. J. Chem. Soc. Faraday Trans. 1·**1987**, *83*, 1469.

102. Khaleel, A.; Dellinger, B. Environ. Sci. Technol. **2002**, *36*, 1620.

103. Vigué, H.; Quintard, P.; Merle-Méjean, T.; Lorenzelli, V. J. Eur. Ceram. Soc. **1998**, *18*, 305.

104. Gracia, E.J.; Miller, J.T.; Kropf, A.J.; Wolf, E.E. J. Catal. **2002**, *209*, 341.

105. Webb, A.N. Ind. Eng. Chem. **1957**, *49*, 261.

106. Strand, P.; Kraus, M. Collect. Czech. Chem. Commun. **1965**, *30*, 1136.

107. Matsuura, K.; Watanabe, T.; Suzuki, A.; Itoh, M. J. Catal. **1972**, *26*, 127.

108. Yusheckenki, V.V.; Antipina, T.V. Kinet. Katal. **1970**, *11*, 134.

109. Scokart, P.O.; Selim, S.A.; Damon, J.P.; Rouxhet, P.G. J. Colloid. Interf. Sci. **1979**, *70*, 209.

110. Kerkhof, F.P.J.M.; Oudejans, J.C.; Moulijn, J.A.; Matulewicz, E.R.A. J. Colloid. Interf. Sci. **1980**, *77*, 120.

111. Holm, V.C.F.; Clark, A. J. Catal. **1967**, *8*, 286.

112. Corma, A.; Fornés, V.; Ortega, E. J. Catal. **1985**, *92*, 284.

113. Matulewicz, E.R.A.; Kerkhof, F.P.J.M.; Moulijn, J.A.; Reitsma, H.A. J. Colloid. Interf. Sci. **1980**, *77*, 110.

114. Hirschler, A.E. J. Catal. **1963**, *2*, 428.

115. Weber, A.B.R. Thesis; Univ. of Delft, Netherlands, 1957.

116. Holm, V.C.F.; Clark, A. Ind. Eng. Chem., Prod. Res. Dev. **1963**, *2*.

117. Clark, A.; Holm, V.C.F.; Blackburn, D.M. J. Catal. **1962**, *1*, 244.

118. Hedge, R.I.; Barteau, M.A. J. Catal. **1989**, *120*, 387.

119. Scokart, P.O.; Rouxhet, P.G. J. Colloid. Interf. Sci. **1982**, *86*, 96.

120. Kowalak, S. Acta. Chim. Acad. Sci. Hung. **1981**, *107*, 27.

121. Kowalak, S. Acta. Chim. Acad. Sci. Hung. **1981**, *107*, 19.

122. O'Reilly, D.E. Adv. Catal. **1960**, *12*, 66.

123. DeCanio, E.C.; Edwards, J.C.; Scalzo, T.R.; Storm, D.A.; Bruno, J.W. J. Catal. **1991**, *132*, 498.

124. Fischer, L.; Harlé, V.; Kasztelan, S.; D'Espinsoede la Caillerie, J.-B. Solid State NMR **2000**, *16*, 85–91.

125. Davis, S.M.; Meitzner, G.B.; Fischer, D.A.; Gland, J. J. Catal. **1993**, *142*, 368.
126. Baird, T.; Bendada, A.; Selougha, M.; Webb, G.; Winfield, J.M. J. Fluorine Chem. **1994**, *69*, 109.
127. McHenry, K.W.; Bertolacini, R.J.; Brennan, H.M.; Wilson, J.L.; Seeling, H.S. *Proc. 2nd Int. Congr. Catal.* (Paris, 1990); 1962; Vol. 2, p. 2293.
128. Drake, L.C. personal communication.
129. Bursian, N.R.; Kogan, S.B.; Davydova, Z.A. Kinet. Katal. **1967**, *8*, 1085.
130. Johnson, M.F.L.; Keith, C.D. J. Phys. Chem. **1963**, *67*, 200.
131. Putanov, P.; Ivanovic, M.; Selakovic, O. React. Kinet. Catal. Lett. **1978**, *8*, 223.
132. Kluksdahl, H.E.; Houston, R.J. J. Phys. Chem. **1961**, *65*, 1469.
133. Ermakova, S.I.; Zaidman, N.M. Kinet. Katal. **1969**, *10*, 1158.
134. Shekhobalova, V.I.; Luk'yanova, Z.V. Russian J. Phys. Chem. **1979**, *53*, 1551.
135. Lietz, G.; Lieske, H.; Spindler, N.; Hanke, W.; Völter, J. J. Catal. **1983**, *81*, 17.
136. Botman, M.J.P.; She, L.Q.; Zhang, J.Y.; Driessen, W.L.; Ponec, V. J. Catal. **1987**, *103*, 280.
137. Botman, M.J.P. Ph.D. thesis; University of Lieden, 1987.
138. Lemaitre, J.L.; Menon, F.G.; Delannay, F. In *Characterization of Heterogeneous Catalysts*; Delannay, F., Ed.; Marcel Dekker: New York, 1984; 299–366.
139. Adams, C.R.; Benesi, H.A.; Curtis, R.M.; Meisenheimer, R.G. J. Catal. **1962**, *1*, 336.
140. Klug, H.P.; Alexander, L.E. *X-ray Diffraction Procedures*; John Wiley and Sons: New York, 1954; Chap. 9.
141. Srinivasan, R.; Rice, L.; Davis, B.H. J. Am. Ceram. Soc. **1990**, *73*, 3528.
142. Ascarelli, P.; Contini, V.; Giorgi, R. J. Appl. Phys. **2002**, *91*, 4556.
143. Rhodes, H.E.; Wang, P.-K.; Stokes, H.T.; Slichter, C.P.; Sinfelt, J.H. Phys. Rev. B **1982**, *26*, 3599.
144. Van Hardevelt, R.; Hartog, F. Surf. Sci. **1969**, *15*, 189.
145. White, D.; Baird, T.; Fryer, J.R.; Freeman, L.A.; Smith, D.J.; Day, M. J. Catal. **1983**, *81*, 119.
146. Gai, P.L.; Goringe, M.J.; Barry, J.C. J. Microsc. **1986**, *142*, 9.
147. Datye, A.K.; Smith, D.J. Catal. Rev.-Sci. Eng. **1992**, *34*, 129.
148. Treacy, M.M.J.; Rice, S.B. J. Microsc. **1989**, *156*, 211.
149. Schwank, J.; Allard, L.F.; Deeba, M.; Gates, B.C. J. Catal. **1983**, *84*, 27.
150. Tesche, B.; Zeilter, E.; Delgado, E.A.; Knözinger, H. *Proc. 40th Ann. Mtg. Electron Microscopy Society of America*; Bailey, G.W., Ed.; San Francisco Press: San Francisco, 1982; p. 682.
151. Huang, Z.; Fryer, J.R.; Park, C.; Stirling, D.; Webb, G. J. Catal. **1996**, *159*, 340.
152. Langmuir, I. J. Am. Chem. Soc. **1916**, *38*, 2221.
153. Benton, A.F.; White, T. J. Am. Chem. Soc. **1932**, *54*, 1820.
154. Podgurski, H.H.; Emmett, P.H. J. Am. Chem. Soc. **1953**, *57*, 159.
155. Brunauer, S.; Emmett, P.H. J. Am. Chem. Soc. **1935**, *57*, 1754.
156. Brunauer, S.; Emmett, P.H. J. Am. Chem. Soc. **1937**, *59*, 310.
157. Via, G.H.; Meitzner, G.; Lytle, F.W.; Sinfelt, J.H. J. Chem. Phys. **1982**, *79*, 1527.
158. Hollaway, P.H. Surf Interf. Anal. **1985**, *7*, 204–205.

159. Benesi, H.A.; Curtis, R.M.; Studer, H.P. J. Catal. **1968**, *10*, 328.
160. Dorling, T.A.; Eastlake, M.J.; Moss, R.L. J. Catal. **1969**, *14*, 23.
161. Dorling, T.A.; Lynch, B.W.J.; Moss, R.L. J. Catal. **1971**, *20*, 190.
162. Christmann, K.; Ertl, G.; Pignet, T. Surf. Sci. **1976**, *54*, 365.
163. Adler, S.F.; Keavney, J.J. J. Phys. Chem. **1960**, *64*, 208.
164. Herrmann, R.H.; Adler, S.F.; Goldstein, M.S.; DeBaun, R.M. J. Phys. Chem. **1961**, *65*, 2189.
165. Boronin, V.S.; Nikulina, V.S.; Poltorak, O.M. Russian J. Phys. Chem. **1963**, *37*, 1174.
166. Boronin, V.S.; Poltorak, O.M.; Turakulova, A.O. Russian J. Phys. Chem. **1974**, *48*, 156.
167. Frennet, A.; Wells, P.B. Appl. Catal. **1985**, *18*, 243.
168. Rabo, J.A.; Schomaker, V.; Pickert, P.E. In *Proceedings 3rd Int. Congr. Catal.*; W.M.H. Sachtler, G.C. Schuit, Zwietering, P. Eds.; North-Holland Pub. Co., Amsterdam, 1964; p. II-1264.
169. Sato, S. J. Catal. **1985**, *92*, 11.
170. McVicker, G.B.; Baker, R.T.K.; Garten, R.L.; Kugler, E.L. J. Catal. **1980**, *65*, 207.
171. Krishnamurthy, S.; Landolt, G.R.; Schoennagel, H.J. J. Catal. **1982**, *78*, 319.
172. Vis, J.C.; Van't Blik, H.F.J.; Huizinga, T.; Van Grondelle, J.; Prins, R. J. Catal. **1985**, *59*, 333.
173. Vis, J.C.; Van't Blik, H.F.J.; Huizinga, T.; Van Grondelle, J.; Prins, R. J. Mol. Catal. **1984**, *25*, 367.
174. Koningsberger, D.C.; Sayers, D.E. Solid State Ionics **1985**, *16*, 23.
175. Kip, B.J.; Van Grondelle, J.; Martens, J.H.A.; Prins, R. Appl. Catal. **1986**, *26*, 353.
176. Kip, B.J.; Duivenvoorden, F.B.M.; Koningsberger, K.C.; Prins, R. J. Catal. **1987**, *105*, 26–38.
177. Sinfelt, J.H.; Lam, Y.L.; Cusumano, J.A.; Barnett, A.E. J. Catal. **1976**, *42*, 227.
178. Sinfelt, J.H.; Via, G.H. J. Catal. **1979**, *56*, 1.
179. Sayari, A.; Wang, H.T.; Goodwin, J.G., Jr. J. Catal. **1985**, *93*, 368.
180. Yang, C.; Goodwin, J.R., Jr. J. Catal. **1982**, *78*, 182.
181. Kramer, R.; Andre, M. J. Catal. **1979**, *58*, 287.
182. Cavanagh, R.R.; Yates, J.T., Jr. J. Catal. **1981**, *68*, 22.
183. Bianchi, D.; Lacroix, M.; Pajonk, G.; Teichner, S.J. J. Catal. **1979**, *59*, 467.
184. Wanke, S.E.; Dougharty, N.A. J. Catal. **1972**, *24*, 367.
185. Konvalinka, J.A.; Scholten, J.J.F. J. Catal. **1977**, *48*, 374.
186. Eberhardt, W.; Greuter, F.; Plummer, E.W. Phys. Rev. Lett. **1981**, *46*, 1085.
187. Yates, J.T., Jr.; Peden, C.H.F.; Houston, J.E.; Goodman, D.W. Surf. Sci. **1985**, *160*, 37.
188. Benson, J.E.; Boudart, M. J. Catal. **1965**, *4*, 704.
189. Menon, P.G. In *Advances in Catalysis Science and Technical (Proc. 7th Natl. Catal. Symp., India, Baroda, 1985)*; Wiley Eastern Ltd.: New Delhi, 1986; L1–L15.
190. Prasad, J.; Murthy, K.R.; Menon, P.G. J. Catal. **1978**, *52*, 515.
191. Isaacs, B.H.; Petersen, E.E. J. Catal. **1984**, *85*, 1.
192. Kobayashi, M.; Inoue, Y.; Takahashi, N.; Burwell, R.L.; Butt, J.B.; Cohen, J.B. J. Catal. **1980**, *64*, 74.
193. Freel, J. J. Catal. **1972**, *25*, 149.

194. O'Rear, D.J.; Löffler, D.G.; Boudart, M. J. Catal. **1990**, *121*, 131–140.
195. Gland, J.L. Surf. Sci. **1980**, *93*, 487.
196. Wilson, G.R.; Hall, W.K. J. Catal. **1970**, *17*, 190.
197. Wilson, G.R.; Hall, W.K. J. Catal. **1972**, *24*, 306.
198. Gallezot, P.; Alarcon-Diaz, A.; Dalmon, J.A.; Renouprez, A.J.; Imelik, B. J. Catal. **1975**, *39*, 334.
199. Netzer, F.P.; Gruber, H.L. Z. Phys. Chem. **1975**, *NF96*, 25.
200. Menon, P.G.; Froment, G.F. J. Catal. **1979**, *59*, 138.
201. Porod, G. *Small-Angle X-ray Scattering*; Glatter, O., Kratky, O., Eds.; Academic Press: New York, 1982; Part II, Chap. 2.
202. Brumberger, H. Trans. Am. Crystallogr. Assoc. **1983**, *19*, 1.
203. Brumberger, H.; Delaglio, F.; Goodisman, J.; Whitefield, M. J. Appl. Crystallogr. **1986**, *19*, 287.
204. Brill, G.L.; Weil, C.G.; Schmidt, P.W. J. Colloid Interf. Sci. **1968**, *27*, 479.
205. Delaglio, F.; Goodisman, J.; Brumberger, H. J. Catal. **1986**, *99*, 383.
206. Hall, P.G.; Williams, R.T. J. Colloid Interf. Sci. *104*, 151.
207. Glatter, O.; Kratky, O. *Small-Angle X-ray Scattering*; Academic Press: New York, 1982.
208. Whyte, T.E., Jr.; Kirklin, P.W.; Gould, R.W.; Heinemann, H. J. Catal. **1972**, *25*, 407.
209. Cocco, G.; Schiffini, L.; Strukul, G.; Garturan, G. J. Catal. **1980**, *65*, 348.
210. Somorjai, G.A.; Powell, R.E.; Montgomery, P.W.; Jura, G. *Small-Angel X-ray Scattering*; Brumberger, H., Ed.; Gordon and Breach: New York, 1967; 449–466.
211. Brumberger, H. Makromol. Chem., Macromol. Symp. **1988**, *15*, 223–230.
212. Goodisman, J.; Brumberger, H.; Cupelo, R. J. Appl. Crystallogr. **1981**, *14*, 305–308.
213. Brumberger, H.; Delaglio, F.; Goodisman, J.; Phillips, M.G.; Schwarz, J.A.; Sen, P. J. Catal. **1985**, *92*, 199–210.
214. Brumberger, H.; Chang, Y.C.; Phillips, M.G.; Delaglio, F.; Goodisman, J. J. Catal. **1986**, *97*, 561.
215. Brumberger, H.; Goodisman, J. J. Appl. Crystallogr., **1983**, *16*, 83.
216. Kronig, R. de L. J. Phys. **1931**, *70*, 317; **1932**, *75*, 468; **1932**, *76*, 468.
217. Winick, H.; Doniach, S. *Synchrotron Radiation Research*; Plenum Press: New York, 1980.
218. Stern, E.A.; Sayes, D.E.; Lytle, F.W. Phys. Rev. B **1975**, *11*, 4825, 4836.
219. Teo, B.K. *EXAFS: Basic Principles and Data Analysis*; Springer-Verlag: Berlin, 1986.
220. Teo, B.K.; Joy, D.C. Eds. *EXAFS Spectroscopy: Techniques and Applications*; Plenum Press: New York, 1981.
221. Lytle, F.W.; Greegor, R.B.; Marques, E.C. In *Proc. 9th Int. Catal. Congr.*; Phillips, M.J., Ternan, M., Eds.; The Chemical Institute of Canada, Ottawa, 1988; Vol. 5, p. 54.
222. Sinfelt, J.H.; Via, G.H.; Lytle, F.W. Catal. Rev.-Sci. Eng. **1984**, *26*, 81.
223. Prins, R.; Koningsberger, D.C. In *Catalysis*; Anderson, J.R., Boudart, M., Eds.; Springer-Verlag, New York, 1987, 1988; Vol. 8, p. 321.
224. Bazin, D.; Dexpert, H.; Lagarde, P. Top. Curr. Chem. **1988**, *145*, 69.
225. Lagarde, P.; Dexpert, H. Advances in Physics **1984**, *33*, 567.

226. Sinfelt, J.H.; Meitzner, G.D. Acc. Chem. Res. **1993**, *26*, 1.
227. Heald, S.M.; Tranquada, J.M. *Physical Methods in Chemistry*, 2nd Ed.; 1990; Vol. 5, p. 189.
228. Guyot-Sionnest, N.S.; Bazin, D.; Lunch, J.; Bournonville, J.T.; Dexpert, H. Physica **1989**, *B158*, 211.
229. Lengeler, B.; Eisenberger, P. Phys. Rev. **1980**, *B21*, 4057.
230. Zhao, J.; Montano, P.A.; Phys. Rev. **1989**, *B40*, 3401.
231. Zhao, J., Ph.D. thesis, West Virginia University; 1991; personal communication.
232. Lytle, F.W.; Greegor, R.B.; Marques, E.C.; Sandstrom, D.R.; Via, G.H.; Sinfelt, J.H. J. Catal. **1985**, *95*, 546.
233. Lytle, F.W.; Greegor, R.B.; Marques, E.C.; Biebesheimer, V.A.; Sandstrom, D.R.; Horsley, J.A.; Via, G.H.; Sinfelt, J.H. *Catalyst Characterization Science*; Deveney, M.L., Gland, J.L., Eds.; ACS Symp. Series; 1985; Vol. 288, p. 280.
234. Marques, E.C.; Sandstrom, D.R.; Lytle, F.W.; Greegor, R.B. J. Chem. Phys. **1982**, *77* (2), 1027.
235. Dalla Betta, R.A.; Boudart, M.; Foger, K.; Löffler, D.G.; Sánchez-Arrieta, J. Rev. Sci. Instrum. **1984**, *55*, 1910.
236. Neils, T.L.; Burlitch, J.M. J. Catal. **1989**, *118*, 79.
237. Dexpert, H. J. Phys., Colloque C8 **1986**, *47* (Suppl. 12), C8-219, C8-226.
238. Guyot-Sionnest, N.S.; Villain, F.; Bazin, D.; Dexpert, H.; Le Peltier, F.; Lynch, J.; Bournonville, J.P. Catal. Lett. **1991**, *8*, 283.
239. Wei, P.S.P.; Lytle, F.W. Phys. Rev. **1979**, *B19*, 679.
240. Lytle, F.W. J. Catal. **1976**, *43*, 376–379.
241. Meitzner, G.; Via, G.H.; Lytle, F.W.; Sinfelt, J.H. J. Phys. Chem. **1992**, *96*, 4960.
242. Lytle, F.W.; Wei, P.S.P.; Greegor, R.B.; Via, G.H.; Sinfelt, J.H. J. Chem. Phys. **1979**, *70*, 4849.
243. Sinfelt, J.H.; Via, G.H.; Lytle, F.W.; Greegor, R.B. J. Chem. Phys. **1981**, *75*, 5527.
244. Meitzner, G. Catal. Today **1998**, *39*, 281.
245. Bazin, D.; Guczi, L. Rec. Res. Dev. Phys. Chem. **1999**, *3*, 387.
246. Newton, M.A.; Dent, A.J.; Evans, J. Chem. Soc. Rev. **2002**, *31*, 83.
247. Bazin, D.; Sayers, D.; Rehr, J.J.; Mottet, C. J. Phys. Chem. B **1997**, *101*, 5332.
248. Bazin, D.; Mottet, C.; Tréglia, G. Appl. Catal. A: Gen **2000**, *200*, 47.
249. de Groot, F.M.F. Top. Catal. **2000**, *10*, 179.
250. Koningsberger, D.C.; Mojet, B.L.; van Dorssen, G.E.; Ramaker, D.E. Top. Catal. **2000**, *10*, 143–155.
251. Rehr, J.J.; Ankudinov, A.; Zabinsky, S.I. Catal. Today, **1998**, *39*, 262.
252. Lagarde, P. Ultramicroscopy, **2001**, *80*, 255.
253. Fernández-García, M. Catal. Rev.-Eng. Sci. **2002**, *44*, 59–121.
254. Bazin, D.; Triconnet, A.; Moureaux, P. Nucl. Instrum. Meth. Phys. Res. B **1995**, *97*, 41.
255. Spieker, W.A.; Liu, J.; Miller, J.T.; Kropf, A.J.; Regalbuto, J.R. Appl. Catal. A: Gen. **2002**, *232*, 219.
256. Ramaker, D.E.; van Dorssen, G.E.; Mojet, B.L.; Koningsberger, D.C. Top. Catal. **2000**, *10*, 157, 167.
257. Yu, I.; Gibson, A.A.V.; Hunt, E.R.; Halperin, W.P. Phys. Rev. Lett. **1980**, *44*, 348.

258. de Ménorval, L.-C.; Fraissard, J.P. Chem. Phys. Lett. **1981**, 77, 309.
259. Rouabah, D.; Benslama, R.; Fraissard, J. Chem. Phys. Lett. **1991**, 179, 218.
260. de Ménorval, L.-C.; Ito, T.; Fraissard, J. J. Chem. Soc. Faraday Trans. **1982**, 78, 403.
261. Dalla Betta, R.A.; Boudart, M. Proc. 5th Int. Congr. Catal.; Hightower, J.W., Ed.; North-Holland: Amsterdam, 1973; Vol. 2, p. 1329.
262. Boudart, M.; Djéga-Mariadassou, G. Kinetics of Heterogeneous Catalysis Reactions; Princeton University Press: Princeton, NJ, 1984.
263. Gallezot, P.; Mutin, I.; Dalmai-Imelik, G. J. Microsc. Spectrosc. Electron. **1976**, 1, 1.
264. Gallezot, P.; Bienenstock, A.I.; Boudart, M. Nouv. J. Chim. **1978**, 2, 263.
265. Weber, R.S. PhD dissertation, Stanford University: Stanford, CA, 1985.
266. Ryoo, R.; Pak, C.; Chmelka, B. Zeolites **1990**, 10, 790.
267. Boudart, M.; Valença, G.P. J. Catal. **1991**, 128, 447.
268. Boudart, M.; Samant, M.G.; Ryoo, R. Ultramicroscopy **1986**, 20, 125.
269. Rivera-Latas, F.J.; Dalla Betta, R.A.; Boudart, M. AIChE J **1992**, 38, 771.
270. Dexpert, H.; Freund, E.; Lesage, E.; Lynch, J.P. Stud. Surf. Sci. Catal. **1982**, 11, 53.
271. Boudart, M.; Ryoo, R.; Valença, G.P.; Van Grieken, R. Catal. Lett. **1993**, 17, 273.
272. Sharma, S.B.; Laska, T.E.; Balaraman, P.; Root, T.W.; Dumesic, J.A. J. Catal. **1994**, 150, 225.
273. Han, O.H.; Larsen, G.; Haller, G.L.; Ziln, K.W. Bull. Korean Chem. Soc. **1998**, 19, 934.
274. Griffiths, J.M.; Bell, A.T.; Reimer, J.A. J. Phys. Chem. **1994**, 98, 1918.
275. Griffiths, J.M.; Bell, A.T.; Reimer, J.A. J. Phys. Chem. **1993**, 97, 9161.
276. Subramanian, S.; Schwarz, J.A. Appl. Catal. **1991**, 68, 131.
277. Defossé, C. Characterization of Heterogeneous Catalysts; Delannay, F., Ed.; Marcel Dekker: New York, 1984; 225–298.
278. Querini, C.A.; Fung, S.C. Catal. Today **1997**, 37, 277.
279. Jess, A.; Hein, O.; Kern, C. Catalyst Deactivation 1999; Delmon, B., Froment, G.F., Eds., Elsevier: Amsterdam, 1999; 81–88.
280. Boudart, M.; de Ménorval, L.-C.; Fraissard, J.P.; Valença, G.P. J. Phys. Chem. **1988**, 92, 4033.
281. Powell, C.J.; Erickson, N.E.; Madey, T.E. J. Electron. Spectrosc. Relat. Phenom. **1979**, 17, 361.
282. Dianis, W.P.; Lester, J.E. Anal. Chem. **1973**, 45, 1416.
283. Contour, J.P.; Mouvier, G. J. Electron. Spectrosc. Relat. Phenom. **1975**, 7, 85.
284. Ogilvie, J.L.; Wolberg, A. Appl. Spectrosc. **1972**, 26, 401.
285. Madey, T.E.; Wagner, C.D.; Joshi, A. J. Electron Spectrosc. Relat. Phenom. **1977**, 10, 359.
286. Hoflund, G.B.; Asbury, D.A.; Corallo, C.F.; Corallo, G.R. J. Vac. Sci. Technol. **1988**, A6, 70.
287. Asbury, D.A.; Hoflund, G.B. Surf. Sci. **1988**, 199, 552.
288. Dwyer, D.J.; Robbins, J.L.; Cameron, S.D.; Dudash, N.; Hardenbergh, J. Strong Metal–Support Interactions; Baker, R.T.K., Tauster, S.J., Dumesic, J.A., Eds.; American Chemical Society: Washington, DC, 1986; 21–33.
289. Dwyer, D.J.; Cameron, S.D.; Gland, J. Surf. Sci. **1985**, 159, 430.

290. Ewertowski, F.; Klimas, J.; Maj, A. Rocz. Chem. **1975**, *49*, 351.

291. Klimas, J.; Ewertowski, F. Pol. J. Chem. **1980**, *54*, 867.

292. Pliskin, W.A.; Eischens, R.P. Z. Phys. Chem. (Frankfurt) NF, **1960**, *24*, 11.

293. Barth, R.; Pitchai, R.; Anderson, R.L.; Verykios, X.E. J. Catal. **1989**, *116*, 61.

294. Lantz, J.B.; Gonzalez, R.D. J. Catal. **1976**, *41*, 293.

295. Sen, B.; Vannice, M.A. J. Catal. **1991**, *130*, 9.

296. Johnson, M.F.L.; LeRoy, V.M. J. Catal. **1974**, *35*, 434.

297. Webb, A. J. Catal. **1975**, *39*, 485.

298. Johnson, M.F.L. J. Catal. **1975**, *39*, 487.

299. McNicol, B.D. J. Catal. **1977**, *46*, 428.

300. Bolivar, C.; Charcosset, H.; Frety, R.; Primet, M.; Tournayan, L.; Betizeau, C.; Leclercq, G.; Maurel, R. J. Catal. **1975**, *39*, 249.

301. Yao, H.C.; Shelef, M. J. Catal. **1976**, *44*, 392.

302. Wagstaff, N.; Prins, R. J. Catal. **1979**, *59*, 434.

303. Isaacs, B.H.; Petersen, E.E. J. Catal. **1982**, *77*, 43.

304. Mieville, R.L. J. Catal. **1984**, *87*, 437.

305. Ambs, W.F.; Mitchell, M.M. J. Catal. **1983**, *82*, 226.

306. Menon, P.G.; Sieders, J.; Streefkerk, F.J.; van Keulen, G.J.M. J. Catal. **1973**, *29*, 188.

307. Eskinazi, V. Appl. Catal. **1982**, *4*, 37.

308. Kirlin, P.S.; Strohmeier, B.R.; Gates, B.C. J. Catal. **1986**, *98*, 308.

309. Adkins, S.R.; Davis, B.H. ACS Symp. Series, **1985**, *228*, 57.

310. Nacheff, M.S.; Kraus, L.S.; Ichikawa, M.; Hoffman, B.M.; Butt, J.B.; Sachtler, W.M.H. J. Catal. **1987**, *106*, 263.

311. Onuferko, J.H.; Short, D.R.; Kelley, M.J. Appl. Surf. Sci. **1984**, *19*, 227.

312. Short, D.R.; Khalik, S.M.; Katzer, J.R.; Kelley, M.J. J. Catal. **1981**, *72*, 288.

313. Kelley, M.J.; Freed, R.L.; Swartzfager, D.G. J. Catal. **1982**, *78*, 445.

314. Meitzner, G.; Via, G.H.; Lytle, F.W.; Sinfelt, J.H. J. Chem. Phys. **1987**, *87*, 6354.

315. Bazin, D.; Dexpert, H.; Lagarde, P.; Bournonville, J.P. J. Physique, **1986**, *47*, C8-293.

316. Hilbrig, F.; Michel, C.; Haller, G.L. J. Phys. Chem. **1992**, *96*, 9893.

317. Vuurman, M.A.; Stufkens, D.J.; Oskam, A.; Wachs, I.E. J. Mol. Catal. **1992**, *76*, 263.

318. Nan-Yu Tøpsoe, J. Catal. **1991**, *128*, 499.

319. Turek, A.M.; Decanio, E.; Wachs, I.E. J. Phys. Chem. **1992**, *96*, 5000.

320. Sibeijn, M.; Spronk, R.; van Veen, J.A.R.; Mol, J.C. Catal. Lett. **1991**, *8*, 201.

321. Hardcastle, F.D.; Wachs, I.E.; Horsley, J.A.; Via, G.H. J. Mol. Catal. **1988**, *46*, 15.

322. Jothimurugesan, K.; Nayak, A.K.; Mehta, G.K.; Rai, K.N.; Bhatia, S.; Srivastava, R.D. AIChE J, **1985**, *31*, 1997.

323. Kelley, M.J.; Short, D.R.; Swartzfanger, D.G. J. Mol. Catal. **1983**, *20*, 235.

324. Prestvik, R.; Tøtdal, B.; Lyman, C.E.; Holmen, A. J. Catal. **1998**, *176*, 246–252.

325. Rønning, M.; Gjervan, T.; Prestvik, R.; Nicholson, D.G.; Holmen, A. J. Catal. **2001**, *204*, 292–304.

326. Bensaddik, A.; Caballero, A.; Bazin, D.; Dexpert, H.; Didillon, B.; Lynch, J. Appl. Catal. A: Gen. **1997**, *162*, 171–180.

327. Biloen, P.; Helle, J.N.; Berbeck, H.; Dautzenberg, F.M.; Sachtler, W.M.H. J. Catal. **1980**, *63*, 112.

328. Sachtler, W.M.H. J. Mol. Catal. **1984**, *25*, 1.

329. Burch, R. Platinum Metals Rev. **1978**, *22*, 57.

330. Sexton, B.A.; Hughes, A.E.; Folger, K. J. Catal. **1984**, *88*, 1566.

331. Lieske, H.; Völter, J. J. Catal. **1984**, *90*, 46.

332. Burch, R. J. Catal. **1981**, *71*, 348.

333. Burch, R.; Garla, L.C. J. Catal. **1981**, *71*, 360.

334. Burch, R.; Mitchell, A.J. Appl. Catal. **1981**, *6*, 121.

335. Müller, A.C.; Engelhard, P.A.; Weisang, J.E. J. Catal. **1979**, *56*, 65.

336. Adkins, S.R.; Davis, B.H. J. Catal. **1984**, *89*, 371.

337. Stencel, J.M.; Goodman, J.; Davis, B.H. Proc. 9th Int. Congr. Catal. **1988**, *3*, 1291.

338. Li, Y.-X.; Stencel, J.M.; Davis, B.H. Reaction Kin. Catal. Lett. **1988**, *37*, 273.

339. Li, Y.-X.; Stencel, J.M.; Davis, B.H. Appl. Catal. **1990**, *64*, 71.

340. Bouwman, R.; Biloen, P. Surf. Sci. **1974**, *41*, 348.

341. Hoflund, G.B.; Asbury, D.A.; Kirszenzatejn, P.; Laitinen, H.A. Surf. Interf. Anal. **1986**, *9*, 169.

342. Hoflund, G. *Preparation of Catalysts III*; Poncelet, G., Grange, P., Jocobs, P.A., Eds.; Elsevier: Amsterdam, 1983; 91–100.

343. Gardner, S.D.; Hoflund, G.B.; Schryer, D.R. J. Catal. **1989**, *119*, 179.

344. Cox, D.F.; Hoflund, G.B. Surf. Sci. **1985**, *151*, 202.

345. Gardner, S.D.; Hoflund, G.B.; Schryer, D.R.; Upchurch, B.T. J. Phys. Chem. **1991**, *95*, 835.

346. Laitinen, H.A.; Waggoner, J.R.; Chan, C.Y.; Kirszensztejn, P.; Asbury, D.A.; Hoflund, G.B. J. Electrochem. Soc. **1986**, *133*, 1586.

347. Hoflund, G.B.; Asbury, D.A.; Gilbert, R.E. Thin Solid Films, **1985**, *129*, 139.

348. Bacaud, R.; Bussiere, P.; Figueras, F. J. Catal. **1981**, *69*, 399.

349. Berndt, V.H.; Mehner, H.; Völter, J.; Meise, W. Z. Anorg. Allg. Chem. **1977**, *429*, 47.

350. Bacaud, R.; Bussiere, P.; Figueras, F.; Mathieu, J.P. C.R. Acad. Sci. Paris Ser. C **1975**, *281*, 159.

351. Bacaud, R.; Bussiere, P.; Figueras, F. J. Phys. Colloq. **1979**, *40*, C2-94.

352. Charlton, J.S.; Cordey-Hayes, M.; Harris, I.R. J. Less-Common Met. **1970**, *20*, 105.

353. Li, Y.-X.; Zhang, Y.-F.-; Klabunde, K.J. Langmuir **1988**, *4*, 385.

354. Klabunde, K.J.; Li, Y.-X.; Purcell, K.F. Hyperfine Interact. **1988**, *41*, 649.

355. Lin, L.; Wu, R.; Zang, J.; Jiang, B. Acta Pertrol Sci. (China) **1980**, *1*, 73.

356. Pakhomov, N.A.; Buyanov, R.A.; Yurchenko, E.N.; Cherynshev, A.P.; Kotel'nikov, G.R.; Moroz, E.M.; Zaitseva, N.A.; Patanov, V.A. Kinet. Katal. **1981**, *22*, 488.

357. Gray, P.R.; Farha, F.E. *Mössbauer Effect Methodology*; Grunerman, I.J., Seidel, Eds.; Plenum: New York, 1976; Vol. 10, p. 47.

358. Yurchenko, E.N.; Kuznetsov, V.I.; Melnikova, V.P.; Startsev, A.N. React. Kinet. Catal. Lett. **1983**, *23*, 137.

359. Zhang, P.; Shao, H.; Yang, X.; Pang, L. Cuihua Xuebao **1984**, *5*, 101.

360. Li, Y.-X.; Zhang, Y.-F.; Shai, Y.-F. Cuihua Xuebao, **1985**, *5*, 311.

361. Zhang, S.; Xie, B.; Wang, P.; Zhang, J. Cuihua Xuebao, **1980**, *1*, 311.
362. Li, Y.-X.; Klabunde, K.J.; Davis, B.H. J. Catal. **1991**, *128*, 1.
363. Kuznetsov, V.I.; Belyi, A.S.; Yurchenko, E.N.; Smolikov, M.D.; Protasova, M.T.; Zatolokina, E.V.; Duplyakin, V.K. J. Catal. **1986**, *99*, 159.
364. Srinivasan, R.; Davis, B.H. Platinum Metal Rev. **1992**, *36*, 151.
365. Srinivasan, R.; De Angelis, R.J.; Davis, B.H. J. Catal. **1987**, *106*, 449.
366. Li, Y.-X.; Chiu, N.-S.; Lee, W.-H.; Bauer, S.H.; Davis, B.H. Charaterization and catalyst development. An interactive approach. ACS Symp. Series, **1989**, *411*, 328.
367. Chiu, N.-S.; Lee, W.-H.; Li, Y.-X.; Bauer, S.H.; Davis, B.H. *Advances in Hydrotreating Catalysts*; Occelli, M.L., Anthony, R.G., Eds.; Elsevier Sci. Pub: Amsterdam, 1989; 147–163.
368. Meitzner, G.; Via, G.H.; Lytle, F.W.; Fung, S.C.; Sinfelt, J.H. J. Phys. Chem. **1988**, *92*, 2925.
369. Srinivasan, R.; Davis, B.H. Appl. Catal. (A General), **1992**, *87*, 45.
370. Le Peltier, F.; Didillon, B.; Jumar, J.-C.; Olivier-Fourcade, J. Eur. Pat. Appl. EP 1181978 A1, (2002).
371. Macleod, N.; Fryer, J.R.; Stirling, D.; Webb, G. Catal. Today **1998**, *46*, 37–54.
372. Srinivasan, R.; Sparks, D.; Davis, B.H. J. Mol. Catal. **1994**, *88*, 325.
373. Srinivasan, R.; Davis, B.H. J. Mol. Catal. **1994**, *88*, 343.
374. Srinivasan, R.; Sparks, D.; Davis, B.H. J. Mol. Catal. **1994**, *88*, 359.
375. Sinfelt, J.H. *Bimetallic Catalysts. Discoveries, Concepts, and Applications*; John Wiley & Sons: New York, 1983.
376. Chan, S.C.; Fung, S.C.; Sinfelt, J.H. J. Catal. **1988**, *113*, 164.
377. Subramanian, S.; Schwarz, J.A. Appl. Catal. **1991**, *68*, 131.
378. Huang, J.-J.; Fung, S.C.; Gates, W.E.; McVicker, G.B. J. Catal. **1989**, *118*, 192.
379. Subramanian, S.; Schwarz, J.A. Appl. Catal. **1990**, *61*, L15.
380. Huang, Y.-J.; Fung, S.C. J. Catal. **1991**, *131*, 378.
381. McVicker, G.B.; Ziemiak, J.J. Appl. Catal. **1985**, *14*, 229.
382. von Brandis, H.; Wagner, F.E.; Sawicki, J.A.; Marcinkowska, K.; Rolston, J.H. Hyperfine Interact. **1990**, *57*, 2127.
383. Sinfelt, J.H.; Via, G.H.; Lytle, F.W. J. Chem. Phys. **1982**, *76*, 2779.
384. Romero, T.; Arenas, B.; Perozo, E.; Bolívar, C.; Bravo, G.; Marcano, P.; Scott, C.; Pérez Zurita, M.J.; Goldwasser, J. J. Catal. La Plata, Argentina, **1990**, *124*, 281.
385. Romero, T.; Tejeda, J.; Jaunay, D.; Bolívar, C.; Charcoset, H. *Proceedings of the VIIth Iberoamerican Congress in Catalysis*; La Plata: Argentina, 1980; p. 453.
386. Goldwasser, J.; Arenas, B.; Bolívar, C.; Castro, G.; Rodriguez, A.; Fleitas, A.; Giron, J. J. Catal. **1986**, *100*, 75.
387. Bouwman, R.; Biloen, P. J. Catal. **1977**, *48*, 209.
388. de Miguel, S.R.; Scelza, O.A.; Castro, A.A. Appl. Catal. **1988**, *44*, 23.
389. de Miguel, S.R.; Martinez Correa, J.A.; Baronetti, G.T.; Castro, A.A.; Scelza, O.A. Appl. Catal. **1990**, *60*, 47.
390. Völter, J.; Lieske, H.; Lietz, G. React. Kinet. Catal. Lett. **1981**, *16*, 87.
391. Palizov, A.; Bonev, Kh.; Kadinov, G.; Shopov, D.; Lietz, G.; Völter, J. J. Catal. **1981**, *71*, 1.
392. Lieske, H.; Völter, J. React. Kinet. Catal. Lett. **1983**, *23*, 403.

393. Bernard, J.R. *Proc. Fifth Int. Conf. Zeolites*, Heyden: London, 1980; p. 686.
394. Besoukhanova, C.; Guidot, J.; Barthomeuf, D. J. Chem. Soc. Faraday Trans. I, **1981**, *77*, 1595.
395. Hughes, T.R.; Buss, W.C.; Tamm, P.W.; Jacobson, R.L. *New Developments in Zeolite Science and Technology*; Murakami, Y., Iijima, A., Ward, J.W., Eds.; Elsevier: Amsterdam, 1986; 725–732 and discussion 120–126.
396. Ward, J.W. J. Catal. **1968**, *10*, 34; *Zeolite Chemistry and Catalysis*; Rabo, J.A., Ed.; 1976; ACS Monograph Series, Vol. 171, 229–234.
397. Newell, P.A.; Rees, L.V.C. Zeolites, **1983**, *3*, 22, 28.
398. Anderson, J.R. *Structure of Metallic Catalysts*; Academic Press: New York, 1975.
399. Rice, S.B.; Koo, J.Y.; Disko, M.M.; Treacy, M.M.J. Ultramicroscopy **1990**, *34*, 108.
400. Miller, J.T.; Sajkowski, D.J.; Modica, F.S.; Lane, G.S.; Gates, B.C.; Vaarkamp, M.; Grondelle, J.V.; Koningsberger, D.C. Catal. Lett. **1990**, *6*, 369.
401. van Zon, J.B.A.D.; Koningsberger, D.C.; van't Blik, H.F.J.; Sayers, D.E. J. Chem. Phys. **1985**, *83*, 5742.
402. Larsen, G.; Haller, G.L. Catal. Today **1992**, *15*, 431.
403. McVicker, G.B.; Kao, J.L.; Ziemiak, J.J.; Gates, W.E.; Robbins, J.L.; Treacy, M.M.J.; Rice, S.B.; Vanderspurt, T.H.; Cross, V.R.; Ghosh, A.K. J. Catal. **1993**, *139*, 48.
404. Iglesia, E.; Baumgartner, J.E. 10th Int. Congr. Catal. Budapest, 1992; paper 0-65.
405. Vaarkamp, M.; Miller, J.T.; Modica, F.S.; Lane, G.S.; Koningsberger, D.C. J. Catal. **1992**, *138*, 675.
406. Tatsumi, T.; Dai, L.-X.; Sakashita, H.; Catal. Lett. **1994**, *27*, 289–295.
407. Jacobs, G.; Ghadiali, F.; Pisanu, A.; Padro, C.L.; Borgna, A.; Alvarez, W.E.; Resasco, D.E. J. Catal. **2000**, *191*, 116–127.
408. Jongpatiwut, S.; Sackamduang, P.; Rirksomboon, T.; Osuwan, S.; Alvarez, W.E.; Resasco, D.E. Appl. Catal. A: Gen. **2002**, *230*, 177–193.
409. Watwe, R.M.; Spiewak, B.E.; Cortright, R.D.; Dumesic, J.A. Catal. Lett. **1998**, *51*, 139–147.
410. Sharma, S.B.; Laska, T.E.; Balaraman, P.; Root, T.W.; Dumesic, J.A. J. Catal. **1994**, *150*, 225–233.
411. Hunger, M.; Weitkamp, J. Angew. Chem. Int. Ed. **2001**, *40*, 2954–2971.
412. Shido, T.; Prins, R. Curr. Opin. Solid State Mater. Sci. **1998**, *3*, 330–335.
413. Le Normand, F.; Borgna, A.; Garetto, T.F.; Apesteguia, C.R.; Moraweck, B. J. Phys. Chem. **1996**, *100*, 9068–9076.
414. Somorjai, G.A. Appl. Surf. Sci. **1997**, *121/122*, 1–19.
415. Knop-Gericke, A.; Hävecker, M.; Schedl-Niedrig, Th.; Schlögl, R. Top. Catal. **2000**, *10*, 187–198.
416. Mojet, B.L.; Miller, J.T.; Ramaker, D.E.; Koningsberger, D.C. J. Catal. **1999**, *186*, 373–386.
417. Zaera, F. Appl. Catal. A: Gen. **2002**, *229*, 75–91.

7
Optimization of Catalyst Pore Structure by Kinetics and Diffusion Analysis

Jerzy Szczygieł
Wrocław University of Technology
Wrocław, Poland

1 INTRODUCTION

The design of the catalyst is of paramount importance to the economics of catalytic processes,[1–5] and the choice of an optimal porous structure is an inherent part not only of catalyst design but also of catalyst selection for the intended technological process. In the assessment of industrial catalysts it is essential to relate catalytic activity on a unit volume basis. This volume-related activity depends on a variety of factors, such as the specific activity of the catalyst relative to the developed surface area, the amount of the surface area in the unit catalyst volume, and the porous structure of the catalyst. The porosity is responsible for the transport of reacting substances and heat in the grains. All these parameters are determined by the conditions of catalyst preparation and the nature of the material.[6,7] A high specific activity can be achieved by controlling the chemical composition of the catalyst under design.[8] However, it is worth remembering that the more active the catalyst, the more difficult becomes the utilization of the catalytic potential because fast reactions induce resistances concomitant with the transport of reagents. It is therefore essential for the designer of a new catalyst to define a priori the required internal surface and the most advantageous pore structure for the intended reaction so as to provide free access of the reacting substances to the surface of the catalyst pores.

The optimal parameters of the pore structure depend on the specific activity of the catalysts, on the kinetics of the process, on the reactant nature, and on the technological conditions. Where elementary reactions are concerned, the (qualitative) effect of catalyst activity on the choice of the optimal structure is quite clear. Thiele[9] was first to construct a mathematical model that incorporated the effect of diffusion on the rate of a chemical reaction. With such a model in hand it was possible to assess the influence of mass transport in the catalyst pores on the progress of the reaction. Although Thiele's model involved mass transport equations for ideal geometrical grains, it approximated real phenomena with good adequacy. The Thiele model may also be of use in estimating how the parameters of the catalyst's porous structure affect the reaction rate.[10] In the absence of diffusion resistance, the rate of the reaction is proportional to the surface developed within the catalyst and does not depend on the pore volume. In the presence of molecular diffusion, reaction rate increases linearly with the square root of the surface area and pore volume developed in the catalyst. Under conditions of Knudsen diffusion, reaction rate does not depend on the catalyst surface but is proportional to the pore volume.

Analysis of the pore structure effect, as well as the choice of an optimal porous structure, raises problems whenever a catalytic process involves many reactions—both simultaneous and consecutive. In that particular case the influence of the porous structure on the efficiency of the catalytic process increases with the increasing difference in rate between individual reactions. Faster reactions are limited by diffusion; slower reactions run in the kinetic region. The effect of diffusional limitation on the total rate and selectivity of the composite reaction depends on the structure of these complex changes.[5] Thus, the optimal parameters of the porous structure for a catalyst of a defined chemical composition should be selected separately, according to the desired kinetic scheme and process conditions, by analyzing the mass transfer and heat transfer phenomena that occur in the catalyst grain. The optimal pore structure is determined by the transfer processes in the grain, as well as by the kinetics of the reactions participating in the process.

The objective of catalytic reforming is to produce high-octane gasoline by increasing the content of aromatic compounds and isoparaffins in an organic material starting with predominantly n-paraffins and naphthenes. The kinetics of the process and the reactions involved, as described by Froment and coworkers,[11–15] are sophisticated. In the course of the process, paraffins are dehydrocyclized to naphthenes, which in turn convert to aromatic compounds, whereas n-paraffins isomerize to form isoparaffins. The undesired reactions of hydrocracking convert some part of the product into light paraffins. There is a substantial difference in the rate between particular reactions. Thus, the hydrogenation of naphthenes to aromatic hydrocarbons operates at a rate that is two orders of magnitude faster than the rate with which paraffins are dehydrocyclized to naphthenes. Hence, under the same process conditions, slower reactions may run in the kinetic region whereas faster reactions may be constrained by diffusion resistance, thus slowing down the rate of the process. Control of transport phenomena via a careful choice of the pore structure can disclose further catalytic potential and upgrade the performance of the catalyst. Recent developments in computer techniques have enhanced preferences for mathematical methods when analyzing the effect of the porous structure on the performance of the catalyst. Such an approach requires assumptions for the physical model of the catalyst grain, description of the porous structure, as well as assumptions for the mathematical description of transport phenomena and reactions in the grain model.

1.1 Models of Porous Structures

Mass transfer and chemical reactions in the catalyst are interrelated. To gain a better understanding of this interrelationship it is advisable to model the porous structure of the catalyst support and to quantify the effective diffusion coefficients. Real porous structures are complex in nature. They form an irregular network of channels, which range in size from micropores (with much smaller radii than the mean free paths of the diffusing components) to macropores (of micrometer dimensions). The pores are randomly interconnected. In chemical reaction engineering use is made of a variety of models that provide a simple description of such a structure. Specialized literature contains many references to the porous structures of a variety of catalysts—from monodispersive,[16] through bidispersive,[17–19] to models including a wide range of pore distributions.[20] A chronological review of such models (from the simple Wheeler model to the models developed in the 1990s) can be found elsewhere.[21,22] Thus, the model developed by Wheeler[23] suffices for the approximation of a monodispersive catalyst,[24] but it fails to describe the topology of the space produced by the pores of a real catalyst. What is more, the coefficient of tortuosity, τ (with a value ranging from 2 to 7), includes too much information and smoothes over the discrepancy between a real porous structure and the one described by the model. The random model constructed by Wakao and Smith[17] allows a description of bidispersive systems; it incorporates Knudsen diffusion and molecular diffusion for narrow pores and wide pores, respectively. The parallel cross-linked pore model established by Steward[25] includes a wide range of pore size distribution, and the calculated effective diffusion coefficient values differ only slightly from those determined experimentally.[26]

In the past few years preference has been given to such models that incorporate not only porosity and pore volume distribution as a function of pore radii, but also pore length distribution and a parameter describing the method of pore interconnection.[27] For the purpose of simplification, all of the models mentioned above make use of the assumption that the pores are cylindrical in shape. Quite recently, a number of publications[28–32] have directed attention to the more likely shapes of the pores described by fractal models. In their description of the catalytic reforming process, Coppens and coworkers[31–33] demonstrate how the fractal roughness of the catalyst pore walls affects the yields and distribution of industrial processes.

An earlier study of ours[34] described a "globular model," which was tested for the support of an Al_2O_3-reforming catalyst. Using this model, the variation of the pore volume can be related to a wide range of variations in the pore radii. The model[34] is based on the assumptions of the Wakao–Smith model,[17] but it contains one more parameter—a variable number of contacts of the molecule with its "neighbors"—which extends the application of the model to the description of a more complex porous structure. The bidispersive "layered model"[19,35] was used to optimize the porous structure of the reforming catalyst, and the values proposed for the parameters of the porous structure were confirmed by experiments.[41] In a previous study,[71] the same catalyst grain model was used to analyze the problem of how the kinetics of the reforming process affects the choice of an optimal porous structure. These models of pore networks apply to a variety of situations—to the prediction of effective diffusion coefficients,[37] to the simulation of diffusion and reactions in the catalyst pores,[19,35,38,39,71] and to the interpretation of experimental results and measured diffusion values.[40]

1.2   Optimization of Porous Structure Reported in Literature

Optimizing the porous structure can be viewed as an increase in the potentiality of upgrading the efficiency of the catalyst unit volume. The porous structure of the catalyst can be optimized using more or less sophisticated models.[19,27,33,42–46] The mathematical methods made use of in such analyses are of particular importance when the interactions of the complex kinetics of the process, diffusion, and porous structure make it difficult to experimentally assess the optimal values for the parameters of the grain pore structure.

Apart from porosity and mean pore radius, the following parameters are to be considered: pore radii distribution, pore length distribution, pore shape, and pore interconnection. At this stage of catalyst design, it is also essential to choose the right shape and size of the catalyst grains so as to match them with the optimal porous structure.

Boreskov[47] seems to have been first to express the need of determining an optimal porous structure, and he specified some general principles for choosing it. The parameters of an optimal porous structure differ according to the adopted objective that is to be achieved. Thus, if the primary objective is to maximize the process rate, then—for a simple reaction and a low-activity catalyst—it is advisable to use a monodispersive catalyst with a developed internal surface. When the catalyst displays a high activity, the application of bidispersive catalysts will be better suited because (apart from the narrow pores that develop the catalyst surface) they include a certain portion of wide (transport) pores.[47] The transport pores facilitate access of the raw material to the surface developed by the narrow pores, thus enabling complete utilization of the internal surface. Dyrin and coworkers[48] corroborated the above-mentioned recommendations. Beskov and coworkers also focused on the optimization of the porous structure.[49,50] Their recommendations pertaining to the choice of the optimal porous structure for a catalyst were based on the analysis of mass transport phenomena inside the grain, and their results were made use of by relevant industries. In their studies, Beskov and coworkers aimed at determining the interaction between the optimal porous structure, grain size, and process parameters.

Owing to the developments in computational techniques, the porous structure can be optimized using a variety of more sophisticated models. Detailed accounts of these models and of relevant optimizing techniques were reported by a number of investigators.[1,22,51–53] Delancey[43] made use of a simple diffusion model to analyze a single, reversible, first-order reaction and proposed an optimal density for a catalyst of a defined composition, an optimal composition for a catalyst of a defined density, as well as an optimal composition and an optimal density. Hegedus[44] applied a random model proposed by Wakao and Smith[17] to the optimization of the porous structure for an automotive catalyst and pointed to a variety of advantages when using a catalyst of an optimal porous structure. Pereira and coworkers[45] used a mathematical model for the design and simulation of a monolithic automotive catalyst. The model, as well as the optimizing procedure, made it possible to predict the optimal values for a variety of factors (washcoat thickness, porous structure, channel size, and wall thickness) needed by the designer of the catalyst to minimize diffusion in the presence of poisons. In the majority of complex processes involving simultaneous and consecutive reactions, an optimal structure must be able to deliver maximal quantities of the defined final product.

Keil and Rieckman optimized the porous structure of a catalyst for the complex hydrodemetallation of crude oil (HDM process) by making use of various theoretical models—the macro-micropore model[46] and the three-dimensional network of interconnected cylindrical pores.[27,55] The porous structure models and the mathematical methods of analysis were found to be a useful tool in determining the optimal parameters for this structure because the interaction between kinetics, diffusion, and geometrical pore structure is too complicated to allow intuitive methods for parameter determination. The methods are useful in optimizing the porous structure for a variety of complex processes.

2 SCOPE OF CHAPTER

In this chapter, we present an algorithm with which one can optimize the porous structure of the reforming catalyst. The algorithm consists of three major steps: analysis of kinetic phenomena in the catalyst grain (Section 3); analysis of diffusion phenomena in the catalyst grain (Section 4), and construction of a mathematical model that makes it possible to optimize the parameter values for the porous structure of the reforming catalyst grain (Section 5). For convenience, the experimental verification of the results obtained with the algorithm has been itemized as Section 6.

In Section 3, the paraffin reaction paths and the kinetic scheme are determined on the basis of experiments in a flow-through reactor and in the presence of a catalyst. Then the relevant mathematical equations describing the time-related changes in substrate concentrations are derived. The proposed kinetic model of the process and the calculated reaction rate constants can be used in further analysis of the model constructed for optimizing the catalyst pore structure.

In Section 4, the attempt is made to assess the effective coefficients of diffusion for the group components of a reforming mixture. The group components originate in the interior of the catalyst grain. The proposed equation makes it possible to calculate the numerical values of these coefficients as a function of the parameters of the porous structure. On the basis of experimental data, a method is suggested for determining the efficiency of grain utilization.

In Section 5, the objective is to construct a mathematical model for optimizing the porous structure of the reforming catalyst. Using a layered model of the catalyst grain and the kinetic scheme adopted for the process, equations are derived to describe the mass transfer in the catalyst grain under conditions of reforming. Particular consideration is given to the problem of how the parameters of the porous structure affect the efficiency of the process.

Finally, in Section 6, the effect exerted by the porous structure of the reforming catalyst on its activity in the reaction of n-heptane dehydrocyclization is determined experimentally, using statistical analysis for the interpretation of the results. Also determined is the minimal radius of pores functioning at the two different temperatures. The results substantiate those obtained by theoretical analysis.

Each of the four steps constitutes an entity with a defined objective and scope, with a description of the mathematical tools, an algorithm, the analysis of results, and partial conclusions. In each of the three major steps, analysis is carried out at two levels—a statistical level and a level involving thorough physicochemical analysis of the investigated phenomena. The statistical approach, which aims at smoothing the experimental results, is helpful in constructing the hypothetical course of the phenomena under study and in interpreting the results obtained. The other approach involves mathematical models that provide a thorough description of the phenomena occurring in the adopted physical models of the catalyst grain. The two approaches are complementary, thus enabling a more detailed analysis of the phenomena examined.

The full range of the tasks performed when optimizing the porous structure of the reforming catalyst, as well as the relations among particular steps, is included in Figure 1, which is a block diagram of the algorithm displaying the

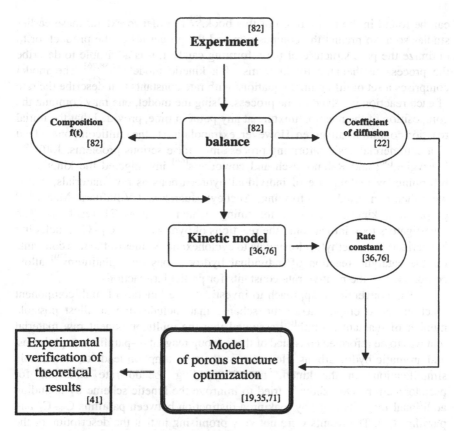

Figure 1 Optimization of porous structure for a reforming catalyst.

accomplishment of the main goal: optimization of the porous structure of the reforming catalyst. As shown by this diagram, analysis of the kinetics on the developed grain surface, along with the analysis of diffusion phenomena inside the catalyst grain, is a prerequisite to generate the input data—reaction rate constants and diffusion coefficients—for the mathematical model, which was built with the aim of optimizing the porous structure.

3 KINETICS AND REACTION PATHS FOR REFORMING

3.1 Previous Kinetic Modeling

Catalytic reforming is a well-known source of high-octane gasoline and aromatic hydrocarbons.[4,56,57] An in-depth review of the process and many related aspects can be found in the first edition of this book.[3] In order to extend these earlier studies so as to predict the composition and the properties of the product, or to optimize the pore structure of the reforming catalyst, it is advisable to describe the process mathematically in terms of a kinetic model.[39,58,76] The model comprises a set of differential equations with rate constants that describe the rate of each reaction involved in the process. Using the model, one may compute the composition of the reaction mixture at any point in time, provided that the initial reaction conditions are given. However, extrapolation to the multicomponent feed in a commercial-scale reforming process may raise serious problems. Joffe,[65] Szczygieł,[36] and Rabinovilsch and coworkers[66] investigated the kinetics of reforming by making use of individual hydrocarbons as raw materials, which were then converted by reforming. Ancheyta-Juarez and Villafuerte-Macias[54] proposed a kinetic model of reforming, which includes 71 reactions. The investigators took into account the reactions of hydrocarbons C_1-C_{11}, including isomerization of methylcyclopentane to cyclohexane. Values of the rate constants for the catalytic reaction of individual hydrocarbons over platinum[59] allow comparison of the relative rate constants for particular reactions.

The conventional approach to investigate the kinetics of multicomponent reactions is to employ a kinetic scheme that includes the smallest possible number of reactants. Smith[60] assumed that the multicomponent raw material that was to be reformed consisted of three group reagents—paraffins, naphthenes, and aromatic hydrocarbons. However, such an assumption leads to undesirable simplifications in the kinetic model, yielding pseudo–rate constants for pseudoreactions Dorozhov[61] tried to imnrove the kinetic scheme by including additional reagents (e.g., by making a distinction between paraffins C_5-C_6 and paraffins C_7). The results were not very promising in that the description of the process was only slightly better but the model itself was more complicated. Zhorov[62,63] incorporated a relationship between the reaction rate constants and the composition of the reagent groupings. Kmak and Stuckey[64] simulated the Powerforming process within a wide parameter range, using data from single components, mixtures of pure components, and a naphtha fraction. This model attempted to describe the performance of a pilot unit over a wide spectrum of operating parameters and starting materials, and to determine the concentration profiles of 22 components in four reactors in series while targeting octane number upgrade.

Ramage et al.[67,68] developed a comprehensive kinetic model, which attempted to capture the reactivity differences between particular raw materials and to modify the process kinetics by incorporating deactivation of the catalyst due to coke formation. The reaction network proposed by Ramage and coworkers was based on kinetic studies of pure components and narrow-boiling fractions of naphthas. The studies led to the construction of the Mobil Kinetic Model of the Reforming Process (KINPtR—"start-of-cycle" and "deactivation" kinetics).

Interconversion between 13 kinetic lumps participating in the reactions of hydrocracking, hydrogenation–dehydrogenation, cyclization, and isomerization was predicted. Reactions of hydrocarbons with equal carbon atom number were assumed to be reversible, whereas those for which carbon content changed were regarded as irreversible. Joshi et al.[69] described the use of software (NetGen) that allows construction of a sophisticated model, simulation of the reforming process, and prediction of the results of the reforming process over a wide range of parameters for various feeds. Up to 79 components may participate in the process, and NetGen simulates up to 464 reactions. The program identifies the components and then selects potential reactions. A model of the process is generated that determines products and rates of change of each product, in good agreement with experimental data.

3.2 Experiments and Methods

The experiments were carried out in a flow-through apparatus (a 3-cm^3 volume microreactor of OL 115/09 type) (Fig. 2). In order to establish a reliable kinetic scheme of the reforming process, experiments were performed to examine the conversion for n-heptane, an n-heptane and methylcyclohexane mixture, and a naphtha fraction with varying reaction times (controlled by the space velocity of the feed) at set temperatures of 470, 490, and 510°C. In addition to following the changes in the feed species with time at different temperatures, this enabled us to trace the variations in diffusion concomitant with the variations in the composition of the reaction mixture, as well as to establish the effect of temperature on the values of the diffusion coefficients. All experiments were run at 1 MPa and an H_2/raw material ratio of 10 : 1 over a Pt-Re catalyst of 0.15 mm grain size, which had been shown in previous experiments to ensure a kinetic limitation of the process rate. Chromatographic analysis was used. The results are listed in Table 1.

Two mathematical methods of analysis were employed: a statistical description of the process, where reforming (as a "black box") was defined in terms of polynomial equations, and second by a description based on the material balance.[76] Both methods were complementary. Statistical description smoothed the experimental results and offered certain suggestions pertaining to the kinetic schemes proposed. The description involving the physicochemistry of reforming made it possible to verify these suggestions and to establish an appropriate kinetic scheme for a process carried out with a different feed component. The use of three feed materials allowed the determination of how the fractional composition of the starting compound affected the course of the process. The kinetic schemes proposed (along with the rate constants and diffusion coefficient values calculated in[22,36]) can be used in optimizing the porous structure of the reforming catalyst by the method described in.[19,35,71] A detailed description of the mathematical tools (along with the graphics generated by them) can be found

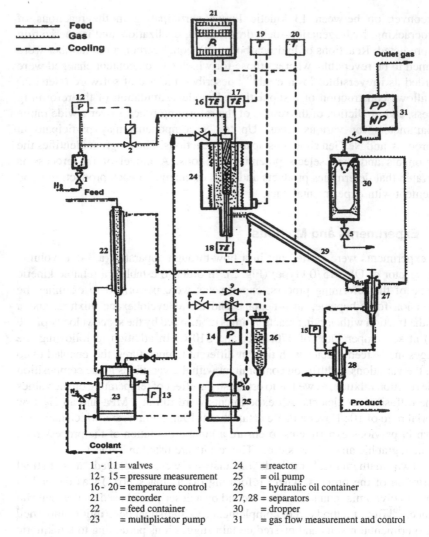

Feed
Gas
Cooling

1 - 11 = valves
12 - 15 = pressure measurement
16 - 20 = temperature control
21   = recorder
22   = feed container
23   = multiplicator pump

24   = reactor
25   = oil pump
26   = hydraulic oil container
27, 28 = separators
30   = dropper
31   = gas flow measurement and control

Figure 2   Experimental setup.[41]

elsewhere.[36,76] Statistical analysis of the results in Table 1 made it possible to generate the coefficients of polynomial equations (Table 2), to plot the isolines of the investigated output quantities in the reaction temperature–reaction time coordinate system, and to assess unequivocally the effect of these parameters on the course of the process. Analysis of the graphics was helpful in establishing the hypothetical kinetic schemes of the process.[76]

Table 1 Result of Reforming Tests for Various Raw Materials

| Process parameters | | n-Heptane | | | | | Heptane + methylcyclohexane | | | | | Naphtha fraction (60–150°C) | | | |
| | | | Yield (wt%) | | | | | | Yield (wt%) | | | | Yield (wt%) | | |
Temp. (°C)	S.vel. (h⁻¹)	Conv. (%)	Gases	Paraffins	Naphthenes	Aromatics	Conv. (%)	Gases	Paraffins	Naphthenes	Aromatics	Gases	Paraffins	Naphthenes	Aromatics
470	1	37.74	14.65	81.47	1.03	2.85	52.78	9.41	55.41	3.22	31.97	14.98	36.51	2.51	46.01
470	2	29.44	14.21	82.82	1.27	1.72	46.96	8.21	58.99	2.18	30.62	10.01	40.25	2.94	46.81
470	4	17.97	8.54	89.32	1.17	0.97	38.57	3.55	62.32	2.61	31.53	7.63	44.81	3.25	44.32
470	8	11.86	6.61	92.11	0.72	0.56	32.98	2.48	62.78	5.68	29.07	4.31	48.62	3.45	43.63
490	1	55.76	22.73	69.2	1.11	6.97	60.97	12.29	51.75	3.71	32.25	21.01	31.41	1.55	46.04
490	2	41.31	17.43	77.4	1.23	3.95	56.24	11.99	53.31	3.11	31.61	14.71	35.39	1.98	47.92
490	4	28.07	13.48	82.38	1.18	2.06	46.83	6.49	59.31	2.09	32.12	10.11	40.34	2.63	46.92
490	8	16.63	6.82	90.66	1.19	1.33	37.21	2.75	63.94	2.49	30.83	6.54	44.22	3.19	46.05
510	1	83.28	37.36	48.08	0.96	13.6	75.79	18.5	42.19	1.31	38.01	24.26	28.24	1.44	46.06
510	2	55.66	25.23	63.45	1.17	8.61	66.7	14.6	49.5	1.71	34.19	18.95	32.63	1.78	46.64
510	4	42.87	16.13	76.76	1.82	5.29	53.72	8.44	55.86	2.73	32.97	13.83	36.47	2.14	47.56
510	8	24.41	10.7	86.64	0.16	2.5	44.12	4.41	60.52	2.23	32.85	10.16	40.51	2.91	46.42

Table 2 Coefficient of Polynomial Equations

Raw material	Y	b_0	b_1	b_2	b_{11}	b_{12}	b_{22}	R	F_T
Heptane and	Conversion	43.71	8.02	−12.55	1.14	−2.84	6.18	0.99	252.54
methylcyclohexane	Gases	5.48	2.44	−5.32	0.32	−1.63	2.79	0.99	59.77
	Paraffins	61.01	−3.38	6.24	−1.13	2.54	−4.11	0.99	59.49
	Naphthenes	2.54	−0.82	0.41	−0.14	−0.52	0.62	0.73	1.39
	Aromatics	30.98	1.77	−1.33	0.95	−0.39	0.69	0.91	4.93
	Reformate	89.93	−2.63	7.36	−3.58	−1.53	−4.41	0.95	10.07
Heptane	Conversion	23.32	12.21	−19.61	2.46	−6.79	12.51	0.99	46.28
	Gases	9.97	4.21	−7.78	0.75	−3.12	5.51	0.93	8.21
	Paraffins	87.21	−6.74	10.92	−1.55	5.19	−7.48	0.96	12.97
	Naphthenes	1.47	−0.03	−0.17	−0.14	−0.11	−0.51	0.75	1.56
	Aromatics	1.36	2.57	−2.97	0.93	−1.97	2.49	0.98	34.55
	Reformate	89.82	−1.01	7.01	0.41	3.81	−12.03	0.91	5.13
Naphtha fraction	Gases	11.92	5.02	−7.71	−2.18	−1.02	4.78	0.99	61.16
(60–150°C)	Paraffins	37.41	−5.17	6.15	0.67	0.84	−2.49	0.99	76.72
	Naphthenes	2.19	−0.41	0.81	0.08	−0.19	−0.21	0.99	51.71
	Aromatics	48.47	0.54	0.74	1.43	0.36	−2.07	0.96	15.27

Source: Ref. [76].

3.3 Discussion of Results

Heptane + Methylcyclohexane

The weight ratio of heptane to methylcyclohexane in the mixture was 67 : 33. Aromatic hydrocarbons are produced primarily from methylcyclohexane after a short reaction time. Even under mild conditions (at 470°C and a reaction time of 8 h⁻¹), a considerable portion of naphthenes (methylcyclohexane) undergoes dehydrogenation to aromatic hydrocarbons, while there is very poor dehydrocyclization of the paraffins (Table 1). The results evidence the well-known findings that the dehydrogenation of naphthenes runs at a very fast rate, while the paraffins that are present in the starting material undergo conversion at a noticeably slower rate. They are extensively converted at higher temperatures and a longer reaction time, but then a considerable portion of the paraffins undergoes cracking to gases. At longer reaction times, the content of aromatic hydrocarbons exceeds that of naphthenes in the raw material, which evidences the dehydrocyclization of heptane. On comparing the plots of Figure 3 with the data of Table 1, it can be seen that, depending on the reaction conditions, the process yields either large amounts of reformate, which are rich in paraffins and much poorer in aromatic hydrocarbons, or reduced amounts of reformate with a high aromatic and much lower paraffin content.

n-Heptane

The use of heptane alone as a raw material for the reforming process has corroborated the well-described fact that aromatic hydrocarbons are formed via cyclization of paraffins to naphthenes and by a very fast dehydrogenation of the latter. This finding is evidenced by the variation in naphthene content with time through a maximum, which is typical of consecutive reactions (Table 1). The effect of reaction time and temperature on the aromatic hydrocarbon content is similar to the heptane–methylcyclohexane case. This suggests that the reforming process involves the same kinetic scheme. From the increased content of gases produced under equivalent conditions (Table 1) it can be inferred that paraffins are the main source of gas production and that naphthene cracking is very slow. The relationship between the quantity of the reformate and the content of aromatic hydrocarbons and paraffins in the reformate is the same as for the heptane–methylcyclohexane mixture when used as raw material.[76]

Naphtha Fraction (60–150°C)

The group composition of this raw material can be itemized as follows: 1.40 wt % of gases, 52.80 wt % of paraffins, 31.96 wt % of naphthenes, and

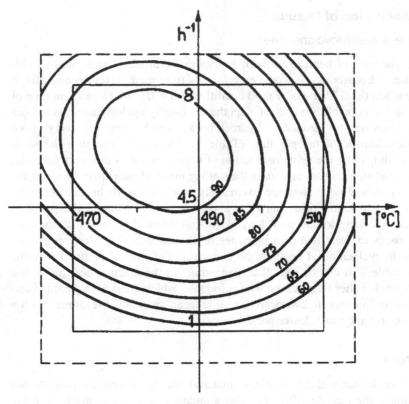

Figure 3 Isolines of reformate content (wt %) in reaction product as a function of temperature (*T*) and time of reaction (h^{-1}); raw material: heptane + methylcyclohexane.[76]

13.84 wt % of aromatic hydrocarbons. As shown by the product data of Table 1, naphthene content is noticeably lower and aromatic hydrocarbon content distinctly higher than in the raw material—even at the onset of the reaction. This is indicative of the very fast naphthenes–aromatic hydrocarbons reactions. At a reaction time of $4\,h^{-1}$, aromatic hydrocarbon content (in statistically smoothed data[76]), approached 60% and was higher than the sum of naphthenes and aromatic hydrocarbons contained in the raw material, indicating that paraffins through dehydrocyclization were the source of aromatic hydrocarbons. At 510°C, with a longer reaction time, the content of paraffins decreased to 20 wt %, and gas content amounted to 35%, which confirms the previous finding that under such conditions paraffins undergo significant cracking.[76]

3.4 Choice and Analysis of Kinetic Schemes

After data processing by statistical methods, and making use of the results from analyzing the plots, potential kinetic schemes were examined:

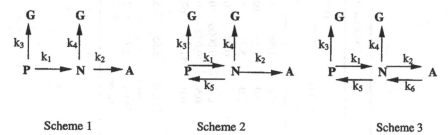

| Scheme 1 | Scheme 2 | Scheme 3 |

where P is paraffins, N is naphthenes, A is aromatic hydrocarbons, and G is gases. Each of the three schemes was tested by the numerical methods described in[76] with respect to the experimental results obtained with each of the three raw materials (Tables 3–5).

n-Heptane

Scheme 1 and Scheme 3 were found to be inadequate, as shown by the estimated values of the rate constants (Table 3). Scheme 1 fails to satisfy the fundamental relationship between the rate constants and temperature (Arrhenius law). The values of the rate constants k_4 at 490°C and 510°C are negative. For Scheme 3, all the rate constants take positive values, but their decrease with increasing temperature (for k_1, k_3, k_4, and k_6) makes the model inadequate. Scheme 2 satisfies all the criteria for an appropriate choice of the model. The small error of the model at positive values of the rate constants supports the scheme. The description allows transition of naphthenes through a maximum (and this should be attributed to the incorporation of the transition of P → N → A in the scheme) and is in agreement with the experiment.

n-Heptane + Methylcyclohexane

The best fitting of the theoretical curves (lowest F_c value) to experimental data is that of Scheme 3 (Table 4). However, the negative values of k_6 and the nonconformance of the k_3 values with the Arrhenius law make the model inadequate and infeasible. The other two schemes generally fulfill the criteria of an adequate description. Under equivalent temperature conditions, the values of the objective function for Scheme 2 are lower than those for Scheme 1, so Scheme 2 is recommended as best suited for the description of the process with an n-heptane + methylcyclohexane mixture as raw material.

Table 3 Rate Constants for the Investigated Schemes of Reforming[a]

Rate constant (min^{-1})	Scheme 1				Scheme 2				Scheme 3			
	Initial value	Temp. (°C)			Initial value	Temp. (°C)			Temp. (°C)			Initial value
		470	490	510		470	490	510	470	490	510	
k_1*10^2	0.5	0.65	1.04	1.42	0.7	0.84	1.28	2.02	1.36	1.2	1.56	0.6
k_2*10^2	50.0	187.7	70.64	10.32	70.0	4.28	10.54	16.71	4.57	10.84	17.86	60.0
k_3*10^2	1.0	5.37	3.42	4.44	3.0	2.99	3.31	4.06	5.91	3.42	4.31	2.0
k_4*10^2	0.5	0.48	−0.72	−1.63	1.0	0.59	0.76	2.23	0.33	0.21	0.24	0.5
k_5*10^2	—	—	—	—	0.5	0.75	1.20	3.00	0.0006	0.0006	0.0003	0.5
k_6*10^2	—	—	—	—	—	—	—	—	0.032	0.041	0.038	0.5
K_F	50.00	80.95	68.07	46.50	45.0	75.54	67.71	46.37	81.24	67.93	46.5	30.0
F_C	Initial value	6.71	6.48	12.63	Initial value	12.18	9.96	5.1	8.65	5.08	8.36	Initial value
	Final value	3.47	3.59	3.54	Final value	3.47	2.9	3.59	2.7	2.88	3.54	Final value

[a]Raw material: *n*-heptane.
Source: Ref. [76].

Table 4 Rate Constants for Investigated Schemes of Reforming[a]

Rate constant (min^{-1})	Scheme 1				Scheme 2				Scheme 3			
	Initial value	Temp. (°C)			Initial value	Temp. (°C)			Initial value	Temp. (°C)		
		470	490	510		470	490	510		470	490	510
k_1*10^2	0.5	0.26	0.34	0.63	0.7	0.41	0.57	0.95	0.6	0.35	0.59	0.87
k_2*10^2	50.0	23.59	30.22	45.76	70.0	23.09	33.71	57.25	60.0	23.05	34.14	60.47
k_3*10^2	1.0	2.41	3.83	3.98	3.0	3.44	4.89	5.61	2.0	3.01	2.41	2.09
k_4*10^2	0.5	0.69	0.74	0.97	1.0	0.54	0.71	0.79	0.5	0.24	1.08	1.37
k_5*10^2	—	—	—	—	0.5	0.25	0.38	0.48	0.5	0.358	0.5	0.67
k_6*10^2	—	—	—	—	—	—	—	—	0.5	-0.0729	-0.13	-0.053
K_F	50.00	52.46	51.27	43.47	45.0	55.26	52.28	46.92	30.0	54.02	48.59	37.06
F_C	Initial value	6.03	6.77	8.46	Initial value	8.92	7.26	6.71	Initial value	8.09	9.63	11.36
	Final value	4.31	4.80	4.70	Final value	4.21	4.70	4.57	Final value	4.14	4.47	3.77

[a]Raw material: n-heptane + methylcyclohexane.
Source: Ref. [76].

Table 5 Rate Constants for Investigated Schemes of Reforming[a]

Rate constants (min^{-1})	Scheme 1 Initial value	Scheme 1 Temp. (°C) 470	490	510	Scheme 2 Initial value	Scheme 2 Temp. (°C) 470	490	510	Scheme 3 Initial value	Scheme 3 Temp. (°C) 470	490	510
k_1*10^2	0.5	0.64	0.71	0.49	0.7	0.58	0.62	0.71	0.6	0.43	1.06	0.79
k_2*10^2	50.0	30.77	48.89	40.15	70.0	30.29	41.81	51.02	60.0	30.07	51.81	42.82
k_3*10^2	1.0	2.71	4.20	5.60	3.0	2.82	4.28	5.79	2.0	2.07	3.42	5.89
k_4*10^2	0.5	0.38	0.45	0.35	1.0	0.02	0.037	0.051	0.5	0.093	0.019	0.074
k_5*10^2	—	—	—	—	0.5	2.78E−05	3.52E−05	4.30E−05	0.5	−3.42E−03	5.01E−03	3.34E−03
k_6*10^2	—	—	—	—	—	—	—	—	0.5	−9.60E−02	−1.80E−01	−9.80E−02
K_F	50.00	34.00	30.22	27.6	45.0	34.08	30.2	27.65	30.0	31.03	29.29	27.79
F_C	Initial value				Initial value				Initial value			
	8.04	10.15	11.4		6.31	8.41	9.89		12.21	13.73	14.73	
	Final value				Final value				Final value			
	4.1	4.37	4.5		4.15	4.31	4.13		4.04	3.95	4.22	

[a]Raw material: naphtha fraction.

Naphtha Fraction

When a naphtha fraction is used as raw material, it is possible to predict the direction of conversion in a real system. The models tested for this system favored Scheme 2 as best suited (Table 5). Scheme 1 was rejected due to the lack of the required relationship between rate constant variations and temperature. Scheme 3 generated negative values of k_6 for each of the investigated temperatures and a negative value of k_5 for the temperature of 470°C. The plots of k_1 and k_4 variations with temperature also raised serious objections. Scheme 2 was found to satisfy all the criteria required, so it can be regarded as suitable for the description of the reforming process for naphtha as raw material. Even though the curve describing the time dependence of naphthene content variations runs below the experimental points for each of the investigated temperatures, the fit is good, as shown by the plots of Figure 4.

3.5 Models Based on an Analytical Description of Time-Related Variations in the Content of Reagents

Analysis of the kinetics identifies the same model (kinetic Scheme 2), irrespective of the starting compound. This finding has been further substantiated by investigations into dehydrocyclization of n-heptane over a more active

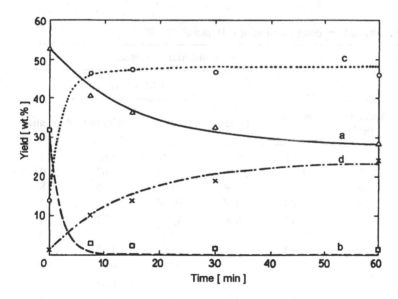

Figure 4 Composition of reaction product (wt %) vs. time for 510°C and 1 MPa. Lines for Scheme 2: a, paraffins; b, naphthenes; c, aromatics; d, gases. Experimental points: △, paraffins; □, naphthenes; ○, aromatics; ×, gases. Raw material: naphtha fraction.[76]

catalyst, as well as by functional descriptions of time-related variations in the reagents for Scheme 1 and Scheme 2.[36]. The results of tests performed with a pressure apparatus (Fig. 2) are listed in Table 6. The use of functional descriptions of time-related reagent variations for the estimation of the rate constants noticeably saves computation time and makes it possible to avoid errors resulting from numerical calculations during selection of an appropriate integration step.

Both models provide reliable profiles of reagent concentration variations; they incorporate the maximum and the decreasing time-related content of naphthenes.[36] According to expectations, both models yield a k_2 value that is considerably higher than the values of other rate constants (Table 7). The model represented by Scheme 2 provided better fitting of the curves to experimental data, which is manifest in the lower value of the objective function F_c (Table 7).

3.6 Summarizing Comments

Two complementary mathematical methods for the analysis of the reforming process were employed. Statistical analysis based on relevant plots (figures from[36,76]) "smoothes" the experimental data, making it possible to predict the content of the substrates at any point of time and temperature. By extension, it

Table 6 Results of Dehydrocyclization of n-Heptane

Process parameters		Raw material: n-heptane				
		Conversion (%)	Yield (wt%)			
Temp. (°C)	Space vel. (h^{-1})		Gases	Paraffins	Naphthenes	Aromatics
470	1	92.81	38.78	45.36	1.63	14.24
470	2	86.14	35.07	53.23	1.67	10.03
470	4	71.3	19.36	69.05	1.78	9.82
470	8	64.64	22.06	70.94	0.7	6.3
490	1	96.17	46.9	38.43	1.27	13.4
490	2	90.62	33.65	52.25	2.03	12.07
490	4	81.14	29.86	56.65	1.27	12.22
490	8	73.33	22.7	70.91	0.61	5.78
510	1	99.17	57.12	28.5	0.45	13.93
510	2	96.6	42.5	40.56	1.08	15.87
510	4	93.06	38.13	46.22	1.24	14.41
510	8	78.35	27.35	67.11	0.95	4.58

Table 7 Rate Constants for Investigated Schemes of Dehydrocyclization[a]

Rate constants (min^{-1})	Scheme 1						Scheme 2					
	Initial value	Temp. (°C)			$\ln k_0$	E/R	Initial value	Temp. (°C)			$\ln k_0$	E/R
		470	490	510				470	490	510		
k_1*10^2	2.11	1.26	1.75	2.15	8.39	7244.04	1.6	1.53	1.89	2.21	9.19	7733.80
k_2*10^2	20.56	25.12	32.04	43.13	9.79	9787.50	30.13	31.18	35.0	53.42	5.03	6138.70
k_3*10^2	6.07	3.36	5.07	6.59	6.49	7931.33	3.77	3.95	5.0	6.04	18.27	17048.90
k_4*10^2	1.74	1.5	2.07	2.59	—	—	0.97	0.94	1.7	3.05	13.75	13776.54
k_5*10^2	—	—	—	—	—	—	0.78	0.86	1.3	2.23	—	—
k_F	35.00	35.02	35.01	28.34	—	—	38.75	39.5	38.07	28.3	—	—
F_C Initial value		38.04	26.63	22.95	—	—		15.14	13.78	11.86	—	—
F_C Final value		17.47	14.93	12.63	—	—		14.92	12.64	11.80	—	—

[a]Raw material: *n*-heptane.

allows a level of optimization of the process and an appropriate choice of the process parameters so as to achieve an anticipated goal, e.g., a maximal yield of the reformate or a maximal content of aromatic hydrocarbons in the reformate. Statistical analysis of the black box led to certain suggestions about the kinetics of the process, which were verified while constructing the kinetic model based on the mass balance and on the physicochemical principles of the reforming process. It was found that irrespective of the raw material used, analysis of the kinetics of the process evidences the applicability of the same model (Scheme 2). This finding was additionally corroborated in other studies using n-heptane.[36] Figure 5 shows the criterion of compliance with the Arrhenius law for the rate constants generated by the adopted model with the use of various raw materials. The numerical Runge–Kutta approach to the solution of the differential equations representing the investigated models was found to be effective, so it may be of equal utility when applied to the analysis of other kinetic schemes.

This part of the study deals with the macrokinetics of the reforming process. Consequently, consideration is given to the rates of pseudoreactions (conversion of the group components), not to the rates of the reactions between individual compounds. That is why not all of the thermodynamic relations are fulfilled, and the rate constants depend on the composition of the raw material and are effective constants. However, for kinetic schemes the Arrhenius law is satisfied quite well and the model describes the experimental values. Hence, Scheme 2 may be of utility in predicting and optimizing the reforming process. The calculated rate constants of conversion will be employed in the analysis of the model to optimize the porous structure of the reforming catalyst.

4 DIFFUSION PHENOMENA IN THE BIDISPERSIVE GRAIN OF THE CATALYST

At standard temperature and pressure, the reforming process runs in the internal diffusion region because of the different rates of the reactions involved. This implies that the rate of the reforming process is limited by the rate of the substrate and product transport inside the catalyst grain. Thus, the porous structure becomes an important factor affecting the rate of the process.[72] To find an optimal porous structure it is necessary to analyze the transport phenomena in the grain interior and to incorporate the transport equations (which include the parameters of the porous structure) into the model of the reforming process.[19,35] Under reforming conditions, the substrates and products occur in gaseous form, requiring the consideration of molecular diffusion and of Knudsen diffusion in the catalyst pores. Whether molecular or Knudsen diffusion becomes dominant depends on the ratio of the pore diameter to the mean free path of molecular motion.[73] At typical reforming conditions, and with a range of pore size for each

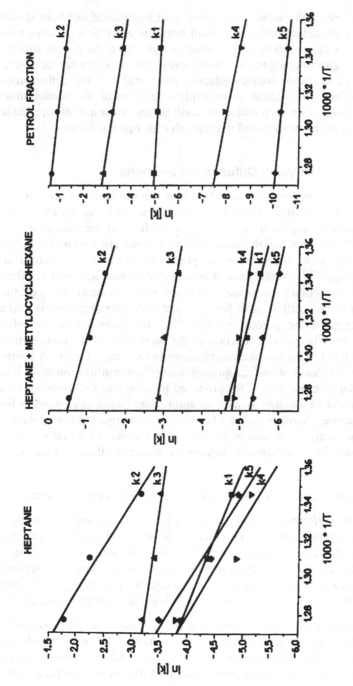

Figure 5   Arrhenius line for reaction rate constant calculated from Scheme 2.[76]

porous structure, both mechanisms are of vital importance and both should be considered when determining the overall diffusion coefficient. When equations are derived for the transport in the entire catalyst grain, the porous structure is regarded as a quasi-homogeneous environment that obeys the law of gases, and the diffusion coefficients are then effective constants.[74,75] The definition of the effective diffusion coefficient as an explicit function of the porous structure parameters will make it possible to analyze the influence of the latter on the efficiency of the process and to establish their optimal values.

4.1 Tasks in Analyzing Diffusion Phenomena

It is necessary to propose a method and an algorithm in order to assess the coefficients of molecular, Knudsen, overall, and effective diffusion for the group components regarded as chemical individuals in the reforming process under conditions of internal diffusion.[22] These values are used in the analysis of the reforming model to optimize the porous structure of the catalyst at the successive stage.[19,35] Making use of the adopted geometrical model of the pore network,[34] a function is proposed. The function relates the effective coefficient of diffusion to the coefficients of Knudsen and molecular diffusion, as well as to the parameters of the porous structure.[22] The expression of the effective coefficients of diffusion as a function of the parameters of the porous structure will dictate the distribution of concentrations in the catalyst grain. A controlled distribution of concentrations can guarantee an effective utilization of the internal surface. After incorporation of the proposed function into the equations of mass transport, it will be possible to make an appropriate choice of parameters for the porous structure. Another major objective is to analyze some properties of bidispersive catalytic structures in the applied range of variations of their structural and kinetic parameters, thereby substantiating their usefulness.

4.2 Assumptions in the Description of Diffusion Phenomena

For use in the reforming process model, the term "chemical individuals" will denote the following group components: paraffin hydrocarbons, naphthenes, aromatic hydrocarbons, and gases.[36,76] For the substrates, the transport phenomena that occur in the cylindrical pores and in the entire grain regarded as a quasi-homogeneous environment are analyzed. The transport phenomena considered here can be defined as macrodiffusion. The assumptions are as follows:

1. Under conditions of reforming, the substrates and the products have a gaseous form. For this reason, Knudsen diffusion in addition to molecular diffusion may occur in both wide and narrow pores. To calculate the overall coefficients of diffusion in both pore types, use

should be made of the relation:

$$D_{iov} = \left[\frac{1}{D_{im}} + \frac{1}{D_{ik}}\right]^{-1} \tag{1}$$

which is valid for the transition zone and limit cases.[77]

2. There are no reasons for the occurrence of mass stream other than diffusion. The effect of surface diffusion is neglected because in a catalytic process surface diffusion would require a high activation energy. The existence of the stream induced by pressure drop in the reactor is neglected because it exists in very large pores only (≥ 100 nm, which account for a very small percentage of the total pore number) or at very high pressure (>100 atm). Also neglected is convectional mass motion induced by the change of the mole number in the reaction.

3. The coefficient of molecular diffusion depends not only on the type of diffusing substance (group component) but also on the composition of the reaction mixture (other group components). To calculate the coefficient of diffusion through the mixture for the ith group component (D_{im}) it is necessary to use the expression derived by Wilke.[78]

4. The coefficient of diffusion of a group component through another pomponent ($D_{i,j}$) is calculated in terms of the correlation established by Hirschfelder et al.[79,80]

5. The effect of hydrogen as a component of the reaction mixture is neglected when calculating the coefficient of molecular diffusion for the group components.

6. Making use of the process balance and of the reported values of the Lennard–Johns force constants (ε/k and σ) for the chemical individuals which are present in the reforming products (Table 8),[79] as well a considering the mole fraction of the compounds that occur in the adopted group components, molecular weights—(M_i), the ε/k ratio, and the σ values of these group components are calculated as additives to the molar fraction of the compounds in these groups. The process balance is used to calculate the molar fractions of the group components y_i.

7. For the transition from D_{iov} to D_{ief}, use is made of a geometrical model of the pore network, which is based on the globular structure of the reforming catalyst grain tested in a previous study.[34] If the contacting particles are identical in size and have a constant density of arrangement (homogeneous monodispersive model), the pore network model will be similar to the bidispersive random model developed by Wakao and Smith. Therefore, the effective coefficient

Table 8 Reported Data for Calculation of Diffusion Coefficients

Component	Group	M_c	ε/k (K)	σ (Å)	T_c (K)	V_c (cm^3/g)	P_c (atm)
Hydrogen	—	1	33.3	2.968	33.2	—	12.8
Methane	Gas	16	136.5	3.822	190.6	99.0	45.8
Ethane	Gas	30	230.0	4.418	305.2	148.0	48.8
Propane	Gas	44	254.0	5.061	370.0	195.0	42.0
i-Butane	Gas	58	313.0	5.341	408.2	263.0	36.0
n-Butane	Gas	58	410.0	4.997	425.2	254.58	37.5
i-Pentane	Paraffin	72	355.0	5.500	461.0	304.85	32.8
n-Pentane	Paraffin	72	345.0	5.769	470.2	312.44	33.0
i-Hexane (2-methylcyclopentane)	Paraffin	86	385.0	5.850	500.0	367.00	30.0
Cyclopentane	Naphthene	70	394.0	5.220	511.8	260.00	44.6
n-Hexane	Paraffin	86	413.0	5.909	508.0	—	29.50
Methylcyclopentane	Naphthene	84	410.0	5.580	532.8	319.00	—
i-Heptane (2-methylhexane)	Paraffin	100	430.4	6.420	559.0	488.00	27.60
n-Heptane	Paraffin	100	282.0	8.880	540.0	426.00	26.80
Cyclohexane	Naphthene	84	313.0	6.143	554.2	308.0	40.0
Methylcyclohexane	Naphthene	98	440.6	5.850	572.2	367.88	34.3
Benzene	Aromatic	78	308.0	6.920	561.8	258.94	48.0
Toluene	Aromatic	92	185.0	12.00	593.8	315.60	41.6
Xylenes (naphthalene)	Aromatic	106	485.1	6.010	630.0	400.00	36.0
Aromatics (n-butylbenzene)	Aromatic	120	508.2	6.460	660.00	497.00	32.3

Source: Ref. [7–9].

of diffusion related to the parameters of the porous structure will be described by the relation resulting from the analysis of the Wakao–Smith model.[17] The grain model, the resulting pore structure model, and the parameters of this structure are shown in Figure 6.

8. To analyze the effect of structural and kinetic parameters on the utilization of micrograms and macropores (f_μ and f_M), the notion of the Φ_S modulus (analogous to the Thiele modulus, h[9] and based on experimental data) is introduced.

$$\Phi = h^2 * f \tag{2}$$

Thus, it is possible to determine the relationships $f_\mu = g(\Phi_\mu)$ and $f_M = g(\Phi_M)$ from experimental data. This method of analysis[81] can be modified by including the effective diffusion coefficient (D_{ief}) calculated with the Wakao–Smith model[17] and by distinguishing the terms responsible for the diffusion in micro- and macropores. A detailed algorithm adopted for the assessment of the D_{if} values and for the analysis of the properties of bidispersive catalytic systems can be found in a previous paper.[22]

4.3 Examples of Calculations

Microreactor results for n-heptane conversion[82] at various reaction times and two temperatures (470°C and 510°C) enabled us to investigate the diffusion variations with each change in the composition of the reaction mixture and to determine the effect of temperature on the values of the diffusion coefficients. Based on the ε/k and σ and M_c values reported in the literature (Table 8), as well as on the results of chromatographic analysis and balance for each experiment, the additive values of these quantities were calculated for the group components. The additive values were used as input data for the equations of the proposed algorithm[22] to calculate the diffusion coefficients in the pores and the effective coefficients of diffusion for each group component (Table 9).

Statistical Analysis

For a reliable qualitative analysis of the influence of porosity ε_M and ε_μ on the effective diffusion coefficients of the group components, the D_{ief} values calculated in terms of the algorithm[22] for 27 points within the ranges ε_M (0.1–0.5) and ε_μ (0.1–0.3) were approximated by adequate polynomial equations, which were analyzed by transformation to the canonical form[83] and by the Hoerle method.[84] The transformation method enables analysis on clear "maps" where variations of the D_{ief} values can be examined in the system of two variables, ε_M

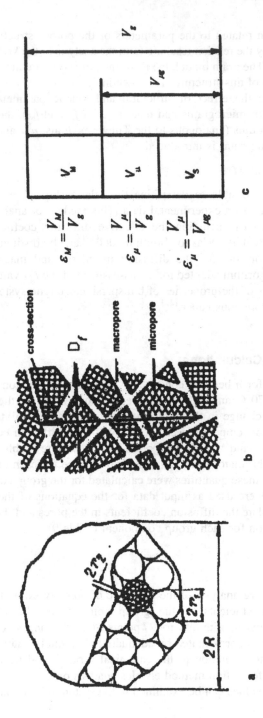

Figure 6 (a) Model of catalyst grain. (b) Model of porous structure of catalyst grain. (c) Parameters of porous structure of catalyst grain.[22]

Table 9 Diffusion Coefficients of Group Components for Different Conditions of Reforming

Group component	Space velocity $(h^{-1}) = 1$						Space velocity $(h^{-1}) = 8$					
	M_c	ε/k	σ	$D_{\mu\mu}$	D_{iM}	D_{ief}	M_c	ε/k	σ	$D_{\mu\mu}$	D_{iM}	D_{ief}
Temperature, 470°C												
Gases	38.12	244.91	4.67	1.291E−02	1.74E−02	8.312E−03	44.66	270.32	4.93	1.129E−02	1.455E−02	7.185E−03
Paraffins	98.65	314.99	8.23	9.437E−03	1.535E−02	6.418E−03	99.69	288.05	9.76	8.471E−03	1.211E−02	5.545E−03
Naphthenes	98.00	440.00	5.85	8.562E−03	1.226E−02	5.608E−03	95.72	423.15	5.81	7.866E−03	1.030E−02	5.026E−03
Aromatics	92.00	185.00	12.00	6.030E−03	6.030E−03	3.704E−03	92.00	185.00	12.00	5.345E−03	5.759E−03	3.255E−03
Temperature, 510°C												
Gases	37.70	243.57	4.66	1.375E−02	1.909E−02	8.926E−03	40.82	258.19	4.77	1.289E−02	1.748E−02	8.314E−03
Paraffins	94.88	367.33	7.08	1.017E−02	1.732E−02	7.017E−03	99.29	299.99	8.53	9.542E−03	1.526E−02	6.455E−03
Naphthenes	98.00	440.00	5.85	1.006E−02	1.728E−02	6.961E−03	98.00	440.00	5.85	8.679E−03	1.227E−02	5.663E−03
Aromatics	92.00	185.00	12.00	7.743E−03	9.684E−03	4.889E−03	92.00	185.00	12.00	6.134E−03	6.835E−03	3.765E−03

Source: From Ref. [22].

and ε_μ (Fig. 7). The other method allows determination of the ε_M and ε_μ, values, which maximize the D_{ief} coefficients, relating them to the radius of the experimental region R. The values of ε_M and ε_μ can be read on the common axis X (Fig. 8) as a result of their standardization in the limits of -1 to $+1$ prior to the generation of the polynomial equations. The radius of the experimental region takes the form $R = \sqrt{X_1^2 + X_2^2}$, where X_1 and X_2 are standardized values of ε_M and ε_μ, respectively. The increase of both macropore and micropore volumes is advantageous to the increase of D_{ief} for all components. The inclinations of the isolines evidences a noticeably greater contribution of ε_M (Fig. 7). Hoerle's analysis (Fig. 8) makes it possible to take readings of the optimal values (i.e., those maximizing D_{ief}) of macropore and micropore porosity for a defined R.

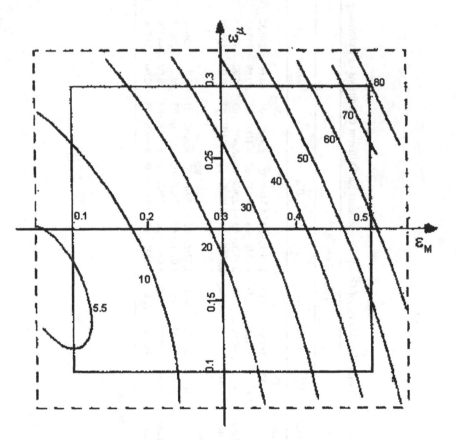

Figure 7 Isolines of $D_{pef}*1000$ value (effective coefficient of diffusion of paraffins) as a function of macroporosity (ε_M) and microporosity (ε_μ) of grain.[22]

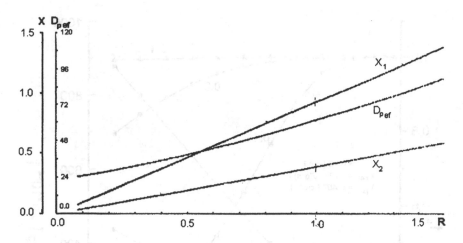

Figure 8 Extreme values of $D_{pef} \ast 1000$ value (effective coefficient of diffusion of paraffins) and corresponding values of standardized macroporosity (X_1) and microporosity (X_2) vs. R.[22]

Properties of Bidispersive Catalytic Systems

With the algorithm proposed[22] it is possible to analyze the effect of r_M and V_M on the utilization and diameter of micrograms in a bidispersive catalyst ($d_{\mu g}$) (Fig. 9). The analysis was performed for a 0.15-cm-diameter catalyst with V_μ of 0.3 cm^3/g and for two n-heptane dehydrocyclization rate constants, K_{exp}: 11 and 22 cm^3/ (g*s). Figure 9 shows the efficiency of micrograin utilization for the macropore volume of 0.4 cm^3/g within the radius range from 0 to 40 μm practically equals 1. At $V_M = 0.2$ and 0.1, the efficiency of micrograin utilization decreases with an increasing value of r_M and is smaller for a smaller macropore volume. The increase in diffusivity may be attributed to the decreased number of macropores (at an increased radius), to their constant volume, and consequently to a more restricted availability of the micrograin pores. For higher rate constant values,[22] the limitation of the process is shifted toward the diffusion region. In both cases the efficiency of micrograin utilization in the applied range of macropore radii is unity, irrespectve of the V_M value. However, the efficiency of macropore utilization f_M increases with increasing V_M. Therefore, it is recommended to use a catalyst characterized by an increased V_M value (0.4 cm^3/g) so as to increase the overall coefficient of grain utilization, f_g. V_M values greater than 0.4 cm^3/g may decrease the mechanical strength of the catalyst.

Making the assumption that in the investigated range of macropore radii, a free transport of mass is provided, then the utilization of the macropores f_M will not depend on their radii, and the overall coefficient of grain utilization at

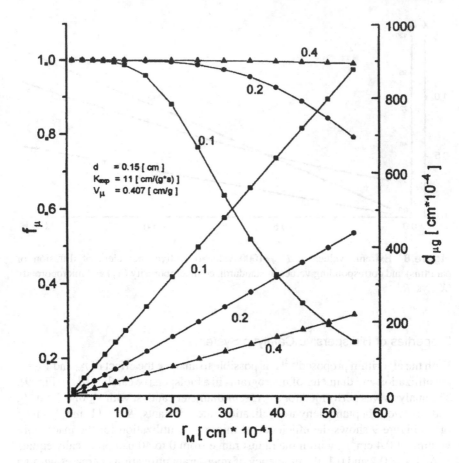

Figure 9 Utilization of micrograins f_μ (left-hand side), micrograins diameter $d_{\mu g}$ (right-hand side) vs. macropore radius r_M for different volume of macropore.[22]

constant V_M will not vary in a wide range of the macropore radius. Knowing the value of the rate constant related to the unit surface K_i (which is calculated in step 10 of the proposed algorithm[22]), one can not only calculate the efficiency of micrograin surface utilization, f_0, in the absence of macropores ($V_M = 0$), but also assess the increase of grain effectiveness resulting from the presence of macropores. The results are demonstrated in Table 10. The data of column 1 and column 2 (for $V_M = 0.1$) pertain to the catalyst which was used in the experiment. It can be inferred that the increase of the macropore volume in the catalyst will increase the utilization of the catalyst surface. The other columns include the results calculated in terms of the algorithm for hypothetical data. These results are utilized to determine their influence on the diffusivity in the grain.

Table 10 Diffusion Parameters for Various Parameters of Porous Structure

	1	2	3	4	5	6
S (cm^2/g)	192.00	192.00	192.00	192.00	192.00	192.00
d (cm)	0.15	0.15	0.15	0.15	0.15	0.10
D_M (cm^2/s)	0.015350	0.012110	0.015350	0.015350	0.015350	0.015350
D_μ (cm^2/s)	0.009437	0.008471	0.009437	0.009437	0.001000	0.001000
K_{exp} (cm^3/g·s)	22.00	22.00	11.00	22.00	22.00	22.00
V_μ (cm^3/g)	0.407	0.407	0.407	0.200	0.407	0.407

Volume of macropores V_M (cm^3/g)

0.1

f_M	0.203	0.181	0.406	0.083	0.033	0.073
f_μ	~1	~1	~1	~1	~1	~1
f_g	0.203	0.181	0.406	0.083	0.033	0.073
K_i	5.56E−05	6.37E−05	1.41E−05	1.38E−04	3.51E−04	1.56E−04
f_o	0.170	0.152	0.341	0.062	0.022	0.049
η	1.20	1.19	1.20	1.34	1.47	1.47

0.2

f_M	0.273	0.240	0.547	0.138	0.069	0.155
f_μ	~1	~1	~1	~1	~1	~1
f_g	0.273	0.240	0.547	0.138	0.069	0.155
K_i	4.19E−05	4.77E−05	1.05E−05	8.28E−05	1.66E−04	7.38E−05
f_o	0.197	0.175	0.394	0.080	0.032	0.072
η	1.38	1.37	1.38	1.72	2.13	2.13

0.4

f_M	0.437	0.374	0.870	0.283	0.179	0.404
f_μ	~1	~1	~1	~1	~1	~1
f_g	0.437	0.374	0.870	0.283	0.179	0.404
K_i	2.63E−05	3.06E−05	6.58E−06	4.04E−05	6.38E−05	2.84E−05
f_o	0.249	0.219	0.498	0.114	0.052	0.117
η	1.75	1.71	1.75	2.46	3.45	3.45

Source: Ref. [22].

On the basis of the data presented in this section, the following generalizations can be made:

- The limitation for a slower reaction shifts towards the kinetic region, which manifests in the increase of surface utilization (Table 10, column 3).
- The decrease of micropore volume at an unchanged surface is concomitant with a decrease of the micropore radius, thus inhibiting the access of the substrate to the grain interior. This manifests in a decrease of the f_g value (Table 10, columns 1 and 4).
- The increase of diffusion coefficients D_M and D_μ enhances the utilization of the catalyst surface, whereas an increase of the grain

effectiveness η (contributed by the presence of macropores) depends on the D_M/D_μ ratio, when V_μ and V_M are constant (Table 10, columns 1, 5, and 6).

- Application of a small grain diameter (d) shifts the process towards the kinetic region, which manifests in an increase of the f_g value (Table 10, column 6).

4.4 Summary Comments on Diffusion Phenomena in the Grain of the Reforming Catalyst

The realization of the presented algorithm made it possible not only to analyze the diffusion phenomenon in a bidispersive grain of the reforming catalyst but also to include experimental results. Making use of the physical model for a bidispersive structure,[34] calculations yielded the following parameters: the coefficients of macrodiffusion for the group components in macro- and micropores, the effective coefficient of macrodiffusion for the entire grain, and the effective coefficients of macrodiffusion for micrograms. All these values are needed when calculating the coefficients for the utilization of macropores (f_M) and micrograms (f_μ).

Analysis of the bidispersive structure revealed a low degree of utilization for macropores and a high degree of utilization for micrograms (approaching unity in the investigated range of macropore radii). The overall utilization of the grain can be raised by the increase of the macropore volume, but such increase is concomitant with the deterioration of the grain strength. This finding calls for a reasonable compromise. This study showed that in a wide range of variation the macropore radius exerted no effect on the overall utilization of the grain.

5 OPTIMIZING THE POROUS STRUCTURE OF THE REFORMING CATALYST

An appropriately chosen pore structure for the reforming catalyst may shift the course of some reactions from the diffusion region to the kinetic one, thus increasing the efficiency of the process.[71] Narrow pores increase the reaction surface. But this increase is paralleled by the deterioration of the access to that surface. The reforming catalyst should therefore contain a certain number of wide pores (transporting pores), which facilitate such an access.[85,86] The expression "optimal structure" denotes a bidispersive catalyst consisting of wide pores (radius ρ_1) and narrow pores (radius ρ_2) in an exactly defined proportion ω, of which provides an optimal efficiency for the reforming process. State-of-the-art technologies of catalyst preparation make it possible to obtain reforming catalysts of various porous structures, so that it is conceivable to implement porous structure parameters proposed in the final part of this section.

5.1 Scope of the Simulation of the Optimal Structure

The last link of the block diagram in Figure 1 includes two tasks—construction and analysis of the mathematical model for the optimization of the porous structure. For the purposes of the model a physical "layered" model of the catalyst grain[19] was assumed and then verified in Section 3. Then, using a probable kinetic scheme (which comprised consecutive, simultaneous, and opposite reactions) the coefficients of diffusion were derived in Section 4. The input data for the model (composition of reagents, coefficients of diffusion, rate constants of conversion) were obtained from the results of various studies. With the goal of raising the efficiency of the process by controlling the transport phenomena in the catalyst grain, the effects exerted by the parameters of the porous structure on the diffusion limitation were analyzed. Following the mathematical analysis of transport phenomena and process kinetics, an optimal porous structure in which the transport of the substrates would not inhibit the rate of the entire process was proposed. Analysis of the model with the aim to optimize the porous structure of the reforming catalyst was performed for the dehydrocyclization of n-heptane under reforming conditions. The rate constants for particular reactions were assessed using an analytical form of description that showed time-related variations in the content of particular reagents (Table 7).

5.2 Topological and Geometrical Representation of the Pore Space

Mathematical analysis of the reaction–diffusion phenomena in the grains of the reforming catalyst was carried out with a layered model (Fig. 10a), which involved the following assumptions:

- The grain has a radius R_g and consists of n adjacent layers, F_j.
- The number of n layers in the grain, which are of the same thickness (R_g/n), must provide convergence of calculations.
- The surfaces limiting the layers pass through the points of pore ramification, and the pore number increases with the increasing sphere radius.
- If the layer thickness is appropriately small, the pores passing through the layer are parallel and have the same length, l, which is equal to the thickness of each layer.
- Each layer contains two types of cylindrical pores: wide pores of radius ρ_1 (macropores) and narrow pores of radius ρ_2 (micropores).

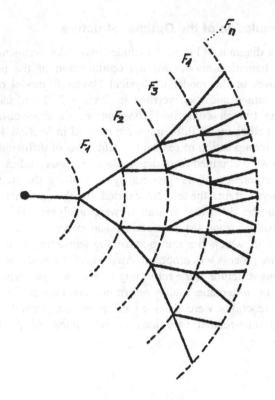

a)

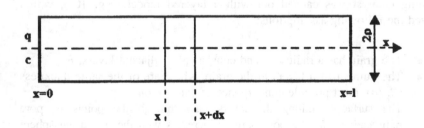

b)

Figure 10 (a) Layered model of catalyst grain.[19] (b) Scheme of cylindrical pore.[19]

- The numbers assigned to the wide (m_1) and narrow (m_2) pores of particular layers are functions of the number of the layer (j); m_1 and m_2 are interrelated by the condition of constant porosity

$$\frac{m_1(j) \cdot \pi \cdot \rho_1^2 + m_2(j) \cdot \pi \cdot \rho_2^2}{4 \cdot \pi(j \cdot l)^2} = \kappa \tag{3}$$

as well as by their ratio, which is constant in all layers:

$$m_1(j) = \omega \cdot m_2(j); \qquad \omega = \text{constant} \tag{4}$$

- The porous structure of the catalyst grain described by the proposed model is characterized by the parameters ρ_1, ρ_2, and ω.

Like the Bethe lattice topology developed by Reyes and Jensen,[37] the pore lattice topology presented in this work (Fig. 10) contains no closed loops. Real pore walls can be tortuous and have a rough fractal surface. As shown by Coppens and Froment,[31,32] the fractal dimension exerts a noticeable effect on the efficiency of the catalyst grain. In our model, the pores are cylindrical in shape, and the fractal dimension effect is reduced by the controlled short length of the pores.

5.3 Physicochemical Phenomena of the Reforming Process

Kinetics

It is on the surface of the cylindrical pore lattice (Fig. 10b) that the chemical reactions of the reforming process proceed and the reactions follow the macrokinetic pattern described earlier as Scheme 2. The reactions adopted in the scheme are first-order reactions. From this, it can be shown that if paraffins are the starting compounds (n-heptane $[P(t_0)] = 100\%$), the content of naphthenes C_N at an arbitrary point of time (except at $t_0 = 0$) can be expressed approximately as a function of paraffin content (C_p) in terms of the following equation:

$$C_N \approx \frac{k_1}{(k_2 + k_4 + k_5) - (k_1 + k_3)} \cdot C_p \tag{5}$$

The rate constant values for the reactions of Scheme 2 (Table 8), which have been used as input data to the derived model, are from a previous study[36] reporting on investigations into the kinetics of the reforming process.

Diffusion

Group components behave in the same way as chemical individuals. Diffusion transport phenomena therefore pertain to the same group components treated as individuals (macrodiffusion). Considering the assumptions in Section 4.2 and making use of the theoretical relations of the algorithm,[22] the overall coefficients of diffusion in micropores and macropores were assessed as follows:

$$D_{i_\mu} = \left\{ \frac{1}{D_{i_m}} + \frac{1}{(D_i)_{K_\mu}} \right\}^{-1} \qquad D_{iM} = \left\{ \frac{1}{D_{i_m}} + \frac{1}{(D_i)_{KM}} \right\}^{-1} \tag{6}$$

where

$$D_{i_m} = \frac{1 - y_i}{\sum_{j=1}^{LSM} (y_j/D_{ij})} \qquad D_{ij} = \frac{0.001858 \cdot T^{1.5} \cdot [(M_i + M_i)/M_i \cdot M_j]^{0.5}}{P \cdot ((\sigma_i + \sigma_j)/2)^2 \cdot \Omega_{i,j}^{ii}} \tag{6a}$$

$$(D_i)_{K\mu} = 2/3 \cdot r_\mu \left\{ \frac{8 \cdot R_G \cdot T}{\Pi \cdot M_i} \right\}^{1/2} \qquad (D_i)_{KM} = 2/3 \cdot r_M \cdot \left\{ \frac{8 \cdot R_G \cdot T}{\Pi \cdot M_i} \right\}^{1/2} \tag{6b}$$

5.4 Method and Scope of Model Analysis

The detailed mathematical analysis[71] yielded relations that can be regarded as a basis for assessing the efficiency of the reforming catalyst grain. The first of these related the performance of the catalyst grain as the ratio of raw material flux—q (directed to the depth of the grain) to raw material concentration (C) on the grain surface, QP/CP. The second related the diffusion resistance (referred to as diffusivity) of the grain with its measure, which is the ratio of the raw material concentration on the catalyst surface to the raw material concentration in the center of the catalyst grain, CP/C_0.

   In a preceding section, we analyzed the effect which the porous structure of the reforming catalyst, specifically the wide pore radius, ρ_1 (transport macropores in the range of 500–900 Å and 3000–7000 Å); the narrow pore radius, ρ_2 (micropores developing the catalyst surface in the range 4–40; 100–200 Å); the proportion of wide pores to narrow pores, ω (in the range of 0.1–0.3 for $\rho_1 = 500$–900 Å; and in the range of 0.001–0.005 for $\rho_2 = 3000$–5000 Å), and the catalyst grain radius, R_g (for 0.035 and 0.05 cm), exerts on the two defined parameters characterizing the efficiency of the grain.

   This was accompanied by optimization of the values of the porous structure parameters in order to upgrade the efficiency of the grain, i.e., to increase the grain performance (QP/CP) without deteriorating diffusivity (CP/C_0), which enables free transport of reagents in the grain (relative optimization for the assumed value of $CP/C_0 \cong 2.0$). Then, the correlation between the parameters of grain efficiency (QP/CP and CP/C_0) and those of the specific surface which is a function of the porous structure parameter values were analyzed.

The effect of the porous structure on the grain efficiency for a fixed kinetic scheme is determined by making use of the plots established from the algorithm.[71] Statistical analysis was used to carry out a conditional optimization which provides advantageous values of the process efficiency parameters for the adopted set of porous structure parameter values. The calculations of input data for the proposed model (kinetic and diffusion constants) (Tables 7 and 9) involved the results of chromatographic analysis and the experimental balance,[82] as well as the algorithms described in a number of previous publications.[22,36,76]

5.5 Discussion of Results Obtained from Analysis of the Mathematical Model

Effect of Porous Structure on Grain Performance

In order to determine the effect that the porous structure of the reforming catalyst exerts on the values of the grain efficiency parameters, the plots obtained from the computer simulations of transport phenomena and reactions in the catalyst grain were analyzed. Grain efficiency is found to be high when its performance, QP/CP, takes the highest possible values at a diffusivity, CP/C_0, enabling free transport of the substrates in the grain. A value of $CP/C_0 \leq 2$ was adopted as the diffusivity value that provides a relatively free transport of the reacting substances in the grain. The use of bidispersive grains enhances the process efficiency. Wide pores facilitate transport of the reacting substances into the narrow pores, with the surfaces available for the reaction.

Figures 11–15 present plots characterizing the efficiency of the grain with an $R_g = 0.035$ (cm) radius at $T = 470°C$. The plots relate the QP/CP (a) and CP/C_0 (b) values to the radius of micropores (ρ_2) for different ranges of the macropore radius (ρ_1) and for different proportions of macro- and micropores (ω). Figures 11, 12 and 13 include calculated results for $\omega = 0.05$ and for two ranges of macropore radii, 3000–7000 Å (Figs. 11 and 13) and 500–900 Å (Fig. 12). From Figure 11 it can be inferred that as the radius of macropores increases, performance decreases (Fig. 11a). This is an indication that the reactions in the wide pores proceed in the "pure" kinetic range. An increase in the micropore radius from 4 to 40 Å produces no distinct changes in diffusivity, and there is a slight concomitant increase in performance. This suggests that the selected radius of narrow pores, ρ_2, is too small to allow the starting compound to penetrate the pore interior. Thus, it is only the 3000-Å macropore radius that brings about a comparatively distinct increase of diffusivity and performance (slope of line 1 in Fig. 11a) with the increasing radius of micropores. This indicates that the contribution of the micropores has increased and the reactions on their surfaces

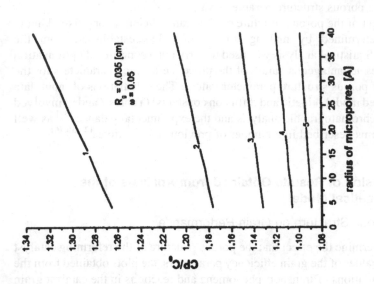

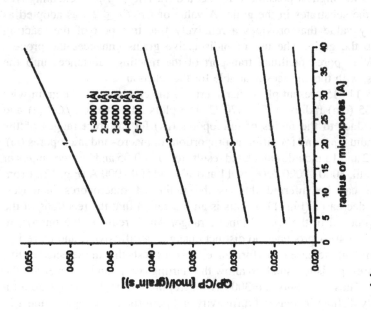

Figure 11 Performance (a) and diffusivity (b) of the grain related to micropore radius (ρ_2) for various macropores (ρ_1); $T = 470°C$, $h^{-1} = 1$.[71]

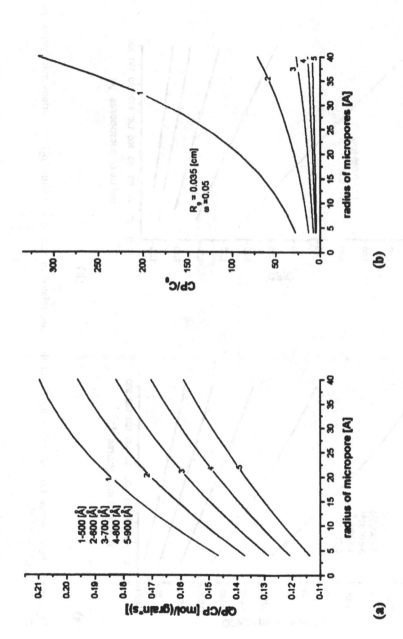

Figure 12 Performance (a) and diffusivity (b) of the grain related to micropore radius (ρ_2) for various macropores (ρ_1); $T = 470°C$, $h^{-1} = 1$.[71]

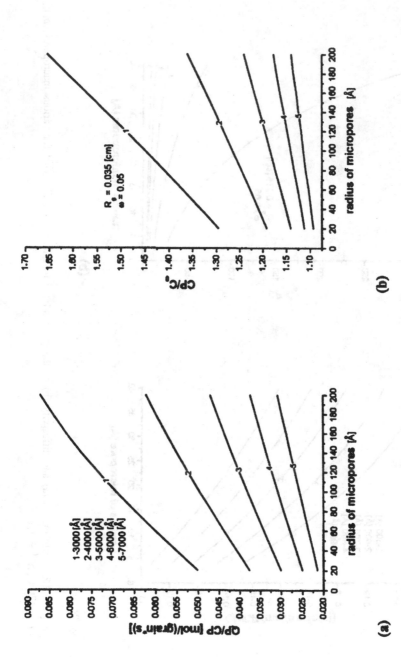

Figure 13 Performance (a) and diffusivity (b) of the grain related to micropore radius (ρ_2) for various macropores (ρ_1); $T = 470°C$, $h^{-1} = 1.$[71]

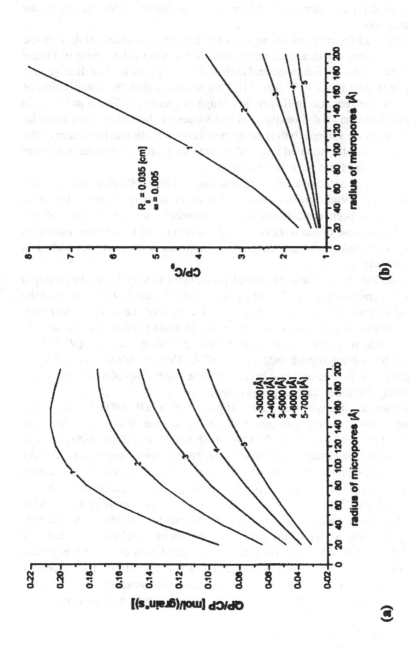

Figure 14 Performance (a) and diffusivity (b) of the grain related to micropore radius (ρ_2) for various macropores (ρ_1); $T = 470°C$, $h^{-1} = 1$.[71]

have accounted for the increase in performance, as the available surface of the catalyst increases.

In Figure 13 the range of micropores has been increased to 200 Å, with the values of the remaining structural parameters being identical to those in Figure 11. The slopes of the performance and diffusivity curves, as well as their values, are all greater than those in Figure 11. This indicates that the contribution of micropores to the reaction in this radius range is greater, reflected not only in performance but also in diffusivity. But the values of the ratio of raw material concentration on the surface to the raw material concentration in the center of the grain are lower than the adopted limit value, which equals 2. This is an indication that the utilization of the grain is satisfactory.

In Figure 12, for the macropore radius range of 500–900 Å, performance is one order of magnitude greater than in the other relevant figures. The high performance has been contributed by the increased internal surface of the catalyst. The increase of the internal surface as compared to previous examples should be attributed to the presence of macropores with a smaller radius at constant porosity.

The diffusivity for these structural parameters is very high, suggesting a low degree of grain utilization. To reduce diffusivity it is advisable to increase the quantity of macropores, i.e., the value of ω. Increase of the ω value decreases diffusivity, but there is a concomitant decrease in grain performance, as seen in Figure 15, which relates performance and diffusivity to ω $(QP/CP(\omega)$; $CP/C_0(\omega))$ for the micropore radius $\rho_1 = 80$ Å. Once a value of $\omega \approx 0.4$ has been achieved, its further increase brings about a slight drop of diffusivity and a comparatively large decrease in performance.

Summing up, it is seen that the use of pores with a 500 to 900-Å radius to act as transport pores in the catalyst grain fails to upgrade the efficiency of the grain. Considering Figure 13 (ρ_1: 3000–7000 Å), it should be noted that diffusivity has values smaller than the limit value across the entire micropore range (ρ_2: 0–200 Å), which is an indication that at $\omega = 0.05$ the porous structure imposes no limitations to the penetration of the starting compound into the grain interior. Owing to the low diffusivity value ($CP/C_0 < 2$), one can increase the content of micropores in the grain (decrease the value of ω) in order to extend the effective surface, thus upgrading the performance of the grain. The results of simulation are plotted in Figure 14 for $\omega = 0.005$. Grain performance obtained with $\omega = 0.005$ is greater than that obtained with $\omega = 0.05$ for diffusivities ≤ 2. Figure 14 suggests several optimal values for the parameters of a porous grain structure with a radius $R_g = 0.005$, which is to work at $T = 470°C$. For $\omega = 0.005$, one predicts

1. $\rho_1 = 3000$ Å; $\rho_2 = 40$ Å
2. $\rho_1 = 4000$ Å; $\rho_2 = 80$ Å
3. $\rho_1 = 5000$ Å; $\rho_2 = 130$ Å

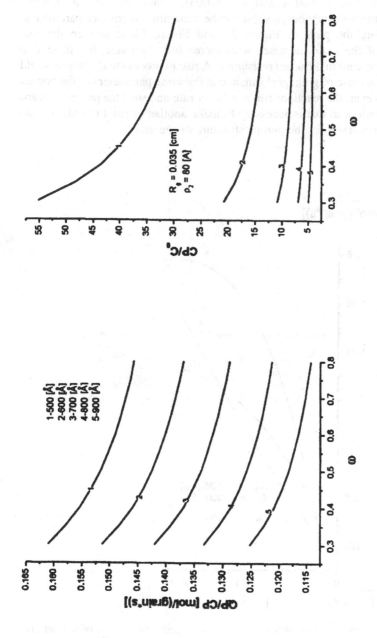

Figure 15 Performance (a) and diffusivity (b) of the grain related to quantitative ratio of wide to narrow pores (ω) for various macropores (ρ_1); $T = 470°C$, $h^{-1} = 1$.[71]

Figure 16 relates the performance, diffusivity, and specific surface of the grain to ρ_2 (for $\rho_1 = 5000$ Å and $\omega = 0.003$), which enables an accurate assessment of the advantageous ρ_2 value for the remaining determined parameters.

Comparing the plots of Figure 17 and Figure 13, it is seen that the performance of the grain increases with increasing grain size, but there is a concomitant increase in diffusion resistance. A rise in process temperature would allow for the increase of grain performance at the same parameters of the porous structure. However, the reactions run at a faster rate and shift the process toward diffusion limitation, so that is necessary to make another attempt in order to find the optimal parameters for the porous structure of the grain.

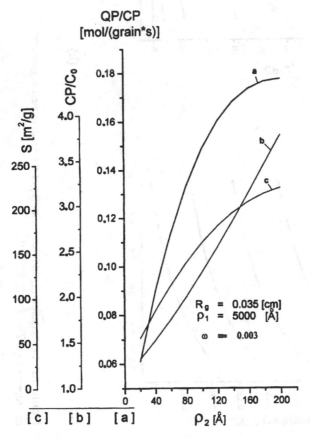

Figure 16 Performance (a), diffusivity (b), and specific surface (c) of a grain as a function of micropore radius ρ_2 for macropore radius $\rho_1 = 5000$ (Å).[71]

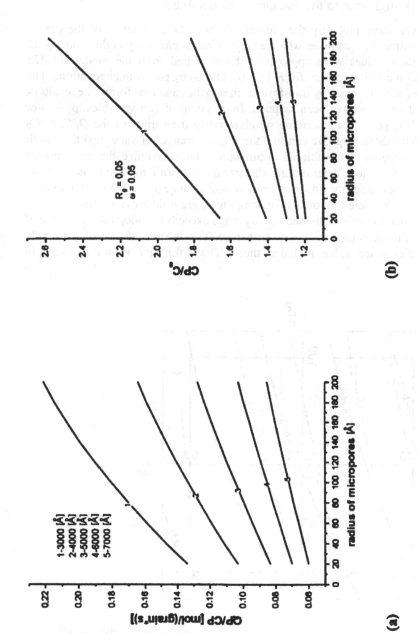

Figure 17 Performance (a) and diffusivity (b) of the grain related to micropore radius (ρ_2) for various macropores (ρ_1); $T = 470°C$, $h = 1$.[71]

Statistical Approach to the Results of Model Analysis

To analyze more precisely the problem of how the parameters of the porous structure affect the quantities which characterize the efficiency of the grain (Q/C, C/C_0), their values were approximated (calculated from the model for 125 points, with actual T, ranges for R_g, ρ_1, ρ_2, and ω) by polynomial equations. The equations were analyzed by transformation into the canonical form. The results of statistical analysis have been mapped. In a system of two variables, $\rho_x - \omega$ or $\rho_1 - \rho_2$, it is possible to examine simultaneously the variations the Q/C, C/C_0 ratios. With these maps we can find the regions (ranges of variables) that fulfill the requirements of conditional optimization, i.e., maximize the performance of the grain without deteriorating the transport conditions. From the plots of Figure 18 we can see that in the macropore radius range $\rho_1 = 500-900$ Å, even at a high ω value (high amount of macropores), grain diffusivity defined by the CP/C_0 value in the investigated $\rho_1 - \rho_2$ range exceeds the adopted limit value of 2. This eliminates the plots from further analysis in spite of the comparatively high performance value included there. The diffusivity values included in

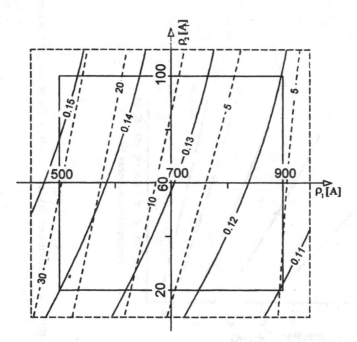

Figure 18   Isolines of grain performance [QP/CP] and diffusion resistance [CP/C_0] as a function of micropore radius (ρ_2) and macropore radius (ρ_1); $T = 470°C$, h$^{-1} = 1$, $R_g = 0.035$ (cm), $\omega = 0.5$.[71]

Figure 19 ($\rho_1 = 3000-7000$ Å) are smaller than 2 in the whole investigated range, which means that the performance of the grain can be increased by reducing the number of transport pores (by decreasing ω).

5.6 Conclusions

The assumptions included in the physical model simplify the real image of the catalyst grain. Even so, the conclusions drawn from the analysis of the model are not inconsistent with the available knowledge, and the model should be of utility in analyzing the bidispersive properties of the catalytic systems for different processes. With the mathematical model it is possible to analyze the effect exerted by the parameters of the porous structure on the performance of the catalyst grain and take into account any change in the kinetic schemes of the investigated processes. Based on the verification of the kinetic schemes,

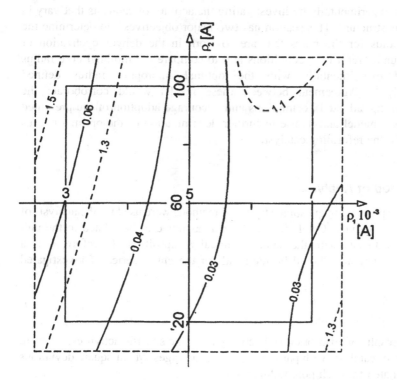

Figure 19 Isolines of grain performance [QP/CP] and diffusion resistance [CP/C_0] as a function of micropore radius (ρ_2) and macropore radius (ρ_1); $T = 470°C$, h$^{-1} = 1$, $R_g = 0.035$ (cm), $\omega = 0.5$.[71]

the results of porous structure optimization for Scheme 2 are the most probable options for the porous structure of the catalyst grain for process of n-heptane dehydrocyclization at 470°C. The next step of the study should include experimental verification of the optimal parameters for the porous structure by incorporation during catalyst preparation.

6 EXPERIMENTAL VERIFICATION OF THEORETICAL RESULTS

Pores of a small radius generate a large surface, but eventually their radii become too small to facilitate access for reagents, with the result being the surface is insufficiently utilized. It is therefore of importance to define the minimal pore radius that will facilitate such access. The value of the minimal radius affects the reaction rate–diffusion rate ratio. The numerical value of the radius can be determined experimentally by investigating the activity of catalysts that vary in their porous structure. This section has two major objectives—to determine the minimal radius for the pores that are to work in the dehydrocyclization of n-heptane under reforming conditions and to compare the minimal pore radius established experimentally with the optimal micropore radius defined theoretically.[41] Agreement between these results would corroborate the adequacy of the adopted assumptions and encourage adoption of the presented mathematical model and its use in further determination of the optimal porous structure for the reforming catalyst.

6.1 Method of Analysis

To solve the problem in question, statistical methods were used for the analysis of pore size and activity. Catalytic activity was assumed to be related to the unit surface of catalysts with the same chemical composition. Therefore, it is a constant quantity and should be identical for the entire series of investigated catalysts.

$$\frac{A_1}{S_1} = \frac{A_2}{S_2} = \cdots = \frac{A_j}{S_j} = a_j \tag{7}$$

Using the results of adsorption and activity tests, the specific activities, a_{ij}, were calculated for catalysts with particular pore radii ranges. Mean-square deviations were calculated for each pore radius range.

$$MSD = \frac{\sum_{i=1}^{nc} (\overline{a}_i - a_{ij})^2}{nc \cdot (nc - 1)} \tag{8}$$

If it is assumed that pore surface distribution as a function of radius for the investigated catalysts differs from one catalyst to another, the minimal value of *MSD* will be equivalent to the critical pore radius, which does not exhibit pore restriction. Knowing the radius of these limit pores and the portion of the specific surface corresponding with the pores of radius $\geq$ the limit pores, one can calculate catalytic activity in terms of the "working" surface unit of the catalyst. On that basis it is possible to predict, for real catalysts of a specified chemical composition, their volume activity, which is a function of the porous structure.

6.2 Experimental

Preparation of Support

The support was prepared using technology developed at the Institute of Chemistry and Technology of Petroleum and Coal, Wrocław University of Technology, which serves for the manufacture of spherical aluminum oxide supports for reforming catalysts.[87,88] There is a general resemblance between these methods and those in patents issued to various catalyst manufacturers. The technology involves the following steps: preparation of an aluminum salt solution (by treating metallic aluminum with hydrochloric acid and nitric acid), mixing the solution with hexamine, formation and conditioning of the support. After drying, the support is subject to calcination for the removal of organic material and for conversion to an active aluminum oxide. Nine supports that varied in their porous structures were selected.

Preparation of Catalyst

The support is first contacted with rhenium from an ammonia perrhenate/2% acetic acid solution. After drying and calcination, platinum is applied from a solution made of hexachloroplatinic acid, hydrochloric acid, and acetic acid. After each operation of metal contacting, the catalyst is dried at 60°C for 12 h and at 120°C for another 12 h prior to calcination according to a selected program. Pore surface within the radius range of 1.5–100 nm is determined by the method of benzene vapor adsorption (Table 11).

Activity Tests: Dehydrocyclization of *n*-Heptane

The tests were carried out as described in Section 3.3 with 5 cm^3 bed for 1 h duration. Each catalyst tested had a grain size varying between 0.3 and 0.75 mm. The activity of the catalysts under test was assessed in terms of the process balance, which served as a basis for calculation of the activity index. The activity

Table 11 Porous Structure of the Investigated Reforming Catalyst

No. of catalyst	\>3.33 (nm)	\>30 (nm)	\>10 (nm)	\>5 (nm)	\>3 (nm)	\>1.5 (nm)	Composition of catalyst
	Pore surface in defined range of radius S_i (m²/g)						
1	0.48	0.67	3.50	16.29	177.40	259.30	
2	0.56	0.78	4.13	27.84	124.13	248.78	Support: Al_2O_3
3	0.55	0.77	3.80	19.19	152.41	241.45	
4	0.45	0.89	2.36	8.60	202.46	275.70	Metals:
5	0.40	0.57	3.47	1.322	140.74	267.45	Pt, 0.6 wt %
6	0.69	0.97	5.99	20.46	165.31	288.74	Re, 0.6 wt %
7	0.82	1.15	6.91	31.95	184.86	262.57	
8	0.66	0.93	5.49	21.70	174.64	271.00	
9	0.56	0.80	3.37	16.37	209.40	310.59	

Source: Ref. [41].

index includes the reaction rate at which toluene forms; it is expressed in moles/ (g_{cat}*sec) and is calculated as follows:

$$V_{TOL} = \frac{M_{TOL}}{G_c * t} \tag{9}$$

Tables 12 and 13 gather the numerical values of catalyst activity and the calculated values of specific activity for defined ranges of pore radii.

6.3 Analysis of Results

The results listed in Tables 11–13 serve as a basis for the analysis and assessment of the effect exerted by the porous structure of the catalyst on the dehydrocyclization of *n*-heptane under reforming conditions. Tables 12 and 13 gather the numerical values of catalyst activity and the calculated values of specific activity for defined ranges of pore radii. Statistical analysis was used to determine the limit (minimal) values of the pore radii at which the particles of the reagent can adsorb on the pore surface. These values are in correspondence with the minimal values of the parameter *MSD* (Fig. 20). The limiting radius of the "working" pores is smaller at 470°C (3 nm) than at 510°C (5 nm). This finding can be attributed to the temperature-enhanced effect of diffusion factors on the process rate. At 470°C, pores of 3-nm radius are accessible to the reagents. However, at 510°C, when the process rate depends to a great extent on the diffusion of reactants, the radius of the pores providing readily available surfaces must be increased. The average specific activity (related to unit surface) of the

Table 12 Total Activity and Specific Activity of Reforming Catalysts (470°C)

No. of catalyst	Activity of catalyst mol/(g_{cat}·S)	Specific activity in defined range of radius a_{ij} [mol/(m²*s)]					
		>33.3 (nm)	>30 (nm)	>10 (nm)	>5 (nm)	>3 (nm)	>1.5 (nm)
1	2.680	5.583	4.000	0.766	0.165	0.015	0.010
2	1.560	2.786	2.000	0.378	0.056	0.013	0.006
3	2.120	3.855	2.753	0.558	0.110	0.014	0.009
4	3.090	6.867	3.472	1.309	0.359	0.015	0.011
5	2.110	5.275	3.702	0.608	0.160	0.015	0.008
6	2.470	3.580	2.546	0.412	0.121	0.015	0.009
7	2.770	3.378	2.409	0.401	0.087	0.015	0.011
8	2.610	3.955	2.806	0.475	0.120	0.015	0.010
9	3.140	5.607	3.925	0.932	0.192	0.015	0.010
MSD		199E−01	5.81E−02	1.06E−02	8.61E−04	8.31E−08	2.64E−07

Source: Ref. [41].

Table 13 Total Activity and Specific Activity of Reforming Catalysts (510°C)

No. of catalyst	Activity of catalyst mol/($g_{cat} \cdot S$)	Specific activity in defined range of radius a_{ij} [mol/($m^2 \cdot s$)]					
		>33.3 (nm)	>30 (nm)	>10 (nm)	>5 (nm)	>3 (nm)	>1.5 (nm)
1	3.560	7.42	5.31	1.02	0.22	0.02	0.01
2	6.080	10.86	7.79	1.47	0.22	0.05	0.02
3	4.180	7.60	5.43	1.10	0.22	0.03	0.02
4	1.870	4.16	2.10	0.79	0.22	0.01	0.01
5	2.880	7.20	5.05	0.83	0.22	0.02	0.01
6	4.460	6.46	4.60	0.74	0.22	0.03	0.02
7	6.960	8.49	6.05	1.01	0.22	0.04	0.03
8	4.730	7.17	5.09	0.86	0.22	0.03	0.02
9	3.560	6.36	4.45	1.062	0.22	0.02	0.01
MSD		3.57E−01	2.50E−01	5.43E−03	1.48E−08	1.52E−05	4.50E−06

Source: Ref. [41].

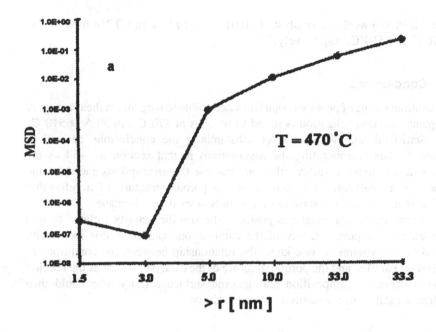

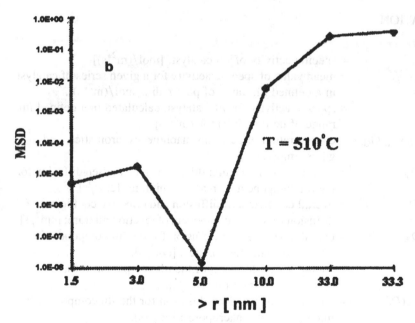

Figure 20 Standard deviation of specific activity for the investigated catalysts vs. different ranges of pore radii at (a) 470°C, (b) 510°C.[41]

0.6 wt % Pt/0.6 wt % Re catalysts is 0.015 mol/(m^{2*}s) and 0.218 mol/(m^{2*}s) at 470°C and 510°C, respectively.

6.4 Conclusions

The minimal radius of pores with surfaces accessible to reagents in the reaction of n-heptane dehydrocyclization is found to be 30 Å at 470°C and 50 Å at 510°C.

Statistical analysis of results substantiates the conclusions derived in Section 5, thus corroborating the assumptions in that section as well as the mathematical model. It follows that one can use the assumptions and the same model when analyzing and optimizing the porous structure of a reforming catalyst that is to process starting compounds other than n-heptane.

Knowing the chemical composition, the specific activity (related to unit surface), and the pore structure of the catalyst, one can determine its activity related to unit volume. If one knew the relationship between the conditions of catalyst preparation and the porous structure of the catalyst, as well as the relation between chemical composition and specific surface activity, one could then produce a catalyst of the desired volume activity.

NOTATION

a_j	= specific activity of jth catalyst; [mol/(m^{2*}s)].
$\bar{a}_i$	= mean value of specific activity for a given series of catalyst in a defined ith range of pore radii; [mol/(m^{2*}s)].
a_{ij}	= specific activity for jth catalyst, calculated in a defined ith range of pore radii; [mol/(m^{2*}s)].
C_P, C_N, C_A, G_G	= concentration of paraffins, naphthenes, aromatics, and gases; [mol/cm^3].
D_{im}, D_{ik}	= coefficient of molecular diffusion and Knudsen diffusion for the ith component of reaction mixture; [cm^2/s].
D_{iov}, D_{ief}	= overall coefficient of diffusion and effective coefficient of diffusion for the ith component of reaction mixture; [cm^2/s].
$D_{i\mu}$, D_{iM}	= overall coefficient of diffusion for the ith component in micropores and macropores; [cm^2/s].
D_{ij}	= coefficient of diffusion of the ith component for a two-component system i,j; [cm^2/s].
$(D_i)_{K\mu}$, $(D_i)_{KM}$	= coefficient of Knudsen diffusion for the ith component in micropores and macropores; [cm^2/s].
d, $d_{\mu g}$	= grain diameter and micrograin diameter; [cm].
F_j	= jth grain layer.

f_M, f_μ, f_g	= utilization factor for macropores, micrograins, and catalyst grain.
G_c	= catalyst weight; [g].
h_0, f_0	= dimensionless Thiele modulus and efficiency of monodispersive catalyst surface utilization.
K_{exp}, K_i	= experimental rate constant per gram of catalyst [cm/m^3/ (g*s)] and rate constant per surface of catalyst; [cm/s].
k_i	= reaction rate constants; [1/s].
l	= length of pore; [cm].
M_i	= molecular weights of group components.
M_{tol}	= mole fraction of toluene in liquid reaction product.
MSD	= mean-square deviations.
$m_1(j)$	= number of macropores in the grain layer F_j.
$m_2(j)$	= number of micropores in the grain layer F_j.
n	= number of layers in the grain.
nc	= number of catalysts tested.
P, P_c	= pressure and critical pressure; [atm].
R	= radius of experimental region.
R_G	= gas constant; [erg/K].
R_g	= radius of grain; [cm].
r_M, r_μ	= mean macropore and micropore; [cm].
S	= specific surface of the catalyst; [m^2/g].
S_j	= "working" surface of jth catalyst; [m^2/g].
T, T_c	= temperature and critical temperature; [$^\circ$C], [K].
y_i	= mole fractions of group components.
q	= feed stream; [mol/s].
q_i	= feed stream for the ith component; [mol/s].
V_{TOL}	= rate of toluene-producing reaction; [mol/(g*s)].
V_M, V_μ	= volume per gram of macropores, micropores; [cm^3/g].
t	= time; [s].
Ω	= collision integral.
Φ	= dimensionless modulus made up of experimental data.
Φ_μ, Φ_M	= modulus for microporous grain and entire grain.
$\varepsilon_M, \varepsilon_\mu$	= macroporosity of grain, microporosity of grain.
$\sigma, \varepsilon/k$	= Lennard–Jones force constants; [Å], [K].
ω	= ratio of number of wide pores (macropores) to number of narrow pores (micropores).
χ	= porosity.
ρ	= radius of pore; [Å].
ρ_1	= radius of wide pores (macropores); [Å].
ρ_2	= radius of narrow pores (micropores); [Å].
η	= coefficient of increase of catalyst surface utilization.

ACKNOWLEDGMENTS

This study was carried out under KBN Grant No. 300359101. This chapter is dedicated to the memory of my mother.

REFERENCES

1. Becker, E.R.; Pereira, C.J. Eds. *Computer-Aided Design of Catalysts*; Marcel Dekker: New York, 1993.
2. Hegedus, L.L. Ed. *Catalyst Design*; John Wiley and Sons: New York, 1987.
3. Antos, G.J.; Aitani, A.A.M.; Parera, J.M. *Catalytic Naphtha Reforming: Science and Technology*; Marcel Dekker: New York, 1995.
4. Little, D.M. *Catalytic Reforming*; PennWell Publishing: Tulsa, OK, 1985.
5. Joffe, I.I.; Reshetov, V.A.; Dobrotvorskii, A.M. *Raschetnye Metody v Prognozir-Ovanii Aktivnosti Geterogennykh Katalizatorov*; Khimiya: Leningrad, 1977.
6. Radomyski, B.; Szczygieł, J.; Marcinkowski, M. Przemysł Chemiczny **1986**, *65* (7), 352.
7. Radomyski, B.; Szczygieł, J.; Marcinkowski, M. Przemysł Chemiczny **1986**, *65* (8), 423.
8. Radomyski, B.; Marcinkowski, M.; Szczygieł, J. Nafta **1984**, *2*, 65.
9. Thiele, E.W. Ind. Eng. Chem. **1939**, *31*, 916.
10. Thomas, J.M.; Thomas, W.J. *Principles and Practice of Heterogeneous Catalysis*; VCH: Weinheim, 1996.
11. Hosten, L.; Froment, G.F. Ind. Eng. Chem. Proc. Des. Dev. **1971**, *10*, 280.
12. Marin, G.B.; Froment, G.F. Chem. Eng. Sci. **1982**, *37*, 759.
13. Van Trimpont, P.A.; Marin, G.B.; Froment, G.F. Appl. Catal. **1986**, *24*, 53.
14. Van Trimpont, P.A.; Marin, G.B.; Froment, G.F. Ind. Eng. Chem. Res. **1988**, *27*, 51.
15. Marin, G.B.; Froment, G.F. *Proc. 1st Kuwait Conf. Hydrotreating Processes*, EFCE Public. Ser. 27(II), 1989.
16. Wheeler, A. Adv. Catal. **1951**, *3*, 249.
17. Wakao, N.; Smith, J.M. Chem. Eng. Sci. **1962**, *17*, 825.
18. Mann, R.; Thomson, G. Chem. Eng. Sci. **1987**, *42*, 555.
19. Szczygieł, J. Comput. Chem. **2000**, *24*, 203.
20. Johnson, M.F.L.; Steward, W.E. J. Catal. **1965**, *4*, 248.
21. Keil, F.J. Catal. Today **1999**, *53*, 245.
22. Szczygieł, J. Comput. Chem. **1999**, *23*, 121.
23. Wheeler, A. Adv. Catal. **1950**, *3*, 250.
24. Zhang, L.; Seaton, N.A. Chem. Eng. Sci. **1994**, *49* (1), 41.
25. Johnson, M.F.L.; Steward, W.E. J. Catal. **1965**, *4*, 248.
26. Feng, C.; Steward, W.E. Ind. Eng. Chem. Fund. **1973**, *12* (2), 143.
27. Keil, F.J.; Rieckmann, C. Chem. Eng. Sci. **1994**, *49*, 4811.
28. Coppens, M.O.; Froment, G.F. Chem. Eng. Sci. **1995**, *50*, 1013.
29. Coppens, M.O.; Froment, G.F. Chem. Eng. Sci. **1995**, *50*, 1027.
30. Coppens, M.O.; Froment, G.F. Chem. Eng. Sci. **1994**, *49*, 4897.

31. Coppens, M.O.; Froment, G.F. Chem. Eng. Sci. **1996**, *51*, 2283.
32. Coppens, M.O.; Froment, G.F. Chem. Eng. J. **1996**, *64*, 69.
33. Coppens, M.O. Catal. Today **1999**, *53*, 225.
34. Szczygieł, J. Inżynieria Chemiczna i Procesowa **1997**, *4*, 693.
35. Szczygieł, J. Inżynieria Chemiczna i Procesowa **1992**, *3*, 489.
36. Szczygieł, J. Inżynieria Chemiczna i Procesowa **1998**, *3*, 621.
37. Reys, S.; Jensen, K.F. Chem. Eng. Sci. **1985**, *40* (9), 1723.
38. Hollewand, M.P.; Gladden, L.F. Chem. Eng. Sci. **1992**, *47* (7), 1761.
39. Sahimi, M.; Gavvalas, G.R. Chem. Eng. Sci. **1990**, *45* (6), 1443.
40. Burghardt, A.; Rogut, J.; Gotkowska, J. Chem. Eng. Sci. **1988**, *43* (9), 2463.
41. Szczygieł, J. Inżynieria Chemiczna i Procesowa **2000**, *21*, 533.
42. Larraz, R. Hydrocarbon Proc. **1999**, (July), 69.
43. DeLancey, G.B. Chem. Eng. Sci. **1974**, *29*, 1391.
44. Hegedus, L.L. Ind. Eng. Chem. Prod. Des. Dev. **1980**, *19*, 533.
45. Pereira, C.J.; Kubsh, J.E.; Hegedus, L.L. Chem. Eng. Sci. **1988**, *43* (8), 2087.
46. Keil, F.J.; Rieckmann, C. Hung. J. Ind. Chem. **1993**, *21*, 277.
47. Boreskov, G.K. Khim. Prom. **1947**, *9*, 253.
48. Dyrin, W.G.; Levinter, M.E.; Loginova, A.N. Neftekhimiya **1976**, *XVI* (2), 215.
49. Slinko, M.G.; Malinovskaya, O.A.; Beskov, B.C. Khim. Prom. **1967**, *9*, 1.
50. Beskov, V.; Flokk, V. *Modelling of Catalytic Processes and Reactors*; Khimiya: Moskva, 1991 (in Russian).
51. Johnson, M.F.L.; Steward, W.E. J. Catal. **1965**, *4*, 248.
52. Wijngaarden, R.J.; Kronberg, A.; Westerterp, K.R. *Industrial Catalysis: Optimizing Catalyst and Processes*; Wiley-VCH: Weinheim, 1998.
53. McGreavy, J.S.; Andrade, K.; Rajagopal, K. Chem. Eng. Sci. **1992**, *47* (9–11), 2751.
54. Ancheyta-Juarez, J.; Villafuerte-Macias, E. Energy Fuels **2000**, *14*, 1032.
55. Rieckmann, C.; Diren, T.; Keil, F.J. Hung. J. Ind. Chem. **1997**, *25*, 137.
56. Rassev, S.D.; Ionescu, C.D. *Reformowanie Katalityczne Benzyn*; WNT: Warszawa, 1965.
57. Maslanskii, G.N.; Shapiro, R.N. *Kataliticheskii Riforming Benzinov*; Leningrad, 1985.
58. Zhorov, J.M. *Modelowanie Procesów Przeróbki Ropy Naftowej*; WNT: Warszawa, 1982.
59. Krane, H.G.; Groh, A.B. *Proceedings of the Fifth World Petroleum Congress*, New York, 1959, Sec. III, p. 39.
60. Smith, I.M. Chem. Eng. Prog. **1959**, *55* (6), 76.
61. Dorozhov, A.P. Moskva, 1971; PhD thesis.
62. Zhorov, J.M.; Panchenkov, G.M. Kinetika i Kataliz **1967**, *8*, 658.
63. Zhorov, J.M.; Shapiro, I.J. Khem. Techn. Topliv i Masel **1973**, *4*, 1.
64. Kmak, W.S.; Stuckey, A.N. *AIChE National Meeting*; New Orleans; March 1973; Paper No. 56a.
65. Joffe, J.J. Neftekhimiya **1980**, *2*, 225.
66. Rabinovilsch, J.P.; et al. Khimija i Khimicheskaya Technol. **1983**, *2*, 214.
67. Ramage, M.P.; Graziani, K.R.; Krambeck, F.J. Chem. Eng. Sci. **1980**, *35*, 41.
68. Ramage, M.P.; Graziani, K.R.; Schipper, P.H.; Krambeck, F.J.; Choi, B.C. Adv. Chem. Eng. **1987**, *13*, 193.

69. Joshi, P.V.; Klein, M.T.; Huebner, A.L.; Leyerle, R.W. *AIChE National Meeting*; March 1997, Paper No. 133d.

70. Dennis, J.E.; Woods, D.J. *New Computing Environments: Microcomputers in Large Scale Computing*; Wouk, Ed.; SIAM, 1987; 116 pp.

71. Szczygieł, J. Computers & Chemical Engineering (submitted).

72. Sokolov, V.; Zaidman, N. React. Kinet. Catal. Lett. **1976**, *6* (3), 329.

73. Froment, G.F.; Bischoff, K.B. *Chemical Reactor Analysis and Design*; John Wiley and Sons: New York, 1990.

74. Zeldovich, B. Zhur. Fiz. Khim. **1939**, *13*, 163.

75. Przezeckii, S. Zhur. Fiz. Khim. **1947**, *21*, 1019.

76. Szczygieł, J. Energy Fuels **1999**, *13*, 29.

77. Satterfield, Ch.N.; Sherwood, T.K. *The Role of Diffusion in Catalysis*; Addison-Wesley: London, 1963.

78. Wilke, C.R. Chem. Eng. Progr. **1950**, *46*, 95.

79. Hirschfelder, J.; Curtiss, F.; Bird, R. *Molecular Theory of Gases and Liquids*; John Wiley and Sons: New York, 1954.

80. Reid, R.C.; Sherwood, T.K. *The Properties of Gases and Liquids*; McGraw-Hill: New York, 1958.

81. Montarnal, P. The porous structure of catalyst and transport processes in heterogeneous catalysis. *4th International Congr. Catal.*, Symposium III Novosibirsk, 1970.

82. Szczygieł, J. KBN Grant No. 300359101, 1993.

83. Davies, O.L. *Design and Analysis of Industrial Experiments*; Oliver and Boyd: London, 1967.

84. Hoerl, A.E. Chem. Eng. Progr. **1959**, *55* (11), 69.

85. Boreskov, G.K. *Geterogennyi Kataliz*; Nauka: Moskva, 1986.

86. Malinovskaya, A.; Boreskov, G.K.; Slinko, M.G. *Modelirowanie Kataliticheskikh Processov na Poristikh Zernakh*; Nauka: Moskva, 1975.

87. Pat. Pol. 234698 (1983).

88. Grzechowiak, J.; Radomyski, B.; Marcinkowski, M. Przem. Chem. **1984**, *63*, 135.

8

The New Generation of Commercial Catalytic Naphtha-Reforming Catalysts

George J. Antos, Mark D. Moser, and Mark P. Lapinski
UOP, LLC
Des Plaines, Illinois, U.S.A.

1 INTRODUCTION

Catalytic reforming maintains its position as a major process in the petroleum refinery. Catalytic naphtha reforming and fluid catalytic cracking are the backbone processes for the production of high-octane gasoline for automotive use. Catalytic reforming also provides a key link between the refining and petrochemical industries through its effective production of aromatics. The high concentration of aromatics in reformate offers high octane ratings, and is a rich source of benzene, toluene, and, particularly, xylenes. The hydrogen produced is a valuable coproduct, particularly as the demands for hydroprocessing in the refinery increase.

The importance of the naphtha reforming process in the refinery in and of itself would generate continuous evolution of the technology. These improvements would be observed in both the processing and equipment pieces of the technology, as well as the catalyst component. The development of the technology through the early 1990s was covered in Chapter 15[1] and in Chapter 13[2] of the first edition. Since publication of the first edition, new drivers for technology evolution have surfaced. Popular concerns for actions on environmental issues have led to a worldwide series of environmental regulations and clean-fuel legislation. Gasoline properties, such as sulfur content, oxygenate content, benzene content, and vapor pressure, have varying levels of restrictions placed on them, all of which have an impact on the reforming operation and the

335

margins available to the refiner. In general, the reformer needs to produce greater quantities of higher octane product.

The second driver is the need to increase profitability by the refiner. Recent analyses have indicated a typical return on capital in the European refining sector of around 4%, compared to 15–20% for highest performing companies.[3,4] This is to be achieved against largely negative market factors such as crude availabilities, crude prices, shifting fuel markets, legislation, a worldwide economic slowdown, and international turmoil. Increasing capacity with new-unit construction to meet local demands for octane barrels frequently does not have enough payback to be an acceptable approach. Thus, maximizing existing asset utilization becomes the strategy employed by the refiner.

Reforming technology vendors have responded. One of the most utilized improvements to naphtha reforming technology that enables the refiner to maintain or increase margins or to respond to market demands with existing assets is to employ new catalysts. Many examples of improvements to the flexibility and profitability of the refiner using catalysts, or catalyst and revamp combinations, are in the literature.[5–8] This chapter extends beyond the catalyst improvements in the first edition[1] to include commercial catalysts available from the main reforming catalyst vendors. Some of these catalysts have been developed to alleviate process issues, which are discussed in Chapter 11 of this edition. This chapter complements the process improvement discussions in Chapter 13 of this edition. The organization of the chapter is according to the usual alignment in the naphtha reforming process. Semiregenerative unit catalysts are treated first followed by cyclic unit and continuous-regeneration unit catalysts. Zeolitic catalysts are reviewed last. Some mention of revamp opportunities and process modifications that assist the refiner are included at various points.

2 SEMIREGENERATIVE REFORMING CATALYSTS

The first platforming units, commercialized in 1949, were designed as semiregenerative (SR) or fixed-bed units, employing monometallic Pt on alumina catalysts. Semiregenerative reforming units are periodically shut down to regenerate the catalyst. To maximize the length of time (cycle) between regenerations, these early units were operated at high pressures in the range of 2760–3450 kPa (400–500 psig). The high reactor pressure minimizes deactivation by coking. Typically, the time between regenerations is a year or more. The platforming process was improved through introduction of platinum–rhenium catalysts to SR reforming units in 1968.[9] These catalysts enabled a lower pressure, higher severity operation: about 1380–2070 kPa (200–300 psig) at 95 to 98 octane while maintaining typical cycle lengths of one year. The Pt-Re catalyst has proven to be reliable and easily regenerable, making it the primary catalyst in use today.

The cycle length of a semiregenerative reforming unit can be determined in a number of ways. It can be defined as a 'normal temperature cycle' of 30–40°F of catalyst activity loss that usually represents 0.5–1.0% C_5+ yield decline. Some refiners may choose to operate to a greater decline in yield in order to obtain a longer run length. Other refiners may dictate run length based on a turnaround schedule, regardless of whether the catalyst has reached the end of run based on a temperature cycle. Refiners that are temperature limited may define the end of cycle when the reactor temperatures reach the maximum reactor design temperature limits. Therefore, high activity, excellent temperature stability, and yield stability are important to obtain long run length and high average yields over the cycle.

The early, monometallic Pt semiregenerative catalysts are much less stable than modern Pt-Re bimetallic catalysts. Figure 1 shows the relative performance for Pt and Pt-Re catalysts. The rhenium is proposed to lengthen cycle length by reducing the deactivating effect of coke on the platinum. This is either by hydrogenating the coke to a less graphitic species[10] or cracking the coke precursors.[11] The stability of the Pt-Re catalyst roughly correlates with the Re/Pt ratio over a limited range of values. An additional component, usually sulfur, is added to reduce the amount of metal activity for the Pt-Re catalysts.[12] This increases the liquid product yield and, depending on conditions, may impact stability. More recent work, described below, has used additional components other than, or in addition to, sulfur and claimed additional benefits.

The semiregenerative catalytic reformer remains one of the most critical units in today's refineries. More than 500 units are operating worldwide, producing approximately 5 million BPSD of reformate. This large installed capital base presents the opportunity for refiners to increase their profits by maximizing the utilization of these existing assets. However, most semiregenerative reforming units are older designs that have been revamped over the years to push the operating parameters to the limit so as to maximize throughput and yields. The older designs may also have lower mechanical design temperature limits than newer reforming units. Given the limitations of the units, improvements have focused on:

- Improved reformate, aromatic and hydrogen yields
- Increased activity and stability to allow increased octane or throughput
- Lower cost of catalyst

The semiregenerative catalyst may be regenerated multiple times before a new load of catalyst is required. At the end of a catalyst's life, it may be

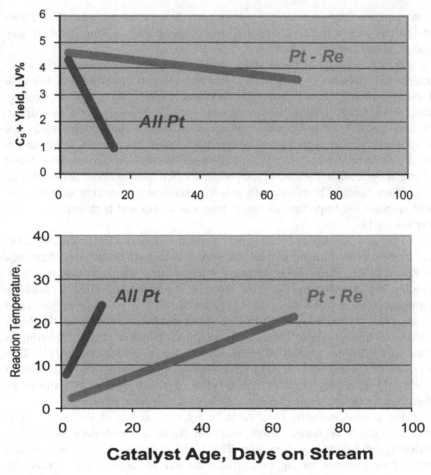

Figure 1 Comparison of process performance for Pt and Pt-Re catalysts.

difficult to completely rejuvenate the catalyst to fresh performance due to low surface area, resulting in lost gasoline yield and short run lengths. In many circumstances, a refiner may consider new catalyst prior to the end of a catalyst's useful life in order to improve process economics. The availability of new catalysts with better activity, yields, and stability may make it attractive to reload with a new catalyst. A change in process requirements to more severe conditions for increased hydrogen production or higher octane may also lead a refiner to consider a reload to a catalyst that maximizes yield, activity, and stability.

2.1   Recent Semiregenerative Catalyst Developments

The desire to increase yields has led Axens to introduce the RG 582 and, more recently, the RG 682 catalysts.[13] The RG 582 catalyst is equal weight percents of platinum and rhenium on an alumina extrudate to which a third component is added to reduce Pt and Re hydrogenolysis activity. Relative to the equal-metals Pt-Re catalyst, the yield of C_5+ product for the trimetallic RG 582 is 1 wt % greater and the cycle length is $0-7\%$ less.[14]

   The RG 682 catalyst uses the same promoter as RG 582 but has a higher Re/Pt ratio. The trimetallic RG 682 catalyst is 0.6 and 0.06 wt % more selective for C_5+ and hydrogen products, respectively, and is more stable and active than RG 582.

   Criterion also has a higher Re/Pt ratio catalyst, PR-29. The support has been modified to increase selectivity and maintain cycle length.[15] More recently, Criterion proposed the use of an equal-metals Pt-Re catalyst, PR-11, in the initial reactors of a semiregenerative unit, to provide increased C_5+ and hydrogen yield in a catalyst system called 'staged loading'.[16]

   In 1994, UOP introduced a staged loading concept with the use of an R-72 catalyst in the initial reactors of a semiregenerative unit, followed by a Pt-Re catalyst in the latter reactors.[17,18] In 2000, UOP introduced a modified extrudate support for Pt-Re catalysts, with a resultant increase of C_5+ selectivity by 1 wt % while maintaining cycle length. This support is 15% lower in density than previous Pt-Re catalysts, saving in the weight of catalyst loaded in the reactors and reducing the amount and cost of Pt and Re that is used. The catalysts are available with equal-metals concentrations, R-88, or higher ratio Re/Pt, R-86.[19]

   The trends in semiregenerative catalysts that have emerged in recent years are the addition of a third component to reduce Pt and Re hydrogenolysis activity, the modification of support acidity to decrease cracking, the use of multiple catalysts in a reforming unit in the concept called staged loading, and a reduction in the density of the catalyst to decrease catalyst cost. The advances have been driven by the need to improve process economics without additional capital investment. Typically, these units are at the maximal severity achievable while maintaining an acceptable catalyst life.

2.2   Process Improvements

The economics of the semiregenerative process unit are typically inferior to that of a continuous reforming unit. Therefore, it is difficult to justify substantial capital investments other than the conversion to a continuous reforming unit. There are instances, however, where smaller capital investments can be justified. Unit modifications designed at improving the flow characteristics and additives used to increase cycle length have been proposed.

Axens has provided a refractory composite cover for the top of a radial flow catalyst bed since 1992.[13] The claimed benefits due to improved flow dynamics are increased activity from better utilization of the existing catalyst, reduced pressure drop, and reduced catalyst loading time. UOP has offered a coverless deck design that claims similar benefits. Conoco-Phillips is offering an improved regeneration procedure and the use of a proprietary additive that is claimed to reduce coking and increase the time between semiregenerative catalyst regenerations by 25%.

2.3 Conclusion

Significant process changes in semiregenerative reforming are unlikely to occur due to the more favorable economics of conversion of these units into CCR platforming units.[8] However, the large installed operating capacity of more than 500 semiregenerative units assures that research and development will continue to improve the economics through advances in catalysts and process modifications. In fact, increased competition among catalyst vendors has accelerated the introduction of improved catalysts in recent years. Semiregenerative reforming will remain a primary process in refineries for many years to come.

3 REGENERATIVE REFORMING PROCESSES AND CATALYSTS

3.1 Cyclic and Continuous Units

The two general types of fully regenerative reforming processes are cyclic and continuous. In both types, a portion of partially aged reforming catalyst is taken from the process and regenerated in a regeneration system and a portion of freshly regenerated catalyst is returned to the process. The on-process catalyst train is maintained in a less deactivated state, resulting in increased activity and no losses in yield as is typically experienced in semiregenerative units when deactivating coke levels become high during a cycle (typically 6–18 months).

By removing coke and maintaining high activity, regenerative units can operate at lower pressures which leads to large increases in C_5+, aromatic, and hydrogen yields. Regenerative units can also recover quickly from upsets such as a compressor trip or a feed upset, which lead to increased coke on the catalyst. For a semiregenerative unit, the increased coke level can drastically shorten the catalyst cycle length or terminate the run.

Cyclic units utilize a swing reactor to substitute for one of the on-line process reactors while that reactor is regenerated in a separate regeneration system. The regeneration sequence and regeneration time are key factors in keeping the coke level low on all reactor positions. The lead reactors are regenerated less often because the large endotherm in those reactor positions leads to lower coke production. The tail reactors have smaller endotherms and are at higher average bed temperatures, which leads to more rapid coking. The tail reactors therefore need to be regenerated more frequently. The total time to regenerate is also an important variable since longer regeneration times result in higher buildup of coke in the reactors that are processing naphtha. The regeneration time can vary from 16 to 72 or more hours for different units. Short regeneration times are essential for operations under conditions that lead to a high rate of deactivation, such as with low pressures, low hydrogen/hydrocarbon ratios, high temperatures, heavy feeds, and monometallic Pt catalysts.

Continuous units move reforming catalyst in spherical form continuously through the reactor train and back and forth between the reactor system and the regenerator system. UOP's CCR platforming process utilizes a stacked-reactor design.[20] The catalyst flows gently by gravity downward from reactor to reactor, which minimizes attrition. After moving through the last reactor, the catalyst flows to a collector vessel and is lifted by nitrogen or hydrogen gas to a catalyst hopper above the regeneration tower. Catalyst then flows to the regeneration tower where it is coke-burned and then reoxychlorinated in separate zones. The regenerated catalyst is returned to the top of the reactor section via a similar system as the reactor-to-regenerator transfer. The Axens Octanizing and Aromizing processes move catalyst between each reactor and to/from the regenerator system via lift gas.[2]

Continuous units have been very successful due to their abilities to operate at very low pressures and produce high octanes and increased hydrogen, C_5+, and aromatic yields. As of 2003, UOP has 175 CCR platforming units operating around the world and Axens has more than 50 licenses awarded for Octanizing.[21] The combination of high octanes, excellent product selectivities, favorable economics, and continual process and catalyst improvements has made continuous units the most favorable for new constructions. Additional process details may be found in[2] and Chapter 13 of this edition.

3.2 Catalyst Advances

The objectives for improved catalysts include improved octane, higher gasoline volume, improved aromatic yields and selectivities, and maximal hydrogen production. Long catalyst life is important in order to maximize the time on stream. Specifically for cyclic units, high surface area stability is desired, whereas for continuous moving-bed operations, both high surface area stability and low catalyst attrition are desired. Due to clean-fuel regulations, lower benzene is important for gasoline blending. For reformate used as a petrochemical feedstock, enhanced production of benzene, toluene, and especially xylene (BTX), is advantageous. Specific catalyst advances for cyclic and continuous units will be discussed below.

Cyclic Catalysts

The original advantages of cyclic units with monometallic platinum catalysts compared to semiregeneration units operating with monometallic catalysts were the higher yields and octanes from operating at lower pressures.[22] With the introduction of Pt-Re bimetallic catalysts in 1968,[9] semiregeneration units could be run at lower pressures and higher severities due to the high activity and stability of the bimetallic catalysts. Pt-Re catalysts were utilized in cyclic units as well, allowing even lower pressure operations; however, revamp costs to operate at lower pressures were substantial for some units.

In addition to monometallic Pt and bimetallic Pt-Re catalysts, bimetallic platinum-tin (Pt-Sn) and various trimetallic catalysts have been used in cyclic service. Some units run a combination of different types in different reactor positions; for example, a Pt-only catalyst may be run in the lead position with Pt-Re in the remaining positions. The Pt-only catalyst in the lead position guards against sulfur upsets, which would more severely deactivate a Pt-Re catalyst. An example of a trimetallic catalyst that has been used in cyclic service is the Axens RG582. Some cyclic units do not utilize Pt-Re catalysts due to difficulties of sulfiding the Pt-Re catalysts after catalyst regeneration and/or due to high feed sulfur levels. For listings of the reforming catalyst offerings from the various vendors, see[23,24] and Chapter 13 of this edition.

Due to the frequent regenerations, especially in the tail positions, surface area stability is an important factor in maintaining long catalyst life. In addition, some cyclic units experience temperature excursions during coke burning, during reduction when using recycle gas, and/or when returning a reactor back to process. Large temperature excursions can lead to alumina phase transitions that convert γ into the θ and α phases that are less desired catalytically. Criterion's catalysts such as P-93, P-96 (Pt-only), and PR-9 (Pt-Re) were designed to be hydrothermally stable for frequent regenerations with little surface area loss. The catalyst supports were modified to inhibit paraffin cracking, potentially providing higher hydrogen and reformate yields.[25] UOP has recently developed new catalysts, R-85H (Pt-only) and R-88H (Pt-Re), specifically for cyclic service with enhanced C_5+ yields, enhanced surface area stability, and increased resistance to alumina phase transitions. Figure 2 shows a significant reduction in phase transitions as a function of the hydrothermal severity for the R-88H catalyst as compared to the R-88 catalyst.

Continuous Catalysts

Continuous catalysts today are mainly bimetallic Pt-Sn on chlorided alumina in spherical form. The advantages include increased activity and selectivity to aromatic formation during paraffin dehydrocyclization, decreased rate

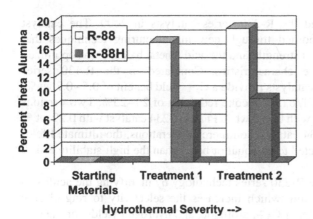

Figure 2 UOP's cyclic catalyst showing increased resistance of γ-alumina phase transitions to θ-alumina from hydrothermal treatments.

of deactivation vs. Pt-only catalysts, ability to attain a high degree of Pt dispersion, and resistance to agglomeration.[26–28] All of the major vendors have come out with new catalysts to improve process performance and/or operations.

Axens has improved their continuous catalysts by stringent control in manufacturing of the Pt and Sn distributions throughout the catalyst particle. The focus is to provide 'nanoscale' control of the homogeneity of the metallic phase. New preparation techniques for controlling the surface reaction of the tin precursor were developed. This is reported to increase activity and C_5+ yields and decrease coke production for the new RG401 and RG405 catalysts vs. the RG200 series catalysts. In addition, two new catalysts, AR-501 and AR-505, were developed and commercialized for aromatics production with the objective of optimizing the catalysts for pressure. For example, AR-501 was optimized for low-pressure units.[28] Both new catalysts require less Pt metal than earlier AR series catalysts.[29]

In 1998, Criterion introduced the PS-40 catalyst. It was specifically designed to reduce paraffin cracking and coke production. Compared to Criterion's older continuous catalyst, PS-10, the PS-40 catalyst gives higher C_5+ yields, higher hydrogen yields, and lower coke. Criterion's catalysts include new manufacturing methods for distributing metals and acidity modification of the alumina support. In addition, the alumina support was designed to reduce attrition losses and increase the surface area stability. The Pt level in PS-40 is 0.30 wt %, and a higher Pt catalyst version, PS-30, is also offered, at 0.375 wt % Pt.[30,31]

UOP commercialized the R-230 series catalysts in 2000. The catalyst support structure was modified through new manufacturing techniques that resulted in improved catalyst strength, and the acid–metal balance was optimized for improved performance characteristics. Compared to the R-130 series catalysts, the R-230 series catalysts provide a C_5+ yield benefit of 0.5–0.7 wt %, a hydrogen benefit of 35 scf/b, and a coke reduction of 20–25%. Two versions are offered: R-232 catalyst with 0.375 wt % Pt and R-234 catalyst with 0.29 wt % Pt. Based on existing R-234 catalyst commercial operations, the ultimate life of the new catalysts are expected to be equal or better than the high-stability R-130 series catalysts.[5,7]

UOP has built on the R-230 series technology by modifying the acidity to decrease the paraffin cracking which increases the selectivity to ring closure reactions. This results in higher C_5+, aromatic, and hydrogen yields for the new R-270 series catalysts. Figure 3 illustrates how the reaction network is modified. Two versions are offered, R-272 catalyst with 0.375 wt % Pt and R-274 catalyst with 0.29 wt % Pt. The R-274 catalyst was commercialized in 2002 and offers 1–2 lv % increase in C_5+ product yield, a 5% increase in hydrogen production, and reduced coke production by 25–40% as compared to R-134 catalyst.[32] The catalyst also shows increased total and C_8 aromatic yields. The highest yield benefits can be attained with lower endpoint feeds, leaner feeds and higher severity operation. The surface area stability and strength are equivalent to those of the R-230 series catalysts.[8]

3.3. Regenerator Advances

Both UOP and Axens have made improvements to their regenerator systems. Compared to previous designs, Axens has increased the catalyst circulation and therefore the regeneration frequency allowing more severe operations. The Axens' Octanizing process uses the RegenC technology for regeneration. UOP commercialized the CycleMax CCR regenerator to enable optimal platinum

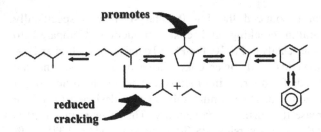

Figure 3   Effect of UOP's R-270 series catalyst on reforming reactions.

redispersion, optimal operating conditions for maximum catalyst life, catalyst 'change-out-on-the-fly' capabilities (no shutdown required), simplified instrumentation and operation to reduce cost, and stainless steel construction with reduced structure height to reduce capital investments.[33]

3.4 Combining Semiregeneration with Continuous Regeneration

The performance of existing semiregenerative units can be enhanced through a revamp that adds a new tail reactor, a heater, and a continuous catalyst regenerator. The catalyst in the new tail reactor is spherical and moves through the reactor to and from the regenerator in a similar fashion as in standard continuous units. Axens licenses two processes: Dualforming and Dualforming Plus. In the Dualforming process, the effluent from the last fixed-bed reactor is passed to a new heater and then to the new tail reactor. The new reactor operates at the same pressure as the semiregenerative reactors. In terms of catalyst circulation, the aged catalyst at the bottom of the new reactor is lifted to the top of the regenerator vessel where it is regenerated and then lifted back to the top of the new reactor. In the Dualforming Plus process, the new reactor gets its feed from the product separator and operates as a second stage. This allows operations at lower pressures than used in the fixed-bed reactors.[6,34]

UOP has developed the CycleX system comprising of a new reactor, interheater and CCR regenerator as a revamp solution for fixed-bed reformers. A new style of regenerator, the sequential regenerator, was developed to address small-capacity (5000–15,000 BPSD) fixed-bed, side-by-side or stacked reactor reforming units. Most operators of these small-capacity reforming units need a low capital cost solution for increasing hydrogen to produce clean reformulated fuels. Based on market assessments, there are three primary situations for the application of the CycleX system: (1) extend the fixed-bed reformer cycle length to match hydrodesulfurization unit cycles, (2) extend the fixed-bed reformer cycle length to match hydrodesulfurization unit cycles and increase hydrogen and reformate production, and (3) maintain the fixed-bed reformer cycle length and increase hydrogen and reformate production. The features of the CycleX system include low capital and operating costs, increased hydrogen and reformate yields, increased fixed-bed catalyst cycle length, and minimal downtime for implementation of the revamp solution.

A diagram of the CycleX system is shown in Figure 4. The only major pieces of equipment that need to be added to an existing fixed-bed reformer are a new heater, a new tail reactor, and a new CCR regenerator. Figure 5 shows the addition of the CycleX system to an existing fixed-bed reformer. As Figure 5 illustrates, the effluent from the last fixed-bed reactor (Rx. 3) is sent to a new

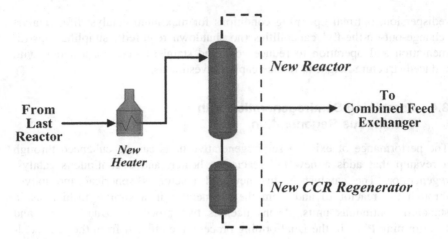

Figure 4 UOP's CycleX system.

heater that provides the proper temperature to the inlet of a new, radial flow reactor (Rx. 4). The spherical catalyst flows by gravity to the CCR regenerator, which is directly below the new reactor for the new sequential style regenerator. The catalyst is fully regenerated and then is lifted back to the top of the reactor.[35]

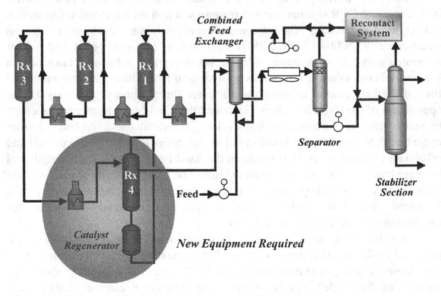

Figure 5 UOP's CycleX system added to a fixed-bed reformer.

4 ZEOLITIC REFORMING CATALYSTS

Molecular sieves demonstrate an amazing range of acidities and, due to their crystalline structure, possess the ability to differentiate the accessibility of reactants to active sites and the desorption of products from active sites by diffusional constraints. These properties have been found useful in a number of petroleum processes. There are many reviews in publication, both from historical and forward-looking perspectives.[36,37]

The use of molecular sieves in naphtha reforming has employed both highly acidic zeolites for cracking and basic zeolites for dehydrocyclization. The diffusional constraint or shape selectivity is used to crack paraffinic species or the side chains of aromatics. Therefore, an emerging use of acidic zeolites is the cracking of reformate to change the endpoint in response to environmental concerns or aromatic distribution to obtain higher valued aromatics. The use of zeolites to dehydrocyclize paraffins has been applied to liquified petroleum gas (LPG) and to raffinates.[38–40] The use of zeolite catalysts for isomerization rather than reforming is not included in this discussion.

4.1 Reformate Processing

Reformate or stabilized liquid product from the reformer may be further reacted to change its composition. Desirable changes are driven by both economic and environmental concerns. Increasing the benzene, toluene, and xylene (BTX) content through cracking ethylbenzene and heavier aromatics has been published.[13] An acidic zeolite is used to crack the aromatics to the BTX range and may be considered economically desirable, although there is some reduction in liquid gasoline yield and hydrogen, according to the reference. The initial name of the process as presented by Mobil was 'Mobil Reformate Upgrading' (MRU).[41] The acidic zeolite is placed at the bottom of the last reactor in a fixed-bed reformer, where it reacts with reformate prior to separation of the light gases. The process with the same objectives, possibly the same process or a derivative thereof, is now marketed under the name BTXtra by Exxon–Mobil.

4.2 Aromatics from LPG

The conversion of LPG to aromatics using gallium-doped zeolite catalysts has been jointly developed by UOP and BP and is called the Cyclar process.[38] This technology uses UOP's moving-bed catalyst regeneration system and has been previously described.[1] The reaction pathway in the Cyclar process is shown below. The product from the Cyclar process can be fractionated without further processing to recover aromatics since there is a negligible nonaromatic content in the BTX distillation temperature range. Sud-Chemie introduced the CPA technology utilizing an MFI catalyst to accomplish similar reaction chemistries in 1999.[23]

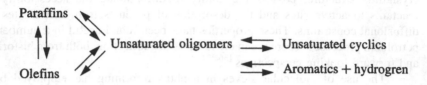

4.3. Aromatics from Naphtha

The use of Pt-containing L zeolite catalyst for the conversion of light paraffins in the C_6-C_9 range to the corresponding aromatics was patented by Elf in the late 1970s.[42] Based on this catalyst, Chevron has commercialized the AROMAX process and UOP has commercialized the RZ platforming process.[39,40] Figure 6 shows the significant selectivity advantage for the RZ platforming process compared to a conventional reforming process converting C_6-C_8 paraffins in raffinate product from an aromatics separation unit.

Many theories have been proposed for the superior selectivity for aromatics production. Some theories espouse the hydrocarbon being preferably oriented for ring closure by the zeolite structure.[43-45] An alternative theory is that the structure protects Pt sites from coke deactivation.[46]

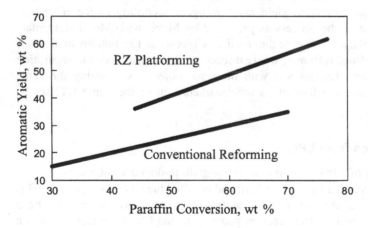

Figure 6 Comparison of aromatics yields for processing highly paraffinic feedstock derived from aromatics separation unit raffinate.

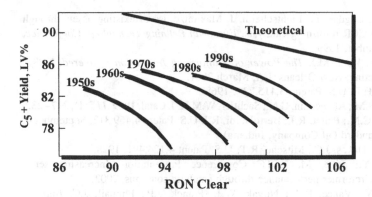

Figure 7 Chronology demonstrating the increase in catalytic reforming performance with catalyst and process innovations.

5 FUTURE DEVELOPMENTS

Catalyst development for cyclic and continuous reforming processes has focused on reducing coke make, allowing longer time before the catalyst needs to be replaced, and increased selectivity for aromatics, hydrogen, and gasoline. Since hundreds of fixed-bed reformers remain in operation, catalyst research for these units is continuing. The primary goals of this research are to develop catalysts that are more selective or are capable of operating under more severe conditions in order to produce greater quantities of aromatics. Product octane has increased as have gasoline, aromatics, and hydrogen yields. The improvement of catalytic reformer performance over time is shown in Figure 7.

REFERENCES

1. Sivasankar, S.; Ratnasamy, P. Reforming for gasoline and aromatics: recent developments. In *Catalytic Naphtha Reforming*; Antos, G.J., et al., Eds.; Marcel Dekker: New York, 1995; p. 483.
2. Aitani, A. Catalytic reforming processes. In *Catalytic Naphtha Reforming*; Antos, G.J., et al., Eds.; Marcel Dekker: New York, 1995; p. 409.
3. Berger, R.; Partner. *Study on Oil Refining in the European Community*, Report prepared for European Commission, DGXVII/B2, 1997; p. 5.
4. Wood McKenzie Consultants Ltd. The Future for European Refining, Multi-Refining Study, 1998.
5. Beshears, D.R. *CCR Platforming Catalyst Selection Improves Unit Flexibility and Profitability*, NPRA Annual Meeting; San Antonio, TX, March 2000.
6. Clause, O.; Mauk, L.; Martino, G. Trends in catalytic reforming and paraffin isomerization. In *Proceedings 15th World Petroleum Congress*, 1998; p. 695.

7. Gautam, R.; Bogdan, P.; Lichtscheidl, J. Maximize use of existing assets through advances in CCR reforming catalysts. *European Refining Technology Conference*, Paris, November, 1999.

8. Sajbel, P.O.; Wier, M.J. *The Power of Platforming Innovation Delivered*, NPRA Annual Meeting; New Orleans, LA, March 2001.

9. Kluksdahl, H.E. U.S. Patent, 3,415,737, 1968.

10. Augustine, S.M.; Alameddin, G.N.; Sachtler, W.M.H. J. Catal. **1989**, *115* (1), 217–232.

11. Sorrentino, C.M.; Pellet, R.J.; Bertolacini, R.J. U.S. Patent 4,469,812, September 4, 1984 (to Standard Oil Company, Indiana).

12. Antos, G.J.; Hayes, J.C.; Mitsche, R.T. U.S. Patent 4,124,491, 1978.

13. LeGoff, P.Y.; Pike, M. ECTC Conference Presentation. Increasing semi-regenerative reformer performance through catalytic solutions, 2002.

14. Morse, R.W.; Vance, P.W.; Novak, W.J.; Franck, J.P.; Plumail, J.C. Improve Reformer Yield and Hydrogen Selectivity with Tri-Metallic Catalyst, Hart's Fuel Reformulation, March/April 1995.

15. Phansalkar, S.S.; Edgar, M.D.; Grey, S.; Lee, B. Catalytic reforming optimization, Criterion Newsletter, 1998.

16. Edgar, M.D. *Recent Developments in Catalytic Reforming*, AIChE 1999 Spring National Meeting; Houston, TX, March 14–18.

17. Moser, M.D.; Wei, D.H.; Haizmann, R.S. Improving the Catalytic Reforming Process, CHEMTECH, October 1996; 37–41.

18. Haizmann, R.S.; Moser, M.D.; Wei, D.Y. *UOP R-72 Staged Loading for Maximum Reformer Yields*, 1995 NPRA Annual Meeting; San Francisco, CA, March 1995.

19. Senn, D.R.; Lin, F.-N.; Wuggazer, T. Improve reforming catalyst performance. World Refining **2000**, *10* (4), 48–49.

20. Dachos, N.; Kelly, A.; Felch, D.; Reis, E. UOP Platforming processes. In *Handbook of Petroleum Refining Processes*, 2nd Ed.; Meyers, R.A., Ed.; McGraw-Hill: New York, 1997; p. 4.19.

21. Octanizing, Axens IFP Group Technologies web site, Jan. 2003.

22. Remsberg, C.; Higdon, H. *Ideas for Rent, The UOP Story*; UOP: Des Plaines, Ill., 1994, p. 348.

23. Armor, J.N. New catalytic technology commercialized in the USA during the 1990's. Appl. Catal. A: Gen. **2001**, *222*, 407–426.

24. Stell, J. Worldwide catalyst rep. Oil Gas J. **2001** (Oct. 8), 56.

25. Criterion Catalysts and Technologies Product Bulletins, P-93, P-96, and PR-9 Reforming Catalysts, 1998–1999.

26. Boitiaux, J.P.; Deves, J.M.; Didillion, B.; Marcilly, C.R. Catalyst preparation. In *Catalytic Naphtha Reforming*; Antos, G.J., et al., Eds.; Marcel Dekker: New York, 1995; p. 98.

27. Murthy, K.R.; Sharma, N.; George, N. Structure and performance of reforming catalysts. In *Catalytic Naphtha Reforming*; Antos, G.J., et al., Ed.; Marcel Dekker: New York, 1995; p. 229.

28. Le Goff, P.-Y.; Le Peltier, F.; Domergue, B. Increasing reformer performance through catalytic solutions. *European Refining Technology Conference*, Paris, France, 2002.

29. AR 501 and AR 505, Axens IFP Group Technologies web site, April 2003.

30. Gray, S.; Fischer, M.; Phansalkar, S.S. *Application of the Criterion PS-40 CCR Catalyst at the ERE Refinery*, Lingren Germany, NPRA Annual Meeting; San Antonio, TX, March 17–19, 2002; Paper AM-02-62.
31. Catalyst and Technology News, Criterion Catalysts and Technologies, Oct–Nov 2002; p. 3.
32. High-yield CCR platforming catalyst commercialized, UOP Technology and More, August 2002; 1–2.
33. UOP Technical Sheet, UOP CycleMax CCR Regenerator, UOP LLC, 1998.
34. Dualforming, Axens IFP Group Technologies web site, Jan. 2003.
35. Peters, S.T. *Platforming Technology Advances: CycleX System for Increased Hydrogen Production from a Fixed-Bed Reforming Unit*, NPRA Annual Meeting; San Antonio, TX, March 2003.
36. Flanigen, E.M. Zeolites and molecular sieves: an historical perspective. Stud. Surf. Sci. Catal. **2001**, *137* (Introduction to Zeolite Science and Practice, 2nd Ed.), 11–35.
37. Rabo, J.A. New advances in molecular sieve science and technology. Periodica Polytechnica Chem. Eng. **1988**, *32* (4), 211–234.
38. Imai, T.; Kocal, J.A.; Gosling, C.D.; Hall, A.H.P. New route to aromatics production. The UOP-BP Cyclar Process, Kikan Kagaku Sosetsu **1994**, *21*, 144–151.
39. O'Rear, D.J.; Scheuerman, G.L. *Zeolite Catalysis in Chevron, Book of Abstracts*, 214th ACS National Meeting; Las Vegas, NV, September 7–11, 1997.
40. Solis, J.J.; Moser, M.D.; Ibanez, F.J. RZ platforming process improves profitability – first new unit at CEPSA, Algeciras, *European Oil Refining Conference & Exhibition*, Cascais, Portugal, June 18–20, 1997.
41. Sorenson, C.M.; Harandi, M.N.; Sapre, A.V.; Freyman, D.G. *Improving Reformate Quality for BTX Using the Mobil Reformate Upgrading Process*, National Petroleum Refiners Association Annual Meeting; San Antonio, Texas, March 16–18, 1997.
42. Bernard, J.R. U.S. Patent 4,104,320, 1978.
43. Tauster, S.J.; Steger, J.J. Molecular die catalysis: hexane aromatization over Pt/KL. J. Catal. **1990**, *25*, 382–389.
44. Derouane, E.G.; Vanderveken, D.J. Structural recognition and preorganization in zeolite catalysis: direct aromatization of hexane on zeolite L-based catalysts. Appl. Catal. **1988**, *45* (1).
45. Lane, G.S.; Modica, F.S.; Miller, J.T. Platinum/zeolite catalyst for reforming hexane: kinetic and mechanistic considerations. J. Catal. **1991**, *129* (1), 145–158.
46. Iglesia, E.; Baumgartner, J.E. A mechanistic proposal for alkane dehydrocyclization rates on platinum/L-zeolite. Inhibited deactivation of Pt sites with zeolite channels. Stud. Surf. Sci. Catal. **1993**, *75* (New Front. Catal. B), 993–1006.

9
Naphtha Reforming Over Zeolite-Hybrid-Type Catalysts

Grigore Pop
S.C. Zecasin S.A.
Bucharest, Romania

1 INTRODUCTION

During the last two decades, aromatization over zeolite catalysts has become an important way to upgrade individual lower alkanes, alkane mixtures, or alkanes contained in naphtha-reforming products. As examples of these technologies, the commercialized Cyclar process for C_2-C_3 aromatization[1] and the M2 forming process[2] are well known. These technologies were described in Chapter 15 of the first edition of this book.[25] However, there is little discussion of reacting the entire naphtha fraction. This is a purpose of the experimental work described in this chapter.

Many papers and patents have been published that claim the best catalysts for these processes are hybrid catalysts composed of two components: typically HZSM-5 zeolite modified with a metal component, such as Ga,[2] Zn,[3] Y,[4] and Ag.[5] The metals are presumed to be located on the cationic sites of the zeolite network where they arrive by ion exchange with multivalent cations, e.g., Ga(III) or Zn(II). In the case of Ga/HZSM-5, temperature-programmed reduction and X-ray diffraction studies[7] show that the active species is the Ga(I) cation. In this instance, Ga^+Z^- (Z^- is the zeolitic anionic site) resulted from two solid–solid reactions between Ga_2O_3 and HZSM-5, namely, reduction of Ga_2O_3 followed by ion exchange. With NaZSM-5, even under a hydrogen stream, the process is stopped at the first step[8] and the material obtained has no catalytic activity. For zinc-containing catalysts, the method for catalyst preparation, whether ion exchange, impregnation, or mechanical admixing of H^+ or Na^+ZSM-5 zeolite with Zn compounds, was shown to have no influence on the catalytic activity and

353

selectivity.[9] Using XANES analysis, it was demonstrated that in all cases Zn species migrate toward the acidic sites, producing a Zn ion-exchange zeolite $Zn^{2+}(Z^-)_2$.[10] Ag^+ ions in ZSM-5 zeolite are also reported to be active species of the catalyst.[6] Under the reducing conditions during the alkanes conversion, the cationic specie Ga^+, Zn^{2+}, or Ag^+ is potentially reducible to the metallic state and, together with coking, may result in deactivation of the catalysts after relatively short time on stream, in some cases 8–24 h. Upon air oxidation, the catalysts can be regenerated to their initial activity and selectivity.

Very selective conversion of light alkanes and olefins to aromatics is known to occur on these hybrid-type catalysts. In these reactions, a competition takes place between the cracking processes that consume hydrogen and the dehydrogenation reactions that release hydrogen. The enhanced production of olefins and aromatics on hybrid catalysts is theorized to be due to a long-distance hydrogen transfer from the zeolite acid sites to the metal cocatalyst surface. This is the so-called *long-distance hydrogen back-spillover* action (LD-HBS mechanism). This concept allows for considerable flexibility in the catalyst formulation since the different components may be modified separately to obtain the desirable catalytic properties for a particular hydrocarbon or hydrocarbon fraction.[11]

If the hydrogen generated over the acid sites of the zeolite is not efficiently removed by discharging it on the metal cocatalyst sites as molecular hydrogen, the reaction continues toward hydrogenolysis and cracking reactions become predominant. The cocatalyst must be in contact or positioned at very close distance to the zeolite particles to favor the hydrogen adsorption. Once adsorbed on the surface of the metal cocatalyst, the hydrogen can desorb molecularly (the sink action) or react with an adsorbed unsaturated molecule, reducing it to the corresponding saturated molecule (the scavenging action). Therefore, in order to obtain high selectivity in alkanes conversion to aromatics, a very careful design of the hybrid catalyst is necessary.

Very little information is available for naphtha reforming over these hybrid catalysts. One would desire the catalyst to have aromatization, isomerization, alkylation, and desulfurization activities and also shape selectivity for the gasoline range, i.e., C_4-C_{10} hydrocarbons. The synthesis of such a catalyst has been carried out using the existing knowledge about lower alkane aromatization and the main principles of reforming and hybrid catalyst formulation.

For the catalyst preparations, HZSM-5 zeolite with different SiO_2/Al_2O_3 ratio that was modified with Zn or Ga was used. The test reaction for catalyst screening was *n*-hexane conversion. This reaction can illustrate the aromatic alkylation processes with higher alkane and has the advantage that the reactant is in the range of gasoline. After conducting these screening tests, the final catalyst was used for naphtha, coking, and cracked gasoline reforming in lab-scale and in demonstration unit facilities.

2 CATALYST PREPARATION AND CHARACTERIZATION

Silica–alumina gel mixtures with desired SiO_2/Al_2O_3 ratios were prepared by precipitation from Na_2SiO_3 and $Al(NO_3)_3$ solutions, followed by filtration and drying at 120°C. The gel was suspended in water. A Zn^{2+} salt was added in order to modify the sample structures. Finally, the template was added. (The template can be di-*n*-propylamine or hexamethylenediamine.) The mixture having a $pH = 12.2$–12.3 was hydrothermally treated at 450 K for 24–150 h. The crystallization time strongly depends on the Zn^{2+} concentration in the reaction mixture. ZSM-5 zeolite is obtained at temperatures greater than 95°C and is characterized with XRD for degree of crystallization and phase purity. After filtration, drying, and a calcination at 825 K, the zeolite was extracted with 2 M HNO_3 until no Zn^{2+} and Na^+ could be found in the extraction liquid. It can be proposed that after treatment the zinc that remained in the zeolite is located by isomorphous substitution of the Al into the framework of MFI structure. The $1614 \, cm^{-1}$ peak in the infrared spectra of the pyridine adsorbed on the samples confirms this supposition.[12] Some characteristics of Zn-modified ZSM-5 zeolite with SiO_2/Al_2O_3 ratio of about 90 are presented in Table 1. Al, Si, and Zn contents in zeolites were measured by the neutron scattering method.[13]

It is interesting to point out that Zn causes an important increase in the number of Lewis sites whereas the number of Brönsted sites remains almost constant at about 0.7–1.0×10^{20} sites/g. As shown in Figures 1 and 2, Lewis acidity (absorbance of 19b Lewis peak PyH^+ at $1540 \, cm^{-1}$) is strictly proportional to the Zn content whereas absorbance of the 8a PyZn peaks at $1614 \, cm^{-1}$ rises exponentially. The $1614 \, cm^{-1}$ peak was assigned to the 8a vibration mode of pyridine coordinated to framework Zn.[12] These Zn-modified ZSM-5 zeolite samples contain strong Lewis sites, which are responsible for specific catalytic activities.[14]

Table 1 Some Characteristics of the ZSM-5 Samples Modified with Zn

Sample no.	Specific surface area (m²/g)	XRD phase purity (%)	IR crystallinity (%)	SiO_2/Al_2O_3 mole ratio XRD	SiO_2/Al_2O_3 mole ratio Analysis	Zn content (wt %)
1	475	100	100	83	87	—
2	460	100	99	91	92	0.6
3	465	95	97	87	89	0.7
4	455	95	95	88	90	1.3
5	450	94	93	89	90	2.0

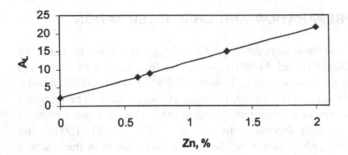

Figure 1 Lewis acidity (absorbance of Lewis peak PyH$^+$ at 1540 cm^{-1}) dependence on Zn content. (From Ref.[12].)

The described zeolites were encapsulated in 40% of an alumina or silica matrix and extruded to cylinders of 2 mm diameter and 2–5 mm in length. The catalysts obtained were tested in a dynamic tubular reactor filled with 10 ml catalyst, using atmospheric pressure, and varying the different temperature and space velocities. The test reaction used was n-hexane conversion.

The results obtained are presented in Figures 3 and 4. As can be inferred from these figures, n-hexane conversion decreases with the increase of the Zn content in catalyst, while liquid C_5^+ fractions increase as percent Zn increases. The aromatic hydrocarbon selectivity passes through a maximum. The same effect is obtained with catalysts prepared by Zn impregnation as those with prepared with Zn during crystallization, although about two times greater Zn concentrations are necessary. Further optimization of the catalyst composition was made using ZSM-5 zeolites with different SiO_2/Al_2O_3 ratios, Zn/Al ratios,

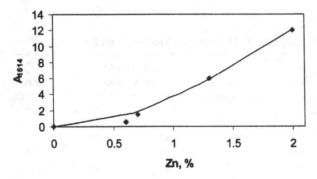

Figure 2 Absorbance of 8a PyZn peak at 1614 cm^{-1}. Influence of Zn content. (From Ref.[12].)

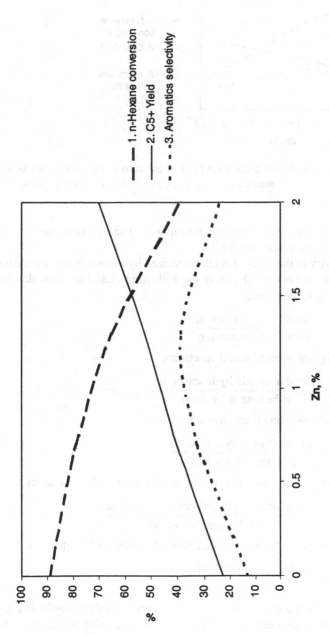

Figure 3 Influence of catalyst Zn content on catalytic properties (Zn modification in the crystallization step). (1) *n*-Hexane conversion. (2) C_5^+ yield. (3) Aromatics selectivity.

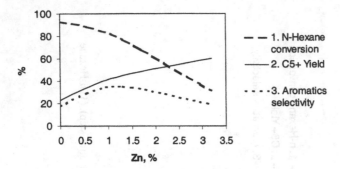

Figure 4 Influence of Zn content in the catalyst on catalytic properties (Zn modification by impregnation). (1) n-Hexane conversion. (2) C_5^+ yield. (3) Aromatics selectivity.

and Zn modifications in the crystallization step. The alumina matrix was also modified with variable Zn content.[15]

The activity of the catalyst is characterized by the n-hexane conversion at a given temperature. For the selectivity the following quantities were defined:

Selectivity for cracking:

$$S_C = \frac{\text{total H to } C_5 \text{ fraction}}{n\text{-hexane conversion}} \tag{1}$$

Selectivity for aromatization reactions:

$$S_A = \frac{\text{total aromatic hydrocarbons}}{n\text{-hexane conversion}} \tag{2}$$

Selectivity for isomerization reactions:

$$S_I = \frac{\text{isoparaffins} + \text{cycloalkanes}}{n\text{-hexane conversion}} \tag{3}$$

Selectivity for aromatic hydrocarbon alkylation to C_9 aromatics:

$$S_{A9} = \frac{\text{total trimethylbenzenes} + \text{ethyltoluenes}}{n\text{-hexane conversion}} \tag{4}$$

Selectivity for aromatic hydrocarbon alkylation to C_{10} aromatics:

$$S_{A10}\frac{\text{total } C_{10} \text{ aromatichydrocarbons}}{n\text{-hexane conversion}} \tag{5}$$

Table 2 illustrates the results obtained with catalyst employing a silica matrix. In Table 3, data refer to a similar catalyst in an alumina matrix. As these results show, unmodified HZSM-5 catalyst is very active but the cracking reactions are predominant. Modification with Zn improves the yields in C_5^+

Table 2 Results Obtained Over Catalysts with Silica Matrix

Zeolite			HZSM-5			
Catalyst sample no.		1		2		3
SiO_2/Al_2O_3		80		79		81
Zn/Al		0		0.5		1.0
LHSV (h^{-1})		1.0		1.0		1.0
Reaction temp. (°C)	340	370	340	370	340	370
n-hexane conversion (%)	49.41	95.97	70.98	71.52	22.30	24.60
H to C_5 hydrocarbon yield (%)	10.41	83.35	2.92	22.63	3.34	7.34
S_C	21.07	86.85	37.43	31.64	14.98	29.84
S_A	10.10	10.48	24.44	41.04	23.59	20.49
S_I	68.13	2.71	38.60	28.22	64.62	48.66
S_{A9}	1.98	1.53	7.47	22.34	3.08	2.62
S_{A10}	0.50	1.46	0.24	3.62	7.10	6.05

hydrocarbons and also the selectivity for aromatization and alkylation reactions. The catalysts with an alumina matrix have a more stable catalytic activity than those employing a silica matrix. The selectivity of the (Zn)-HZSM-5 zeolites could be further improved without changing the overall activity by matrix modification with Zn. Some experimental tests are presented in Table 4 for catalysts with 4% Zn in an alumina matrix combined with an HZSM-5 zeolite with 0–1.0% framework Zn. The best results were obtained with HZSM-5 zeolite containing 0.5–0.9% Zn. These compositions manifest especially good activity in isomerization reactions while C_{10} aromatic hydrocarbon content remains very low.

Table 3 Results Obtained Over Catalysts with Alumina Matrix

Zeolite			IIZSM-5			
Catalyst sample no.		4		5		6
SiO_2/Al_2O_3		80		79		81
Zn/Al		0		0.5		1.0
LHSV (h^{-1})		1.0		1.0		1.0
Reaction temp. (°C)	340	370	340	370	340	370
n-hexane conversion (%)	92.98	93.64	34.61	73.54	75.71	79.40
H to C_5 hydrocarbon yield (%)	69.02	77.46	27.79	27.32	49.39	62.89
S_C	74.23	82.69	80.29	37.03	65.25	79.21
S_A	17.04	12.40	6.70	36.21	11.08	19.32
S_I	5.47	2.77	13.01	25.98	7.46	13.12
S_{A9}	3.44	0.43	2.20	16.49	2.23	0.74
S_{A10}	3.64	6.65	0.22	11.96	1.25	5.96

Table 4 Results Obtained Over Catalyst Double-Promoted with Zn[a]

Catalyst	ZSM-5, $SiO_2/Al_2O_3 = 125$									
Zn in alumina matrix (%)					4.0					
Catalyst sample no.	7		8		9		10		11	
Zn in HZSM-5 framework (%)	0		0.25		0.50		0.90		1.0	
Reaction temperature (°C)	340	370	340	370	340	370	340	370	340	370
n-hexane conversion	84.9	85.36	64.65	77.92	76.89	89.00	39.37	69.94	55.96	65.64
H to C_5 hydrocarbons yield (%)	56.82	67.65	34.80	48.82	30.32	35.21	17.65	39.43	22.96	38.80
S_C	66.92	79.25	53.83	62.65	39.43	39.56	44.82	56.38	41.07	59.11
S_A	12.42	16.99	15.92	21.34	12.50	12.79	18.57	26.79	10.00	19.17
S_I	12.05	7.65	30.26	15.99	48.08	47.40	41.07	19.17	46.25	21.72
S_{A9}	2.32	3.30	1.95	2.87	2.66	2.88	4.09	4.71	0.97	0.97
S_{A10}	0.83	0.6	0.32	0.47	0.34	0.41	1.73	2.26	0.24	0.31

[a]Normal pressure; LHSV = $1\,h^{-1}$.
Source: Ref.[15].

Table 5 shows the influence of the SiO_2/Al_2O_3 ratio in H-ZSM-5 zeolites modified with Zn in the framework. The catalysts are formulated in an unmodified alumina matrix. The SiO_2/Al_2O_3 ratio in the zeolite has a great influence upon all the reactions that occur during n-hexane transformation. Total conversion has a minimal value corresponding to the SiO_2/Al_2O_3 ratio value of 80. The aromatization and alkylation reaction contribution grows continuously with increased SiO_2/Al_2O_3 ratio. The isomerization reactions pass through a maximal value, corresponding to the SiO_2/Al_2O_3 ratio value of approximately 80. Unfortunately, the selectivity for cracking reactions is higher for all compositions.

Table 6 shows a comparison between catalysts with the same modification level for two HZSM-5 zeolites with high SiO_2/Al_2O_3 ratios. Over these catalysts, aromatization and isomerization reactions combined are dominant at a more acceptable cracking selectivity. All of these data show that an improved hybrid catalyst for hydrocarbon transformation is obtained by manipulation of the combination of three main parameters: SiO_2/Al_2O_3 ratio in the zeolite, Al substitution by Zn in the zeolite framework, and Zn modification of the alumina matrix. Catalyst samples 19 and 20 represent, at 80–90% conversion, more favorable selectivity for aromatization and isomerization reactions in n-hexane transformation. Zeolites with high SiO_2/Al_2O_3 ratios, i.e., with stronger acid sites, favor isomerization processes whereas zeolites having SiO_2/Al_2O_3 ratios less than 100 promote aromatization and aromatic hydrocarbon alkylation. In fact, the resultant reformate composition is very near that of reformulated gasoline, namely, medium aromatic content, benzene under 2%, low olefin and n-paraffin content, and high isoparaffins and cycloalkanes content.

The reaction parameter dependence over catalyst sample 20 was studied using a statistical model with two variables: temperature and space velocity. The experimental data used for the mathematical derivation of the program are presented in Table 7. A second-degree polynomial fits the experimental data well. Numerical values of the parameters in the statistical model are shown in Table 8. The graphical representations of the regression equations enable the following conclusions (Fig. 5a–d):

Maximal n-hexane conversion (90%) is situated at an intermediate value of the space velocity ($2 \, h^{-1}$) and high temperature (713 K).

The maximal yield of aromatic hydrocarbons (30 wt %) is observed in the same parameter area.

Minimal cracking activity, represented by propane yield, is reached in two regions, corresponding to higher temperature and space velocity and to lower temperature and space velocity.

Hydrogen yield increases continuously with increasing temperature and increasing space velocity.

Table 5 Influence of SiO_2/Al_2O_3 Ratio in HZSM-5 Zeolite[a]

Catalyst sample no.	12	13	14	15	16	17	18
SiO_2/Al_2O_3 ratio in zeolite	46	60	60	60	80	120	125
Zn/Al ratio in the zeolite framework	1.0	0	0.4	0.75	0.90	0.90	0.89
n-hexane conversion (%)	89.23	92.10	91.02	78.668	79.40	79.45	85.87
H to C_5 hydrocarbons yield (%)	78.18	72.42	73.82	66.00	62.89	60.59	66.08
S_C	76.41	78.63	81.10	83.88	79.21	76.26	76.95
S_A	12.50	13.03	13.18	14.48	19.32	19.75	22.14
S_I	6.80	7.37	5.63	17.35	13.12	6.39	0.91
S_{A9}	1.66	1.94	1.66	1.45	1.74	3.48	4.30
S_{A10}	0.59	1.12	1.13	1.46	1.96	2.42	2.66

[a]Catalyst in unmodified alumina matrix.
Source: Ref.[15].

Table 6 Influence of SiO_2/Al_2O_3 Ratio in HZSM-5 Zeolite Catalyst in Zn-Alumina Matrix, LHSV = 1 h^{-1}

Catalyst type	HZSM-5; 5% Zn in alumina matrix; 0.50% Zn in zeolite framework			
Catalyst sample no.		19		20
SiO_2/Al_2O_3 ratio in zeolite		80		125
Reaction temp. (°C)	340	370	340	370
n-hexane conversion (%)	73.63	91.47	79.98	86.50
H to C_5 hydrocarbons yield (%)	26.46	37.11	19.78	41.79
S_C	35.94	40.57	24.73	67.29
S_A	14.60	21.47	9.16	10.07
S_I	49.45	37.76	66.10	41.20
S_{A9}	0.98	4.05	1.54	1.85
S_{A10}	0.14	0.16	0.36	1.46

Source: Ref.[15].

3 EVIDENCE OF LD-HBS MECHANISM

This section presents the available evidence for the *long-distance hydrogen back spillover* (LD-HBS) mechanism in the *n*-hexane transformation over Zn- and Ga-HZSM-5 hybrid catalyst. Dufresne et al.[11] have demonstrated that the enhanced aromatic production on a bifunctional catalyst is due to a long-distance hydrogen transfer (*back spillover*) from the zeolite acid sites to the cocatalyst surface. Roesner et al.[16] showed that in cyclohexane dehydrogenation, activated hydrogen supplied by the Pt is necessary. The isomerization to methylcyclopentane requires both activated hydrogen and acidic sites. Using different arrangements of acidic and metallic sites, the action of the hydrogen spillover was observed far from the activated center.[16] In butane aromatization over Pt-Ga ZSM-5 catalyst, it was demonstrated that a balance of direct and reverse hydrogen spillover is necessary to provide both rapid dehydrogenation and selective aromatization.[17] Selective *n*-pentane isomerization to isopentane over Pt-HZSM-5 catalysts also requires spillover hydrogen.[18]

If the spillover hydrogen is necessary for selective hydrocarbon transformation, the production of hydrogen could be evidence for the LD-HBS action in the alternative mechanism for cracking processes with light paraffin formation on the catalyst acid sites. Evidence for these two mechanisms was found in *n*-hexane transformation over HZSM-5 zeolite and hybrid catalysts

Table 7 Experimental Data for the Statistical Model for *n*-Hexane Conversion on Catalyst Sample No. 20

Exp. no.	1	2	3	4	5	6	7	8	9	10
Temp. (°C)	335	405	335	405	320	440	370	370	370	370
Space velocity (h^{-1})	1.07	1.28	2.24	2.40	2.00	1.87	1.90	0.68	2.82	1.80
H$_2$–C$_5$ yield (wt %)	48.39	62.70	31.67	51.67	24.03	59.76	46.30	51.58	29.74	46.84
C$_6{}^+$ yield (wt %)	51.61	37.30	68.33	48.33	75.97	40.26	53.70	48.41	70.26	53.16
Aromatics yield (wt %)	11.68	26.39	15.12	30.69	7.40	27.43	21.38	17.20	10.97	16.99
Product distribution (wt %)										
Hydrogen	0.58	1.24	0.50	1.48	0.27	2.02	0.66	0.84	0.62	1.00
Methane	0.24	0.668	0.13	0.95	0.10	1.43	0.44	0.39	0.21	0.53
Ethane	1.08	0.15	0.50	2.99	0.38	4.89	1.61	2.41	1.02	2.08
Ethene	0.34	0.59	0.42	1.69	0.26	1.97	0.53	0.52	0.62	0.54
Propane	17.59	28.84	11.39	17.22	12.47	25.23	22.55	23.58	15.55	17.17
Propene	1.19	3.23	1.23	3.61	0.72	5.86	1.44	1.51	2.37	2.82
iso-butane	8.09	10.58	4.81	6.84	3.98	5.67	6.04	6.84	3.38	7.07
n-butane	8.09	6.05	4.73	5.02	4.16	1.19	5.43	7.06	3.36	5.99
Butenes	0.55	1.83	0.94	1.44	0.79	1.52	2.78	0.65	0.82	1.32

iso-pentane	1.05	1.42	2.23	1.06	0.45	0.92	0.82	1.00	0.52	1.19
n-pentane	0.68	0.57	0.41	0.48	0.31	0.37	0.40	0.52	0.30	0.55
C_5-olefins	8.81	7.10	1.75	8.63	1.86	8.82	4.33	4.42	2.85	7.18
n-hexane	27.99	11.03	42.17	17.74	51.94	12.78	27.07	27.00	48.21	30.56
iso- and cycloalkanes C_7^+	0.46	1.97	0.66	1.29	0.31	1.72	0.92	0.83	0.64	1.02
Benzene	0.46	1.97	0.66	1.29	0.31	1.72	0.92	0.83	0.64	1.02
Toluene	3.88	12.06	4.75	12.09	2.06	9.59	7.07	6.34	4.44	7.39
Ethylbenzene	0.57	1.03	0.72	0.65	0.36	1.28	0.99	0.82	0.50	0.76
p-xylene	1.65	2.64	2.39	4.06	1.43	3.65	3.05	1.96	1.55	1.94
m-xylene	2.35	4.44	2.53	5.69	1.33	5.16	4.13	3.40	1.78	2.78
o-xylene	0.98	1.70	0.84	2.26	0.46	2.29	1.63	1.41	0.55	1.05
Ethyltoluene	0.98	1.63	2.63	2.32	0.76	1.94	2.05	1.87	0.83	1.20
1,3,4-trimethylbenzene	0.25	0.36	0.33	0.58	0.20	0.55	0.46	0.40	0.18	0.25
1,2,3-trimethylbenzene	0.21	0.19	0.31	0.32	0.16	0.37	0.37	0.87	0.16	0.21
Durenes	0.01	0.02	0.07	0.03	0.03	0.04	0.05	0.04	0.01	0.02
C_{10}^+ hydrocarbons	0.37	0.64	0.48	0.56	0.14	0.73	0.38	0.52	0.05	0.17

Table 8 Numerical Values of the Coefficients in the Statistical Model $Z = a + bT + cS + dT^2 + eS^2 + fTS$

Variable, Z	Coefficients					
	a	b	c	d	e	f
Z_1, n-hexane conversion (%)	−116.558	0.962027	−47.4627	-1.25898×10^{-3}	−7.78293	0.174411
Z_2, aromatic hydrocarbons yield (wt %)	−187.761	0.944257	−0.171666	-1.13379×10^{-3}	−5.33661	4.90265×10^{-2}
Z_3, hydrogen yield, mole/mole n-hexane converted	2.51011	-1.24601×10^{-2}	−0.736844	1.81924×10^{-5}	-1.44429×10^{-2}	-2.24042×10^{-3}
Z_3, cracking activity, propane yield, mole/mole n-hexane converted	−2.62207	1.29288	0.704512	-0.05099×10^{-5}	3.50011×10^{-2}	-2.35027×10^{-3}

T, temperature; S, space velocity.

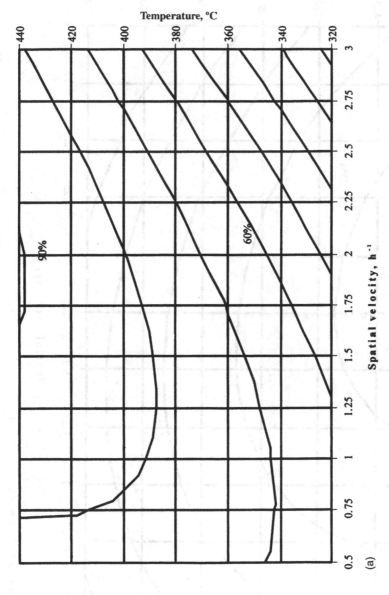

Figure 5 Graphic representation of the statistical model for *n*-hexane transformation over Zn-modified HZSM-5 hybrid catalyst. (a) *n*-hexane conversion. (b) Aromatic hydrocarbons yield. (c) Cracking activity (propane yield). (d) Hydrogen yields.

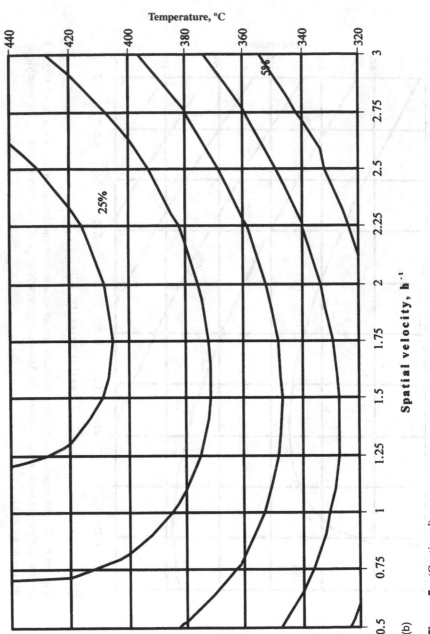

Figure 5 (Continued).

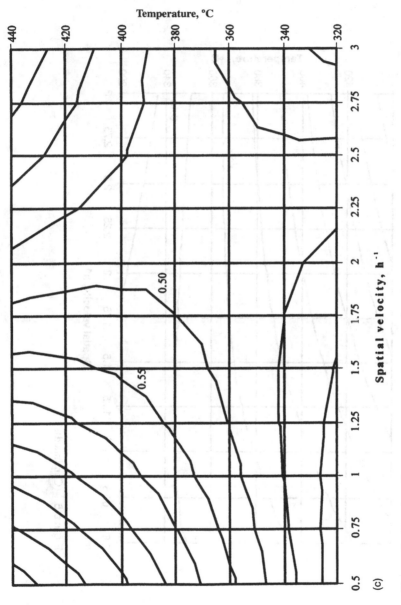

Figure 5   (Continued).

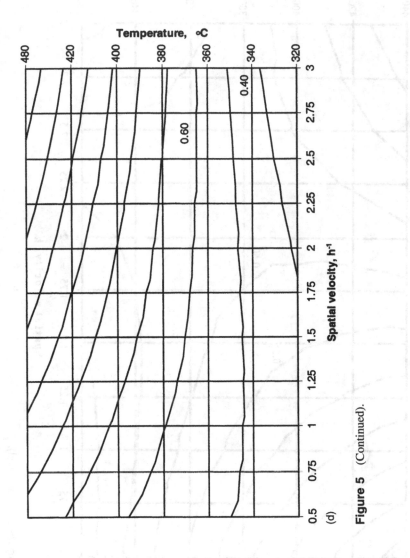

Figure 5 (Continued).

Zn-ZSM-5 and Ga-ZSM-5.[19] Some results that confirm these mechanisms are shown in Table 9. The high concentrations of propane and butanes in the reaction products are ascribed to the high concentration of Al in the zeolite lattice, which is evidence for a large number of acid sites. In order to increase the hydrogen spillover action, more metallic centers and a decreased number of acidic sites are needed. Increasing the SiO_2/Al_2O_3 ratio of the zeolite could do this. The zeolite acidity also becomes stronger and the isomerization–alkylation reactions are favored.

Table 9 Reaction Product Distribution in *n*-Hexane Transformation Over H-, Zn-s and Ga-Modified ZSM-5 Zeolite (Mole/Mole *n*-Hexane Converted; $T = 370°C$; $LHSV = 1\ h^{-1}$)

Sample	—	$SiO_2/Al_2O_3 = 40$ Zn- and Ga-modified ZSM-5 during crystallization stage	
		Zn/Al = 0.5	Ga/Al = 0.5
n-hexane conversion, %	63.93	39.89	43.91
Hydrogen	0.99	8.82	8.21
Methane	0.12	0.14	0.49
Ethane	—	0.38	0.32
Ethylene	—	0.24	0.14
Propane	44.32	45.60	38.78
Propylene	—	0.59	0.29
iso-butane	19.20	16.64	20.17
n-butane	17.78	16.17	17.89
1 + *iso*-butene	0.06	0.32	0.26
2-butene	2.99	2.27	0
iso-pentane	1.13	1.28	2.37
n-pentane	—	—	1.54
Pentenes	1.02	2.01	2.04
Benzene	0.50	0.22	0.28
Toluene	3.57	1.57	1.97
Ethylbenzene	0.60	0.23	0.34
p-xylene	1.11	0.72	0.75
m-xylene	2.13	1.15	1.34
o-xylene	0.87	0.45	0.63
Ethyltoluenes	1.37	0.74	0.91
Trimethylbenzenes	0.54	0.38	0.65
C_{10} aromatics	1.06	0.43	0.63

4   REACTION PATHWAYS IN *n*-HEXANE
TRANSFORMATION

Paraffin transformation over Zn-HZSM-5 hybrid catalyst is a very complex process. If one presumes that the LD-HBS action is valid, the following reaction pathway could be accepted:

At the metallic site of the catalyst, paraffins are converted to olefins and molecular hydrogen is released.

Catalyst acid centers promote the cracking reactions. Propane, butanes, and low molecular weight olefins are the most important reaction products.

Olefin dehydrocyclization to cycloalkanes and aromatization reactions are catalyzed by the metal active centers with hydrogen spillover formation.

At the acid centers of the zeolite, many other processes are possible, e.g., paraffin and olefin isomerization, and paraffin and aromatic alkylation with olefins, polycondensation, isomerization, and transalkylation of aromatic hydrocarbons.

These complex reaction pathways over Zn-HZSM-5 hybrid catalyst were confirmed by the analysis of the reaction product distribution obtained in *n*-hexane transformation at 375°C and atmospheric pressure.[20] The data presented in Table 10 demonstrate the good catalyst activity in cracking, aromatization, isomerization, and alkylation reactions. The high proportion of *n*-paraffins in the reaction product is due to propanes and butanes resulting from the cracking reactions and to unreacted *n*-hexane. Low benzene concentration in the aromatic fraction demonstrates high alkylation activity of the catalyst.

One can imagine that the optimization of the catalyst formula could be carried out by modification of the SiO_2/Al_2O_3 ratio in the zeolite, in order to control the number and strength of the acid centers. In addition, one could increase the concentration and repartition of the metallic centers by suitable Zn modification. This type of optimization would be required, since a good catalyst for naphtha reforming must have low cracking activity, high hydrogen spillover action, and high alkylation and isomerization activities. Catalyst samples 19 and 20 in Table 6 can be considered examples of improved hybrid catalysts for hydrocarbon fraction reforming. As the experiments with *n*-hexane show, the catalysts are active in reforming processes at relatively low temperature (375–440°C) and at atmospheric pressure. These capabilities were demonstrated in naphtha reforming experiments at lab scale and in a demonstration unit.

Table 10 Product Distribution in *n*-Hexane Transformation over Zn-HZSM-5 Hybrid Catalyst

No.	Component	Distribution					
		H_2	*n*-Paraffins	*iso*-Paraffins	Olefins	Cycloparaffins	Aromatics
1.	Hydrogen	0.30					
2.	Methane		0.13				
3.	C_2		0.45		0.19		
4.	C_3		15.22		1.02		
5.	C_4		12.15	11.19	0.53		
6.	C_5		3.17	3.07	0.52		
7.	C_6		22.12	2.92	0.14	1.31	0.90
8.	C_7		0.19	0.51	0.21	0.35	7.92
9.	C_8			0.07		0.25	
10.	Ethylbenzene						1.11
11.	$m + p$-xylenes						7.48
12.	o-xylene						1.85
13.	C_9						3.3
14.	C_{10}						0.41
15.	$C_{10}+$						1.02
	Total	0.30	53.43	17.76	2.61	1.91	23.99

Source: Ref.[20].

5   COMMERCIAL NAPHTHA REFORMING OVER HYBRID CATALYSTS

Experimental work has been carried out using Zn-HZSM-5 hybrid catalysts for reforming of commercial naphtha. The experiments were made with the catalyst sample 19, in a tubular reactor of 20 mm ID and 30-ml catalyst load, at atmospheric pressure. The characteristics of the feedstock were:

$d_4^{20} = 0.736$
Initial boiling point, 45°C
Final boiling point, 171°C
Sulfur content, 0.21 wt %
RON, 70.1.

The starting reaction temperature was 380°C and the corresponding space velocity. LHSV, was $1 \, h^{-1}$. Temperature was raised periodically as shown in Figure 6. At 4 h on stream, all aromatic hydrocarbons passed through a maximal value, as shown in Figure 6. After 12 h on stream, the active cycle of the catalyst is completed with values of the aromatics back at baseline. The reformate average RON was 83 and the liquid yield 91.6 wt %.

The stepwise increase of the temperature to 430°C apparently had no effect on the catalyst activity. Figure 7 shows that the experiment with the same feedstock at 450°C had a longer active cycle. The reformate composition became constant after 16 h on stream but at low conversion values. The RON value for the reformate obtained in the first 16 h was 89.4, with 80 wt % gasoline yield.

The catalyst deactivation is reversible. By burning the catalyst in air for 2 h at 550°C, the catalytic activity is completely restored, as shown in Figure 8. Since HZSM-5 zeolite is very resistant to coking, one might expect that levels of 5% coke would have little influence on the catalytic performance. Therefore, it is presumed that deactivation takes place by blocking the active centers with sulfur compounds. In order to confirm this hypothesis, a desulfurization study was performed over hybrid catalyst 19. The results obtained are presented in Figure 9. The results suggest an adsorptive sulfur effect rather than a chemical one [21].

A sulfur-free hydrocarbon fraction was processed under the same conditions as for the sulfur-containing naphtha. The results obtained changed dramatically. The hydrocarbon fraction composition was:

i-C$_6$	4.94 wt %
n-C$_5$	41.01 wt %
n-C$_6$	32.58 wt %
i-C$_6$	4.04 wt %
C$_7$+	1.93 wt %
Aromatics	2.89 wt %

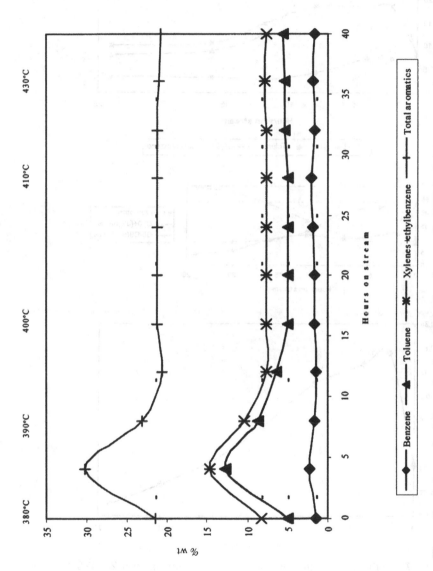

Figure 6 Active cycle of the catalyst in naphtha reforming over Zn-HZSM-5 hybrid catalyst with progressive temperature increase.

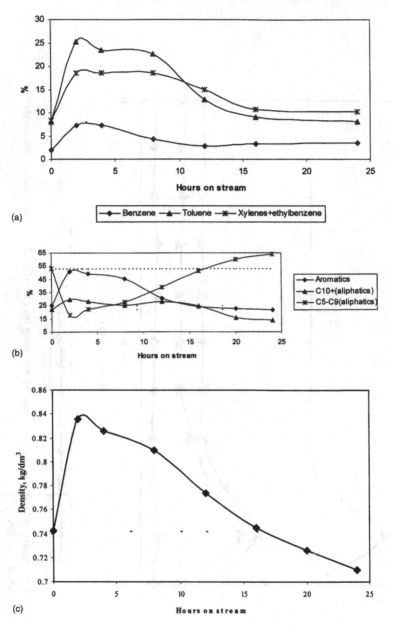

Figure 7 Active cycle of the catalyst in naphtha reforming over Zn-HZSM-5 hybrid catalyst at 450°C. (a) Benzene, toluene, and xylene + ethylbenzene concentration. (b) Total aromatics, heavy hydrocarbons C_{10}^+, C_5–C_9 hydrocarbons. (c) Density (ρ) variation of the liquid fraction.

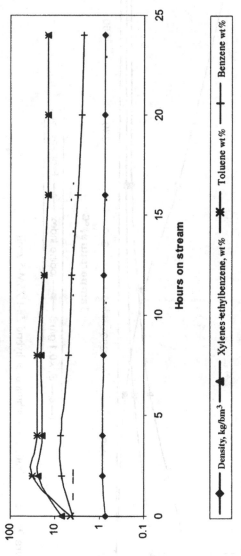

Figure 8 Active cycle of the regenerated catalyst.

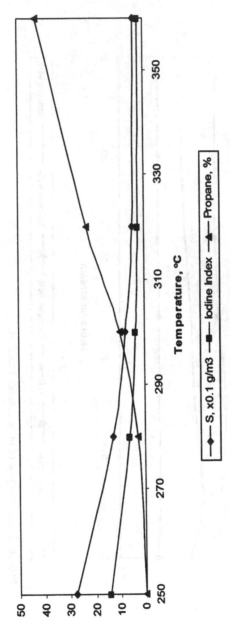

Figure 9 Desulfurization of naphtha over hybrid Zn-HZSM-5 zeolite.

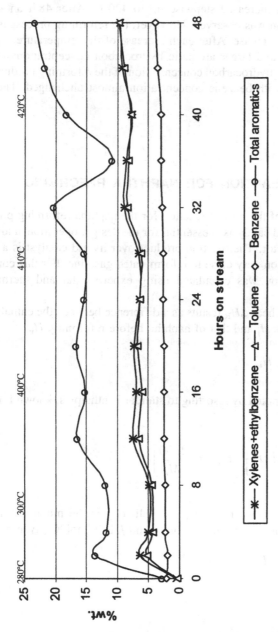

Figure 10 Sulfur-free feedstock reforming over Zn-HZSM-5 hybrid catalyst.

Results are presented in Figure 10. From an initial reaction temperature of 380°C, this parameter was increased stepwise up to 420°C. After 48 h on stream, no catalyst deactivation was observed. However, the reforming process is sensitive to the temperature increase. After each increase of the temperature, a plateau is achieved at larger and larger aromatic hydrocarbon concentrations. At 420°C, after 48 h, aromatic hydrocarbon concentration of the reforming product gasoline reached 24 wt %, with benzene concentration almost unchanged. The gasoline yield was 81 wt %.

6 HEAT OF REACTION FOR NAPHTHA REFORMING

Exact calculation of the heat of reaction for the naphta reforming process over Zn-HZSM-5 hybrid catalysts is essential for the design, operation, and control of the reforming reactor. The reaction product over hybrid catalyst is a reformate with the composition very close to reformulated gasoline. For this composition, the heat of reaction was calculated using experimental and thermodynamic data available.[22]

The reaction heat ΔH_R results as a difference between the enthalpies of the reforming products H_f and that of naphtha before reforming, H_i.

$$\Delta H_R = H_f - H_i \tag{6}$$

H_f and H_i are computed by resorting to standard enthalpies, known for 298.15 K as follows:

$$H_s = \sum_{j=1}^{c} \left[y_j \left(H_j^o \int_{T_0}^{T} C_{p,j} dT \right) \frac{1000}{M_j} \right] \tag{7}$$

where subscript s stands for i or j, as well. $C_{p,j}$ is the molar specific heat at constant pressure for the chemical compound j, in J/mol.K, given by[23]:

$$C_{p,j} = a_j + b_j T + c_j T^2 + d_j T^3 \tag{8}$$

Therefore:

$$\Delta H_R = \sum_{j=1}^{c} \left[(\delta y_j) \left(H_j^o + \int_{T_0}^{T} C_{p,j} dT \right) \right] \frac{1000}{M_j} \tag{9}$$

Table 11 Thermodynamic Data for Chemical Species Included in Heat of Reaction Calculations[a]

Chemical compound	Molecular weight (kg/kmol)	Standard enthalpy of formation for the ideal gas (J/mol)	Isobaric heat capacity of the ideal gas (J/mol.K)				Mass fraction (%)	
			a	b	c	d	Inlet, $y_{i,j}$	Outlet, $y_{e,j}$
Hydrogen	2.016	0	2.714E+01	9.274E−3	−1.381E−5	8.645E−9	0	2.26*
Methane	16.043	−7.49E+4	1.925E+1	5.213E−2	1.197E−5	−1.132E−8	0	3.14*
Ethylene	28.054	3.806E−1	5.234E+4	1.566E−1	−8.348E−5	1.755E−8	0	0.90*
Ethane	30.070	−8.474E+4	5.409E+0	1.781E−1	−6.938E−5	8.713E−9	0	3.97*
Propylene	42.081	2.043E+4	3.710E+0	2.345E−1	−1.160E−4	2.205E−8	0	2.14*
Propane	44.094	−1.039E+5	−4.224E+0	3.063E−1	−1.586E−4	3.215E−8	0.04	0.08 47.37*
trans-2-butene	56.108	−1.118E+4	1.832E+1	2.564E−1	−7.013E−5	−8.989E−9	0	2.41*
n-butane	58.125	−1.262E+5	9.487E+0	3.313E−1	−1.108E−4	−2.822E−9	0.38	1.79 13.04*
iso-butane	58.124	−1.346E+5	−1.390E+0	3.847E−1	−1.846E−4	2.895E−4	0.10	0.64 17.00*
n-pentane	72.151	−1.465E+5	−3.262E+.	4.893E−1	−2.580E−4	5.305E−8	1.43	2.78
2-methylbutane	72.151	−1.546E+5	−9.525E+0	5.066E−1	−2.729E−4	5.723E−8	1.01	2.42
Benzene	78.114	8.298E+4	−3.392E+1	4.739E−1	−3.017E−4	7.130E−8	1.43	2.68
Methylcyclopentane	84.162	−1.068E+5	−5.011E+1	6.381E−1	−3.642E−4	8.014E−8	2.10	2.28
2-methyl-2-pentene	84.162	−6.653E+4	−1.475E+1	5.669E−1	−3.341E−4	7.963E−8	0	1.09
n-hexane	86.178	−1.673E−5	−4.413E+0	5.820E−1	−3.119E−4	6.494E−8	2.88	2.62 7.77*
2-methylpentane	86.178	−1.744E+5	−1.057E+1	6.187E−1	−3.573E−4	8.085E−8	2.26	6.04
3-methylpentane	86.178	−1.717E+5	−2.386E+0	5.690E−1	−2.870E−4	5.033E−8	1.21	1.73
Toluene	92.141	5.003E+4	−2.435E+1	5.125E−1	−2.765E−4	4.911E−8	2.36	7.26
1,1-dimethylcyclopentane	98.189	−1.384E+5	−5.789E+5	7.670E−1	−4.501E−4	1.010E−7	1.56	1.55
1,2-dimethylcyclopentane, *trans*	98.189	−1.368E+5	−5.452E+1	7.591E−1	4.480E−4	1.017E−7	1.15	1.42
Methylcyclohexane	98.189	−1.549E+5	−6.192E+1	7.842E−1	−4.438E−4	9.366E−8	6.28	3.99
n-heptane	100.205	−1.879E−5	−5.146E+0	6.762E−1	−3.651E−4	7.658E−8	4.10	2.01
2-methylhexane	100.205	−1.951E+5	−3.939E+1	8.642E−1	−6.289E−4	1.836E−7	1.53	1.60

(*continues*)

Table 11 Continued

Chemical compound	Molecular weight (kg/kmol)	Standard enthalpy of formation for the ideal gas (J/mol)	Isobaric heat capacity of the ideal gas (J/mol.K)				Mass fraction (%)	
			a	b	c	d	Inlet, $y_{i,j}$	Outlet, $y_{e,j}$
3-methylhexane	100.205	−1.924E+5	−7.046E+0	6.837E−1	−3.734E−4	7.834E−8	2.15	2.40
2,3-dimethylpentane	100.205	−1.994E+5	−7.046E+0	6.837E−1	−3.734E−4	7.834E−8	0.71	1.01
o-xylene	106.168	1.900E+4	−1.585E+1	5.962E−1	−3.443E−4	7.528E−8	1.07	1.84
p-xylene	106.168	1.796E+4	−2.509E+1	6.402E−1	−3.374E−4	6.820E−8	3.91	7.00
Ethylbenzene	106.168	2.981E+4	−4.310E+1	7.072E−1	−4.811E−4	1.301E−7	1.22	1.99
1,1-dimethylcyclohexane	112.216	−1.811E+5	−7.211E+1	8.997E−1	−5.020E−4	1.030E−7	0.40	1.13
1,2-dimethylcyclohexane, *trans*	112.216	−1.801E+5	−6.848E+1	9.123E−1	−5.355E−4	1.181E−7	1.67	2.47
1,4-dimethylcyclohexane, *cis*	112.216	−1.768E+5	−6.415E+1	8.826E−1	−5.016E−4	1.068E−7	2.25	2.97
Ethylcyclohexane	112.216	−1.719E+5	−6.389E+1	8.893E−1	−5.108E−4	1.103E−4	4.25	4.30
n-propylcyclopentane	112.216	−1.482E+5	−5.597E+1	8.447E−1	−4.924E−4	1.117E−4	0.27	1.19
3-methylheptane	114.232	−2.128E+5	−9.215E+0	7.859E−1	−4.400E−4	9.697E−8	2.26	1.81
3-methyl-3-ethylpentane	114.232	−2.151E+5	−9.215E+0	7.859E−1	−4.400E−4	9.697E−8	0.53	1.89
1-methyl-3-ethylbenzene	120.195	−1.930E+3	−2.900E+1	7.293E−1	−4.363E−4	9.998E−8	0.54	1.34
1,2,4-trimethylbenzene	120.195	−1.394E+4	−4.668E+0	6.238E−1	−3.263E−4	6.376E−8	1.33	1.74
1,3,5-trimethylbenzene	120.195	−1.608E+4	−1.959E+1	6.724E−1	−3.692E−4	7.700E−8	1.02	0.93
n-propylcyclohexane	126.243	−1.934E+5	−6.252E+1	9.889E−1	−5.795E−4	1.291E−7	7.09	4.81
1-nonene	126.243	−1.036E+5	3.718E+0	8.122E−1	−4.509E−8	9.705E−8	0.18	2.58
2-methyloctane	128.242	−2.292E+5	−1.011E+1	8.805E−1	−4.936E−4	1.083E−7	1.58	1.51
2,2,-dimethylheptane	128.242	−2.470E+5	−2.089E+1	9.668E−1	−6.120E−4	1.570E−7	3.65	1.21
3,3,5-trimethylheptane	142.286	−2.587E+5	−7.037E+1	1.232E+0	−8.646E−4	2.455E−7	2.44	1.15
2,2,3,3-tetramethylhexane	142.286	−2.587E+5	−5.883E+1	1.231E+0	−8.834E−4	2.585E−7	2.41	1.32

[a]Coefficients given in scientific exponential notation, e.g., 17.013E−5 = 17.013 × 10^{−5}.

Note: (*) represents gas mass fraction (20% of the total reaction products). Outlet species with less than 1% mass fraction have not been listed in the table.

Source: Ref.[22].

where M_j is the molecular weight for the chemical species j, kg/kmol, and T, the reaction temperature. Δy_j is the difference between the outlet and inlet of the reactor, for the enthalpy, in J/kg mixture, given by the following equation:

$$\Delta y_j = 0.8 y_{e,lq,j} + 0.2 y_{e,g,j} - y_{i,j} \tag{10}$$

in which the assumption is made that 80% of the reaction products are obtained as liquid fraction and 20% as gas product, as shown by the experimental data. The final relation for ΔH_R is therefore:

$$\Delta H_R = \sum_{j=1}^{c} (0.8 y_{e,lq,j} + 0.2 y_{e,g,j} - y_{i,j})$$

$$\times \left[H_j^0 + a_j(T - T_o) + b_j \frac{T^2 - T_o^2}{2} + c_j \frac{T^3 - T_o^3}{3} + d_j \frac{T^4 - T_o^4}{4} \right] \tag{11}$$

The thermodynamic data and the chemical compositions of the fluxes used for computation are presented in Table 11.

The composition of the naphtha and of the reaction product presented in Table 11 were obtained by GC-MS technique, The reaction was conducted at atmospheric pressure, 370°C and $1.2 \, \text{h}^{-1}$ LHSV. In the derivation of ΔH_R, 89 chemical species were considered, representing more than 95% of the mass quantity for feed and for the reaction product. The result was a positive value: $\Delta H_R = 42.34$ kcal/(kg reaction mixture) for naphtha reforming over hybrid catalyst at 370°C reaction temperature. This value confirms the assumption that this process could be carried out in a single fixed-bed catalytic, adiabatic reactor. The heated naphtha feed (450°C, temperature acceptable for the zeolite catalyst) has an approximate specific heat of 0.5 kcal/kg.K; therefore, the temperature will cool to no less than 350°C, due to the endothermic effect of the reforming reaction.

7 DEMONSTRATION UNIT FOR NAPHTHA REFORMING

The demonstration unit for naphtha reforming over hybrid catalyst[15,24] is schematically presented in Figure 11. Naphtha is heated to 450°C in the furnace F-1 and then fed into the reactor R, which is filled with 100 kg hybrid catalyst. In the lower section of the reactor, a heating coil is installed, providing the extra heat needed to compensate for the heat due to reaction and losses. The reaction products are evacuated from the upper end of the reactor, cooled and condensed

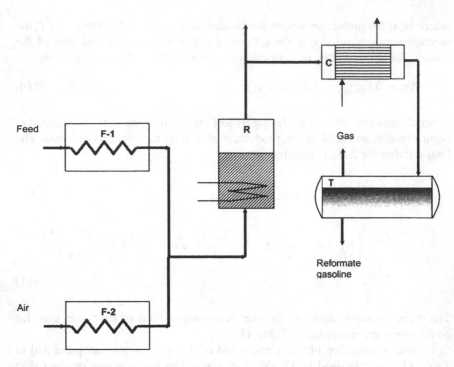

Figure 11 Experimental demonstration unit scheme. F-1, F-2, furnaces; R, catalytic reactor; C, condenser; T, tank. (From Ref.[24].)

in condenser C, and collected in tank T, where the uncondensed fraction is separated from the liquid reformate gasoline.

Some experimental results obtained in the demonstration unit, using 100–150 kg naphtha/h feed rate are illustrated in Table 12. The analyses were made using GC-MS techniques. The data show a good *n*-paraffin conversion, while iso- and cycloparaffin contents remain practically unchanged. The aromatic hydrocarbon concentration in the reformate increases in level by 10–15 wt %. Benzene concentration does not exceed 3–4 wt %. Therefore, the composition of the gasoline obtained over zeolite hybrid catalyst is very near that of a reformulated gasoline. Desulfurization level is 79–90%. The typical gaseous fraction composition in wt % is:

H_2	2.07
CH_4	2.54
C_2H_6	4.12
C_2H_4	4.39

Table 12 Results Obtained in the Demonstration Unit. Naphtha feed: 100 kg/h; temperature 340–360°C; pressure: 1 atm

Raw material		Sulfur	Composition (wt %)				
			n-Paraffins	iso-Paraffins	Olefins	Cycloparaffins	Aromatics
Light naphtha	Initial	0.038	22.01	29.00	1.49	32.81	14.65
	Reformate	0.009	9.28	26.98	4.32	30.41	29.00
Heavy naphtha	Initial	0.0165	20.14	25.74	2.8	29.22	22.12
	Reformate	0.0019	3.41	21.36	3.12	37.22	34.88
Hydrotreated heavy naphtha	Initial	—	19.38	26.82	1.20	34.31	17.59
	Reformate	—	2.76	20.82	2.90	37.33	36.39

Source: Ref.[24].

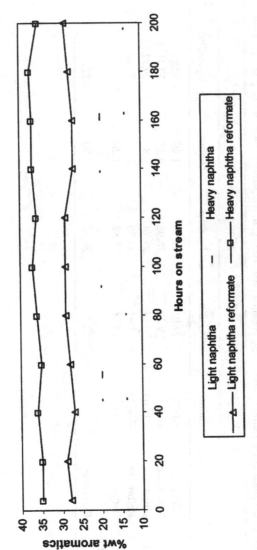

Figure 12 Time on stream catalytic stability for hybrid catalyst.

C_3H_8	44.88
C_3H_6	10.74
$iso\text{-}C_4H_{10}$	13.73
$n\text{-}C_4H_{10}$	12.81
C_4H_8	4.72.

Figure 12 shows a very stable activity of the hybrid catalyst during time on stream.

8 CONCLUSIONS

Zeolitic hybrid catalysts, based on HZSM-5 modified with Zn, exhibit good activity in the naphtha reforming process. The main functions of the catalyst—n-paraffin conversion, isomerization, alkylation, and desulfurization—lead to a reformate with a composition near that of a reformulated gasoline. Burning the deposited coke in air at 500–650°C can easily restore the catalyst activity. One can propose that in a scheme with continuous circulating catalyst, the procedure and catalyst may compete with the classical reforming process based on conventional platinum-type catalyst.

NOMENCLATURE

S_C	selectivity for cracking reactions
S_A	selectivity for aromatization reactions
S_I	selectivity for isomerization reactions
S_{A9}	selectivity for aromatic hydrocarbon alkylation to C_9 aromatics
S_{A10}	selectivity for aromatic hydrocarbon alkylation to C_{10} aromatics
RON	research octane number
ΔH_R	reaction heat
H	enthalpy
$y, \Delta y$	molar fraction
T	temperature
C_p	molar specific heat at constant pressure
M	molecular weight
a, b, c, d	constants for each component in C_p

SUBSCRIPTS

i, 0	initial value
f	final value
s	standard value

c   number of components involved
j   summation index
lq   liquid
g   gas

REFERENCES

1. Mowry, J.R. Arab J. Sci. Eng. **1985**, *10*, 367.
2. Chen, N.Y.; Yan, Y. Ind. Eng. Chem. Proc. Des. Dev. **1986**, *25*, 151.
3. Meriaudeau, P.; Naccache, C. J. Mol. Catal. **1990**, *59*, L31.
4. Mole, T.; Anderson, J.R.; Creer, G. Appl. Catal. **1985**, *17*, 141.
5. Angelescu, E.; Gurau, P.; Pogonaru, G.; Musca, G.; Pop, G.; Pop, E. Rev. Roumaine Chim. **1990**, *35*, 229.
6. Ono, Y.; Osako, K.; Kim, G.-J.; Inoue, Y. *Studies in Surface Science and Catalysis*; Elsevier: Amsterdam, 1994; Vol. 84, 1773 pp.
7. Price, G.L.; Kanazirev, V. J. Catal. **1990**, *66*, 267.
8. Price, G.L.; Kanazirev, V. J. Mol. Catal. **1991**, *66*, 115.
9. Schulz, P.; Baerns, M. Appl. Catal. **1991**, *78*, 15.
10. Hennig, G.; Thiel, F.; Halimeier, K.-H.; Szargan, R.; Hagen, A.; Roessner, F. Spectrochim. Acta **1993**, *49A*, 1495.
11. Dufresne, L.; Yao, J.; Le Van Mao, R. *Chemical Industries, Vol 46: Novel Production Methods for Ethylene, Light Hydrocarbons and Aromatics*; Marcel Dekker: New York, 1992; Paper No. 28.
12. Frunza, L.; Pop, E.; Pop, G.; Ganea, R.; Birjega, R.; Milea, L.; Fota, I.; Ivanov, E.; Plostinaru, D. *Proc. 7th Symp. Heterogeneous Catalysis*; Bulgarian Academy of Science, Bourgas: Bulgaria, 1991; Vol. 1, 301 pp.
13. Musa, M.; Tarina, V.; Stoica, A.D.; Ivanov, E.; Plostinaru, D.; Pop, E.; Pop, G.; Ganea, R.; Birjega, R.; Musca, G.; Paukshtis, A. Zeolites **1987**, *7*, 427.
14. Minacev, M.K.; Charson, M.S. Dokl. Akad. Nauk SSSR **1988**, *300*, 300.
15. Pop, G.; Ganea, R.; Ivanescu, D.; Naum, N.; Birjega, R.; Boeru, R.; Ichim, S.; Natu, N.; Constantin, C.; Olaru, I.; Ionescu, M.; Negroiu, M.; Eparu, E.; Anghelache, C.; Jercan, E.; Juganaru, T. RO Patent $110000B_1$ (13.07.1995).
16. Roesner, F.; Mroczek, U.; Hagen, A. In *New Aspects of Spillov Effect in Catalysis*; Inui, T., Fujimoto, K., Uchijima, T., Masai, M., Eds.; Elsevier: Amsterdam, 1993; 151 pp.
17. Shapiro, E.; Shevchenko, D.P.; Dimitriev, R.V.; Tkachenko, O.P.; Minachev, K.M. In *New Aspects of Spillover Effect in Catalysis*; Inui, T., Fujimoto, K., Uchijima, T., Masai, M., Eds.; Elsevier: Amsterdam, 1993; 159 pp.
18. Aimoto, K.; Fujimoto, K.; Maeda, K. In *New Aspects of Spillover Effect in Catalysis*; Inui, T., Fujimoto, K., Uchijima, T., Masai, M., Eds.; Elsevier: Amsterdam, 1993; 165 pp.
19. Pop, G.; Ivanescu, D.; Ganea, R. In *New Aspects of Spillover Effect in Catalysis*; Inui, T., Fujimoto, K., Uchijima, T., Masai, M., Eds.; Elsevier: Amsterdam, 1993; 137 pp.

20. Pop, G.; Ivanescu, D.; Ganea, R.; Tomi, P. Progr Catal (Bucharest) **1995**, *4*, 15.
21. Business Wire, October 21, 1999.
22. Tsakiris, C.; Tomi, P.; Pop, G. Progr. Catal. (Bucharest) **1994**, *3*, 83.
23. Reid, R.; Poling, B. *The Properties of Gases and Liquids*; McGraw-Hill: New York, 1968; Appendix A.
24. Pop, G.; Ichim, S.; Tomi, P.; Ganea, R.; Ivanescu, D. Progr. Catal. (Bucharest) **1996**, *5*, 31.
25. Sivasanker, S.; Ratnasamy, P. In *Catalytic Naphtha Reforming: Science and Technology*; Antos, G.J., Anitani, A.M., Parera, J.M. Eds.; Marcel Dekker: New York, 1995; Chapter 15.

10

Deactivation by Coking

Octavio Novaro and Cheng-Lie Li*
National University of Mexico
Mexico City, Mexico

Jin-An Wang
National Polytechnic Institute
Mexico City, Mexico

1 INTRODUCTION

Catalytic naphtha reforming processes are accompanied by side reactions leading to carbonaceous deposits both on metallic sites and on the support of the catalyst, causing catalyst deactivation. The carbonaceous deposits on catalysts generally can be divided into different groups: (1) a constant amount of residual carbon; (2) reversible carbon formed instantaneously during operation at working conditions; and (3) irreversibly adsorbed carbon accumulating during reaction after several hours and eventually forming graphitic-like structures.[1,2] Some of the carbon deposits may be beneficial, whereas most of these materials are harmful, depending on the reactions and catalysts.[3] Under working conditions, there exist several different surface compositions and structures on a Pt/Al_2O_3-reforming catalyst: predominantly uncovered ensembles of clean Pt sites, two-dimensional overlayer carbonaceous deposits, and three-dimensional carbon islands.[4a] The presence of coke with different natures, compositions, and structures significantly influences the catalyst activity and frequently alters selectivity and, finally, lifetime. Fortunately, deactivation by coking is usually reversible and the coked catalyst can be refreshed by employing a controlled coke burnoff operation. Thus, the activity and selectivity may be completely or partially restored.

Current affiliation: East China University of Science and Technology, Shanghai, China

Coking processes are complex. They are affected by various factors, including (1) the active metals (metal content, crystallite size, and dispersion); (2) catalyst support (textural properties and acidity); (3) additional promoters (Sn, Re, Ir, Ge, Cl, S); (4) operation conditions (reaction temperature, pressure, H_2/oil ratio, and time on stream) and (5) feedstock properties (molecular weight and structure, hydrocarbon basicity). Clarification of these influences on coke formation is necessary to illuminate coking mechanisms and to facilitate choice of regeneration parameters and optimization of operations. Some important remarks on coke formation are found in the literature.[4b]

In the present chapter, coke locations on the surface of the catalyst and coke distribution across the catalyst pellets and along the catalyst bed are presented. Coke characterization techniques and the influencing factors for coke formation are summarized. Coking mechanisms and kinetics are also discussed. Finally, the effects of coke deposition on the activity and selectivity of the different reactions occurring in the catalytic naphtha reforming process are shown.

2   COKE CHARACTERIZATION TECHNIQUES

One of the major difficulties in the study of deactivation by coking during the reforming process is the limitation in determining the composition and nature of the carbonaceous deposits. The coke usually contains complex mixtures of carbonaceous compounds and unconverted reactants or products. They cannot be related to any single chemical structure, and in most cases it is difficult to analyze their distribution. For overcoming these problems, advanced instrumental techniques are highly necessary. Among those, temperature-programmed oxidation (TPO), ^{13}C NMR, ^{1}H NMR, Fourier transform infrared spectroscopy (FTIR), Raman spectroscopy, X-ray diffraction (XRD), and transmission electron microscopy (TEM) are valuable in the analysis of coke amount, type, composition and structure. Additional descriptive information on some of these techniques can be found in other sources,[4b,4c] and in the chapter on characterization.

2.1   Temperature-Programmed Oxidation

The TPO peak assignments, such as peak temperature and integrated peak area in the exit CO_2 profile, have been widely used to study coke types, location, and combustion behavior. Generally the TPO spectrum is well resolved and consists of low- and high-temperature peaks. The former corresponds to oxidation of the coke formed on the metal function and the other is due to combustion of the coke formed on the acid sites on the support. Figure 1 is a set of TPO profiles of a series of coked bimetallic catalysts.[5]

When coke content is lower than 0.1 wt %, it is difficult to characterize it by a routine TPO technique. However, Querini and Fung[6a] used a modified TPO

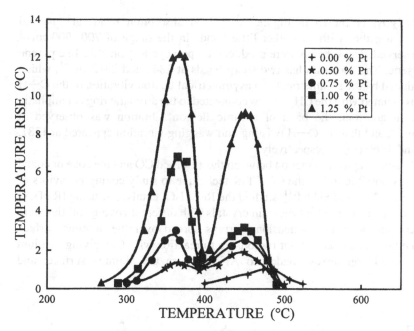

Figure 1   A set of TPO spectra of a series of coked reforming catalysts with various Pt loading. (From Ref.[5].)

method to identify coke type and burnoff behavior by feeding the exit gaseous mixture of coke combustion into a methanator where the Ru catalyst was used to catalyze the conversion of CO_2 to CH_4. This greatly enhanced the sensitivity and stability of the baseline and the resolution of the conventional TPO technique, making it possible to quantify the coke content as low as 0.01 wt %.

Because the coke particle size and morphology show a great influence on TPO profile, by a proper design of TPO experiment in combination with a kinetic model, some information concerning the coke morphology and particles can also be obtained. It was found that the coke deposits with a three-dimensional structure exhibit a coke reaction order increasing from almost 0 to approaching 1 as the oxidation reaction proceeds. By using a linear combination of power law kinetic expressions, coke concentration and coke particle size and number are roughly determined in terms of the TPO behavior.[6b]

2.2   Infrared Spectroscopy

IR has been used to characterize the coke structure and compositions.[7–10] A typical FTIR spectrum of the coke formed on an industrial reforming catalyst

(Pt-Sn/Al$_2$O$_3$) is shown in Figure 2.[10] Two absorption bands at 744 and 871 cm^{-1} together with two other little bands in the range of 700–900 cm^{-1} were observed. These bands were produced by polycyclic aromatics like pyrene or chrysene. The spectrum has two sharp bands at 2846 and 2912 cm^{-1}, which are produced by the symmetrical and asymmetrical flexion vibration of the C—H bonds associated with the CH$_2$ groups connected to the aromatic rings or aliphatic groups. In addition, the band of olefinic flexion vibration was observed at 1606 cm^{-1}, and that for C—H twisting and wagging vibration appeared at 1381, 1374, and 1460 cm^{-1}, respectively.

By studying the interaction between the adsorbed CO and the coke or coked metal, it is possible to use the CO-FTIR technique to study coking behaviors of the catalysts. Yao and Shelef[11] studied the Re/Al$_2$O$_3$ catalyst by using IR of CO adsorption, and two kinds of rhenium crystals with different coking abilities were observed: one with two-dimensional arrays spread over the alumina surface might be less readily coked but responsive to reaction with CO, giving rhenium multicarbonyl complexes bonded to Al^{3+} ions in the alumina surface; and

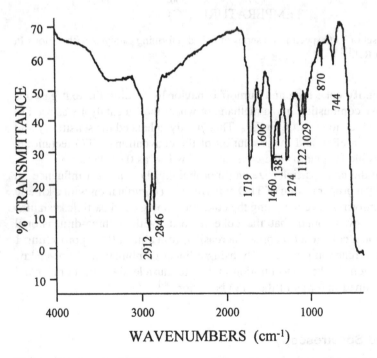

Figure 2 A typical FTIR spectrum of the coke formed on an industrial Pt-Sn/Al$_2$O$_3$ reforming catalyst. (From Ref.[10].)

another with three dimensions, which would be most likely to form linearly adsorbed CO band, which was heavily coked. Anderson et al.[12] reported the IR results of CO adsorption on Pt, Re, and Pt-Re/Al$_2$O$_3$ catalysts before and after coking. About 92% of the CO adsorption was inhibited by coking after 2 min of CO adsorption on coked Pt/Al$_2$O$_3$ compared to the uncoked catalyst. However, on the coked Re/Al$_2$O$_3$, the intensity of CO adsorption band was 52% of that shown in the uncoked sample, indicating that 42% of Re^0 sites for the linear adsorption of CO had been poisoned by coking. When CO adsorbed on the coked Pt-Re/Al$_2$O$_3$, the intensity of the dominant IR band was reduced by 87%, which is slightly lower than that for Pt alone but bigger than that for Re alone. They also found that small patches of uncoked Pt were enlarged after addition of CO due to the mobility of the carbonaceous layer induced by CO adsorption.

2.3 ^{13}C NMR and ^{1}H NMR

^{13}C NMR and ^{1}H NMR techniques are useful for coke characterization because they do not require the separation of the coke deposits from the catalysts. The fairly wide range of chemical shift of the ^{13}C nucleus makes it easy to distinguish among different compounds present in the coked catalysts. Moreover, it is also possible using this technique to obtain information such as mobility of the coke and related host–guest interaction.[13]

Carbon atoms with an sp^2 hybridized state usually present a chemical shift between 82 and 160 ppm, whereas those with an sp^3 hybridized state have a shift between 0 and 82 ppm in the NMR spectrum. Figure 3 shows that the NMR spectrum of coke deposited on an industrial reforming Pt-Sn/γ-Al$_2$O$_3$ catalyst consists of one large absorption peak at around 125 ppm, showing that the coke contains components having polynuclear aromatic and graphite-like structures with an sp^2 hybridized state and negligible amounts of —CH$_3$ side chains with an sp^3 hybridized state.[10] Moreover, by using ^{1}H NMR, different hydrogens (H$_\alpha$, H$_\beta$, and H$_\gamma$) in saturated groups linked in the condensed aromatic rings in the extracted coke are identified.[14–16]

2.4 Raman Spectroscopy

It is well known that in the beginning stage of the reforming process, with a coke amount less than 1 wt %, most of the coke is formed on the metals and some deposits on acid sites in high dispersion with poor crystallization. In this case, it is difficult to apply XRD for coke structure study. However, the laser Raman spectroscopic technique is powerful for coke characterization in the cases of low coke content and graphitization of amorphous coke due to its high sensitivity. It is not necessary to separate the coke from the coked catalysts. Well-crystallized

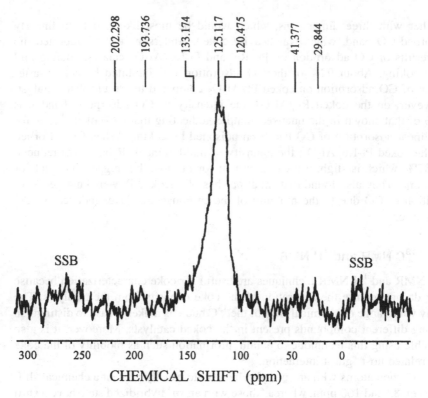

Figure 3 An NMR spectrum of coke deposited on an industrial Pt-Sn/γAl$_2$O$_3$ reforming catalyst. (From Ref.[10].)

graphite usually presents only two Raman-active fundamental modes: one is an interlayer mode at about 42 cm^{-1} and another is an intralayer mode at around 1581 cm^{-1}. Pregraphitic carbons show another broad composite band with a maximum at about 1355 cm^{-1}. The ratio I_{1355}/I_{1581} has been correlated to the size of graphite crystallites or, more precisely, to the average diameter of the aromatic layer.[17,18]

Espinat et al.[19] reported an interesting work concerning the coke formed on both mono- and bimetallic reforming catalysts by laser Raman spectroscopic technique. The coke is shown in a wide range between 0.29 and 27.3 wt % to exhibit a highly aromatic nature and very likely consists of a two-dimensional carbon structure. The alteration of the spectra with the feed composition, H$_2$/oil ratio, and second metal addition clearly reflects the changes in structure, composition, and average crystallite sizes of the coke (Fig. 4).

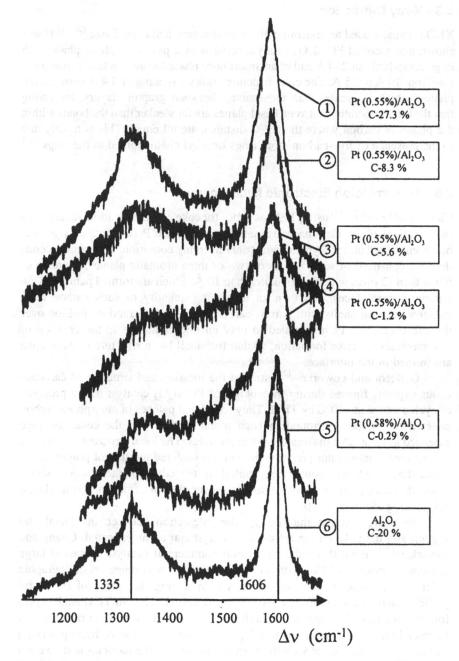

- ① Pt (0.55%)/Al₂O₃ C-27.3 %
- ② Pt (0.55%)/Al₂O₃ C-8.3 %
- ③ Pt (0.55%)/Al₂O₃ C-5.6 %
- ④ Pt (0.55%)/Al₂O₃ C-1.2 %
- ⑤ Pt (0.58%)/Al₂O₃ C-0.29 %
- ⑥ Al₂O₃ C-20 %

1335   1606

1200   1300   1400   1500   1600

$\Delta\nu$ (cm^{-1})

Figure 4   A set of Raman spectra for different coked catalysts. (From Ref.[19].)

2.5 X-ray Diffraction

XRD is usually used to determine the coke structure and crystal size.[20,21] It was shown that a coked Pt/Al_2O_3 catalyst consists of a pseudographitic phase with large crystals about 250 Å and of an amorphous phase located on small clusters of size from 10 Å to 15 Å. The coke structure shows a spacing of 3.4 Å between the planes, which is greater than the spacing between graphite layers, indicating that the forces operating between basal planes are far weaker than the bonds within the planes of carbon where the C—C distance are all equal. This is mostly due to the presence of five-carbon atom rings or alkyl chains joined to the rings.[22]

2.6 Transmission Electronic Microscopy

Cabrol and Oberlin,[23] using TEM as a tool for coke characterization, determined two locations of the carbonaceous materials on a spent Pt/alumina catalyst that had been treated under a typical industrial reforming condition. The carbonaceous deposits consisted of small stacks of two or three aromatic planar structures of fewer than 12 rings with a size smaller than 10 Å, which are formed parallel to the alumina crystal faces, and then an increasing quantity of such carbon units gathered to form shells with porous carbon particles in a random fashion over the entire catalyst. Pt is assumed to play an important role in the creation of intermediates for coke formation, so that the small basic structure units of coke are formed in the interface.

Gallezot and coworkers[24] studied the location and structure of carbonaceous deposits formed during the coking of Pt/Al_2O_3 catalyst in the presence of cyclopentane at 440°C by TEM. They found that patches of amorphous carbon covered the support surrounding each metal particle and the coke coverage extended as far as 20 nm from the given particles. The local structure of the coke was neither graphitic nor pregraphitic, but a disordered mixture of polyaromatic molecules, which was sometimes limited by the edges of the alumina sheets where the metal particles were located. However, the coke frequently spilled onto neighboring alumina sheets.

By using electron microscopy, the interaction between the metal and support and the coke on the reforming catalyst can be investigated. Chang and coworkers[25] found that although a used commercial catalyst contained large amounts of coke (7–12%) with a characteristic of short-range crystallographic order, metal particles were still observed, indicating that some of the metal particles were relatively free of carbonaceous deposits. In situ TEM studies performed on a model Pt-supported catalyst system showed that upon reaction in a hydrocarbon environment, the metal particles were capable of hydrogasifying carbon in their immediate vicinity. During this process the particles underwent a wetting and spreading action at the deposit interface in a narrow temperature range between 480°C and 600°C, which was believed to be a key factor in the maintenance of the small metal particle size during the reforming reactions.[25]

3 LOCALIZATION AND DISTRIBUTION OF COKE

3.1 Locations and Distribution of the Coke on the Catalyst Surface

Coke may deposit on different locations on the catalyst. As shown in Figure 1, two different types of coke with different combustion behaviors formed on different locations are identified: one burned off at low temperature range (200– 400°C) and another at high temperature range (400–550°C), which individually corresponds to the coke produced on the metal and acidic support.[5,26]

On the metal crystals with various coordinations, coke formation is also different. Arteaga and coworkers[27] used CO as probe to monitor the effect of Pt site character on the coke formation on Pt/Al_2O_3 and $Pt-Sn/Al_2O_3$ catalysts that were coked by heat treatment in heptane/hydrogen. Four types of Pt sites with different coordinations on Pt/Al_2O_3 were distinguished: (1) the lowest coordination Pt atoms [$v(CO) < 2030$ cm^{-1}] were free from coke, confirming that these sites, which are at corner or apex atoms on the small Pt particles, were more resistant to coking than the other types of Pt site present; (2) the highest coordination sites in large ensemble of Pt atoms [$v(CO) = 2080$ cm^{-1}] whose exposed low-index planes in Pt/Al_2O_3 were heavily poisoned by coke; (3) the intermediate coordination sites [$v(CO) = 2030-2060$ cm^{-1}] that are possibly at edges or steps of Pt particles were also heavily coked; and (4) the sites in smaller two-dimensional ensembles of Pt atoms exhibiting a bridging bond with CO (2060–2065 cm^{-1}) were partially poisoned by coke.

Espinat et al.[21] observed three zones on a coked catalyst. In the coke-free zone, the sample was identical in all aspects to the fresh catalyst. In the slightly coked zone, a layer of pregraphitic carbon with a thickness of a few atoms was observed. This kind of coke is generally poorly organized. In the highly coked zones, the support is buried in the three-dimensional coke with several hundred angstroms of thickness. These observations confirm that the carbonaceous deposits are greatly heterogeneous. In the case of low coke content, the carbon units gather into the pores of the catalyst and uniformly cover the individual catalyst surface to form a mono- or multilayer of carbon shell. When the coke content is too high, it blocks the mouth of the channel or pores of the catalyst.

3.2 Coke Distribution Across the Catalyst Pellet

Coke is not only formed on the surface of the catalyst but also distributes inside the catalyst pellets. In most cases, the distribution of the coke across the catalyst particles is nonuniform. By using electron microprobe and ion microprobe techniques, the carbon distribution from surface to subsurface or inner layer of the catalyst can be determined. On a $0.67\%Pt-1.47\%Cl/Al_2O_3$ catalyst, two different regions across a coked catalyst pellet were roughly distinguished (Fig. 5): in zone A, which is in the range of $10-100$ μm in catalyst radii, the coke signal is very high, showing that coke concentration reaches the maximum; in zone B, which is in the range of $150-250$ μm, some regions of higher carbon concentration are observed, and others are quite free from coke.[21]

3.3 Coke Distribution Along the Catalyst Bed

Coke buildup along the catalyst bed is mainly determined by coking mechanisms and reaction conditions. Although changes of reaction conditions along the catalyst bed, such as hydrogen partial pressure, temperature, and concentration of each component, usually lead to difficulties in prediction of coke profile, analyses of relationship between the coke profile and influencing variables can provide

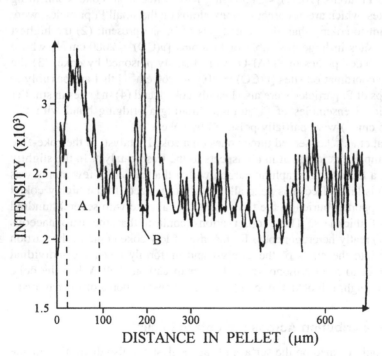

Figure 5 Coke distribution across a catalyst pellet. (From Ref.[21].)

valuable information concerning the coking mechanism and its effects on the catalytic behaviors. Generally, a series mechanism produces an increasing amount of coke throughout the bed, whereas a parallel mechanism usually yields a decreasing coke profile. In more complicated reaction networks, coke deposits may result from a combination of pathways. If the main source of coke deposits is an intermediate compound, the coke profile exhibits a maximum, whereas in a consecutive system A → B → C, if coke is produced from both A and C, there might be a minimum in the coke along the bed.[28a−c,29]

Querini and Fung[29a] studied the coke profile formed along the catalyst bed during n-heptane reforming on both nonsulfided and sulfided Pt and Pt-Re/Al_2O_3 catalysts in a fixed-bed multioutlet reactor under different conditions. They found that the coke profiles strongly depended on the reaction conditions, sulfidation, and catalyst types. The distribution of coke concentration influenced the toluene concentration profile through the catalyst bed. At low pressure (105 kPa), the coke content on the nonsulfided Pt catalyst increases in the catalyst bed from top to bottom; however, at higher pressure (1225 kPa), coke content decreases from inlet to outlet. In a moderate pressure between 105 and 1225 kPa, a maximum in the coke profile is observed near the top one-fourth of the catalyst bed. When Re was added to the Pt catalyst, the maximum of coke profile shifted from the bed bottom to upper section. On the other hand, on the sulfided catalysts, it was found that the effect of sulfur on the coke distribution is opposite to that of Re. Addition of Re to Pt/Al_2O_3 moves the maximum in coke toward the top of the bed; however, addition of sulfur to Pt-Re moves this maximum in the other direction in the catalyst bed.[29b] As shown in Table 1, at the same temperature and pressure, the coke content linearly increases with time on stream in the section of the catalyst bed. At the same reaction condition, the coke content decreases along the catalyst bed from top to bottom.

Coke profile in the catalyst bed is related to the gas-phase composition contacted with catalyst bed. In the upper section of the catalyst bed, the catalysts are exposed mainly to the highest concentration of reforming feed and lowest product fraction. As these fractions change along the bed, this results in a nonuniform coke distribution along the reactor. Since sulfur is slowly stripped from the sulfided catalysts, there is also a sulfur concentration profile that increases along the bed, which leads to a shift toward the bottom of the reactor in the maximum of the coke profile.[29b]

It is noteworthy that coke is formed not only on the catalyst particles but also on the wall surfaces of the reactor. It was found by TEM, TPO, and XRD that this kind of coke is rather different from the one separated from the catalyst surface. This coke is formed from filament-like carbons. At the front end metal particles containing Fe, Cr and Ni have been detected, and Fe is present in large amount.[15] The coke formed on the inner surface of the reactor was catalyzed by the metal reactor itself, which was difficult to remove in a normal regeneration

Table 1 Coke Content as a Function of Time On Stream in Various Sections of the Catalyst Bed

Temp. (K)	Pressure (kPa)	Time (h)	Coke (%)			
			Bed 1.9	Bed 22.1	Bed 55.1	Bed 84.9
772	525	2	0.31	0.34	0.33	0.30
772	525	24	0.99	0.95	0.69	0.58
772	525	34	1.19	1.08	0.83	0.68
772	525	72	2.35	2.18	1.21	0.84
772	525	91	2.43	1.92	1.09	0.97
772	525	120	2.87	2.57	1.53	1.10
772	525	213	4.16	2.81	1.44	1.12
755	525	330	2.06	1.48	1.18	1.05
755	525	240	2.07	1.66	1.32	1.10
755	1225	24	0.57	0.54	0.46	0.40
755	1225	123	1.03	0.74	0.54	0.42

Source: Ref.[29b]

condition. To avoid this kind of coke formation, the wall surfaces of the reactor may require special pretreatment, depending on the nature of the operation.

4 EFFECTS OF CATALYSTS AND OPERATING PARAMETERS ON COKING

4.1 Metallic Platinum Content, Dispersion, and Ensemble Size

Metallic platinum content affects both the amount and nature of coke. Increasing platinum content usually results in an increase in the amount of coke on the catalyst, given the same or similar metal dispersion.[30–32] This is because the metallic function is primarily responsible for the production of the coke precursor species like methylcyclopentadiene.[33]

Coke deposition on metals is a metal structure–sensitive reaction. Some studies have reported that the catalysts with low metal dispersions are more sensitive to autodeactivation by coke deposition than well-dispersed catalysts.[34,35] This is possible as coke formation is preferential on planes rather than on corners and edges of the metallic crystallites.[36–38] From the electronic effect perspective, highly dispersed metals with small crystallites on a support usually show an electron-deficient nature due to the metal–support interaction. When the size of the metallic crystallites decreases, electrons transfer from metal to support causes an electron-deficient character to appear on the metal. This is less favorable to the adsorption of coke precursors like cyclopentadiene, thus producing less coke during the reaction.[34,39,40] Therefore, the small crystallites have greater coke resistance. Coke formation usually requires multiple bonded species, thus requiring high coordination number metal atoms and large ensembles. The most active sites for coking are the face atoms of the metallic crystallites, although corner atoms are also partially covered by coke.[41–45]

Coke formed on the metallic sites can be divided into two types: reversible coke (H/C atomic ratio of 1.5–2.0) and irreversible coke (H/C atomic ratio of about 0.2).[46,47] The reversible coke is more easily removed by hydrogen treatment, whereas the irreversible coke is more graphitic and is much harder to remove with hydrogen. The removal rate by hydrogen for the reversible coke is at least 1000 times higher than that of the irreversible coke.

4.2 Metallic Additives

It has been proven that the addition of a second metal promoter, such as Sn, Re, Ir, In, Ge, and others, improves Pt catalyst stability and enhances the selectivity for the formation of high-octane products.[48a–f] Beltramini and Trimm[48f] found that carbon formation on Pt, Pt-Sn, Pt-Ir, Pt-Re, and Pt-Ge supported alumina catalysts increased with time on stream; however, the carbon amount varied with different catalysts during n-heptane reforming. In the first 2 h, carbon deposition increased in the order Pt-Sn < Pt-Re < Pt-Ge < Pt-Ir < Pt (Fig. 6). Figure 6 shows that bimetallic catalysts present increased ability to resist coking. In a recent work, the deactivation rate for n-hexane conversion caused by both sulfur and coke deposition was found to increase in the order Pt-Ge < Pt ≪ Pt-Sn ≤ Pt-Sn.[49] Table 2 shows the effect of Re and Ir addition on the coke formation and nature.[36]

Burch and Mitchell[50] summarized the various viewpoints presented by different groups regarding the origin of these improvements related to the second metal: (1) the formation of an alloy like Pt-Sn, and Pt-Re, which exhibit different properties either as a result of ensemble or of electron modifications;[51,52] (2) stabilization against sintering;[53] (3) interaction with metal ions of the second element stabilized in the surface of the support;[54,55] (4) increasing hydrogenolysis activity;[56] (5) decreasing the hydrogenolysis activity;[57] (6) suppression of surface carbiding;[58] (7) improving hydrogenation activity;[59] and (8) hydrogenation of coke residues.[60] These conclusions are all directly or indirectly related to coking or decoking processes. The following section mainly focuses on the effects of tin and rhenium additives on coke formation and the interaction between the additives and platinum.

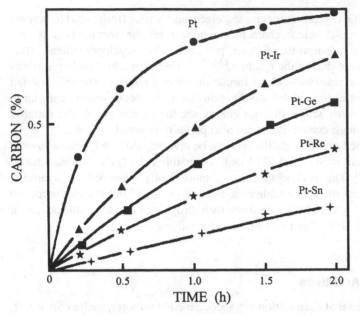

Figure 6 Effects of the second metal additives on the coke formation. (From Ref.[48f].)

Tin Addition

Pt-Sn system is particularly resistant to coke deactivation. The formation of coke precursors is inhibited over Pt-Sn catalyst in n-C_6 reforming because the sites display low activity for the transformation of methylcyclopentene (MCPe), which is the key intermediate for producing coke via consecutive condensation reactions, to methylcyclopentadiene (MCPde), and to further MCPde dehydrogenation. This is consistent with the results obtained by Beltramini and Trimm

Table 2 Effect of Re and Ir Addition on Coke Formation and Nature

Catalyst	Carbon (%)	Graphitic coke (%)	Extractable coke (%)	Products in extractable coke (%) (mol wt < 202)
Pt/Al$_2$O$_3$	1.18	72	28	30
Pt-Re/Al$_2$O$_3$	0.85	78	22	35
Pt-Ir/Al$_2$O$_3$	0.92	75	25	48

Source: Ref.[36].

who postulated that the ability to decrease coke formation results from enhanced gasification of coke precursors by tin.[48f,61]

The coke coverage on tin-doped platinum catalyst is distinctly less than on the monometallic one, although in some cases the total amount of coke is almost the same on both catalysts.[61,62] It is assumed that coke precursors are less strongly adsorbed on the tin-doped platinum sample due to an ensemble effect of tin. Consequently, these coke precursors are more mobile and can more easily migrate to the alumina where they are finally deposited as coke. This "drain-off" effect guarantees that a larger portion of active Pt sites remain free from blocking coke precursors and provide higher activity for the bimetallic samples.

The addition of Sn can also block the lowest coordination Pt sites and destroy large ensembles of Pt by a geometrical dilution effect. Völter and Kürschner[63] reported that the addition of tin strongly modifies the Pt catalyst by forming a Pt-Sn alloy that causes inhibition of hydrogenolysis due to a geometrical effect and retards the deactivation of the catalyst by modifying the coke deposition. The addition of tin did not diminish the amount of coke but increased the ability of Pt-Sn sites to resist coke poisoning by reducing adsorption of coke precursors.

Rhenium Addition

When rhenium is added to platinum, the catalytic stability is significantly improved, extending the operating period of the unit between regenerations. This can be explained by several effects of rhenium: stabilization of the metallic phase on the support and higher resistance to deactivation by coke deposition.[64,65]

Addition of Re may divide the large platinum ensembles into smaller ones, achieving a high metal dispersion and thus inhibiting the coking reaction.[66–69] Addition of Re could also destroy coke precursors by hydrogenolysis, reducing coke formation.[70] This differs from the conclusion stated by Parera et al. that the role of Re is to decrease the dehydrogenation capacities of Pt, and therefore Re addition decreases the coke formation.[71] Table 2 shows that addition of Re not only induces a decrease in the amount of coke deposit by a transformation of the coke precursors into other species, but also induces a modification of the coke nature, leading to more dehydrogenated coke formation and increasing the coke toxicity.[36,72]

Interaction Between Platinum and Additives

Interaction between platinum and the second metal additive produces important influences on both coke formation and its toxicity. The greater the Pt-Re interaction is, the lower the deactivation by coking on the metallic function.[73] During cyclopentane reforming on a nonsulfided Pt-Re/Al$_2$O$_3$ catalyst, the coke deposition for the metallic function and its toxicity decreases when the Pt-Re interaction increases. However, after sulfiding, the deactivation by coking depends on both the Pt-Re interaction and the working pressure. Under normal pressure, the toxicity of the formed coke increases with the degree of the Pt-Re interaction. However, at higher pressure (15 bar) the same catalyst shows less sensitivity to coke deposition.[74]

The interaction between metallic Pt and Re provides activated hydrogen that could migrate and hydrogenate coke precursors on the acidic sites.[75] Also, a strong Pt-Re interaction may inhibit the reverse spillover of hydrogen which was proposed for the elimination of hydrogen from coke precursors adsorbed on the acidic function.[76]

4.3 Catalyst Support

Most of the coke is formed on the acid function of the support. When the acidity of the support decreases in the sequence $Al_2O_3 > TiO_2 > SiO_2 > MgO$, for example, the amount of coke formed on these catalysts shows a similar decreasing sequence from Al_2O_3 to MgO.[76] When the alumina support is modified by other promoters, such as chloride,[77-79] phosphorus,[80] boric acid,[81] or potassium hydroxide,[82] its acidity is enhanced or reduced, resulting in changes in the amount and types of coke produced.

Aranda and coworkers,[83] investigating coking on a supported Pt-niobia catalyst, found that more hydrogenated and lighter coke was formed on the Pt/Nb_2O_5 catalyst than on the Pt/Al_2O_3 due to the reduced density of acidic sites on the former. Table 3 shows the amount of coke and its type for three different Pt-supported catalysts. In addition, effects of support modified by rare earth oxides, such as CeO_2, were also studied.[84,85]

Acidic zeolites with 12 ring pores, such as ZSM-3, ZSM-20, β-zeolite, Y-zeolite, and USY, have also been used as catalyst supports for reforming reactions.[86-90] Recently, platinum-supported basic KL or Ba^{2+}-doped KL

Table 3 Data Relative to the Amount of Coke on Different Catalysts

Type of coke	% wt/wt		
	Pt/Nb_2O_5	Pt/Al_2O_3	Pt-Sn/Nb_2O_5
Total coke	3.7	3.1	1.2
Insoluble	1.9	2.2	—
Soluble	1.8	0.9	—

Source: Ref.[83].

zeolites have been considered as reforming catalysts.[91–95] In comparison with the acidic zeolites, zeolite L provides the highest final activity during reforming of a paraffinic industrial feedstock, indicating that coke is formed preferentially over the acidic sites rather than over the platinum clusters. The reforming activity decreases as $Pt/L > Pt/\beta > Pt/USY$.[96] It was observed that after reaction the Pt/BaKL showed a gray color while the Pt/β and Pt/USY zeolite were black. Under identical conditions, the deactivation rate of USY was higher than β-zeolite due to its more numerous acidic sites.[96] The generation of bulky coke deposits in the supercages of the USY zeolite results in the blocking of its pore, thus preventing the access of the reactants to any internal active sites.

4.4 Operating Conditions

The effect of operating conditions on the performance of reforming catalysts has been discussed in Chapter 2 of this book. Enhancement of gasoline octane number or BTX aromatics concentration in reformates is conveniently achieved in a commercial reforming unit under more severe operating conditions by lowering space velocity (expressed as LHSV) or raising reactor inlet temperature (WAIT) at constant pressures. Experience from commercial reforming units indicates that doubling LHSV requires a WAIT increase of 10°C to keep reformate research octane number (RONC) unchanged in the range of 90–100.[97] Changes in operating parameters also exert an important influence on catalyst coking behaviors. High severity would produce a decrease in reformate yield because of undesired hydrocracking as well as hydrogenolysis reactions. Coke formation is also accelerated and it grows preferably on the support. The coke is more dehydrogenated and graphitic at higher temperatures and lower LHSV under constant pressure and hydrogen-to-oil ratio. The same trend has been observed in increasing time on stream.[98–100]

A reduction in pressure is one other way to increase severity in reforming reactions. In Figure 7, it is shown that doubling of relative severity (RS) from 2 to 4 would boost the reformate RONC from 92.2 to 96.5. Of course, higher severity concomitantly increases catalyst coking rate, shortening the cycle life from 12 months to 6 months. Likewise, at a RONC of 96.2 reducing pressure from 300 to 160 psig will double RS and thus also produce a lowering of the cycle life by half because of deactivation by coke deposition on the catalyst. Contrary to the impact of reaction temperature, Figure 8 shows that a reduction in pressure, which also increases RS, offers benefits of an increase in both reformate and hydrogen yields because it may increase the rate of dehydrogenation and dehydrocyclization while concomitantly decreasing hydrocracking reactions. In addition, it has been demonstrated[36,101,102] that coke is deposited predominantly on the support under increasing pressures that only can be burned off at higher temperatures, while coke deposited under low pressure is more easily removed by regeneration.

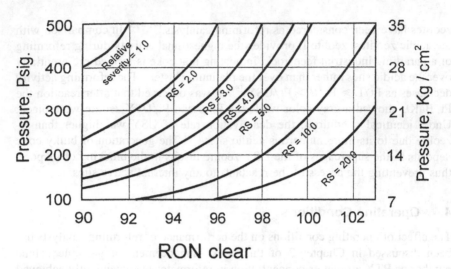

Figure 7 Effect of operating pressure and octane number on catalyst deactivation. (From Ref.[97].)

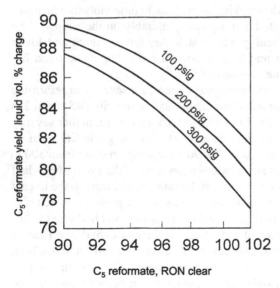

Figure 8 Effect of pressure on reformate yield. (From Ref.[97].)

Therefore, in the development of reforming technology, the initial challenge was to operate under reduced pressure continually. The first approach to accomplishing this lies in the improvement of the catalyst formulation so that reforming processes can still be performed in a semiregenerative fixed bed, but under lower pressure for a cycle time longer than one year. The widely used catalyst in this approach is Pt-Re/Al$_2$O$_3$, which is deactivated at a rather slow rate. A comparison in the coking behaviors between Pt and Pt-Re catalysts is shown in Table 4.[103] It is seen that coke content on the Pt-Re catalysts is somewhat lower than that on the Pt catalyst under the same operating conditions and Re/Pt ratio also has an influence on the coke amount. In addition, it is reported in the literature that Pt-Re catalyst has a much higher coke tolerance.[104] This improvement has also been observed in practice for industrial Pt-Re reforming unit. After 430 days on stream, the amount of coke on the catalyst reached 12%, whereas the activity was maintained at an expense of 8°C temperature increase only.[105]

The other approach is to use a moving-bed or continuous catalyst regeneration system, where a small amount of the coked catalyst is continually drawn out and sent to a regenerator for coke burnoff. In the industrial unit of this system, the residence time of each catalyst particle in the reactor system is only about 4–6 days and the coke content on the coked catalyst increases eventually up to 4–6 wt %. It can be reduced to 0.1 wt % or less in the regenerator and then fed back to the top of the reactor for reuse after some further treatment. In this way the catalyst is always kept at a high activity level. The amount of coke on the catalyst is not a controlling factor in the cycle life in this technology, so that it can be operated at high severity levels. In such a system, WAIT can be enhanced up to 520–535°C and the pressure may be lowered to 0.4 MPa.[106] This system is commonly used for a paraffin-rich feedstock because the reaction converting paraffin to aromatics, dehydrocyclization, is the most difficult to perform in reforming process and is only competitive with other reactions like hydrocracking under the conditions of low pressure and high temperature given above. A test was performed for a paraffin-rich feedstock with 52% carbon number attributed to paraffin. If the test pressure was in the range of 0.35–1.2 MPa,

Table 4 Coke Amount and Re/Pt Ratio on a Sulfided Pt–Re Reforming Catalyst Using n-Heptane as Feed

Re/Pt ratio	0.00	0.82	1.42	1.81	2.30	2.70
Coke amount (wt %)	0.89	0.32	0.21	0.11	0.12	0.12

Source: Ref.[103].

which is commonly used in a moving-bed reforming system, the reaction temperature can be 16–17°C lower to keep the same space velocity and product octane number than that in the range of 1.4–4.2 MPa, which is used in a semiregenerative fixed-bed reforming process. Therefore, continuous reforming is a good choice for such a paraffin-rich feedstock from the viewpoint of liquid yield and coke formation. In fact, a moving-bed approach can be implemented at higher temperature to obtain high aromatic concentration reformate. A Pt-Sn catalyst is preferably used in this approach because incorporation of Sn improves catalyst stability, especially under low pressure and higher temperatures, by hindering coking on activated sites of Pt particles. Of course, increasing coke deposition has also been observed on the alumina support of this catalyst.[107,108] All new reforming units built in 1990–1995 belong to one of these two approaches, about one-half each.[109]

4.5 Reaction Environment Factors

It is very important for a reforming catalyst to have the proper environmental atmosphere to achieve all the required reactions and maintain the catalyst performance. First, recirculating hydrogen is necessary to provide an environment with reasonable hydrogen-to-hydrocarbon ratio (HHC) so as to avoid catalyst deactivation by coking. Second, presulfiding at startup of a run is a key operation for bimetallic catalysts like Pt-Re catalysts in a semiregeneration unit. In addition, the feed must contain a very low concentration of sulfur because it is necessary to control the H_2S content of the hydrogen atmosphere. Third, chlorine-containing compounds should be injected in the feed to maintain the chlorine content at a certain level so as to provide a reasonable water–chlorine equilibrium environment, so that ultimately the reforming catalyst has optimal acidic properties. All of these reaction environmental factors are relevant to deactivation by coking, and therefore will be discussed in the following sections.

Circulating Hydrogen

The main reforming reactions produce hydrogen by dehydrogenation and dehydrocyclization to form aromatics. A hydrogen-rich gas stream must be recycled and passed over the catalyst with the feedstock to reduce coke formation and preserve catalyst activity to permit longer runs between regenerations. In this way, a hydrogen-rich environment is provided to increase hydrogen partial pressure in the reactors under constant operating pressure. Hydrogen reacts with coke precursors, removing them from the catalyst before they can form polycyclic aromatics, which ultimately convert to coke and deactivate the catalyst. Reducing HHC from 8 to 4 increases coke amount by 75%, whereas it increases 3.6 times if HHC ratio changes from 4 to 2 and thus shorten the life cycle remarkably. The decrease in HHC would also affect the nature of coke, which is in line with the reduction in space velocity and total operating pressure.[110]

Sulfur and Sulfurization

Sulfur-containing compounds are poisons for reforming catalysts. When a catalyst is deactivated by poisoning, the reaction temperature must be raised to keep RON of the reformate unchanged. As shown in Figure 9, the presence of

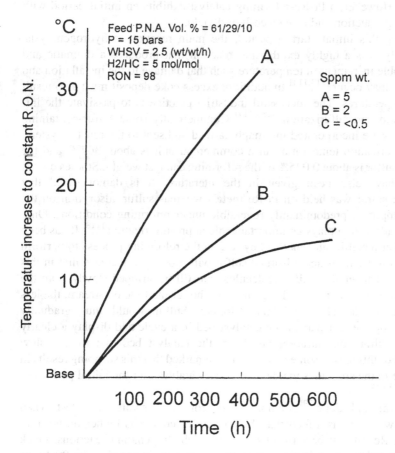

Figure 9 Rate of deactivation versus sulfur content of the feed on Pt/Al_2O_3 catalyst at approximately 50°C. (From Ref.[123].)

sulfur necessitates a higher reaction temperature for the same performance and thus results in a higher coke formation.[110] The simultaneous deactivation by coke and sulfur of bimetallic reforming catalysts was also studied. It was demonstrated that Pt-Sn catalysts showed high resistance to coke deactivation but were severely poisoned by sulfur, whereas for the Pt-Re catalyst, sulfur produces a significant decrease in both.[111]

The maximal sulfur concentration permitted in the naphtha feed to facilitate a stable operation is limited to 20 ppm for Pt/Al_2O_3 and 1 ppm for Pt-Re catalysts, mainly because the aromatic yield of the latter is more sensitive to sulfur.[29,112] It has been mentioned above that the Pt-Re-reforming catalyst is now widely used in modern fixed-bed reforming units because of its improved stability. It allows a much longer cycle life between regeneration under pressure of 15 bars. However, a Pt-Re-reforming catalyst exhibits an initial period with greater gas production and a reduced liquid yield.

During this initial startup period, the main reaction is hydrogenolysis. Hydrogenolysis is a highly exothermic reaction that produces a dramatic and uncontrollable increment in temperature such that damage to the installation and the catalyst may occur.[113,114] In addition, excess coke deposit must also result at high temperatures. The successful industrial practice is to passivate the initial activity by presulfurization.[115–117] Commercially some sulfur-containing compounds were incorporated into naphtha feed and sent to the reactor system. The presulfurization temperature in a commercial unit is about 367°C and the amount of sulfur is about 0.025% of the reforming catalyst weight. Studies on this treatment have also been given in the literature. It is demonstrated that irreversible sulfur was held on Pt-Re metals whereas sulfur adsorption on the alumina support is predominantly reversible under reforming conditions. Only the former adsorption plays an important role in presulfurization.[118] It has been accepted that metallic sites have a key role in the reforming process to perform aromatization reactions and adsorbed sulfur will produce a decrease first in this reaction.[122] Irreversible sulfur molecules can still be stripped slowly from the catalyst during the course of a long run. The shorter the time on stream, the less pronounced is the effect of this stripping. Sulfur would thus gradually accumulate on the rear part of the catalyst bed in a cycle and display a clearly increasing sulfur concentration profile in the catalyst bed due to this slow stripping. For this reason some authors have remarked that this stripping results in an increase in the aromatics yields for a Pt-Re catalyst after the initial periods of start-up.[119,123]

Querini and Fung[29] demonstrated from the results in a test when n-heptane was used as a feed that the amount of coke was higher on the presulfided Pt-Re catalyst because of increasing dehydrogenating reactions. Pieck et al. concluded that presulfurization produced an increase in coke on a Pt/Al_2O_3 catalyst whereas the reverse effect was observed on the Pt-Re catalyst.[74]

On the contrary, other investigators showed that coke decreased after sulfurization.[119–121] The nature of coke is also still a matter of debate. The reasons for these conflicting results and explanations were discussed by Frank and Martino.[110]

Tin has the ability to inhibit hydrogenolysis, so that this presulfurization treatment is not necessary for Pt-Sn catalyst. However, a very low-level sulfur-containing environment is also important in a Pt-Sn catalyst containing moving-bed reforming system. A moving-bed unit is always operated at higher temperature with WAIT up to 530°C and reduced pressure of 0.4 MPa to produce high BTX concentration reformates. Under these severe conditions, it might be possible to produce coke agglomeration on the hot metal surface of reforming reactors, heaters, and catalyst conveying line, if a sulfur-containing hydrogen environment is not provided. Coke amount could increase in this system and thus produce negative effects to reforming reactions. Under the direst of conditions, one can imagine the Pt-Sn-reforming catalysts not able to move smoothly or even blocked in the reactor–regenerator system by this coke formation on the metal surface of the installation. This problem was once experienced in an industrial moving-bed unit in which a deep hydrogenated naphtha was used as the feedstock. Serious coke clogging was observed in reactors and other hot parts of that unit, which made it necessary to shut down immediately.[122] To elucidate the cause of this kind of coke formation, catalysts and coke samples were discharged and studied. It was observed this coke did not deposit on the catalyst but was mainly formed on the metal surface of different parts. Large coke blocks appeared in the bottom of reactors. It was seen from TEM observation that this coke was composed of carbon filaments. Coaxial hollow channels were present within the filaments and the metal particles at their tips were pear shaped. These results are in line with studies on coke formation on bare metal surfaces reported in the literature.[122] In this reforming unit, the naphtha feed had been prehydrogenated in a hydrogenation guard reactor to further remove sulfur-containing compounds. Therefore, the sulfur content in that feed was only 0.02 ppm and not enough to passivate the surface of the metal walls. In such a sulfur-deficient environment with high reforming temperatures, hydrocarbons would be adsorbed on the metal surface and then convert to carbon and encapsulate the metal particles. This carbon deposited and dissolved around the metal particles, and carbon filaments were formed. Coke blocks were eventually produced by further agglomeration of the latter. From the above results it was determined to remove the deep prehydrogenation guard reactor and to incorporate sulfur-containing compounds into the feed to keep 0.5 ppm sulfur content. In this way, this kind of carbon formation problem on the metal surface was solved. This is the other reason that an H_2-H_2S atmosphere environment must also be provided in the whole circuit of a moving-bed reforming unit.

Chlorine and Water Content

Naphtha reforming reactions occur on the bifunctional catalyst. Acidity provided by alumina is not enough and must be adjusted by adding chloride compounds. Therefore, chlorine content is the other important environmental factor and will influence catalyst coking behavior in the reforming process.[124] Ardiles et al. reported the presence of chlorine produces more uniform deposition of Pt over alumina in the Pt/Al_2O_3 catalyst and thus improves the dispersion of the metallic phase.[125] Metal redispersion by chlorine under an oxidizing condition after coke burnoff in the regenerator has been industrially used to refresh activity of the catalyst and was discussed previously.[126] Highly dispersed reforming catalysts are more resistant to deactivation by coke deposition because small platinum crystallites present an electron deficiency due to metal–support interaction resulting in a lower stability of the intermediate adsorbed unsaturated hydrocarbons. On the other hand, coking is connected with catalyst acidic properties that are related to chloride content that promotes polymerization of coke precursors.[127] Therefore, Parera et al. concluded that there is an optimal chloride content on the Pt/Al_2O_3 that allows a reduction of the coke deposit through a mechanism involving the hydrogen spillover from metal to the support.[128]

In the literature, it was also observed that the addition of Cl into Pt catalysts produces a decrease in the first peak on TPO spectra that is attributed to coke combustion on metallic sites.[73] It could be due to a decrease in coke amount deposited on Pt sites or to a decrease in the oxidizing capacity of Pt produced by an inhibiting effect of chloride. Chloride also has an impact on coke deposition for bimetallic catalysts. These authors revealed that chlorine would inhibit Pt-Re interaction. The greater the Pt-Re interaction, the lower the deactivation by coking on metallic sites. Taking cyclopentane as a feedstock in this test, it was also reported that incorporation of Re in the reforming catalyst produces a decrease in coke content as shown in Figures 10 and 11. This change is in line with the concomitant decrease in the percentage of cyclopentene (CP—) and cyclopentadiene (CP=) production, which are considered to be coke precursors. It is also seen that amounts of CP— and CP= change in the same way as amounts of coke at constant chloride content. At different chloride levels, however, coke deposition may be enhanced under lower concentrations of those unsaturated intermediates for the catalyst with 1.1% of Cl. These results elucidate that the acid function is especially important under higher chloride content. As for Pt-Sn-reforming catalysts, it was reported that the effect of chlorine in improving dispersion is less pronounced than that of the monometallic Pt catalyst, but it has the additional effect on Pt-Sn alloy formation and thus lower platinum site deactivation by coking.[129]

Residual water must be maintained in the industrial naphtha feed. An alumina-supported reforming catalyst requires moisture to activate the acid

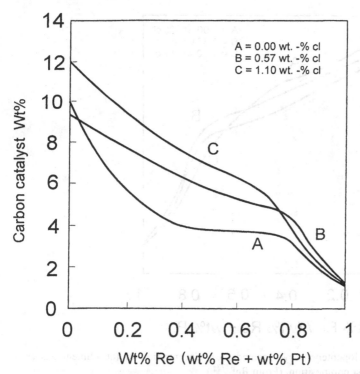

Figure 10 Carbon deposition on catalysts after the run as function of catalyst composition. (From Ref.[73].)

function and provide homogeneous chloride content over the whole catalyst bed. When the environmental atmosphere is too wet, however, chloride in the catalyst may be leached off and thus deteriorate its acidic behavior. Therefore, water control must be performed along with chloride control to maintain a proper chloride–water balance in the environmental atmosphere. From industrial practice, the chloride content on the catalyst should be kept in the range of 0.9–1.2 wt% for the modern bimetallic catalysts. To meet this requirement an environmental atmosphere of 1–5 ppm of hydrogen chloride and 10–20 ppm of water should be provided in the circulating gas over the bimetallic reforming catalysts.[130] At lower Cl level, WAIT will need to be raised to maintain octane number level and hence increase coke deposition. When chloride content is higher than 1.2%, catalyst acidity would be too strong and thus drastically increase the coking rate and shorten the cycle life of a reforming catalyst. Because of the importance of maintaining balance, relationships were given to predict and adjust the amount of chloride that should be injected in the reactor to

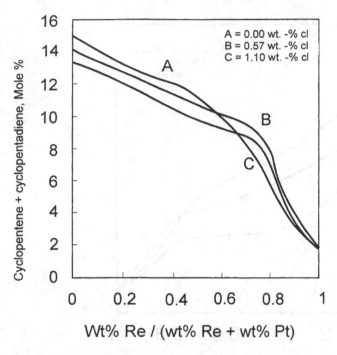

Figure 11 (Cyclopentene + cyclopentadiene) concentration in the effluent gas as a function of catalyst composition. (From Ref.[73].)

provide an optimal chloride-containing environment in various stages of an industrial unit.[131]

4.6 Feed Compositions

Coking is dependent on the structure, boiling point, and basicity of the feedstocks and reformates. The usual feed to the naphtha reforming unit contains 45–70% paraffins, 20–50% naphthenes, 5–15% aromatics, and less than 2% olefins. During the reforming process, the amount of aromatics increases to 60–75%, paraffins and naphthenes decrease to 20–45% and 1–10%, respectively. The olefins almost disappear. Due to different properties, the hydrocarbons produce different effects on coke deposits. Generally, heavy cuts produce more coke.

   The n-paraffins are the main components of naphtha feed, and their coking capacity is important in studying the effects of feed nature on coking process. Parera et al.[132,133] made a detailed measurement of coke formation on pure hydrocarbons and naphtha doped with several series of hydrocarbons over different catalysts. They found that light paraffins produce a small amount of cyclopentanes, which are great coke precursors. The heavy paraffins, if they produce bicyclic compounds with an indenic structure, also served as coke producers. Coke formation on several catalysts fed with naphtha and naphtha doped with 10% of different hydrocarbons is shown in Table 5.

Beltramini et al.[134] reported that when the naphtha is doped with paraffins, e.g., n-heptane, a minimum in coke deposition is formed. Coking deactivation increases, in general, when the number of carbon atoms is higher than 8. Considering aromatics, catalyst deactivation increases with the length of the paraffinic chain linked to the aromatic ring. The ring with five carbon atoms is much more easily polymerized than the one with six carbon atoms; therefore, the 5C atoms ring is an important coke precursor. If the feed contains high concentration of aromatic compounds with 5C atoms ring, more coke will be produced and the catalyst deactivation is rapid.

Olefins are usually poor coke precursors. However, diolefins with conjugated double bonds have greater coking capacity than that of paraffins and monoolefins but are still smaller than that of cyclopentadiene ring. If naphtha contains a ring with 5C atoms, coke formation and catalyst deactivation are higher than for a ring with 6C atoms due to the higher reactivity of 5C atom ring with dienes to form heavy aromatics.

Ginosar and Subramaniam[135] found that oligomers are important coke precursors during the 1-hexene reforming on a Pt/Al_2O_3 catalyst. When the 1-hexene feed was diluted with n-pentane and the mixture was run at supercritical conditions, oligomer concentration was significantly reduced, diminishing coke formation.

Since most coke on the support is formed by acid–base–catalyzed reactions, the basicity of the feedstock will have an important influence on coke formation. The relationship between these factors was established many years ago. Figure 12 shows the coking rates of a series of aromatic feedstocks as a function of their basicity.[36]

5 COKING MECHANISMS

Several assumptions concerning coking mechanisms appeared in the earlier literature. Some of these results contradict others particularly in the arguments on determination of the long-term deactivation by coking. Shum et al.[137] stated that affecting the metallic function by coke controls the long-term deactivation of a reforming catalyst, which is in agreement with the results of Lieske[69] who believed that coking is controlled by a function of metal and not alumina support since the coke formation on alumina without metal remains negligible during the 1-hexane adsorption. Lieske suggested two coke formation routes on the metal function: (1) C_1 route: coke formed from C_1 species adsorbed on a small part of Pt

Table 5 Coke Formation on Several Catalysts Fed with Naphtha and Naphtha Doped with 10% Different Hydrocarbons

Doping agent	Catalyst					
	Pt/Al$_2$O$_3$	Pt-Re-S/Al$_2$O$_3$	Pt-Ge/Al$_2$O$_3$	Al$_2$O$_3$-Cl	SiO$_2$-Al$_2$O$_3$	Pt/SiO$_2$
None	3.36 (−)	2.55 (−)	2.60 (−)	0.20 (−)	0.70 (−)	1.01 (−)
CP	3.75 (+)	3.08 (−)	3.20 (−)	0.20 (−)	0.68 (−)	1.20 (−)
CPe	9.20 (++)	5.68 (+)	5.96 (+)	0.54 (−)	1.88 (−)	2.50 (−)
CPde	19.12 (+++)	17.10 (++)	18.30 (++)	11.50 (+)	20.28 (+)	5.83 (−)
CPde	3.50 (+)	3.52 (+)	3.50 (+)	0.20	1.65	
MCPe	7.10 (++)	5.13 (+)	5.94 (+)	1.09	3.15	
MCPde	18.20 (+++)	17.06 (++)	18.61 (+)	8.04	17.38 (+)	
CH	3.31	2.87	2.49	0.43	1.47	1.10
CHe	3.40	3.70	4.35	0.99	4.10	
CHde	3.40	3.84	2.61	4.83	1.47	1.10
Bz	3.20	1.94	1.58	0.36	0.80	
n-C5	4.09	3.15	2.58	0.22	0.88	
1-n-C$_5$e	4.35	3.15	2.58	0.22	0.88	
1,3-n-C$_6$de	13.10			6.00	9.21	
n-Propyl Bz	9.75 (++)	8.80 (+)			1.09	
1,2,4-TM Bz	3.80	3.75			1.00	
Indene	29.78	22.34			24.12	
Indane	9.30	4.45			2.76	
Naphthalene	4.24	4.94			4.37	

CP, cyclopentane; CPe, cyclopentene; CPde, cyclopentadiene; M, methyl; CH, cyclohexane; CHe, cyclohexene; CHde, cyclohexadiene; Bz, benzene; n-C5, normal pentene; n-C$_5$e, normal pentene; n-C$_6$de, normal hexadiene; TM Bz, trimethylbenzene. (+) or (−), positve or negative detection of heavy aromatics (indene; ArC$_{10}$$^+$).
Source: Ref.[132].

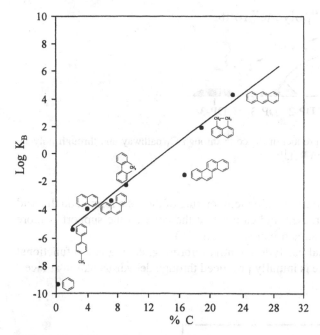

Figure 12 Effect of the basicity of the hydrocarbons on coke formation. (From Ref.[136].)

atoms with high coordination number is mostly formed in the early stage of the reaction and it is a relatively slow process. Nevertheless a very small of amount of such coke can significantly alter the selectivity of the catalyst. (2) Polyene route: more coke was formed on Pt from C_6 hydrocarbons in a fast process that leads to significant Pt coverage with coke. During the steady state of coke formation, the coke transformation from Pt to support is the determinant step. A typical model for production of coke through C_1 pathway on platinum and through polyene pathway is shown in Figure 13.[41,138,139]

The viewpoint of metallic sites controlling long-term deactivation was questioned by Margitfalvi who found the long-term deactivation of Pt/Al_2O_3 was greatly influenced by the deactivation of the acid sites.[140] Parera and coworkers claimed that the long-term deactivation is due to deactivation of the acid function, which determines the length of the cycle.[141] During an initial lineout period, the metal function is partially covered by coke and deactivated. Figure 14 shows that the amount of coke on the metal remains constant along all of the operational cycle after the lineout period, and afterward the coke continues to deposit on the acid function, causing final deactivation.[142] The nature of coke formed on the metal also remains the same during the reaction

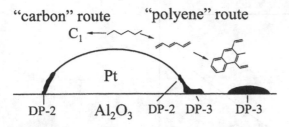

"carbon" route   "polyene" route

Figure 13 A model for production of coke through C_1 pathway and through polyene pathway on platinum. (From Ref.[137].)

of 208 days; however, the combustion temperature of the coke formed on the acid support is gradually increased, indicating that the coke on the support is more graphitic as the time on stream increases (Fig. 15).

In fact, in industrial catalytic naphtha reforming, coking is a bifunctional control process. The coke is initially produced through dehydrogenation at faces,

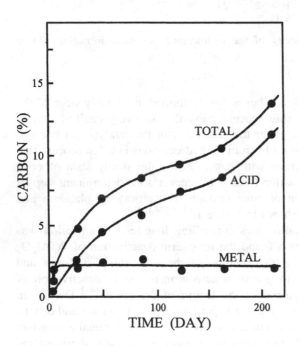

Figure 14 The amount of coke formed on the metal and acid functions. (From Ref.[141].)

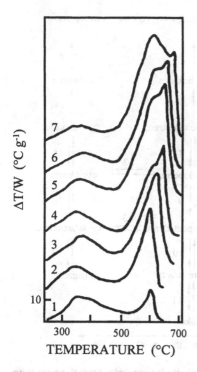

Figure 15 The combustion behaviors of the coke formed on the metal and acid functions. (From Ref.[141].) 1) 4 days; 2) 26 days; 3) 49 days; 4) 87 days; 5) 12 days; 6) 161 days, and 7) 208 days.

edges, and corners of the metallic particles and then diffuses over the support surface in the form of carbide until reaching the interface between platinum and support. These carbonaceous materials can progress to acid sites to begin the polymerization reaction that transforms it into pseudographite. The aromatic products on the reforming catalyst could further dehydrogenate and polymerize to polynuclear aromatic structures and act as precursors in coke formation.[143] A relatively complete reaction network involving coke formation on both metallic and acidic functions is shown in Figure 16.[71]

6 COKING KINETICS

As stated earlier, coke formation is not only governed by the catalyst composition and morphology but also by operating variables, at the level of both the catalyst

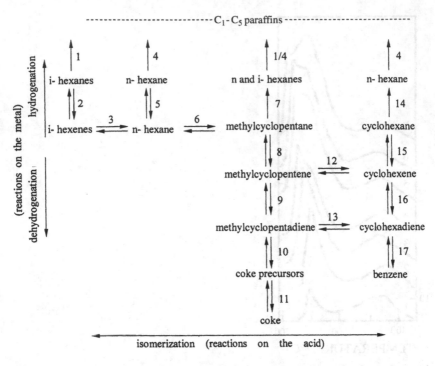

Figure 16 A relatively complete reaction network including coke formation on both metallic and acidic functions. (From Ref.[71].)

particles and the reactors. These complexities cause additional difficulties in kinetic studies and limit the wide application of some models based on relatively simplified assumptions.

Froment[144] developed a general and fundamental kinetic framework on the coke deactivation, that can be used to study site coverage, coke growth and blockage in pores and networks of pores, and diffusional limitations in catalysts. By using a gravimetric reactor, Mipville[145] found that after an initial period, coking on an equilibrated Pt/Al_2O_3 catalyst obeys the following rate law:

$$r_{coke} = A \cdot P_{H2}^{-1} \cdot P_{feed}^{0.75} \cdot C_{coke}^{-1} \cdot \exp\left(\frac{-37000}{RT}\right)$$

where C is the amount of coke, A is a preexponential term, P_{H2} is partial pressure of H_2, P_f is partial pressure of feed, and T is reaction temperature in Kelvin units. Coking kinetics allows evaluation of coke level for any given time, feedstock, pressure, and reaction temperature.

Under an accelerated deactivation condition, the relationship between the amount of coke formed on Pt/Al_2O_3 catalyst and several operational conditions could be expressed as the following:[146]

$$\%C = 4.99 \times 10^6 e^{-8955/T} P^{-0.94} \text{WHSV}^{-1.28} \left(\frac{H_2}{\text{naphtha}}\right)^{-1.33}$$

This equation shows that the coke amount increases by increasing the temperature and by decreasing pressure, space velocity, and H_2-to-naphtha ratio.

The coking rates of the reversible and irreversible coke on metallic function, r_{rev} and r_{irre}, are different, which may be expressed as follows:[46]

$$r_{rev} = k_{rev} \cdot PM \cdot f(a)\varphi_{\frac{1}{3}}(a) = \frac{[M_0]k_p[P]a\varphi_{\frac{1}{3}}(a)f(a)}{1 + K_H[H_2]^{1/2} + K_A[A] + K_B[B] + K_p[P]}$$

$$r_{irre} = k_{irr} C_{rev}\varphi_{\frac{1}{3}}(a)$$

where the function $\varphi_{\frac{1}{3}}(a)$ accounts for the probability of finding platinum ensembles containing at least 2–4 Pt atoms, which is the minimal site requirement for typical coking mechanisms. Function $f(a)$ accounts for the higher coking rate on the more energetic sites, which are covered initially. This accounts for the effect of the metal atom ensembles on coking.

From the practical viewpoint, lumping kinetic models may be more suitable for application in the naphtha reforming process.[47,148] Recently, some lumping kinetic models, which were extended to include coking deactivation, have been developed.[149–151] Jiang et al. developed a coking deactivation model based on a 16-lump kinetic model. The authors claimed that this model could be used to predict the product distribution before and after accelerated coking deactivation and to determine the effects of the operation parameters on coke deactivation.

Ramage et al. developed a 13-lump kinetic model in which the catalyst deactivation by coking was taken into account.[152] This model included hydrocarbon conversion kinetics on fresh catalysts and the modification of the kinetics with time on stream due to catalyst deactivation. A set of three reactors were used to test this kinetics model and results show that it can be used to predict the product distribution, activity, and selectivity.

Tasker and Riggs[153] reported a detailed kinetic scheme involving 35 pseudocomponents connected by a network of 36 reactions within C_5–C_{10} range in a semiregenerative catalytic naphtha reformer, which was modeled by using Hougen–Watson Langmuir–Hinshelwood reaction rate expressions. Deactivation of the catalyst was modeled by including the corresponding equations for coking kinetics. This model was benchmarked with the industrial data to ensure that it adequately represented the actual plant variables at a base operating point.

7 EFFECTS OF COKING ON ACTIVITY AND SELECTIVITY

The overall reforming process involves various reactions. Isomerization, cyclization, and aromatization are desired but the hydrocleavage reactions are not. The metallic function is mainly responsible for hydrogenation–dehydrogenation and hydrogenolysis activity, while the acidic support provides activity for the cracking, isomerization, and aromatization.[20,154] Coking on the metallic function and on the acidic function will directly affect the selectivity of the above reactions.

Since the reactions of dehydrocyclization and hydrogenolysis are mainly controlled by the metallic functions; therefore, the selectivity of these reactions must relate to the coking process on metal. The rates of both reactions usually undergo a rapid initial deactivation and then remain constant up to the end of the operation cycle. However, the relative activity of n-heptane dehydrocyclization to toluene is found to be a simultaneous combination of monometallic mechanism and a bifunctional one controlled by acid function. The conversion rate is initially deactivated by a small amount of coke on metallic sites, and the subsequent linear deactivation should correspond to the bifunctional acid–controlled mechanism.[155]

Isomerization is a typical bifunctional reaction controlled by the acid function of the catalyst. The catalytic activity of the n-pentane isomerization linearly decreases with the amount of carbon deposited on the acid function of the catalyst. Hydrocracking reaction usually shows a large deactivation resulting from a small amount of deposited carbon and then deactivates linearly with carbon accumulation. The initial deactivation coincides with loss of the metallic function and the coking of the acid function causes linear deactivation.[133]

An important chemical property of the carbonaceous deposit on the catalyst is its ability to store and exchange hydrogen with reacting surface species, producing hydrogen spillover from metallic sites to adsorbed species or the reverse from adsorbed species. This also influences the selectivity and activity. Electronic or geometrical effects may explain the coke effect on activity and selectivity. Coke can donate electron charge to Pt that enhances electron density of Pt, and this favors reactions of weaker secondary C—H bonds more than stronger primary C—H bonds in hydrocarbon reactions, altering the isomerization products distribution.[129,156] However, the geometrical effect of coke may be dominant in comparison with the electronic one. The blocking of low-coordination Pt sites by coking decreases the aromatization selectivity.[129]

8 CONCLUSIONS

Although coking is a very complex process, some important conclusions can be obtained based on the studies of model and industrial reforming catalysts and processes. These can be summarized as follows:

Coke can be formed on both the metallic function and on the acid sites on the support. During the lineout stage, coke is mainly deposited on the metal particles. Then it transfers to the acid sites on the support where a more graphitic-like structure is produced through complex acid-base catalyzed reactions. Coke can be formed on the metal function through C_1 or/and polyene pathways.

Coke distributions on the metal crystallites, catalyst support across the catalyst pellet, reactor inner surface, and within the catalyst bed are greatly heterogeneous. Coke buildup along the catalyst bed is mainly determined by reaction conditions and coking mechanisms.

Metallic content, crystallite size, and dispersion show important influences on coke formation. Increasing metal loading generally leads to an increase in coke amount. The high metallic dispersion can reduce the coke amount. Coke is preferentially formed on the larger metallic crystallites with high coordination. The metallic crystallites with low coordination show more resistance to coking than the others.

The structure, surface area, and acidity of the catalyst support are important factors in coke formation. Increasing acidity of the support may increase coke amount and alter coke combustion behaviors.

The coke formation is affected by the addition of promoters. The second metal additives, such as tin, rhenium, and others, can inhibit coke accumulation and modify the coke nature by means of electronic or geometrical effects. Metal addition may destroy the coke precursors and induce a preferential desorption of coke on the support and thus improve the catalyst stability.

The presence of chlorine produces more uniform deposition of Pt over alumina in the Pt/Al_2O_3 catalyst and thus improves the dispersion of the metallic phase. However, as chloride on the support it increases coke deposition. There is an optimal chloride content on the reforming catalyst, which allows a reduction of the coke deposit through a mechanism involving the hydrogen spillover from metal to the support and better metallic dispersion. The presence of sulfur poisons catalyst active sites, altering the resultant coke nature and location. Presulfurization is an effective and necessary method to passivate the hydrogenolytic hyperactivity of the reforming catalyst in the initial periods of startup. Compositions of the naphtha-reforming feedstock are another factor affecting coke formation. Generally heavier cuts produce more coke, and the higher the basicity of the feed hydrocarbons, the more the coke amount on the catalysts.

Reaction parameters such as temperature, pressure, time on stream, and hydrogen-to-oil ratio are closely related to coke depositions. Increasing reaction temperature usually causes an increase in coke amount. When the reaction is under high pressure, coke shows a more dehydrogenated and graphitic nature. Decreasing space velocity, similar to decreasing the H_2/HC ratio at constant pressure, produces more coke deposition on the support.

SYMBOLS

A	pre-exponential term
$[A]$	concentration of reactant in gas phase
$[B]$	concentration of product in gas phase
C	coke amount
k_{rev}	coking rate constant for the reversible coke
k_{irre}	coking rate constant for the irreversible coke
K_H	equilibrium chemisorption constant for reactant
K_A	equilibrium chemisorption constant for product
K_B	equilibrium chemisorption constant for hydrogen
K_p	equilibrium chemisorption constant for coke precursors
P_{H2}	partial pressure of hydrogen
P_f	partial pressure of feedstock
$[P]$	gas concentration of the coke precursor
$[M_0]$	initial concentrations of the metal sites
r_{coke}	coking rate
r_{rev}	rate of reversible coke formation
r_{irr}	rate of irreversible coke formation
T	reaction temperature, K
$\varphi 1/3(a)$:	the probability of finding platinum ensembles containing at least 2–4 Pt atoms
$f(a)$	function describing exponential decay of activity with increasing metal surface coverage

REFERENCES

1. Matusek, K.; Wootsch, A.; Zimmer, H.; Paál, Z. Appl. Catal. A: Gen. **2000**, *191*, 141.
2. Garin, F.; Maire, G.; Zyade, S.; Zauwen, M.; Frennet, A.; Zielinski, P. J. Mol. Catal. **1990**, *58*, 185.
3. Menon, P.G. J. Mol. Catal. **1990**, *59*, 207.

4. (a) Dives, S.M.; Zaera, F.; Somorjai, G.A. J. Catal. **1982**, *77*, 439; (b) Marécot, P.; Barbier, J. In *Catalytic Naphtha Reforming: Science and Technology*; Antos, G.J., Aitani, A.M.; Parera, J.M., Eds.; Marcel Dekker: New York, 1995; 279 pp; (c) Davis, B.H.; Antos, G.J. In *Catalytic Naphtha Reforming: Science and Technology*; Antos, G.J., Aitani, A.M., Parera, J.M., Eds.; Marcel Dekker: New York, 1995; 113 pp.

5. Augustine, S.M.; Alameddin, G.N.; Sachtler, W.M.H. J. Catal. **1989**, *115*, 217.

6. (a) Querini, C.A.; Fung, S.C. Appl. Catal. **1994**, *117*, 53; (b) Querini, C.A.; Fung, S.C. Catal. Today **1997**, *37*, 277.

7. Eberly, P.E.; Kimberlin, C.N.; Miller, W.H.; Drushel, H.V. Ind. Eng. Chem. **1966**, *5*, 193.

8. Best, D.A.; Wojciechowski, B.W. J. Catal. **1977**, *47*, 11.

9. Parmaliana, A.; Frusteri, F.; Nesterov, G.A.; Paukshtis, E.A.; Giordano, N. In *Catalyst Deactivation*; Delmon, B., Froment, G.F., Eds.; Elsevier: Amsterdam, 1987; 198 pp.

10. Li, C.L.; Novaro, O.; Bokhimi, X.; Muñoz, E.; Boldú, J.L.; Wang, J.A.; Lopez, T.; Gómez, R.; Batina, N. Catal. Lett. **2000**, *65*, 209.

11. Yao, H.C.; Shelef, M. J. Catal. **1976**, *140*, 392.

12. Anderson, J.A.; Chong, F.K.; Rochester, C.H. J. Mol. Catal. A: Gen. **1999**, *140*, 65.

13. Kalinina, N.G.; Poluboyarov, V.A.; Anufrienko, V.F.; Ione, K.G. Kinet. Catal. **1986**, *27*, 215.

14. Parera, J.M.; Figoli, N.S.; Beltramini, J.N.; Churin, E.J.; Cabrol, R.A. *Proceedings of the 8th International Congress on Catalysis*; Berlin, 1984; 93 pp.

15. Liu, F.; Shen, G.Q.; Li, C.L. J. East China Inst. Chem. Technol. **1993**, *19* (2), 138.

16. Li, C.L.; Novaro, O.; Muñoz, E.; Boldú, J.L.; Bokhimi, X.; Wang, J.A.; Lopez, T.; Gómez, R. Appl. Catal. A: Gen. **2000**, *199*, 211.

17. Tuinstra, F.; Koenig, J.L. J. Chem. Phys. **1970**, *53* (3), 1126.

18. Nakamizo, M.; Honda, H.; Inagaki, M. Carbon **1978**, *16*, 281.

19. Espinat, D.; Dexpert, H.; Freund, E.; Martino, G.; Couzi, M.; Lespade, P.; Cruege, F. Appl. Catal. **1985**, *16*, 343.

20. Barbier, J. Appl. Catal. **1986**, *23*, 225.

21. Espinat, D.; Fround, E.; Dexpert, H.; Martino, G. J. Catal. **1990**, *126*, 496.

22. Figoli, N.S.; Beltramini, J.N.; Querini, C.A.; Parera, J.M. Appl. Catal. **1986**, *26*, 39.

23. Cabrol, R.A.; Oberlin, A. J. Catal. **1984**, *89*, 156.

24. Gallezot, P.; Leclercq, C.; Barbier, J.; Marecot, P. J. Catal. **1989**, *16*, 164.

25. Chang, T.S.; Rodriguez, N.M.; Baker, R.T.K. J. Catal. **1990**, *116*, 164.

26. Beltramini, J.N.; Wessel, T.J.; Datta, R. In *Catalyst Deactivation*; Bartholomew, C.H., Butt, J.B., Eds.; Elsevier: Amsterdam, 1991; 119 pp.

27. Arteage, G.J.; Anderson, J.A.; Rochester, C.H. Catal. Lett. **1999**, *58*, 189.

28. (a) Butt, J.; Petersen, E.E. *Activation, Deactivation and Poisoning of Catalyst*; Academic Press: London, 1998; (b) Froment, G.F.; Bishoff, K.B. Chem. Eng. Sci. **1962**, *17*, 105; (c) Acharya, D.R.; Ghassemi, M.R.; Hughes, R. Appl. Catal. **1990**, *58*, 53.

29. (a) Querini, C.A.; Fung, S.C. J. Catal. **1993**, *141*, 389; (b) Querini, A.; Fung, S.C. J. Catal. **1996**, *161*, 263.

30. Beltramini, J.N.; Datta, R. React. Kinet. Catal. Lett. **1991**, *44* (2), 345.
31. Beltramini, J.N.; Churin, E.J.; Traffano, E.M.; Franck, J.P. Appl. Catal. **1985**, *19*, 169.
32. Barbier, J. Appl. Catal. **1985**, *23*, 225.
33. Datka, J.; Eischens, R.P. In *Catalyst Deactivation*; Bartholomew, C.H., Butt, J.B., Eds.; Elsevier: Amsterdam, 1991; 127 pp.
34. Barbier, J.; Gorro, G.; Zhang, Y. Appl. Catal. **1985**, *13*, 245.
35. Marecot, P.; Churin, E.; Barbier, J. React. Kinet. Catal. Lett. **1988**, *37*, 233.
36. Barbier, J.; Chuin, E.; Maécot, P. J. Catal. **1990**, *126*, 228.
37. Davis, S.M.; Zaera, F.; Somorjai, G.A. Appl. Catal. **1982**, *77*, 439.
38. Somorjai, G.A.; Blakely, D.M. Nature **1975**, *258*, 580.
39. Gallezot, P., Datka, J.; Massardier, J.; Primet, M.; Imelik, B. *Proceedings of 6th International Congress on Catalysis*; Hightower, J.W., Ed.; London, 1976; 696 pp.
40. Barbier, J.; Marecot, P. Appl. Catal. **1986**, *102*, 21.
41. Trimm, D.L. Appl. Catal. **1983**, *5*, 263.
42. Christmann, K.; Ertl, G. Surf. Sci. **1976**, *60*, 366.
43. Van Broekhoven, E.H.; Schoonhoven, J.W.F.M.; Ponec, V. Surf. Sci. **1985**, *156*, 899.
44. Ponec, V. Adv. Catal. **1983**, *32*, 149.
45. Van Broekhoven, E.H.; Schoonhoven, J.W.F.M.; Ponec, V. Surf. Sci. **1985**, *156*, 899.
46. Biswas, J.; Gray, P.G.; Do, D.D. Appl. Catal. **1987**, *32*, 249.
47. Salmeron, M.; Somorjai, G.A. J. Phys. Chem. **1982**, *86*, 341.
48. (a) Standard Oil Company. US Patent 2,848,377, 1953; (b) French Patent 2,372,883, 1976; (c) UOP. US Patent 3,511,888, 1970; (d) CFR (Compagnie Française de Raffinage), French Patent 2,031,984, 1969; (e) Chevron, US Patent 3,415,777, 1968; (f) Beltramini, J.; Trimm, D.L. Appl. Catal. **1987**, *32*, 71.
49. Borgna, A.; Garetto, T.F.; Apesteguía, C.R. Appl. Catal. A: Gen. **2000**, *197*, 11.
50. Burch, R.; Mitchell, A.J. Appl. Catal. **1983**, *6*, 121.
51. Baucaud, R.; Barbier, J.; Blanchard, G.; Charcosset, H. J. Chem. Phys. **1980**, *77*, 387.
52. Rice, R.W.; Lu, K. J. Catal. **1982**, *77*, 104.
53. Zhdan, P.A.; Kuenestsov, B.N.; Shepelin, A.P.; Kovalchuk, V.I.; Yermakov, Yu.I. React. Kinek. Catal. Lett. **1981**, *18*, 267.
54. Muller, A.C.; Engelhard, P.A.; Weisang, J.E. J. Catal. **1979**, *56*, 65.
55. Burch, R.; Carla, L.C. J. Catal. **1981**, *71*, 348.
56. Carter, J.L.; McVicker, G.B.; Weissman, M.; Kmak, W.S.; Sinfelt, J.H. Appl. Catal. **1982**, *3*, 327.
57. Bacaud, R.; Bussiere, P.; Figueras, F. J. Catal. **1979**, *69*, 399.
58. Rasser, J.C.; Beindorff, W.H.; Scholten, J.J.F. J. Chem. Res. **1979**, 62.
59. Zhorov, Yu.M.; Panchenkov, G.M.; Kartashev, Yu.N. Kinet. Catal. **1981**, *22*, 1058.
60. Johnson, M.F.L.; LeRoy, V.M. J. Catal. **1974**, *35*, 434.
61. Beltramini, J.; Trimm, D.L. Appl. Catal. **1987**, *31*, 113.
62. Lieske, H.; Sárkány, Völter, J. Appl. Catal. **1987**, *30*, 69.
63. Völter, J.; Kürschner, U. Appl. Catal. **1983**, *8*, 167.
64. Barbier, J. Stud. Surf. Sci. Catal. **1987**, *34*, 1.
65. Grau, J.M.; Parera, J.M. Appl. Catal. **1991**, *70*, 9.

66. Tennison, S.R. Chem. Br. **1981**, 536.
67. Dowden, D.A. Chem. Soc. Spec. Publ. Catal. **1978**, 1.
68. Shum, V.K.; Butt, J.B.; Sachtler, W.M.H. J. Catal. **1986**, *96*, 371.
69. Shum, V.K.; Butt, J.B.; Sachtler, W.M.H. J. Catal. **1986**, *96*, 126.
70. Carter, J.L.; McVicker, G.B.; Weisshan, K.; Hmak, W.S.; Sinfelt, J.H. Appl. Catal. **1982**, 327.
71. Parera, J.M.; Beltramini, J.N.; Querini, C.A.; Martinell, E.E.; Churin, E.J.; Aloe, P.E.; Figoli, N.S. J. Catal. **1986**, *99*, 39.
72. Barbier, J.; Marecot, P.; Pieck, C.L. In *Catalyst Deactivation*; Bartholomew, C.H., Fuentes, G.A., Eds.; Elsevier: Amsterdam, 1997; 327 pp.
73. Pieck, C.L.; Marecot, P.; Prera, J.M.; Barbier, J. Appl. Catal. A: Gen. **1995**, *126*, 153.
74. Pieck, C.L.; Marecot, P.; Barbier, J. Appl. Catal. A: Gen. **1996**, *145*, 323.
75. Traffano, E.M.; Parera, J.M. Appl. Catal. **1986**, *28*, 193.
76. Barbier, J. Appl. Catal. **1986**, *23*, 225.
77. Artraga, G.J.; Anderson, J.A.; Rochester, C.H. J. Catal. **1999**, *187*, 219.
78. Augustine, S.M.; Alameddin, G.N.; Sachtler, W.M.H. J. Catal. **1989**, *115*, 217.
79. Gallezot, P.; Leclercq, C.; Barbier, J.; Marécot, P. J. Catal. **1989**, *116*, 164.
80. Parmaliana, A.; Frusteri, F.; Nesterov, G.A.; Paukshtis, E.A.; Giordano, N. In *Catalyst Deactivation*; Delmon, B., Froment, G.F., Eds.; Elsevier: Amsterdam, 1987; 197 pp.
81. Barbier, J.; Elassal, L.; Gnep, N.S.; Guisnet, M.; Molina, W.; Zhang, Y.R.; Bournonville, J.P.; Franck, J.P. Bull. Soc. Chim. Fr. **1984**, *9–10*, 1250.
82. Svajgl, O. Int. Chem. Eng. **1972**, *12* (1), 55.
83. Aranda, D.A.G.; Alfonso, J.C.; Frety, R.; Schmal, M. In *Catalyst Deactivation*; Bartholomew, C.H., Fuentes, G.A., Eds.; Elsevier: Amsterdam, 1997; 335 pp.
84. Fan, Y.-N.; Xu, Z.-S.; Zang, J.-L.; Lin, L.-W. In *Catalyst Deactivation*; Bartholomew, C.H., Butt, J.B., Eds.; Amsterdam, 1991; 683 pp.
85. Parmaliana, A.; Frusteri, F.; Nesterov, G.A.; Paukshtis, E.A.; Giordano, N. In *Catalyst Deactivation*; Delmon, B., Froment, G.F., Eds.; Elsevier: Amsterdam, 1987; 197 pp.
86. Dossi, C.; Tsang, C.M.; Sachtler, W.M.H. Energy Fuels **1989**, *3*, 468.
87. Martens, J.A.; Tielen, M.; Jacobs, P.A. Stud. Surf. Sci. Catal. **1985**, *46*, 49.
88. Smirniotis, P.G.; Ruckenstein, E. Catal. Lett. **1993**, *17*, 341.
89. Smirniotis, P.G.; Ruckenstein, E. J. Catal. **1993**, *140*, 526.
90. Smirniotis, P.G.; Ruckenstein, E. Chem. Eng. Sci. **1993**, *48*, 3263.
91. Fang, X.-G.; Li, F.-Y.; Luo, L.-T. Appl. Catal. A: Gen. **1996**, *146*, 297.
92. Sugimoto, M.; Katsuno, H.; Hayasaka, T.; Hirasawa, K.; Ishiawa, N. Appl. Catal. **1993**, *106*, 9.
93. Hughes, T.R.; Buss, W.S.; Tamm, P.W.; Jacobson, R.J. Stud. Surf. Sci. Catal. **1986**, *28*, 725.
94. Tamm, P.W.; Mohr, D.H.; Detz, C.R. Energy Progress **1987**, *7* (4), 215.
95. Hill, J.M.; Cortright, R.D.; Dumesic, J.A. Appl. Catal. A: Gen. **1998**, *168*, 9.
96. Smirniotis, P.G.; Ruckenstein, E. Appl. Catal. A: Gen. **1995**, *123*, 59.
97. Little, D.M. *Catalytic Reforming*; Penn Well Publishing: Tulsa, OK, 1985; 88 pp.

98.   Barbier, J.; Churin, E.; Marecot, P.; Menezo, J.C. Appl. Catal. **1988**, *36*, 277.
99.   Bakulin, R.A.; Levinter, M.E.; Unger, F.G. Neftekhimiya **1974**, *14* (5), 707.
100. Parera, J.M.; Figoli, N.S.; Traffano, E.M.; Beltramini, J.N.; Martinelli, E.E. Appl. Catal. **1983**, *5*, 33.
101. Barbier, J. In *Catalyst Deactivation*; Delmon, B., Fromont, G.F., Eds.; Elsevier: Amsterdam, 1987; 1 p.
102. Pieck, C.L.; Parera, J.M. Ind. Eng. Chem. Res. **1989**, *28*, 1785.
103. Liang, W.J. *Petroleum Chemistry*; Petroleum University Publisher: China, 1995; 342 pp.
104. (a) Zhorob, Y.M.; Panchenkov, G.M.; Kertashev, Y.N. Kinet. Catal. **1986**, *22*, 1058; (b) Biswas, J.; Bickle, G.M.; Gray, P.G.; Do, D.D.; Barbier, J. Catal. Rev. Sci. Eng. **1988**, *30* (2), 161.
105. Wang, C.D.; Zhang, D.Q.; Sun, Z.L. Petroleum Proc. Petrochem. (China) **2000**, *31* (7), 13.
106. Huang, G.H.; Shan, Q.S.; Zhu, Y.Q.; Yang, S.H. Ibid. **2000**, *31* (7), 5.
107. Barias, O.A.; Holmen, A.; Blekkan, E.A. J. Catal. **1996**, *158*, 1.
108. Larsson, M.; Hulten, M.; Blekkan, E.A.; Anderson, B.D. J. Catal. **1996**, *164*, 44.
109. Ji, L. Petroleum Proc. Petrochem. (China) **1997**, *28* (9), 32.
110. Marecot, P.; Barbier, J. In *Catalytic Naphtha Reforming*; Antos, G.J., Aitani, A.M., Parera, J.M., Eds.; Marcel Dekker: New York, 1995, 279 pp.
111. Borgna, A.; Garetto, T.F.; Apesteguia, C.R. Appl. Catal. A: Gen. **2000**, *197*, 11.
112. Van Trimpont, P.A.; Marin, G.B.; Froment, G.F. Appl. Catal. **1985**, *17*, 161.
113. Parera, J.M.; Verderone, R.J.; Pieck, C.L.; Traffano, E.M. Appl. Catal. **1986**, *23*, 15.
114. Apesteguia, C.R.; Barbier, J. J. Catal. **1982**, *78*, 352.
115. Kluksdahl, H.E. US Patent No. 3,617,520, 1971.
116. Jacobson, R.L. US Patent No. 3,578,582, 1971.
117. Lowell, P.F. U.S. Patent No.3,565,789, 1971.
118. Apesteguia, C.R.; Barbier, J. J. Catal. **1982**, *78*, 252.
119. Augustina, S.M.; Alameddim, G.N.; Sachtler, W.M.H. J. Catal. **1989**, *115*, 217.
120. Coughlim, R.W.; Hasan, A.; Kwakami, K. J. Catal. **1988**, *88*, 163.
121. Wilde, M.; Stolz, T.; Feldhamsa, R.T.; Ander, K. Appl. Catal. **1987**, *31*, 99.
122. Figueiredo, J.L. *Progress in Catalyst Deactivation*; Figueiredo, J.L., Ed.; 1982; 45 pp.
123. Shum, V.K.; Butt, J.B.; Sachtler, W.M.H. Appl. Catal. **1984**, *11*, 151.
124. Gjervan, T.; Prestvik, R.; Totdal, B.; Lyman, C.E.; Holmen, A. Catalyst Today **2001**, *65*, 163.
125. Ardiles, D.R.; de Miguel, S.R.; Castro, A.A.; Scelza, O.A. Appl. Catal. **1986**, *24*, 175.
126. Fernandez-Garcia, M.; Chong, F.K.; Anderson, J.A.; Rochester, C.H.; Haller, G.L. J. Catal. **1999**, *182*, 199.
127. Cooper, B.J.; Trimm, D.L. In *Catalyst Deactivation*; Delman, B., Froment, G.F., Eds.; Elsevier: Amsterdam, 1980; 63 pp.
128. Parera, J.M.; Figoli, N.S.; Jablonski, F.L.; Sad, M.R.; Beltramini, J.N. Stud. Surf. Sci. Catal. **1980**, *6*, 571.
129. Arteaga, G.J.; Anderson, J.A.; Rochester, H. J. Catal. **1999**, *187*, 219.

130. Little, D.M. *Catalytic Reforming*; Penn Well Publishing: Tulsa, OK, 1985; 108 pp.

131. Dai, C.Y. Acta Petrolei Sinica (Petroleum Processing Section) **1995**, *11* (3), 96.

132. Parera, J.M.; Verderone, R.J.; Querini, C.A. In *Catalyst Deactivation*; Delmon, B.; Froment, G.F., Eds.; Elsevier: Amsterdam, 1987; 135 pp.

133. Parera, J.M.; Querini, C.A.; Beltramini, J.N.; Figoli, N.S. Appl. Catal. **1987**, *32*, 117.

134. Beltramini, J.N.; Cabrol, R.A.; Churin, E.J.; Figoli, N.S.; Martinelli, E.E.; Parera, J.M. Appl. Catal. **1985**, *17*, 65.

135. Ginosar, D.M.; Subramaniam, B. J. Catal. **1995**, *152*, 31.

136. Appleby, W.G.; Gibson, J.W.; Good, G.M. Ind. Eng. Chem. Proc. Des. Dev. **1962**, *1*, 102.

137. Shum, V.K.; Butt, J.B.; Sachtler, W.M.H. Appl. Catal. **1984**, *11*, 151.

138. Sarkany, A.; Lieske, H.; Szilagyi, T.; Toth, L. *Proceedings of the 8th International Congress on Catalysis*; Berlin, 1984; Vols. 1 and 2, 613 pp.

139. Luck, F.; Aejyach, S.; Marie, G., *Proceedings of the 8th International Congress on Catalysis*; Berlin, 1984; 234 pp.

140. Margitfalvi, J.; Göbölös, S. Appl. Catal. **1988**, *36*, 331.

141. Parera, J.M.; Querini, C.; Figoli, N.S. Appl. Catal. **1988**, *44*, L1-L8.

142. Parera, J.M. In *Catalyst Deactivation*; Bartholomew, C.H., Butt, J.B., Eds.; Elsevier: Amsterdam, 1991; 103 pp.

143. Parera, J.M. In *Catalyst Deactivation*; Delmon, B., Froment, G.F., Eds.; Elsevier: Amsterdam, 1987; 135 pp.

144. Froment, C.F. In *Catalyst Deactivation*; Bartholomew, C.H., Butt, J.B., Eds.; Elsevier: Amsterdam, 1991; 53 pp.

145. Mipville, R.L. In *Catalyst Deactivation*; Bartholomew, C.H., Butt, J.B., Eds.; Elsevier: Amsterdam, 1991; 151 pp.

146. Figoli, N.S.; Beltramini, J.N.; Martinelli, E.E.; Sad, M.R.; Parera, J.M. Appl. Catal. **1983**, *5*, 19.

147. Ramage, M.P.; Graziani, K.R. Chem. Eng. Sci. **1980**, *35* (1), 41.

148. Froment, G.F. AIChE J. **1975**, *21* (6), 1041.

149. Jiang, H.-B.; Wong, H.-X. J. East China Univ. Sci. Technol. **1998**, *24* (1),42.

150. Jiang, H.-B.; Wong, H.-X. J. East China Univ. Sci. Technol. **1998**, *24* (1), 46.

151. Xie, X.-A.; Ping, S.-H.; Lui, T.-J. China Refinery Design **1996**, *26* (2), 44.

152. Ramage, M.P.; Graziani, K.R.; Schipper, P.H.; Krambeck, F.J.; Choi, B.C. Adv. Chem. Eng. **1987**, *13*, 193.

153. Tasker, U.; Riggs, J.B. AIChE J. **1997**, *43*, 740.

154. Molina, W.; Guisnet, M.; Barbier, J.; Gnep, N.S.; Elassal, L. *First Franco-Venezuelan Congress*, 1993.

155. Gates, B.C.; Katzer, J.R.; Schuit, G.C. *Chemistry of Catalytic Processes*; McGraw-Hill: New York, 1979; 1984 pp.

156. Hlavathy, Z.; Tétényi, P. Surf. Sci. **1998**, *410*, 39.

11

Catalyst Regeneration and Continuous Reforming Issues

Patricia K. Doolin, David J. Zalewski, and Soni O. Oyekan
Marathon Ashland Petroleum, LLC
Catlettsburg, Kentucky, U.S.A.

1 INTRODUCTION

Catalytic reforming technologies and processes have been the subject of numerous discussions and reviews over the past 50 years.[1–8] Since the pioneering work of Vladimir Haensel and UOP[9] on platinum on alumina catalyst and platforming in the 1940s, catalytic reforming technology has evolved from semiregenerative, fixed-bed processes to the more energy efficient, highly reliable, low operating cost, continuous catalyst reforming units.[10] Typical process objectives are the upgrading of low octane naphthas to high octane gasoline blending components, production of aromatics and chemical feedstock, and production of hydrogen. Over the past five decades, numerous advancements have been made with respect to key reforming processes and in the development of high-performance catalysts. In the current environment of low-sulfur gasoline and ultralow-sulfur diesel regulations for reducing air pollutants,[11] catalytic reformers and in particular, continuous reformers have become indispensable as support to refiners' plans for installation of enhanced processes for hydrotreating catalytic cracker feed, distillates, and catalytic cracker gasoline. Refiners will be relying on their catalytic reformers for reliable hydrogen supply for the various desulfurization processes required to achieve near-zero-sulfur gasoline and ultralow-sulfur diesel.

To ensure that catalytic reformers perform reliably, both reactor and catalyst conditions must be maintained. Catalyst activity is largely determined by the effectiveness of the regeneration process. Poor catalyst regeneration will lead to low product yields and increased operating expense. Good regenerations can improve catalyst performance, reduce operating costs, and lower catalyst management costs. This chapter will review key features of commercial catalytic reformers and relate specific catalyst properties that influence unit performance.

2 CATALYTIC REFORMING PROCESSES

Reforming processes are usually classified as semiregenerative, cyclic (fully) regenerative, and continuous (moving-bed) regenerative processes based on their mode of operation, severity, and mode of regeneration. A short review of the reforming processes is provided as reference for the topics on catalyst regeneration issues covered in this chapter.

2.1 Semiregenerative Process

The semiregenerative (SR) reforming process is characterized by operations at relatively low severity as defined by a combination of unit operating conditions, naphtha quality, and catalyst management to extend the cycle length or the time between catalyst regenerations.[12,13] Semiregenerative reformers as shown in Figure 1 contain three to five reactors in series with intermediate heaters to provide necessary heat requirements for the endothermic reforming reactions. Cycle lengths of semiregenerative, fixed-bed reformers are usually determined by one of the following end-of-cycle criteria. The end of cycle can be determined by the reactor metallurgy temperature limit, a prescribed weighted average inlet temperature (WAIT) increase, a specified amount of C_5+ yield decline, a specified amount of hydrogen decline, and refinery and reformer economics. At the end of the cycle, naphtha-reforming operations are discontinued, oil is taken out of the unit, and the catalyst is regenerated in situ in an optimal fashion using a specific regeneration and startup procedure. Ex situ catalyst regenerations are feasible, but most refiners typically conduct in situ catalyst regenerations. Since catalysts can be optimally regenerated using specific procedures for the semiregenerative reformers, a variety of catalysts can be used in the semiregenerative reformers. Catalysts used in the reformers include platinum, platinum/rhenium, platinum/ tin, platinum/germanium, platinum/iridium, staged catalytic systems, and trimetallic catalysts.[14–16] In addition, a variety of catalyst shapes have been used successfully with no detrimental catalytic performance since catalysts are not transported from one reactor to another in a semiregenerative reformer.

2.2 Cyclic Regenerative Process

Cyclic regenerative reformers are operated under more severe process conditions relative to semiregenerative reformers and can process more paraffinic or "harder

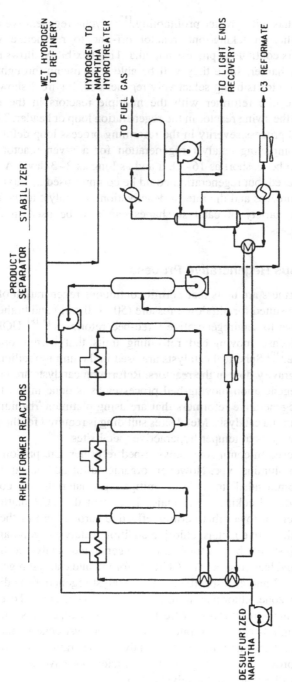

Figure 1 Semiregenerative reformer: Chevron's Rheniformer II.

to reform" naphthas with greater profitability.[17] Cyclic regenerative reformers have the flexibility of taking one reactor off-line to regenerate while the remaining reactors continue to process naphtha. This flexibility allows refiners to push the catalyst harder, since they will be able to maintain throughput while catalyst in one reactor is being selectively regenerated. Figure 2 shows the key features of the cyclic reformer with the multiple reactors in the reforming process loop and the swing reactor in the regeneration loop or header. The type of catalyst used and process severity in the reforming process loop define the time permitted for completing catalyst regeneration for a given reactor. Catalyst regenerations can be as short as 16–18 h and as long as 2–3 days! A variety of catalysts can be used, but regeneration could be compromised due to constraints imposed by the process and the rate of deactivation of catalyst in the reforming process loop. A variety of catalyst shapes can also be used in the cyclic regenerative process.

2.3 Continuous Regenerative Process

Continuous reformers are units that permit continuous regeneration of catalysts and are therefore suited for ultralow-pressure (50–150 psig) and higher severity operations relative to semiregenerative fixed-bed reformers.[4,18] UOP and IFP (now Axens) license moving-bed reforming units that permit operating at ultralow pressure.[19] Spherical catalysts are used in the unit to facilitate catalyst circulation and gravity flow in the reactors. Reforming catalysts are not sulfided in continuous regeneration moving-bed processes as is done in most cases for fixed-bed semiregenerative reformers that are using platinum/rhenium alumina and platinum alumina catalysts. Metal sites sulfiding is required for the platinum/rhenium-type catalysts to temper hyperactive metal sites.

Catalysts used in continuous moving-bed reformers are predominantly of the platinum/tin alumina type. However, organic sulfur addition in the feed is sometimes recommended to aid in mitigating materials of construction (metallurgy)–induced coking.[20] A major feature of the CCR platformer is a dedicated regenerator tower that permits effective carbon burn of the catalysts, redispersion of the active metals, chloride addition, catalyst drying, and catalyst environment adjustment before transporting regenerated catalyst to the catalyst reduction zone and lead reactor. The CCR platformer and catalyst regenerator are shown in Figures 3 and 4, respectively. The catalyst regenerator is divided into the carbon burn zone, chlorination zone, and the drying zone. To ensure that catalyst activity in the reactor is near fresh catalyst activity, catalyst circulation is maintained as high as per design rates and spent catalyst coke is maintained at less than 7 wt %. The time required for catalyst regeneration may vary from 6 to 8 h, and improvements in regeneration efficiencies have led to achieving excellent metals dispersion and catalyst activity.

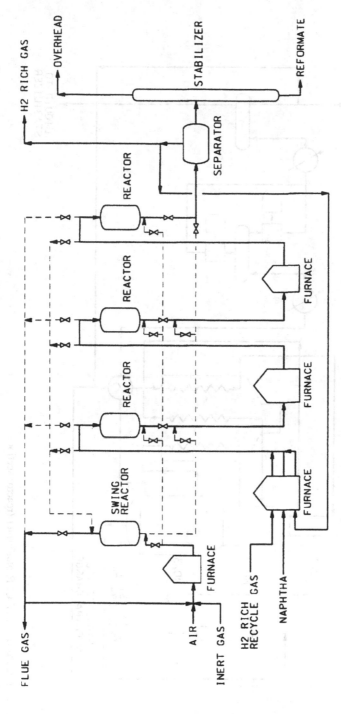

Figure 2 Cyclic regenerative reformer: Amoco's Ultraformer.

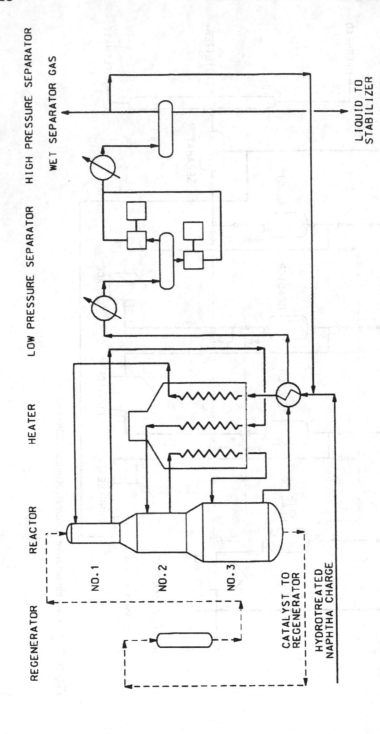

Figure 3 UOP CCR platformer (reactor section).

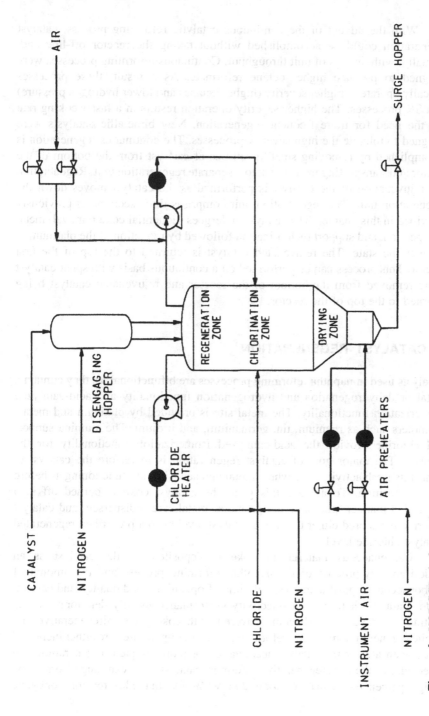

Figure 4 UOP CCR platformer (regenerator).

With the advent of the continuous catalytic reforming process, catalyst regeneration could be accomplished without taking the reactor off-line and, normally, without loss of unit throughput. Continuous reforming processes were designed to produce higher octane reformate. As a result, these processes typically operate at higher severity (higher octanes and lower hydrogen pressure) than SR processes. The higher severity operation results in a faster coking rate and the need for more frequent regeneration. New bimetallic catalysts were designed to tolerate the high-severity processes. The continuous regeneration is accomplished by removing small quantities of catalyst from the bottom of the reactor and transporting the catalyst to a separate regeneration unit. Regeneration and rejuvenation of the catalyst is performed as the catalyst moves down the regeneration unit. The regeneration unit comprises three sections as previously described in this chapter. The catalyst undergoes sequential coke removal, metal redispersion, and support rechlorination followed by reduction of the platinum to the metallic state. The rejuvenated catalyst is returned to the top of the first reactor. This process can be performed on a continuous basis with spent catalyst being removed from the bottom of the reactor and rejuvenated catalyst being returned to the top of the reactor.

3 CATALYST REGENERATION

Catalysts used in naphtha reforming processes are bifunctional, as they contain a metal or dehydrogenation and hydrogenation functionality and acid-catalyzed isomerization functionality. The metal site is provided by platinum and metals promoters such as rhenium, tin, germanium, and iridium. The alumina support and chloride provide the acid-catalyzed isomerization functionality for the catalyst. The major goal of catalyst regeneration is to restore the catalyst to almost its fresh activity state where metal and acid sites are functioning as before coke deposition. To achieve this goal, the catalyst coke is burned off in a controlled manner, platinum and promoter metals are redispersed, and catalyst chloride is restored either to a fresh catalyst level or to a prescribed regenerated catalyst chloride level.

Carbonaceous material or coke is deposited on the catalyst as an undesirable by-product of the naphtha reforming process and the amount of carbonaceous deposit increases with time of operation, feed quality, and catalyst state. Catalytic activity and selectivity performance usually deteriorate as the platinum and acid sites become covered with coke. In semiregenerative and cyclic regenerative fixed-bed reformers, reactor temperatures are either increased to compensate for the catalyst deactivation or more frequent regenerations are effected. For the semiregenerative reformer, catalyst coke can range from a few weight percent on the lead reactor to 20+ wt % coke in the last reactor. For cyclic regenerative reformers, the decision to regenerate a reactor or reactors is usually based on economics dictated by a decline in catalytic performance for one or more of the reactors in the reforming process loop. In the case of the continuous reformer, operating with a platinum/tin catalyst, the spent catalyst coke is usually restricted to less than 7 wt % to permit more stable regenerator tower operation and more economical operation of the reformer. The nature of deposited coke and its chemistry have been discussed extensively.[21,22] Baker et al.[23] and Weisz and Goodwin[24] reported on the burnoff characteristics of the coke in the presence of platinum and transition metal oxides, respectively.

In this chapter, critical issues and challenges of in situ catalyst regeneration are discussed with some emphasis on some of the specific differences for the three reforming processes. The basic steps required for regenerating deactivated platinum–containing catalysts are similar; however, some critical differences exist with respect to the specific catalyst in use, process flexibility, and concerns about unit safety issues.

To regenerate a reforming catalyst, variations of the following steps are conducted:

1. Reactor/catalyst purge using recycle gas and/or nitrogen to remove hydrocarbon and hydrogen to render the catalyst batch, reactor, or reactors safe for handling and subsequent coke burn
2. Controlled coke burn at temperatures higher than 750°F in a mixture of nitrogen and air
3. Oxidation or oxychlorination for metals dispersion and chloride addition
4. Nitrogen purge/drying
5. Metals reduction
6. Metals sulfiding.

3.1 Reactor/Catalyst Purge

In a semiregenerative reformer, catalyst regenerations are usually performed in the reactor vessel without moving the catalyst. The feed to the reformer is stopped and recycled hydrogen gas is used to cool the reactor (<400°F) and act as a purge to remove any residual hydrocarbons.[30] Hydrogen is replaced with a nitrogen purge to ensure a nonreducing environment in the reactor. This step is necessary as a safety measure to guarantee subsequent safe and uniform burn of the catalyst coke and preparation for heat-up of the catalyst to the desired carbon burn temperature. The nitrogen purge step is standard procedure for semiregenerative and cyclic regenerative fixed-bed reformers.

For UOP CCR platformers, a fixed amount of catalyst moves continuously through the unit from the reactors to the regenerator section as determined by the catalyst circulation rate. Spent catalyst from the last reactor flows to a catalyst collector and a hydrogen purge of the spent catalyst is conducted to remove entrained hydrocarbons and gases. The catalyst is then transferred through a number of vessels, pipings, and valves to the regenerator tower via a disengaging hopper. During catalyst transportation from the catalyst collector, the transition is made from a reductive hydrogen atmosphere to one conducive to safe catalyst regeneration in the regeneration tower. Before the final transfer of catalyst to the regenerator, there is provision for substantial reduction of catalyst fines in the disengaging hopper.

3.2 Coke Burn

In an SR unit, spent catalyst is heated in nitrogen at a rate of about 50–100°F per hour to a target 750°F temperature. Primary burn is usually initiated at 750°F and the reactor inlet maintained there while monitoring the temperature rise at any thermocouple in the reactor bed to ensure that it does not exceed 850°F or 100° greater than the reactor inlet temperature of 750°F. A small controlled amount of oxygen is introduced to the reactor system to initiate coke burning. Air is often used as an oxygen source. Oxygen concentrations should be maintained at 0.5–2.0% in nitrogen. Temperatures must be carefully maintained to prevent a runaway exothermal reaction that could result in damage to the catalyst support and excessive sintering of the metal crystallites. Catalyst coke is removed in a series of steps where the temperature or oxygen concentration is increased until there is no evidence of coke on the catalyst.

In cyclic regenerative reformers, coke burn temperatures are maintained over a broad temperature range of 750–1050°F. Catalyst coke burnoff at temperatures greater than 1050°F should be avoided so as to minimize rapid surface area losses due to alumina phase transitions. γ-Alumina transition to the δ phase is known to be initiated at temperatures in excess of 850°C (1562°F), and further transitions can occur to the θ (starting at about 1900°F) and α (starting at 2084°F)[25] phases. Alumina phase transition can occur readily at temperatures in the core of the catalyst particles as the intraparticle temperatures could be 200–300°F higher than bulk gas phase temperatures due to exothermic temperature rise associated with the coke burn. Care is usually taken to control the heat-up rate so as not to damage the catalyst and reactor internals. Oxygen is metered into the nitrogen gas to a concentration of 0.3–1.0%. During primary burn, the exothermic temperature rise is minimized by controlling the amount of oxygen and heater operations.

In the 1970s, with the introduction of platinum/rhenium catalysts, an addition was made to the coke burning procedure by introducing continuous addition of organic chloride to help mitigate metals agglomeration in semiregenerative fixed-bed units. The effluent gas containing high chloride from the last reactor has to be continuously neutralized via circulation of an alkaline caustic solution to minimize corrosion of piping and equipment. It is not clear if additional organic chloride at this coke burn stage provides any significant reduction of metals agglomeration. However, potential benefit of the organic chloride addition could be ensuring that adequate chloride is added to bring the catalyst chloride to the desired level at the end of the regeneration. For most semiregenerative reformer catalysts the target chloride content is in the range of 0.9–1.1 wt %. In the case of regenerated catalysts in cyclic regenerative reformers, the target could range from 0.7 to 1.0 wt %, which is determined in part by corrosion and fouling concerns in the reformer.

For the CCR platformer, the spent catalyst, after catalyst fines removal, is transferred from the disengaging hopper to the regenerator zone where temperature (950–1100°F), air flow, and oxygen concentration (0.8–1.3 wt %) are regulated to permit controlled combustion of the catalyst coke. One of the features of the CCR platformer is the use of outer and inner screens to contain the catalyst and its flow in the reactors and regenerator annuli. Airflow to the regenerator is reduced as the screens begin to plug with catalyst fines leading potentially to incomplete or poor coke burns, screen cleanings, and catalyst damage. However, regenerator screens can be cleaned and the regenerator retstarted without disruption in reformate and hydrogen production in the reactors.

3.3 Oxidation/Chlorination

The now carbon-free reforming catalyst must then be rejuvenated to return the metal(s) to their original dispersed state. Even with controlled coke removal, the metal crystallites may become slightly sintered during regeneration. As metal crystallites become larger in diameter, the activity of the naphtha-reforming catalyst decreases.[31] Therefore, the next step is the redispersion of the metal via an oxychlorination process.[32] This is accomplished by the addition of chlorine under a full-air atmosphere to the reactor at a temperature of 932°F (500°C). Redispersion reduces the size of the metal crystallite particles and thereby produces an overall increase in specific metal surface area.

In an SR process, catalyst regeneration and rejuvenation must be performed sequentially in the fixed catalyst reactor. This results in many days of downtime for the reactor. In most cases, the cyclic-regenerated processes have multiple reactors, allowing one reactor to be taken off-line for regeneration at any given time.

After secondary and proof burns the oxygen partial pressure is increased to 5–15 psia and the temperature raised to 950–975°F. Organic chloride is added to provide the required catalyst chloride and to improve metals dispersion since the catalyst is usually agglomerated in the coke burn step. It is known that metals redispersion can be effected by the appropriate amounts of oxygen, chloride, temperature, and time in semiregenerative, cyclic regenerative, and continuous catalyst regenerative reformers.

Safety concerns in the cyclic regenerative reformers with respect to operating below a hydrocarbon/air explosion limit sometimes constrain operators to conduct the metals dispersion under less than optimal oxychlorination conditions. Typically, the oxygen concentration of the chlorinating gas is limited to less than 2.5 vol %.

For the continuous reformers, the dedicated regenerator and special regenerator metallurgy permit operating the catalyst regeneration with high chloride and high oxygen content. The oxygen partial pressures in the oxychlorination zones are usually limited to less than 10 psia in atmospheric CCR units. UOP has, however, recently licensed higher pressure CCR units that can operate at greater than 10 psia oxygen partial pressures.

3.4 Nitrogen Purge/Drying

Catalyst is purged with nitrogen in preparation for changing from an oxidizing to a reducing atmosphere after the oxychlorination step. The nitrogen purge also aids in drying the catalyst before the metals reduction step. Water is a product of the metals reduction reactions as shown by the simplified platinum and rhenium reduction equations in the next section. To optimize metals reduction, it is beneficial to reduce the amount of water in the regeneration unit or reduction zone before and during reduction of the metals. In semiregenerative reformers, this is accomplished by nitrogen purge and periodic draining of low points in the regeneration circuit.

For the CCR platformers, a specific drying zone is provided in the regenerator tower for drying the catalyst with nitrogen as shown in Figure 4.

3.5 Metals Reduction

Reduction of the platinum and promoter metals on catalysts is accomplished at about 700–900°F using recycle gas from a reformer or high-purity electrolytic hydrogen. To facilitate metals reduction, it is necessary to conduct the reduction at appropriate temperatures, to use as high hydrogen-containing reduction gas as economically feasible, and to allow adequate time. Various platinum/rhenium catalyst studies have shown that reduction of the platinum in reforming catalysts is facile and occurs essentially at temperatures around 600°F for Pt/Al_2O_3 catalyst. It has also been shown that in bimetallic Pt/Re catalysts, platinum is reduced to the zero valence state at about 600°F and that the reduction of rhenium approaches completion if the reduction temperature is maintained at greater than 900°F for 1–4 h as shown by the equations below.[25–29]

$$PtO_2 + 2H_2 \rightarrow Pt + 2H_2O$$

$$Re_2O_7 + 7H_2 \rightarrow 2\,Re + 7H_2O$$

Over the years, refiners have used recycle gas of varying hydrogen purity with mixed success in metals reduction. In some cases, the reduction of platinum-containing catalysts has been hampered by high-temperature rise due to hydrocracking of hydrocarbons contained in the recycle gas. To achieve better metals reduction, the refiner is usually advised to use a reduction gas with as high a hydrogen purity as economically feasible.

3.6 Metals Sulfiding

Platinum/rhenium catalysts often require some sulfiding to temper hyperactive metal sites, and this is usually conducted by sulfiding to levels commensurate with the rhenium content of the catalyst. For balanced, equimolar Pt/Re catalysts, about 0.05–0.06 wt % sulfur is recommended. Sulfur can be more optimally applied to the catalyst after reduction if the unit has historical catalyst sulfur data collected via use of an on-line catalyst sampler.

Cyclic regenerative reformer operators using monometallic platinum catalysts who choose not to sulfide their reduced catalysts sometimes experience excessive hydrogenolysis over hyperactive metal sites when the regenerated reactor is brought into the naphtha-reforming processing loop. Catalyst metals sulfiding is not required in continuous reformers since Pt/Sn catalysts are used in that reforming process.

4 CATALYST PROPERTIES INFLUENCING REFORMING CATALYST ACTIVITY AND PERFORMANCE

The ability to continuously regenerate a controlled quantity of catalyst is a significant innovation of the UOP CCR Platforming unit. The regeneration and reactor sections of the unit can be isolated to permit a shutdown of the regeneration system for normal inspection and maintenance without interruption of unit throughput. Along with these advantages, the CCR design also poses unique challenges to the reforming catalyst and the regeneration process. Since catalyst is circulated through the unit, it become relatively easy to sample and characterize the catalyst as it ages. This allows refiners to pinpoint operational problems that limit reforming activity and reduce catalyst lifetime.

4.1 Surface Area Stability

Reforming catalysts are heterogeneous and composed of active metals supported on a support material (normally Al_2O_3). Thermal stability of the support is a very important criterion for the reforming catalyst. The catalyst must withstand high localized temperatures during the regeneration process caused by coke burning on the catalyst surface. A support is used to produce a high specific metal surface area. Thermal stability or resistance to sintering of the reforming catalyst is impacted by the support manufacture. The exact mode of alumina sintering is complicated by the fact that these high surface area materials are formed with an essentially highly unordered structure. Sintering is impacted by the points of contact within the structure and the presence of small amounts of impurities or foreign ions. Mild sintering occurs within the reforming catalyst over time as it undergoes the regeneration process. Even under the most careful regeneration process, a gradual loss of surface area and pore volume is inevitable. Figure 5 displays the gradual loss in surface area for commercial reforming catalyst as a function of successive regeneration cycles. Curves depicted in Figure 5 represent 4 years of service. Surface area loss is greater with the first few regenerations and then levels off as the catalyst ages and approaches its end of life. Early continuous reforming catalyst formulations would drop

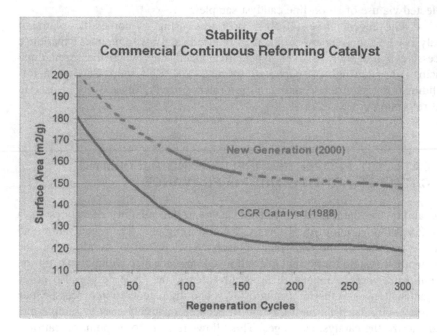

Figure 5 Surface area stability of commercial reforming catalysts.

from an original surface area of $180 \, m^2/g$ to an end-of-run value of $120 \, m^2/g$. Depending on the severity, continuous reforming catalyst would be in service for 2–4 years before being replaced. Advances in the alumina technology have resulted in improved surface area stability. All major catalyst manufacturers offer catalyst that maintains surface areas $20–30 \, m^2/g$ higher than formulations used 20 years ago. The improved surface area retention has allowed refineries to extend run lengths while maintaining high product yields. The extent of surface area that can be lost and still maintain good reforming activity is dependent on the catalyst formulation. Sufficient surface area must be maintained by the catalyst support to allow for highly dispersed platinum atoms while maintaining chloride retention that provides the catalyst acidity.

An extreme form of sintering can occur when conditions are so severe as to result in a phase transformation of the alumina support. Impurities can either accelerate or decelerate phase transformations in alumina.[33] γ-Alumina is widely used as the support material in reforming catalysts because of its relatively high surface area ($150–300 \, m^2/g$), its acidic properties, and its thermal stability. The most stable phase of alumina is α-alumina with a surface area of $2 \, m^2/g$. It is obvious that phase transformation of γ-alumina to α-alumina results in an extensive loss of surface area. Fortunately, alumina phase transformation is significant only at high temperatures. Normal conditions used for the regeneration of reforming catalysts in an SR or a continuous regeneration unit are mild enough to prevent the transition of the γ-alumina phase. However, in the event of runaway regenerator temperatures or excessive localized temperatures, phase transformation of the alumina support can occur.

Temperature excursions can occur when the carbon on catalyst becomes so high that a controlled burn is difficult to maintain. This is particularly critical in a continuous unit since the catalyst flows down through the regeneration zone and into the chlorination/metals redistribution zone. If carbon remains on the catalyst when it enters the oxygen-rich chlorination zone, a temperature runaway situation can arise. Temperature excursions can lead to the formation of α-alumina. In severe cases, formation of shrunken "dwarf" particles occurs. The major alumina phase present in catalyst dwarfs is α-alumina. α-Alumina is more dense than γ-alumina, so that the dwarf particles are smaller in diameter and contain very low surface areas ($<10 \, m^2/g$). The surface area of a commercial catalyst dropped from 154 to $129 \, m^2/g$ during a temperature excursion that raised the regenerator temperature in excess of 1800°F (980°C). During this excursion, 17% of the catalyst located in the regenerator was converted to dwarf material.

When phase transformation of the catalyst support occurs in a commercial unit, costly fresh catalyst makeup will be necessary to replace the damaged catalyst. In some cases, a unit turnaround would be required to effectively remove the damaged catalyst and recharge the unit with fresh catalyst. Active metals may be trapped inside the collapsed alumina structure and be unavailable for metal recovery. In the continuous unit, the presence of dense α-alumina particles can result in increased attrition of less dense γ-alumina particles.

4.2 Catalyst Attrition

The most unique challenge of the continuous catalyst, in contrast to SR, is the importance of particle integrity. Unlike SR catalysts, continuous catalysts are spherical in nature to allow for the smooth movement of the particles from the reactor to the regenerator on a continuous basis. Although continuous regeneration has many advantages, the catalyst particles must withstand grinding and impingement forces caused by the continual movement of catalyst in the reactor and regenerator.

Catalyst must be transported from the reactor to the regenerator on a continuous basis. This movement requires that attrition resistance be built into the catalyst spheres. Commercial catalyst undergoes 200–400 regeneration cycles before being replaced with fresh material. During its lifetime, various attrition mechanisms act on the catalyst as it cycles through the reactor. Lift lines are used to transport catalyst from the bottom of the reactor to the top of the regenerator. Nitrogen gas is used to control the rate of catalyst lift. As the catalyst leaves the lift line, it is subjected to collisions with tube walls and other catalyst particles as it falls into the lift hopper. Gravity is the principal driving force for catalyst movement through the reactor and regenerator section. As the catalyst moves down through the reactor, it experiences compressive and shear stresses as catalyst particles are forced against each other and against the reactor walls. Catalyst can also be crushed in a lock hopper valve if it is not functioning properly. In addition to mechanical stresses, the catalyst is subjected to high-temperature oxidation and reduction cycles, which may weaken or crack the particles.

Catalyst fines and chips generated as the catalyst is circulated can eventually plug the reactor screens and create a maldistribution of gas and catalyst flows. When this occurs, reforming activity drops resulting in a loss in product yield. Poor air distribution at the front of the regenerator can delay the carbon burn and create temperature excursion as carbon is burned off in the oxygen-rich zone of the regenerator.

Attrition can be divided into two classifications based on the type of particles produced. Either a "fracture" or a "grinding" type of mechanism can occur. Catalyst fracture involves breaking whole spheres into smaller pieces. This mechanism occurs when particles collide with a reactor wall or with other catalyst particles. The fracture mechanism creates a variety of different-sized fragments or catalyst chips. The grinding mechanism creates spheres that are slightly smaller along with a large amount of very fine particles. This type of catalyst wear occurs as the catalyst moves through the reactor. Both catalyst–catalyst and catalyst–wall contact contribute to this mechanism. An increase in the amount of small fines suggests that the catalyst is soft and attriting by the grinding mechanism.

Since a variety of attrition mechanisms operate on a catalyst in continuous reformer, a multilevel approach is needed to evaluate how a catalyst will perform in a commercial unit. Particle crush strength measurements can identify catalyst that would likely fail via the fracture mechanism. A rotating drum test would be more useful in identifying material that will fail by a grinding mechanism.

Particle Crush Strength

Particle crush strength measurements can identify weak spheres that are likely to fracture as they are cycled through the continuous reforming reactor. Thermal stresses placed on the catalyst during the regeneration process can be the major factor leading to catalyst fracture. Exposing a catalyst to extreme temperatures can result in alumina phase transformations that create fault lines along the catalyst surface. Fault lines weaken the catalyst structure and lead to an increased rate of fracture. Catalyst pills possessing crush strengths of less than 2.5 pounds per sphere (11 newtons per sphere) show increased tendency to fracture.

Figure 6 compares catalyst samples obtained before and after a unit experienced a temperature excursion in its regenerator. The temperature excursion resulted in some alumina phase damage and created a large number of weak particles. After the excursion, the unit fines make increased by a factor of 10. The undamaged catalyst displayed an average crush strength of 6.7 pounds per sphere (30 newtons per sphere) vs. 2.6 pounds per sphere (12 newtons per sphere) for damaged material collected after the temperature excursion. Particles possessing a crush strength below 2.5 pounds increased from 0% to 54%. In general, all of the spheres present in fresh catalyst possess crush strengths in excess of 2.5 pounds. The presence of particles possessing a crush strength below this value indicates that some catalyst damage has occurred.

Dwarf particles can be created when a catalyst is exposed to extreme temperatures ($>1832°F$, $>1000°C$). At this temperature, alumina is transformed from the γ phase into the α phase, and the catalyst spheres shrink in size. α-Alumina has a low surface area and displays little, if any, reforming activity. If the phase transition is incomplete, stress points can form, which produce weak particles. These weak particles are likely to break apart as they are circulated through the unit and are eventually removed with the unit fines. It is typical to see increased fines make after experiencing a temperature excursion. If conditions in the regenerator allow catalyst dwarfs to anneal, the stress points are removed and very strong particles are created. The hard particles remain intact and act as an abrasion source for the remaining catalyst inventory. The crush strength

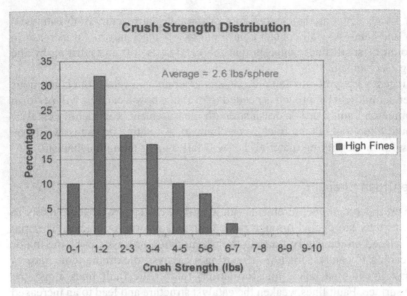

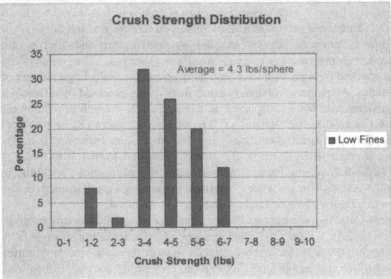

Figure 6 Comparison of particle crush strengths of a commercial continuous reforming catalyst.

distributions of catalyst dwarfs are displayed in Figure 7. The dwarf material displayed an average crush strength of 11.0 pounds per sphere, with 52% of the particles possessing a crush strength in excess of 10 pounds. In comparison, fresh catalyst displayed a crush strength of 6.7 pounds per sphere with only 6% possessing a crush strength in excess of 10 pounds.

Rotating Drum Test

Crush strength measurements are helpful in determining if the particles can withstand impacts or collisions with reactor walls or other catalyst particles. However, crush strength does not address the grinding issues a catalyst faces as it circulates through the unit. The grinding forces can be simulated in the laboratory through a drum test.[34]

A modified attrition drum (MAD) test developed by Ashland Petroleum consists of two concentric rotating drums.[34] The space between the outer and inner drums is set at 11 mm, which can accommodate approximately seven layers of catalyst spheres. Drums are rotated in opposite directions to force catalyst spheres to roll over each other. Both catalyst–catalyst and catalyst–wall interactions are duplicated in this test.

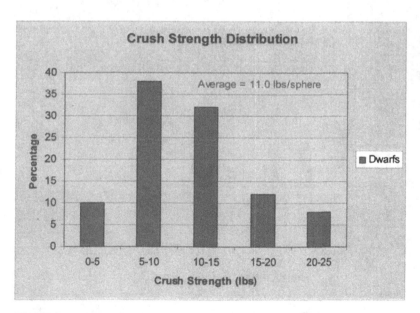

Figure 7 Crush strength distribution of dwarf particles obtained from commercial CCR unit.

Fines obtained in the MAD test display a similar size distribution as those obtained from a commercial unit fines collection system (Fig. 8). The similarity indicates that attrition mechanisms occurring in the MAD tester are mimicking those that occur in a commercial unit. Furthermore, catalysts that display low unit fines have shown a low fines make in the MAD tester.

4.3   Metal Redispersion

The object of catalyst rejuvenation and activation is to return the regenerated catalyst to an active state with properties similar to fresh catalyst. During regeneration by coke burning, some sintering of the metal phase and the alumina support occurs even under the most controlled conditions. Coke deposition, discussed in detail in the previous chapter, occurs by dehydrogenation initiated at the metal site. The catalyst support has an important role in coke deposition. In bifunctional reforming catalysts, two types of coke are formed: (1) those on metallic sites and (2) those on support.[35,36] During regeneration, coke burning on the metal and the support leads to the production of high temperatures and steam. As a result, some sintering of the metal crystallites and support will occur. To return the catalyst to its active state, the sintered metal crystallites must be

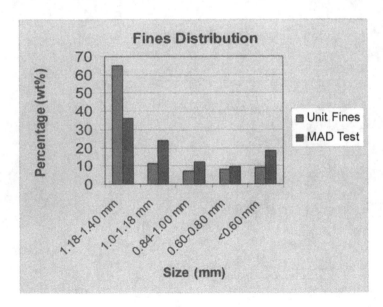

Figure 8   Comparison of catalyst fines produced by the modified attrition drum (MAD) tester to a commercial unit.

returned to a highly dispersed and reduced state. The loss of support surface area is irreversible, but it can continue to provide an active catalyst until surface area is no longer sufficient to allow complete metal redispersion and good chloride retention.

In an air atmosphere, platinum surface area decreases as a function of temperature and time as does chloride content.[37] High chloride content reduces the impact of sintering. This phenomenon is due to inhibition of sintering or to an equilibrium between sintering and redispersion. A transmission electron microscopy study by Harris with Pt/Al_2O_3 catalysts in the absence of chloride suggests that the sintering mechanism involves a combination of migration and coalescence of whole particles.[38] Promoters, such as Sn and Re, are incorporated into the catalyst to help minimize platinum clustering. The reverse reaction, the redispersion of crystalline platinum, is impossible with chloride-free catalysts. Researchers have shown that the oxychlorination reaction involves the formation of $[Pt^{IV}O_xCl_y]_s$. This species is mobile and can migrate on the surface of the alumina. The first step in the redispersion process is the oxidation of surface Pt atoms followed by chlorination of the oxidized Pt sites.

Platinum redispersion on a regenerated (coke-free) catalyst is accomplished in the chlorination zone. During the coke burning period, the metal(s) are converted to a highly oxidized state and some chloride is removed from the support. In the chlorination zone, catalyst is subjected to an oxidizing atmosphere rich in an organic chloride source (typically ethylene dichloride) and a small amount of water vapor at a temperature between 930°F and 1110°F (500°C and 600°C). High concentrations of water vapor are extremely detrimental to the redispersion process. The chloride ions react with the oxidized platinum atoms to form the mobile $[Pt^{IV}O_xCl_y]_s$ species, which redistributes the platinum over the catalyst support. Chloride ions are not only required for redispersing the platinum atoms, but they also help maintain a high platinum dispersion during processing. In addition, chloride ions interact with the alumina support to create acidity needed for the reforming reactions. Reforming catalysts are bifunctional, requiring both acidic and metallic sites. The acid function of the catalyst must be balanced with the metal activity for good reforming activity. Industrial reforming catalysts typically maintain chloride content between 1.0 and 1.1 wt % depending on the catalyst and unit. An excessive amount of chloride is detrimental as it leads to undesirable cracking reactions and increased coke formation.

After drying to remove moisture from the catalyst, hydrogen-rich gas is used to reduce platinum to its metallic state. Approximately 2 h is required to complete the reduction step. The reduced catalyst is then purged with nitrogen to remove any residual hydrogen before returning it to reforming service.

4.4 Regenerator Troubleshooting

Continuous reforming units have proved to be extremely reliable and functional, and the catalysts, properly designed for these operations, can undergo hundreds of regeneration cycles during their lifetime. Commercial units can run for extended periods without the need for shutdown when good regeneration efficiency, clean feeds, effective fines removal, and chloride and sulfur management programs are in place.

The efficiency of the continuous regenerator should be evaluated on a periodic basis. Efficiency is defined as percent increase in platinum dispersion of the regenerated catalyst in comparison to the spent catalyst. Coke on the spent catalyst must be carefully removed in the laboratory without impacting the Pt dispersion. Metal dispersion of catalysts with known platinum content can be measured by volumetric chemisorption techniques such as hydrogen chemisorption, oxygen titration, or hydrogen titration (see Chapter 6). Low or negative efficiencies indicate poor regenerator operation that requires immediate investigation.

Low regeneration efficiency can be caused by restrictions in the regenerator that retard airflow or chloride injection in the chlorination zone. Sufficient quantities of oxygen and ethylene dichloride are required for efficient platinum redispersion. If the oxygen or chloride is reduced, the platinum will not redisperse properly. Figure 9 illustrates the decline in dispersion that occurred on a commercial continuous catalyst during a period when plugged screens prevented the proper balance of air and chloride in the regenerator. Over the course of one year, platinum dispersion dropped to 30% of its original value. At this point only one-third of the platinum atoms in the catalyst were accessible for reforming reactions. In order to correct the problem, a regenerator shutdown and screen cleaning was performed. Reestablishing the proper air/chloride distribution in the regenerator greatly improved its operation. Two months after the screen cleaning, platinum dispersion was restored. Prior to the screen cleaning, the measured regenerator efficiency dipped to -10%. Each time the catalyst passed through the regenerator, the platinum dispersion was lowered. After the reactor screens were cleaned, the regenerator efficiency increased to 15%. Each time the catalyst passed through the regenerator platinum dispersion increased until it reached normal levels. Figure 9 illustrates that the present-day catalysts are robust and can recover from unit upsets.

Regenerators are designed to maintain the performance of the continuous reforming catalyst. Problems created by excessive fines can reduce the efficiency of the regenerator. Often screen plugging occurs gradually over time and goes unnoticed. Since Pt dispersion is very sensitive to regenerator conditions, it can serve as early indicator of potential problems. In general, a drop in Pt dispersion occurs before any operational problems are noticed. By observing a gradual

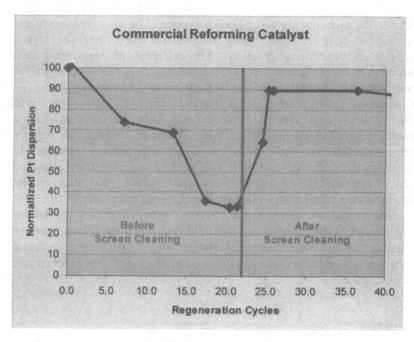

Figure 9   Loss of Pt dispersion due to plugged screens.

decline in platinum dispersion, a refiner can identify and correct potential problems before being forced into unexpected shutdown.

   This emphasizes the importance of a good fines management program. A small quantity of catalyst fines are formed due to the movement of the catalyst through a continuous reformer unit as discussed in the previous section on catalyst attrition. Catalyst fines and broken particles must be removed from the catalyst inventory before entering the regeneration zone. Fines are removed by elutriation gases in the disengaging chamber at the top of the regnerator. Proper gas flow must be established and maintained through a fines management program to ensure effective operation.

5 CONCLUSIONS

Regeneration of bifunctional naphtha-reforming catalysts is a multifaceted process. The regeneration process consists of the elimination of coke by controlled burning followed by a catalyst rejuvenation process. Rejuvenation involves the restoration of the dispersion of the metal(s) by an oxychlorination process over the high surface area alumina support. The acid function of the support must also be restored by chloride adsorption to allow the proper balance between the acid and metal functions. The dispersed platinum must then be reduced in a hydrogen atmosphere to the metallic state. For some catalysts, the regenerated catalyst must be passivated by sulfiding to moderate activity.

Continuous reformers are becoming the workhorses of modern refineries to meet the stringent requirements of reformulated fuels. Although the unique continuous regeneration design provides many advantages to the refiner, it poses new demands on the catalyst and on the regeneration process. Advances in catalyst design have allowed the continuous catalyst to successfully withstand hundreds of regeneration cycles. Refiners who practice good management programs to address sulfur and metals contamination and fines control can expect to have long periods of uninterrupted service. With future new regulations requiring ultralow-sulfur gasoline and diesel, there is still a significant incentive for researchers to find better and more efficient ways to produce these future generation fuels. Naphtha reforming will undoubtedly remain an important part of fuels production.

REFERENCES

1. Ciapetta, F.G.; Wallace, D.M. Catalytic naphtha reforming. Catal. Rev. **1971**, *5*, 67.
2. Gates, B.C.; Katzer, J.R.; Schuit, G.C.A. Reforming. In *Chemistry of Catalytic Processes*; McGraw-Hill: New York, 1979; 184 pp.
3. Sinfelt, J.R. Catalytic reforming of hydrocarbons. In *Catalysis Science and Technology*; Anderson, J.R., Boudart, M., Eds.; Springer-Verlag: Berlin, 1981; Vol. 1, 257 pp.
4. Weiszmann, J.A. Catalytic re-reforming. In *Handbook of Petroleum Refining Processes*; Meyers, R.A., Ed.; McGraw-Hill: New York, 1986; 3-3 pp.
5. Ciapetta, F.G.; Dobres, R.M.; Baker, R.W. Catalytic reforming of pure hydrocarbons and petroleum naphthas. In *Catalysis*; Emmett, P.H., Ed.; Reinhold: New York, 1958; 495 pp.
6. Sivasanker, S.; Ratnasamy, P. Reforming the gasoline and aromatics. In *Catalytic Naphtha Reforming*; Antos, G.J., Aitani, A.M., Parera, J.M., Eds.; Marcel Dekker: New York, 1995; 483 pp.
7. Little, D.M. *Catalytic Reforming*; Pennwell Books: Tulsa, OK, 1985.
8. Edgar, M.D. Catalytic reforming of naphtha in petroleum refineries. In *Applied Industrial Catalysis*; Leach, B.E., Ed.; Academic Press: New York, 1983; Vol. 1, 123 pp.
9. Haensel, V. (a) U.S. Patent 2,479,109, 1949; (b) U.S. Patent 2,479,110, 1949.
10. Sutton, E.A.; Greenwood, A.R.; Adams, F.H. A new processing concept for continuous Platforming. Oil Gas J. **1973**, *71* (20), 136.

11. Meeting the challenges of cleaner fuels, June 2001, *World Refining*; Chemical Week Associates Production; 12 pp.

12. McClung, R.G.; Kramer, R.; Oyekan, S.O. Paper AM-89-45 Reformer feedstock pretreatment; Liquid phase vs. vapor phase sulfur removal systems, 1989, NPRA, San Francisco.

13. McClung, R.G.; Oyekan, S.O. Sulfur sensitivity of Pt/Re catalysts in naphtha reforming, 1988, AIChE spring national meeting, New Orleans.

14. Kluksdahl, H.E. U.S. Patent 3,415,737, 1968.

15. Oyekan, S.O.; Swan, G.A. U.S. Patent 4,436,612, 1984.

16. Antos, G.J. U.S. Patent 4,101,418, 1981.

17. *The Petroleum Handbook*; Elsevier: Amsterdam, 1983; 268 pp.

18. Prins, R. Modern processes for the catalytic reforming of hydrocarbons. In *Chemistry and Chemical Engineering of Catalytic Processes*; Pins, R., Schuit, G., Eds.; Sitjhoff and Noordhoff: Alp hen aan den Rijn, The Netherlands, 1980; 389 pp.

19. *IFP Technology Symposium: Oil Refining and Petrochemicals Production*; IFP Publications: Paris, 1981; 3 pp.

20. 2000 NPRA Q&A Discussions, San Francisco.

21. Aitani, A.M. Catalytic reforming processes. In *Catalytic Naphtha Reforming*; Antos, G.J., Aitani, A.M., Parera, J.M., Eds.; Marcel Dekker: New York, 1995; 409 pp.

22. Beltramini, J.N. Regeneration of reforming catalyst. In *Catalytic Naphtha Reforming*; Antos, G.J., Aitani, A.M., Parera, J.M., Eds.; Marcel Dekker: New York, 1995; 365 pp.

23. Baker, R.T.K.; France, J.A.; Rouse, L.; Waite, R.J. J. Catal. **1976**, *41*, 22.

24. Weisz, P.B.; Goodwin, R.B. J. Catal. **1966**, *6*, 227.

25. Oyekan, S. U.S. Patent 4,539,307, 1985.

26. Johnson, M.F.L.; LeRoy, V.M. J. Catal. **1977**, *35*, 434.

27. Wagstaff, N.; Prins, R. J. Catal. **1979**, *59*, 434.

28. Bertolacini, R.J.; Pellet, R.J. In *Catalyst Deactivation*; Delmon, B., Froment, G.F., Eds.; Elsevier: Amsterdam.

29. Sachtler, W.M.H. J. Mol. Catal. **1984**, *25*, 1.

30. Greenwood, A.R.; Vesely, K.D. U.S. Patent 3,647,680, 1972.

31. Herrman, R.H.; Adler, S.F.; Goldstein, M.S.; deBaun, R.M. J. Phys. Chem. **1965**, *65*, 2189.

32. Lieske, H.; Lietz, G.; Spindler, H.; Volter, J. J. Catal. **1983**, *81*, 8.

33. Bartholomew, C.H.; Butt, J.B. Eds. *Catalyst Deactivation*; Elsevier: Amsterdam, 1991; 29 pp.

34. Doolin, P.K.; Gainer, D.M.; Hoffman, J.F. J. Test. Eval. **1993**, Nov., 481.

35. Parera, J.M.; Figoli, N.S.; Traffano, E.M. J. Catal. **1983**, *79*, 484.

36. Beltramini, N.N.; Wessel, T.J.; Datta, R. In *Catalyst Deactivation*; Bartholomew, C.H., Butt, J.B., Eds.; 1991; 119 pp.

37. Bournonville, J.P.; Martino, G. In *Catalyst Deactivation*; Delmon, B., Froment, G.F., Eds.; Elsevier: Admsterdam.

38. Harris, P.J.F. J. Catal. **1986**, *97*, 527.

12
Precious Metals Recovery from Spent Reforming Catalysts

Horst Meyer and Matthias Grehl
W.C. Heraeus GmbH & Co.
Hanau, Germany

1 INTRODUCTION

Given the importance of reforming as a refinery process, significant quantities of noble metal–containing catalysts are installed worldwide in the operating oil refineries. The value of these catalysts is quite significant, so it is without any doubt that at the end of a catalyst life cycle, when the oil refineries produce a spent catalyst, the precious metals recovery and other related aspects have to be handled with corresponding attention [1]. The modern petroleum-processing industry would not exist without precious metals catalysts, but the associated investment value brings with it special concerns. Most oil refineries already have installed a special care program incorporating guidelines beginning with the purchase of the fresh catalyst until the metals are recovered from the spent one. For each of the precious metals refiners, dealing with spent catalysts goes beyond showing the best possible recovery performance, but also acting as a consultant for the customer in respect to all precious metals–related technical procedures, including fine-metal and precious metals compounds handling, analytical procedures, sampling statistics, logistics, and book-keeping. This chapter explains some of the practical aspects to this high-value part of the naphtha-reforming operation. Certain perspectives were presented in the first edition of this book [2]. Further perspectives are added in this chapter.

2   MATERIALS HANDLING AT THE OIL REFINERY

When it comes to the point where an oil refinery has decided to perform a catalyst change-out, certain in-house procedures come into force. This is especially true for precious metals–containing catalysts, where a variety of important procedures have to be observed. Even though each oil refinery has its own procedures, a general chronological pattern of steps for execution for a catalyst change-out can be established.

1. Decision about the timing of a change-out
2. Choice of the manufacturer for the fresh catalyst (if applicable)
3. Choice of the precious metals refiner for the spent catalyst
4. Reactor unloading/skimming/packaging, etc.
5. Shipping the spent catalyst for further treatment and metals recovery

At the time of the basic decision for a change-out, it will be already known if the catalyst has to go through a regeneration process. Depending on the carbon and/ or hydrocarbon and/or halides concentration on the spent catalyst, regeneration might become necessary to provide a safe handling through all further steps, such as sampling and metals refining. A properly regenerated catalyst will guarantee an accurate sampling operation, resulting in representative samples that will show the actual precious metals content in the catalyst delivered. In addition, possible surcharges that are assessed in the case of excessive carbon, hydrocarbon, or moisture can be avoided. In case of an ex situ regeneration it will be very important to make contact with a corresponding regeneration company as early as possible. Planning ahead for such a step will avoid unnecessary delays which later on would affect the precious metal availability. When it comes to step 3 above, several important pieces of information have to be available:

Type of catalyst
Quantity
Catalyst properties
Expected availability
Kind of change-out
Regeneration in situ or ex situ (if applicable)
Availability of sample

Based on this information a precious metals refiner will be able to calculate the requested terms and conditions for the recovery.

At the time when a specialized company arrives to take care of the unloading of a reactor, certain aspects have to be observed to guarantee smooth handling later on. Depending on the kind of change-out, the catalyst to be unloaded can consist of various fractions, such as:

Inert balls, or support material
Dust and broken pellets or extrudates
Heel catalyst (material containing excessive carbon)
Clean catalyst
Contaminated catalyst

Decisions about classifying catalyst as clean or contaminated depend on the maximum limits of various contaminants given by the precious metals refiner. Unloaded material has to be dumped into suitable drums, flow bins, or big bags. All fractions have to be clearly and carefully labeled and stored in such a way that they can be identified later on. This is important to avoid confusion and intermixing of catalyst types, which might ultimately cause problems during precious metals refining.

Each drum should be labeled as follows:

Name and address of the oil refinery.
Type of catalyst.
Gross and net weight.
Name and address of the precious metals refiner.
If applicable, drums have to be marked according to the appropriate
    transport and waste regulations.

As an additional safety feature, it is recommended to seal each drum before shipping. This ensures that through all the handling steps, catalyst (and precious metals) will not be lost without notice. The supplier of the spent catalyst has to employ very careful control to ensure that everything is properly handled and all relevant safety and transport regulations are obeyed when the catalyst is shipped for reclamation. The requirements are not just a formalism; they are necessary for the safety and protection of value during transport and metals recovery of the catalyst. When it comes to shipping, it will be the refinery's decision either to use their own in-house logistics or to subcontract to the precious metals refiner. In the final analysis, the least price per truck and the possible convenience for the refinery will be the driving force in the decision.

At the end, the paperwork should consist of

Purchase order
Invoice
Packing list
Material declaration (tracking document/AVV number)
Declaration if material is classified as hazardous or nonhazardous
If necessary, any declaration in accordance with transport regulations

In case of the material declaration, assistance should be available from the precious metals refiner.

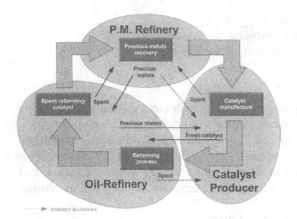

Figure 1   Business transactions in catalyst refining and catalyst life cycle.

Figure 1 gives an overview of possible business transactions during the catalyst life cycle. There are various options that can be chosen depending on the needs of the customer. The most common options are:

Fine metal will be delivered to a catalyst manufacturer.
Fine metal will be kept on an weight account.
Fine metal will be sold at actual rates.

3   DESCRIPTION OF SPENT CATALYSTS

3.1.   Form

Spent reforming and similar alumina-supported catalysts are typically spheres or extrudates with dimensions of 1mm or more. Typically spheres are used for moving-bed reactors and extrudates for fixed-bed reactors.

3.2.   Support Material

The support material is made from high-surface-area alumina (mainly γ-alumina) of various strengths and porosites. During the lifetime of the catalysts, overheating occurs at some areas in the reactor and phase transitions can take place converting alumina phases of high solubility to other alumina phases (γ to α) with low solubility in caustic soda or sulfuric acid. Noble metal catalysts for other processes are also sent for metals recovery. As a result, α-alumina-, silica-, or zeolite-supported catalysts are also common [3,4]. Figure 2 presents typical forms of reforming catalysts in spherical, pellet, and extrudate shapes.

Figure 2 Catalysts—pellets, extrudates, and spheres.

3.3. Precious Metal Content and Promoters

The precious metals content including the rhenium content of fresh reforming catalysts varies from about 0.2% to more than 0.6%. Typical precious metals concentrations of fresh reforming catalysts are 0.3% or 0.375% of each. At least one of the precious metals is contained in any catalyst. Besides platinum and rhenium, some petrochemical processes use palladium, ruthenium, and iridium. Depending on the usage and the physical properties of a reforming catalyst, the precious metals concentration in the spent catalyst should be nearly the same as the corresponding fresh one.

Besides rhenium and iridium, other promoters such as tin, lead, arsenic, cadmium, and germanium are used. Depending on the usage of a given catalyst these metals are used, but they are not considered important enough for metals reclaiming. Catalysts are typically identified by the manufacturer's code using a combination of letters and numbers like R-62, RG-482, and E-603.

3.4. Distribution of the Precious Metals

Starting with the unloading procedure from the reactor, the catalyst is usually separated into fractions such as fines, oversize, and whole catalyst. The precious metals content for the fines often shows variations of $\pm 50\%$ compared to the catalyst, whereas the oversize frequently does not contain any precious metals. The percentage of the fines related to whole catalyst depends on how the catalyst was originally produced and its properties. Also, the use of the catalyst during the cycle in the reforming reactor can influence the percentage of fines and their precious metals content.

3.5. Impurities

Typical impurities on the catalyst are iron and other components of steel such as nickel and chromium. Metals typically contained in the crude oil, such as vanadium and manganese, may also be present, particularly if some upset has occurred. These elements and the additional catalyst promoters can influence the metals refining process. An example of an especially deleterious effect would be contamination of the aluminum solutions resulting from the metals refining process, which could prevent further usage of these solutions. Such a restriction would influence the economics of the refining process.

Other impurities, such as halides, carbon, and hydrocarbons, have to be observed as they could have a direct influence on the metals refining process. Halides usually result from the catalyst manufacturing process of the alumina carrier or the finished catalyst production. For the case of isomerization catalysts, chlorine is continuously added to the process resulting in much higher halide concentrations than for reforming catalysts, where chloride maintenance is of minor extent. Carbon and hydrocarbons result from the carbon or coke deposited during the reforming process in the reactor.

4 WEIGHING, SIEVING, AND SAMPLING PROCEDURES

Because of the considerable value of the precious metals on the catalyst, sampling is the first very important step. Taking into consideration the accuracy of a precious metals analysis and the value involved, the lot size for sampling has to be limited to a certain quantity in order to obtain the most accurate measurement of noble metal value. It is without any question that each sampling step can be witnessed by either the customer or a representative.

During receipt of the material at the precious metals refiner's warehouse an inspection of the containers is performed. Containers will be opened only after release by the customer or the customer's representative. Gross, tare, and net weight of each individual drum, big bag, or container is determined during the sampling operations by means of a calibrated scale. Simultaneously, the customer's packing list is checked to ensure that the delivery is in compliance with the actual material shipped. Any differences or other issues, such as damaged containers, are reported to the production management and customer service department immediately for appropriate action.

For the sampling of spent catalysts, specially designed equipment for screening and sampling is necessary. The objective of the screening and sampling operation is to obtain representative samples of the spent whole catalysts and other precious metals–containing fractions, e.g., the fines fraction. The sampling method will be explained using the Hanau sampling equipment as an example.

After dumping the catalyst into a movable container, the container is placed on top of a sieving device where the catalyst is uniformly fed to a vibrating screener. A continuous separation process now takes place, dividing the material into four fractions, namely, fines, pellets (middle fraction), magnetics, and oversize. In two steps a cut is taken for the pellets and a first intermediate sample of approximately 0.5% is obtained. The mesh width of the sieves is selected according to the pellet shape and size to ensure optimal separation from oversize. Fines of each lot are usually accumulated to form a fines lot. Such a lot will be kept separate from the pellet lots. Figure 3 presents equipment used in screening and sampling of catalysts.

At the end of the sampling operation, a second intermediate sample (reduced in size) is deived from the first intermediate sample. This sample is divided into two portions of approximately 3 kg each. One portion remains in an "as is" state whereas the other portion is ground to a particle size less than 200 μm. Each of these 3-kg portions is then split by a rotary divider into eight samples. A representative 3-kg sample is also taken from the fines, which are then further ground and similarly split into eight samples. These representative samples of ground pellets, original pellets, original as-is pellets, and fines can then be distributed to the analytical laboratories of the refiner, the customer, and/ or an independent representative assigned by the customer. Such samples are usually sealed by both parties using a seal pad that will not allow an uncontrolled opening of the samples. For all necessary sampling steps, such as grinding and dividing, it is important to ensure that all the equipment involved is of sufficient quality to the high value of the catalyst and the required precision. Figure 4 shows a schematic diagram of the screening and sampling device.

5 ASSAYS OF PRECIOUS METALS AND RHENIUM

5.1. Sample Handling and Analysis

Because of the statistics and inhomogeneity of the as-is material, ground material has to be used for the precious metals analysis. During handling of the samples humidity could be absorbed or desorbed resulting in a change of weight. Therefore, a reproducible mass is needed as a basis for the calculation of the

Figure 3 Screening and sampling equipment.

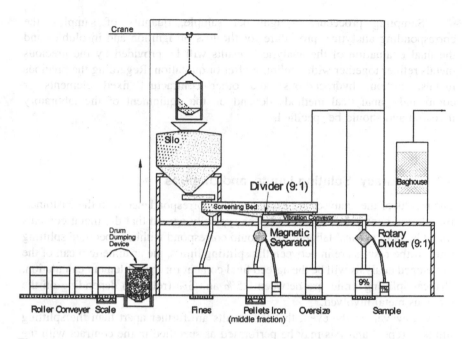

Figure 4   Schematic diagram—screening and sampling device.

precious metals and rhenium contents, which is the ignited material. All the precious metals assays in the Heraeus laboratories and the customers' laboratories are related to the ignited material. The loss on ignition (LOI) is determined gravimetrically using an electric furnace and platinum crucibles. The LOI for the settlement weight is determined for the as-is sample.

Without going into actual analytical details, the steps in the analyses of the precious metals and rhenium are as follows. The alumina support material will be digested in the analytical lab. Traces of precious metals and rhenium that are dissolved will be precipitated. After separation of the precious metals–free solution, the residue from the digestion step, the precipitate, and any nondissolved support material will be totally dissolved in several steps.

Precious metals and rhenium are determined by the ICP method using an internal standard and suitable dilutions. Usually three assays from each lot are executed. For more accuracy, a test sample with known content (which has been assayed by gravimetric methods) is determined in parallel to the actual samples. The procedure is a complete analogue from the digestion of the support until the final ICP determination. The results should not deviate from the theoretical content by more than $\pm 0.002\%$.

Sampling procedure, weight per sample, quantity of samples, the corresponding analytical procedure for the loss on ignition and insolubles and the final evaluation of the analytical results will be provided by the precious metals refiner together with a refining offer or quotation. Regarding the precious metals, carbon, hydrocarbons, and other contractual fixed elements or compounds, analytical methods depend on the equipment of the laboratory involved and should be specified.

5.2. Accuracy, Splitting Limits, and Umpires

The results of the analyses are exchanged by correspondence with the customer. Based on a contractual agreement, it is typically agreed that the metal contents found by both parties' laboratories should correspond within a specified splitting limit. If the results are in between the splitting limits, the arithmetic mean of the exchanged analysis will be the agreed final content for the settlement calculation. Typical splitting limits are between 0.5% and 3% (relative) depending on the precious metal involved.

However, in the event that the results are further apart than the splitting limits, a repeat analysis may be performed as specified in the contract with the customer. If even the repetition is outside the splitting limits, a technical consultation between the responsible analytical person of both parties should take place to resolve the differences. In case the differences are still not resolved, a set of sealed samples will be submitted to an umpire laboratory, as outlined in the contractual agreement.

The accuracy of the sampling and analytical procedures is very important. As an example, we consider an order of 100 metric tons of reforming catalyst. This order contains 350 kg platinum with a total value of US $8 million. A mistake of 0.5% relative is equal to US $40,000. Such a high value of the noble metals content is one of the reasons that the sampling lot is small.

6 TREATMENT ROUTES

6.1. Removal of Halides, Hydrocarbons, and Carbon

Depending on the concentration of carbon and/or hydrocarbons and/or halides and the process involved, the catalyst typically must go through a pretreatment. In most cases, such a treatment is executed by specialized companies offering their services to either oil refineries or precious metals refiners. Heraeus provides it own in-house equipment for carbon and hydrocarbon burnoff.

6.2. Acidic and Alkaline Leach Processes

To extract the precious metals and rhenium, only the catalysts with high surface alumina phases can be treated using the wet chemical route. The alumina support has to be dissolved leaving the platinum, palladium, or iridium as a residue. Alternatively, the metals could be dissolved off the support. However, due to the physical characteristics of the porous support material, precious metals and rhenium are somewhat strongly fixed to the support, so that removal from the alumina is not possible at a satisfying percentage of recovery. There have been processes to remove the platinum by solid-state chlorination or HCl leaching, but these methods did not progress beyond laboratory or pilot scale [5]. Also, the direct reaction of the ground catalyst with concentrated sulfuric acid had no economical success. A continuous autoclave process for the dissolution of alumina has not been used for the treatment of spent reforming catalysts [4].

For spent catalysts with high surface area alumina and reasonable contents of insolubles and carbon, hydrometallurgical approaches are most commonly employed. Alumina support is dissolved with either sodium hydroxide or sulfuric acid [4,6]. If adequate precautions are taken, platinum as a noble metal is not dissolved in sodium hydroxide or in sulfuric acid. Alumina dissolution takes place as shown in the following two reactions:

Alkaline leach process:

$$Al_2O_3 + 2NaOH + 3H_2O \rightarrow 2Na\,Al(OH)_4 \tag{1}$$

Sulfuric acid leach process:

$$Al_2O_3 + 3H_2SO_4 \rightarrow Al_2(SO_4)_3 + 3H_2O \tag{2}$$

After the leaching process, rhenium is in the aqueous phase and the platinum, palladium, or iridium is in the solid residue. The choice of either the alkaline or sulfuric acid leach process depends on the infrastructure to dispose of sidestreams, the equipment, and past experiences in the refiner's environment.

Both processes are operated at elevated temperatures. The alkaline process is also operated at elevated pressure during the first leach step. It is easily understood that elevated temperatures and pressures improve the process in regard to the dissolution time and quantity of nondissolved material. At Heraeus the alkaline pressure leach is used in Hanau, Germany, whereas sulfuric acid leach is used in Santa Fe Springs, California as shown in Table 1 and Figure 5.

6.3. Utilization of Leach Products

Aluminum sulfate and sodium aluminate are used in wastewater purification processes. Industrial customers purchase this material for their wastewater treatment plants. Having these industrial outlets is highly desirable because such

Table 1 Alkaline and Sulfuric Leach Conditions

	Heraeus sulfuric acid leach	Heraeus alkaline pressure leach
Loading	0.75–0.9 metric ton spent catalyst	1 metric ton spent catalyst
Leach chemical	H_2SO_4	NaOH 50%, 45%
Temperature	<100°C	180–220°C
Pressure	1 bar	8–24 bar

usage contributes to the economics of the overall recovery process. Three factors are important for the subsequent sale and use of aluminum solutions. These are impurities outside the specification levels, the availability of competing replacement chemicals like $FeCl_3$, and logistic costs for transporting the solutions.

6.4. Smelting

Smelting of the spent catalyst to a slag and a metallic phase is another option. This approach is typically used if the carrier material is nonsoluble. In case of rhenium-containing catalysts, rhenium goes into the slag phase and will be lost

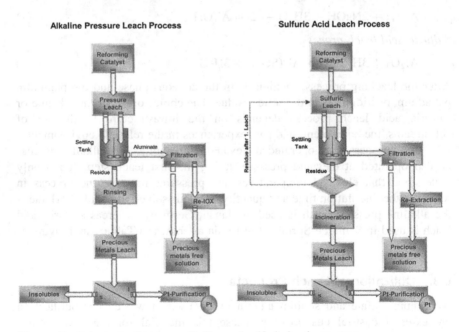

Figure 5 Schematic diagram—alkaline and sulfuric leach process.

for recovery. The recovery yield for iridium would be low compared to the results achieved utilizing the wet chemical route.

7 PRECIOUS METALS SEPARATION AND PURIFICATION

7.1. Dissolution of Leach Residue

To recover platinum, the leach residue is dissolved, in batch reactors. This is typically done with aqua regia or hydrochloric acid and chlorine [reaction (3)] or other hydrometallurgical treatments [7]. For platinum purification, solvent extraction or classical precipitation and redissolution [reaction (4)] as well as ion-exchange methods can be applied.

7.2. Purification Process

At Heraeus, the hexachloroplatinate salt is dissolved, followed by reprecipitation. This results in a very good separation from all potential impurities [8].

7.3. Platinum Metal Production

The conversion of platinum salt to platinum metal can be done by wet chemical reduction, by pyrolytic decomposition [reaction (5)] of the ammonium hexachloroplatinate, or by electrowinning processes [reaction (6)]. At Heraeus Hanau, the electrowinning process is the most important route for metallic platinum (Fig. 6). The electrolysis is operating with a daily capacity of 80 kg. The platinum is deposited as fine black powder on water-cooled titanium anodes. The cathodic chambers are separated by a membrane from the anode chambers. The electrolysis is carried out continuously for 22 h, and then the harvesting process takes place. This is followed by filtration, washing, drying, and incineration to the final sponge form. The sponge is quality controlled by analytical methods to guarantee catalytic grade purity [8]. Common commercial and ASTM grades of platinum are 99.95% and 99.99%. The following set of reactions shows the sequence of platinum conversion from salt to metal.

Platinum dissolution:

$$Pt + 2HCl + 2Cl_2 \rightarrow H_2PtCl_6 \tag{3}$$

Platinum salt precipitation:

$$H_2PtCl_6 + 2KCl \rightarrow K_2PtCl_6 + 2HCl \tag{4}$$

Platinum salt incineration:

$$(NH_4)_2PtCl_6 \rightarrow Pt^0 + 2NH_4Cl + 2Cl_2 \tag{5}$$

Figure 6 Platinum electrowinning process.

Platinum electrowinning:

$$H_2PtCl_6 + 4e^- \rightarrow Pt^0 + 2HCl + 4Cl^- \tag{6}$$

7.4. Recovery Schemes for Palladium and Iridium

In the case of palladium-containing spent catalysts, the palladium is also precipitated as the hexachloropalladium salt. The palladium purification is done by the common ammonia purification process. At the end of the purification process, the yellow salt is reduced and then dried and treated at higher temperatures to obtain the final palladium sponge.

In the case of platinum- and iridium-containing catalysts, the alkaline leach residue is dissolved in a multistep operation. After the HCl/chlorine leach a second step is needed to dissolve the remaining alumina. Then the residue is leached again with HCl/chlorine. Solvent extraction is used for both platinum- and iridium-containing mother liquors to separate the platinum and the iridium. The costs for this process, the cycle time, and especially the yield of the iridium depends mainly on the insolubles of the spent catalysts, the iron content of the spent catalyst, and the iridium loss into the aluminate solution. Iridium purification is done by hydrometallurgical method including solvent extraction, ion exchange, and electrolysis.

8 RHENIUM SEPARATION AND PURIFICATION

8.1. Solvent Extraction or Ion Exchange

Various methods can be used to separate the rhenium from the aluminum sulfate or sodium aluminate solutions. Solvent extraction and ion exchange are the most common. At Heraeus both methods are applied. Heraeus favors the anion-exchange method. The hot sodium aluminate is passed through ion-exchange columns, so that the perrhenate is fixed and chloride is released into solution. To release the perrhenate, the column has to be treated with sodium hydroxide, hydrochloric acid, and then iron chloride solution.

8.2. Purification of Ammonium Perrhenate and Perrhenic Acid

To obtain ammonium perrhenate, the perrhenic acid is converted to ammonium perrhenate, which is crystallized as shown in reactions (7) and (8). The ammonium perrhenate is then purified by recrystallization. Further purification is obtained by passing the ammonium perrhenate solution over a cation-exchange column as shown in reaction (9). The catalyst grade ammonium perrhenate is obtained by precipitation with ammonium hydroxide. After drying and analytical control the ammonium perrhenate is ready for packaging. Common catalyst grade ammonium perrhenate contains 69.4% Re according to the customer's or refiner's specification. Rhenium contained in more dilute mother liquors of precipitation and recrystallization processes is precipitated as potassium perrhenate, since this is much less soluble.

Ammonium perrhenate precipitation and perrhenic acid:

$$HReO_4 + NH_4Cl \rightarrow NH_4ReO_4 + HCl \tag{7}$$

$$HReO_4 + NH_4OH \rightarrow NH_4ReO_4 + H_2O \tag{8}$$

Ion-exchange purification to perrhenic acid:

$$NH_4ReO_4 + Res\text{-}H \rightarrow HReO_4 + Res\text{-}NH_4 \qquad (9)$$

9 COMMERCIAL ASPECTS

In order to meet customers' expectations and to cover their needs on a worldwide basis, Heraeus operates two refineries for spent precious metals catalysts (Hanau, Germany and Santa Fe Springs, California), where a leading share of the available market is treated. We estimate that about 10,000–15,000 metric tons reforming and other refining catalysts containing Pt, Pt/Re, Pt/Ir, and Pd are available worldwide for metals reclamation per year. Taking into consideration a typical value of approximately US $70 per kilogram of spent catalyst (based on average precious metals prices, May 2001), the total worldwide value for precious metals recovery is approximately US $0.7–1.0 billion.

The high value per kilogram for spent catalyst must always be kept in mind, particularly when evaluating treatment and possible pretreatment charges (like carbon burnoff or hydrocarbon stripping). Losses must be minimized. Costs increase considerably in cases when a metal lease is needed. Therefore, it is very important to have all technical parameters of the spent catalyst in hand prior to a pickup. This will enable the oil refinery and precious metals refiner to work on a fixed schedule, thus reducing time for a given metals lease. The resulting advantages will be to avoid unexpected additional costs and to reduce the overall costs of using these noble metal catalysts in reforming.

REFERENCES

1. Gallmeier, J. The recovery of platinum and rhenium from petroleum catalyst. In *Proceedings of a Seminar at the International Precious Metals Institute*, Las Vegas, Nevada, 1997.
2. Rosso, J.P.; El Guindy, M.I. Recovery of Pt and Re from Spent Reforming Catalysts. In *Catalytic Naphtha Reforming: Science and Technology*; Antos, G. J., Aitani, A. M., Parera, J. M., Eds.; Marcel Dekker: New York, 1995; p. 395.
3. Feige, R.; Winkhaus, G. Aluminumoxid—Rohstoff für Ingenieurkeramik. Metall, **1986**, *40*, 598.
4. Hudson, L.K.; Misra, C.; Wefers, K. Aluminum oxide. In *Ullmann's Encyclopedia of Industrial Chemistry*, A1, 1985, 557.
5. Thomas, M.; Grosbois, J. Verfahren zur Rueckgewinnung von Platin und Iridium aus edelmetallhaltigen Katalysatoren, Deutsches Patentamt No. DT 2454647,1976.

6. Hoestra, J.; Michalco, E. Recovery of Platinum from Deactivated Catalyst Composites. US Patent No. 2950965, 1960.
7. Meyer, H.; Grehl, M.; Stettner, M. Verfahren zum Lösen von Edelmetallen aus edelmetallhaltigen Scheidgütern. Deutsches Patentamt No. 19928029, 1999.
8. Meyer, H.; Grehl, M. Paper presented at the Dechema European Workshop on Spent Catalysts: Recycling and Disposal. Frankfurt/Main, Germany, 1999.

13
Licensed Reforming Processes

Abdullah M. Aitani
King Fahd University of Petroleum and Minerals
Dhahran, Saudi Arabia

1 INTRODUCTION

Catalytic naphtha reforming is the technology that combines catalyst, hardware, and process to produce high-octane reformate for gasoline blending or aromatics for petrochemical feedstocks. Reformers are also the source of much needed hydrogen for hydroprocessing operations. Several commercial processes are available worldwide, and the licensing of technology for semiregenerative and continuous reforming is dominated by UOP and Axens (formerly IFP) technologies.

The main difference between commercial reforming processes are catalyst regeneration procedure, catalyst type, and conformation of the equipment. Currently, there are more than 700 commercial installations of catalytic reforming units worldwide, with a total capacity of about 11.0 million barrels a day. About 40% of this capacity is located in North America followed by 20% each in western Europe and the Asia–Pacific region. Table 1 presents a regional distribution of catalytic reforming capacity worldwide [1].

This chapter presents an overview of latest developments in reforming technology, describes major licensed processes, and includes recent introductions of reforming catalysts.

2 PROCESS CLASSIFICATION

Catalytic naphtha reforming processes are generally classified into three types:

1. Semiregenerative
2. Cyclic (fully regenerative)
3. Continuous regenerative (moving bed)

Table 1 Regional Distribution of Catalytic Naphtha Reforming by Capacity [1]

Region	Crude capacity (1000 b/d)	Reforming capacity (1000 b/d)	Reforming as % of crude capacity
N. America	20,030	4075	20.3
W. Europe	14,505	2135	14.7
Asia–Pacific	20,185	2000	10.0
E. Europe	10,680	1430	13.4
Middle East	6075	570	9.4
S. America	6490	400	6.1
Africa	3200	390	12.1
Total	81,165	11,000	13.6

This classification is based on the frequency and mode of regeneration. The semiregenerative requires unit shutdown for catalyst regeneration, whereas the cyclic process utilizes a swing reactor for regeneration in addition to regular in-process reactors. The continuous process permits catalyst replacement during normal operation. Worldwide, the semiregenerative scheme dominates reforming capacity at about 57% of total capacity followed by continuous regenerative at 27% and cyclic at 11%. Table 2 presents a regional distribution of catalytic reforming capacity by process design [1]. Most grassroots reformers are currently designed with continuous catalyst regeneration. In addition, many units that were originally built as semiregenerative units have been revamped to continuous regeneration units. A list of commercial reforming processes with a summary of key process features is presented in Table 3.

Table 2 Regional Distribution of Catalytic Reforming Capacity by Process Type [1]

Region	Total reforming (1000 b/d)	Percentage share of total reforming (%)			
		Semiregenerative	Continuous	Cyclic	Other
N. America	4075	46.4	26.8	22.2	4.6
W. Europe	2135	54.0	31.5	11.0	3.5
Asia–Pacific	2000	42.4	44.8	1.6	11.2
E. Europe	1430	86.4	11.0	1.1	1.5
Middle East	570	63.0	23.1	7.2	6.7
S. America	400	80.4	9.3	0.6	3.5
Africa	390	81.9	0.0	1.8	16.3
Total	11,000	56.8	26.9	11.1	5.2

Table 3 Summary of Naphtha Reforming Processes

Process name	Licensor	Process type and key features	Installations
Platforming	UOP	Semiregenerative and continuous reforming; CycleMax regenerator; product recovery system	Over 800 units with 8 million b/d
Octanizing; Aromizing	Axens	Semiregenerative and continuous reforming; dualforming for conventional process revamp	Over 100 licensed units
Houdriforming	Houdry Div. Air Products	Semiregenerative; high-octane gasoline and aromatics	0.3 million b/d
Magnaforming	Engelhard	Semiregenerative or semicyclic	1.8 million b/d
Powerforming	ExxonMobil	Semiregenerative or cyclic	1.4 million b/d
Rheniforming	Chevron	Semiregenerative; low-pressure operation	1 million b/d
Ultraforming	Amoco	Semiregenerative or cyclic	0.5 million b/d
Zeoforming	SEC Zeosit	Semiregenerative; zeolite-based catalyst	Few small units

2.1. Semiregenerative Process

The semiregenerative process is characterized by continuous operation over long periods, with decreasing catalyst activity as a result of coke deposition. Eventually, as the reactor temperatures reach end-of-cycle levels, the reformers are shut down to regenerate the catalyst in situ. Regeneration is carried out at low pressure (approximately 8 bar) with air as the source of oxygen. The development of bimetallic and multimetallic reforming catalysts with the ability to tolerate high coke levels has allowed the semiregenerative units to operate at 14–17 bar with similar cycle lengths obtained at higher pressures. It is believed that all reforming licensors have semiregenerative process design options.

The semiregenerative process is a conventional reforming process that operates continuously over a period of typically up to one year. As the catalytic activity decreases, the yield of aromatics and the purity of the byproduct hydrogen drop because of increased hydrocracking. Semiregenerative reformers are generally built with three to four catalyst beds in series. The fourth reactor is usually added to some units to allow an increase in either severity or throughput while maintaining the same cycle length. The longer the required cycle length, the greater the required amount of catalyst. Conversion is maintained more or less constant by raising the reactor temperatures as catalyst activity declines. Sometimes, when the capacity of a semiregenerative reformer is expanded, two existing reactors are placed in parallel, and a new, usually smaller, reactor is added. Frequently, the parallel reactors are placed in the terminal position. When evaluating unit performance, these reactors are treated as though they are a single reactor of equivalent volume.

Research octane number (RON) that can be achieved in this process is usually in the range of 85–100, depending on an optimization between feedstock quality, gasoline qualities, and quantities required as well as the operating conditions required to achieve a certain planned cycle length (6 months to 1 year). The catalyst can be regenerated in situ at the end of an operating cycle. It is not unheard of that the catalyst inventory can be regenerated 5–10 times before its activity falls below the economic minimum, whereupon it is removed and replaced. Equally likely is the intent to replace the catalyst with a newer market introduction that will offer economic advantage to the refiner.

2.2. Cyclic (Full Regeneration) Process

The cyclic process typically uses five or six fixed catalyst reactor beds, similar to the semiregenerative process, with one additional swing reactor, which is a spare reactor. It can substitute any of the regular reactors in a train while the regular reactor is being regenerated. In this way, only one reactor at a time has to be taken out of operation for regeneration, while the reforming process continues in operation. Usually, all of the reactors are the same size. In this case, the catalyst in the early stages (or front-end reactors) is less utilized; therefore, it will be regenerated at much longer intervals than the later stages. The cyclic process may be operated at low pressures, may utilize a wide boiling range feed, and may operate with a low hydrogen-to-feed ratio. Coke lay-down rates at these low pressures and high octane severity (RON of 100–104) are so high that the catalyst in individual reactors becomes exhausted in time intervals of from less than a week to a month.

The process design of the cyclic process takes advantage of low unit pressures to gain a higher C_5+ reformate yield and hydrogen production. The overall catalyst activity, conversion, and hydrogen purity vary much less with time than in the semiregenerative process. However, a drawback of this process is that all reactors alternate frequently between a reducing atmosphere during normal operation and an oxidizing atmosphere during regeneration. This switching policy needs a complex process layout with high safety precautions and requires that all the reactors be of the same maximal size to make switches between them possible.

2.3. Continuous Catalyst Regeneration Process

The continuous reforming process is characterized by high catalyst activity with reduced catalyst requirements, more uniform reformate of higher aromatic content, and high hydrogen purity. The process can achieve and surpass reforming severities as applied in the cyclic process but avoids the drawbacks of the cyclic process. The continuous process represents a step change in reforming technology compared to semiregenerative and cyclic processes. Since its introduction in the early 1970s, it has gained wide acceptance by the refining and petrochemical industries worldwide.

In this process, small quantities of catalyst are continuously withdrawn from an operating reactor, transported to a regeneration unit, regenerated, and returned to the reactor system. In the most common moving-bed design, all the reactors are stacked on top of one other. The fourth (last) reactor may be set beside the other stacked reactors. The reactor system has a common catalyst bed that moves as a column of particles from top to bottom of the reactor section. Coked catalyst is withdrawn from the last reactor and sent to the regeneration reactor, where the catalyst is regenerated on a continuous basis. However, the final step of the regeneration, i.e., reduction of the oxidized platinum multimetallic catalyst, takes place in the top of the first reactor or at the bottom of the regeneration train.

Fresh or regenerated catalyst is added to the top of the first reactor to maintain a constant quantity of catalyst in the reactor train. Catalyst transport through the reactors and the regenerator is by gravity flow, whereas the transport of catalyst from the last reactor to the top of the regenerator and back to the first reactor is by the gas lift method. Catalyst circulation rate is controlled to prevent any decline in reformate yield or hydrogen production over time onstream.

In another design, the individual reactors are placed separately, as in the semiregenerative process, with modifications for moving the catalyst from the bottom of one reactor to the top of the next reactor in line. The regenerated catalyst is added to the first reactor and the spent catalyst is withdrawn from the last reactor and transported back to the regenerator.

The continuous reforming process is capable of operation at low pressures and high severity by managing the rapid coke deposition on the catalyst at an acceptable level. Additional benefits include elimination of downtime for catalyst regeneration and steady production of hydrogen of constant purity. Operating pressures are in the 3.5- to 17-bar range and design reformate octane number is in the 95–108 range.

3 MAJOR REFORMING PROCESSES

UOP and Axens are the two major licensors and catalyst suppliers for catalytic naphtha reforming. The processes differ in the type of operation (semiregenerative or continuous), catalyst type, and process engineering design. Both licensors agree on the necessity of hydrotreating the feed to remove permanent reforming catalyst poisons and to reduce the temporary catalyst poisons to low levels. Numerous process design modifications and catalyst improvements have been made in recent years.

3.1. UOP Platforming

Platforming was the first process to use platinum-on-alumina catalysts. The first UOP semiregenerative Platforming unit went onstream in 1949. UOP technology is used in more than 50% of all reforming installations with more than 750 units in service worldwide [2]. The individual capacities of these units range from 150 to 63,000 b/d facilities.

The Platforming process has been adapted to bimetallic catalyst and to both semiregenerative and continuous operation. UOP offers semiregenerative units that use catalyst-staged loading for increased production. In particular, UOP's R-72 staged loading system generates the highest C_5+ yields. Recently, advanced reforming catalysts were introduced, such as R-86 for semiregenerative units and R-270 for CCR applications. New catalysts are continuously aimed at maximizing hydrogen yield, increase operating flexibility and maximizing C_5+ product yield.

As of September 2001, more than 170 stacked-reactor Platforming units were operating with continuous regeneration compared to 550 semiregenerative units. The total capacity of the semiregenerative units exceeded 5.0 million b/d whereas CCR^{TM} units reached 3.8 million b/d. Table 4 presents a regional distribution of UOP Platforming units according to process design and capacity. The simultaneous use of CCR technology and bimetallic catalysts has given UOP a unique position in the field of catalytic reformer process licensing. Recent multimetallic catalyst formulations have improved both aromatic and reformate yields.

Table 4 List of Commissioned UOP Reforming Units [2]

Region	Semiregenerative		CCR	
	Units	1000 b/d	Units	1000 b/d
Americas	294	2713	51	1379
Europe	123	1256	34	747
Far East	86	672	66	1393
Middle East	27	273	13	278
Africa	20	183	4	74
Total	550	5097	168	3871

CCR Platforming

To meet the demand of increased severity, UOP has improved the performance of the conventional Platforming process by incorporating the CCR system. The process uses stacked radial flow reactors and a CCR section to maintain a steady-state reforming operation at optimum process conditions: fresh catalyst performance, low reactor pressure, and minimal recycle gas circulation. The flow pattern through the Platforming unit with CCR is essentially the same as with conventional fixed-bed units. The effluent from the last reactor is heat exchanged against combined feed, cooled, and phase split into vapor and liquid products in a separator. A schematic flow diagram of the CCR Platforming process is presented in Figure 1.

Catalyst flows vertically by gravity down the stack, while the feed flows radially across the annular catalyst bed. The catalyst is continuously withdrawn from the last reactor and transferred to the regenerator. The withdrawn catalyst flows down through the regenerator where the accumulated carbon is burned off. Regenerated catalyst is purged and then lifted in hydrogen to the top of the reactor stack, maintaining nearly fresh catalyst quality. Because the reactor and regenerator sections are separate, each operates at its own optimal conditions. Typical operating conditions for the current design of the UOP CCR process are: reactor pressure 6.8 bar; LHSV 1.6 h^{-1}; H$_2$/HC molar ratio 2–3; and RON-clear 100–107. Table 5 shows the relative operating severities of the UOP semiregenerative and CCR units. The CCR unit operates at higher severity and lower reactor catalyst inventory. In addition, the CCR unit runs continuously compared to 12 months semiregenerative cycle lengths. Typical product yields for the CCR and semiregenerative Platforming units operating at the same conditions are presented in Table 6 [3]. Many of the benefits of CCR include higher hydrogen yield and purity as well as higher octane barrels.

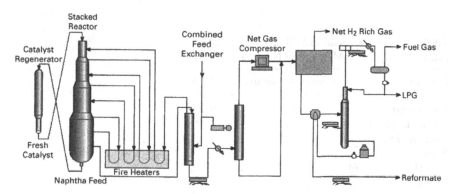

Figure 1   Schematic of UOP CCR Platforming process.

Table 5 Relative Severities for the Semiregenerative and CCR Platforming Units [3]

Parameter	Semiregenerative	CCR
Catalyst type	R-56	R-134
Charge rate, b/d	20,000	20,000
LHSV, h^{-1}	Base	Base × 1.8
H_2/HC	Base	Base × 0.5
RONC	97	102
Pressure, kPa	Base	Base—1035
Cycle life, months	12	Continuous

One recent development in CCR technology is second-generation CCR Platforming with several modifications in the reactor and regenerator sections. The high-efficiency regenerator design resulted in an increased coke burning capacity with reduced regeneration severity and complexity. CycleMax regenerator provided easier operation and enhanced performance over other regenerator designs. The operation at ultralow pressure (3.4 bar) and the use of low-platinum R-34 catalyst ensured the highest yield of the reformate and aromatic product with more cost-effective process operation. The net gas recovery schemes maximized the yields of reformate and hydrogen. Moreover, the new regenerator allowed higher regeneration rates to support the coke generation of the low-pressure operation and high conversion levels.

RZ Platforming

The RZ Platforming process and the RZ-100 catalyst offer constant aromatics selectivity, in the range of 80% or higher. RZ-100 catalyst differs greatly from

Table 6 Typical Yields of UOP Semiregenerative and CCR Units for a Middle East Naphtha Feed [4]

Parameter	Semiregenerative	CCR
Catalyst type	R-72	R-274
Stream factor, d/y	330	360
Pressure, kPa	1380	345
Yield		
Hydrogen, scfd	1270	1690
C_5+ wt %	85.3	91.6
RONC	100	100

conventional reforming catalysts in the production of light aromatics, benzene and toluene. The selectivity of conventional reforming for benzene and toluene is significantly lower than for the C_8 aromatics. Although UOP CCR Platforming is the most efficient means for producing xylenes from heavy naphtha fractions, its conversion of C_6 and C_7 paraffins to aromatics is normally below 50%, even at low pressure.

In general, the RZ Platforming configuration is consistent with other UOP Platforming systems. The process employs adiabatic, radial flow reactors that are arranged in a conventional side-by-side pattern. An interheater is used between each reactor to reheat the charge to reaction temperature. Treated naphtha feed is combined with recycled hydrogen and sent to the reactor section. The effluent from the last reactor is heat exchanged against combined feed, cooled, and phase split into vapor and liquid products in a separator. The liquid from the recovery section is sent to a stabilizer where light saturates are removed from the C_6+ aromatic product. Typical cycle lengths are 8–12 months and the units are designed for either efficient in situ or ex situ catalyst regeneration.

3.2. Axens Reforming Technology—Octanizing and Aromizing Processes

The Axens catalytic reforming technology is based on IFP and Procatalyse reforming expertise for the upgrade of various types of naphtha to produce high-octane reformate, BTX, and liquefied petroleum gas (LPG). The reforming process can be supplied in either semiregenerative or continuous operation. Table 7 presents a compilation of Axens reforming catalysts for conventional and continuous applications [4–6]. Axens semiregenerative version is a conventional reforming process in which the catalyst is regenerated in situ at the end of each

Table 7 Compilation of Axens Reforming Catalysts for Semiregenerative and Continuous Applications [6]

Process	Catalyst type	Remarks
Semiregenerative		
	RG 452	Max. LPG
	RG 492; RG 582	High H_2, C_5+, max stability
	RG 582	High H_2, C_5+
	RG 682	Max. H_2, C_5+, max stability
Continuous reforming		
	CR 401, CR 201/201A	Max. H_2, C_5+
	CR 701/CR 702	For all types of CCR
	AR 501, AR 405	Max. aromatics

cycle. The operating pressure of this process is in the range of 12–25 bar (170–350 psig) with low-pressure drop in the hydrogen loop. The product RON-clear is in the range 90–100. Multimetallic catalyst formulations for semiregenerative applications offer higher selectivity and stability.

The decrease in the reformate yield during the run cycle of the semiregenerative version (in spite of improvement in catalyst stability) has led Axens to also develop a catalyst moving-bed system that allows continuous regeneration of the catalyst. This version, the Octanizing process, is an advanced design that reflects the results of several decades of research and development efforts. Aromizing is Axens continuous reforming process for the selective production of aromatics. It is the petrochemical complement to the Octanizing process. The technology employs an advanced catalyst formulation to achieve high BTX aromatics yield. The technology offers high aromatics yields, low investment and operating costs, and high onstream factor. Table 8 presents a summary of key features of the continuous reforming processes. The heart of the Octanizing technology differentiates itself in its catalyst circulation and continuous catalyst regeneration systems. A schematic flow diagram of the Octanizing process is presented in Figure 2 [7]. The overall process comprises the following:

A conventional reaction system consisting of a series of four radial flow reactors that use a stable and selective catalyst suitable for continuous regeneration.

A catalyst transfer system using gas lift to carry the catalyst from one reactor to the next and finally to the regenerator.

A catalyst regeneration section, which includes a purge to remove combustible gases, followed by catalyst regeneration.

In the Octanizing or Aromizing processes, treated naphtha is mixed with recycle hydrogen, preheated, and passed through a series of adiabatic reactors

Table 8 Key Features of Octanizing and Aromizing Processes [7]

Key feature	Description
Reactor configuration	Easy construction and maintenance
	Minimal thermal expansion problems
	Less than 1% nonflowing heal catalyst
Continuous regenerator	Smooth operation; no pulsing lift system
	Low catalyst attrition
	Easy maintenance
Advanced regenerator	Two distinct burning zones
	Optimized oxychlorination parameters
	Reduced corrosion and low catalyst cost

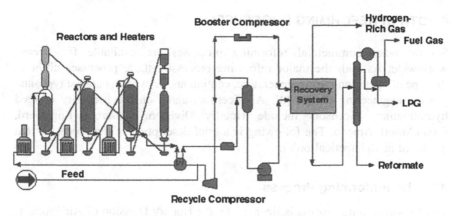

Figure 2 Schematic of Axens Octanizing and Aromizing processes.

and heaters where it is converted to rich-aromatics stream and hydrogen. The effluent is cooled by heat exchange and liquid product is separated from recycle and hydrogen gases. Axens regenerative technology has been improved to allow faster circulation of the catalyst and, as a consequence, increased regeneration frequency as required by the more severe operating conditions (low pressure, low H_2-to hydrocarbon ratio). The Octanizing process features high on-stream efficiency, flexibility, and reliability. Major improvements, compared with previous designs, are the development of catalysts of increased activity, selectivity, and hydrothermal stability along with substantial increases in the yields of $C_5 +$ reformate and hydrogen. Table 9 presents typical yields of the Axens conventional and regenerative process. The total number of installations using the Axens technology is 90 licensed units; of these, 30 units are designed for continuous regenerative technology.

Table 9 Typical Yields of Axens Semiregenerative and Octanizing Processes for a 90–170°C Cut Light Arabian Feedstock [4]

Parameter	Semiregenerative	Octanizing process
Pressure, kg/cm^2 (psig)	10–15	<5 (71)
Yield, wt %		
Hydrogen	2.7	3.8
C$_5$+	83	88
RONC	100	102
MONC	89	90.5

4   OTHER REFORMING PROCESSES

Several other commercial reforming processes are available for license worldwide. As with the major reforming processes, these processes differ in the type of operation (semiregenerative, continuous, or cyclic), catalyst type, and process engineering design [8]. All licensors agree on the necessity of feed hydrotreating. Licensors include Houdry Division, Chevron, Engelhard, ExxonMobil, Amoco. The following is a brief description of the processes that are listed in alphabetical order.

4.1.   Houdriforming Process

The Houdriforming process is licensed by the Houdry Division of Air Products and Chemicals, Inc. The process is used to upgrade various naphthas to aviation blending stocks, aromatics, and high-octane gasoline in the range of 80–100 RON clear. The process operates in a conventional semiregenerative mode with four reactors in series for BTX production, compared with three reactors for gasoline. The catalyst used is usually Pt/Al_2O_3 or may be bimetallic. A small "guard case" hydrogenation pretreater can be used to prevent catalyst poisons in the naphtha feedstock from reaching the catalyst in the reforming reactors. The guard case reactor is filled with the usual reforming catalyst but operated at a lower temperature. It is constructed as an integral stage of the Houdriforming operation when required for the feedstock.

At moderate severity, the process may be operated continuously for either high-octane gasoline or aromatics, without provision for catalyst regeneration. However, operation at high severity requires frequent in situ catalyst regeneration. Typical operating conditions are temperature 755–810 K, pressure 10–27 bar, LHSV $1-4\,h^{-1}$, and H_2/HC ratio 3–6. The total capacity of Houdriforming units is about 250,000 b/d.

4.2.   Magnaforming Process

The Magnaforming process, which is licensed by Engelhard Corporation, is used to upgrade low-octane naphtha to high-octane reformate. The product is a premium blending component or aromatic hydrocarbon source. The feature that most distinguishes the Magnaforming process from other reforming processes is its use of a split hydrogen recycle stream to increase liquid yields and improve operating performance. About half of the recycled gas is compressed and recycled to the first two reactors, which are operated at mild conditions. The greater portion of the recycle gas is returned to the terminal reactors, which operate under severe conditions. It is believed that substantial compressor power saving can be achieved by splitting the recycle.

Engelhard does not offer a continuous design, but it does offer a semiregenerative design and a combination of semiregenerative and cyclic regeneration. The combination design is made by supplementing the terminal reactors with a swing reactor that can alternate with the terminal reactors. It can also operate in parallel with the terminal reactors, permitting these reactors to be regenerated without unit shutdown. Smaller Magnaforming units use a conventional three-reactor system, compared with four reactors in large units.

The process was initially designed to operate with established monometallic platinum catalysts but was adapted to include the newer platinum-rhenium-based catalyst of the E600 and E800 series. The catalysts provide greater activity and stability, enabling use in units where high-severity operation and long cycle lengths are required. A wide range of catalysts has been used in the Magnaforming process in order to optimize operating performance and to produce desired product specifications. Many units incorporate Sulfur Guard technology to reduce sulfur in reformer feed to ultralow levels. There are approximately 150 units totaling 1.8 million b/d using the Engelhard reforming technology.

4.3. Powerforming Process

Powerforming is offered by ExxonMobil to produce gasoline blending stocks from low-octane naphthas. Alternatively, the process may be operated to give high yields of benzene or other aromatics or to produce aviation blending stocks. The process also produces large quantities of hydrogen, which can be used to hydrotreat or improve other products. Powerforming features a semiregenerative or cyclic configuration and proprietary catalyst system tailored to the client's specific needs. Advantages of Powerforming include cost-competitive installations for units less than 12,000 b/d, 98 RON products, mild severity operations, and long cycle lengths.

The staged catalyst system is represented to have a high stability, good selectivity, and adequate selectivity maintenance. It uses the high-activity KX-130 catalyst with very high stability and high benzene/toluene yields. KX-130 is an appropriate debottlenecking catalyst or BTX-producing catalyst. The dual-catalyst system offers high activity catalyst, high benzene/toluene yields, and a C_5+ yield advantage relative to the KX-130 catalyst system. ExxonMobil's catalyst management techniques for hot flue gas regeneration in cyclic units, on-stream chlorination, and analytical tools for monitoring catalyst help enhance the unit's performance. Cyclic Powerformers are designed to operate at low pressure, with a wide range of feed boiling point and low hydrogen-to-feed ratio. The unit has four reactors in series plus a swing reactor. The use of the swing reactor allows any of the on-stream reactors to be taken out of service for regeneration while maintaining continuous operation of the unit. The frequency of regeneration can be varied to meet changing process objectives as well as operating under high-severity, low-pressure conditions, since coke deposition is maintained at low levels. Regeneration is generally performed on a predetermined schedule to avoid having two or more reactors regenerated at the same time. Usually the terminal reactors are scheduled for more frequent regenerations than the early-stage reactors. Run lengths of up to 6 years between shutdowns could be achieved. Powerforming has a wide range of potential refining applications and has been commercially applied in nearly 50 semiregenerative units and more than 30 cyclic units ranging from 1000 b/d to 65,000 b/d.

4.4. Rheniforming Process

The Rheniforming process is used to convert naphthas to high-octane gasoline blendstock or aromatics plant feedstock. The process has gained wide acceptance since Chevron patented the bimetallic Pt/Re catalyst in 1968. Rheniforming is basically a semiregenerative process that comprises a sulfur sorber, three radial flow reactors in series, a separator, and a stabilizer. The process is characterized by the sulfur control step, which reduces sulfur to 0.2 ppm in the reformer feed. A new Rheniforming F/H catalyst system has been used that permits low-pressure operation. The high resistance to fouling of the catalyst system increases the yields of aromatic naphtha product and hydrogen due to the long cycle lengths, which reach 6 months or more. Optimized operating techniques permit maintenance of high catalyst activity throughout each cycle and return to fresh activity after each regeneration. The increased resistance to fouling also provides for expansion of existing plants by using higher space velocities, lower recycle ratios, or increased product octane. Converted units are operating with H_2/HC ratios of 2.5–3.5 and long cycles between regeneration. It is believed that a total of 73 Rheniformers are on stream with a total capacity of more than 1 million b/d.

4.5. Ultraforming Process

The Amoco Ultraforming process is used to upgrade low-octane naphthas to high-octane blending stocks and aromatics. The process is a fixed-bed cyclic system with a swing reactor incorporated in the reaction section, which is usually specified for aromatic (BTX) production. The system can be adapted to semiregenerative operation with the conventional three radial flow reactors in series. The process uses rugged, proprietary catalysts permitting frequent regeneration and high-severity operations at low pressures. The catalyst system has relatively low precious metals content, and its estimated life is perhaps 4 years for cyclic operation vs. 8 years for semiregenerative operation. The swing reactor in cyclic operation replaces any reactor while the catalyst bed in this reactor is being regenerated. Normally, the reactors are all the same size; however, the first reactor is loaded with half the usual amount of catalyst.

Ultraformers may be designed to produce high-purity xylenes and toluene, which can be separated by straight distillation before the extraction step. The benzene fraction can be recovered by extractive distillation. High yields of C_5+ reformate and hydrogen have been reported for the Ultraforming process. The total capacity of Ultraforming is more than 530,000 b/d for 39 commercial units worldwide. However, no new Ultraformers have been licensed in recent years.

4.6. Zeoforming Process

Zeoforming process is a new reforming technology for high-octane gasoline production from hydrocarbon raw materials of various origins (straight-run naphtha fractions from both gas condensates and crude oils, condensates of accompanying gases, olefin-containing gases, secondary hydrocarbon fractions of refinery and petrochemical plants), boiling up to 200–250°C. The process has been developed by the SEC Zeosit Center of the Siberian branch of Russian Academy of Sciences. A special stable and selective IC-30 catalyst based on modified pentasil zeolite is used in the process [9].

Due to acid shape–selective mechanism possibilities for the hydrocarbon conversions in the Zeoforming process, the product gasoline has a lower aromatics content than the reformate with the same octane number. Reactions proceeding under Zeoforming process conditions allow operation to reach benzene concentration under 1 vol % and to produce gasoline with low sulfur content without naphtha hydrotreating owing to the tolerance of the catalyst to sulfur compounds in the feedstocks. The process is based on the catalytic conversion of linear paraffins and naphthenes into isoparaffins and aromatics over zeolite-containing catalysts, which allows an increase of the octane number of naphtha from 45–60 MON up to 80–85 MON.

Stepanov et al. [9] reported that a small scale plant of 120 b/d of naphtha was operated successfully in the north of Siberia, Russia in 1992. Catalyst life reached more than 1.5 years. A new industrial plant based on the Zeoforming process of 1000 b/d capacity has been operating at Glimar Refinery in Poland since 1997. Commercial products of the plant are Eurosuper-95 gasoline and liquefied gas with total product yields up to 92–95%. The first catalyst load had worked more than 1.7 years. Other units of capacities of 120–1000 b/d are under various stages of design and construction in Russia, Kirghizia, Ukraine, and Georgia.

5   COMMERCIAL REFORMING CATALYSTS

Beginning in the 1950s, commercial naphtha-reforming catalysts have been essentially monometallic heterogeneous catalysts composed of a support material

Table 10   List of Commercial Catalytic Naphtha Reforming Catalysts[10]

Catalyst designation	Type	Application	Active agents on alumina
Axens			
AR 405, -501	Bimetallic spherical	Aromatic production	PtSn
CR 201, -301, -401	Bimetallic	Continuous	PtSn
CR 502	Monometallic	Cyclic, semiregen.	Pt
CR 701, -702	High-stability bimetallic	Continuous	PtSn
RG-412	Monometallic	Semiregen.	Pt
RG-534	Bimetallic	Cyclic	PtRe
RG-492	Skewed bimetallic	Semiregen.	PtRe
RG-534	Monometallic	Cyclic	Pt
RG-582, -682	Bimetallic	Semiregen.	PtRe
Criterion Catalyst Co.			
P-15	Mono high activity	Semiregen.	PtCl
P-93, -96	Monometallic	Semiregen.	PtCl
PHF-43, 46	Monometallic	Semiregen.	PtCl
PR-9, -11	Multimetallic	Continuous	PtSnCl
PR-28, -291	Multimetallic	Continuous	PtSnCl
PS-7, -10, -20, -30, -40	Multimetallic	Continuous	PtSnCl
Exxon Research & Engineering			
KX-120	Multimetallic	Semiregen., cyclic	PtReCl
KX-130	Multimetallic	Semiregen., cyclic	PtIrCl
KX-160, 170	Multimetallic	Semiregen.	PtReCl
KX-190	Multimetallic	Cyclic	PtSn

Indian Petrochemicals Corp.

IRC-1001	Monometallic	Semiregen.	PtCl
IRC-1002	Monometallic low Pt	Semiregen.	PtCl
IPR-2001	Bimetallic	Continuous	PtReCl
IPR-3001	Multimetallic	Continuous	PtRe

Instituto Mexicano del Petroleo (IMP)

RNA-1	Bimetallic	Semiregen.	PtRe
RNA-1(M)	Bimetallic wider range	Semiregen.	PtRe
RNA-2	Bimetallic trilobe	Aromatics	PtRe
RNA-4	Bimetallic	Continuous	PtSn

UOP

R-50	Bimetallic	Semiregen.	PtRe
R-55	Monometallic	Semiregen.	Pt
R-56	Bimetallic	Semiregen.	PtRe
R-62	Bimetallic spherical	Semiregen.	PtRe
R-72	Multimetallic spherical	Semiregen.	Pt/Promoters
R-85	Monometallic	Semiregen.	Pt
R-86	Bimetallic	Semiregen.	PtRe
R-132, -134	Bimetallic	Continuous	PtSn
R-162, -164	High density; Bimetallic	Continuous	PtRe
R-232, -234	Low coke; Bimetallic	Continuous	PtSn
R-272, -274	High yield; Bimetallic	Continuous	Pt/Promoters
RZ-100	Monometallic	Aromatics	Pt

(usually chlorided alumina) on which platinum metal was placed. These catalysts were capable of producing high-octane products; however, because they quickly deactivated as a result of coke formation, they required high-pressure, lower octane operations. In the early 1970s, bimetallic catalysts were introduced to meet increasing severity requirements. Platinum and another metal (often rhenium, tin, germanium, or iridium) account for most commercial bimetallic reforming catalysts. The catalyst is most often presented as 1/16-, 1/8-, or 1/4-in. Al_2O_3 cylindrical extrudates or spheres into which Pt and any other metal have been deposited. In commercial catalysts, platinum concentration ranges between 0.3% and 0.7% and chloride is added (0.1–1.0%) to the alumina support (η or γ) to provide acidity.

At present, there are six international manufacturers of reforming catalysts producing more than 80 different types of catalysts suitable for different applications and for a variety of feedstocks. The current demand for reforming catalysts is mainly a replacement market with about 75–80% of it as bimetallics. Reforming catalyst manufacturers continue to develop new catalyst formulations designed to meet a wide array of challenges. Many of these challenges involve environmental regulations that refiners have been, and will be, required to meet during the coming years. Table 10 presents a compilation of commercial reforming catalysts that are available by sale or license to refiners [10]. The list provides information on catalyst supplier, catalyst type, and other selected catalyst properties. In addition, new catalyst introductions have been discussed in Chapter 8 of this book.

6 CONCLUDING REMARKS

The refining industry worldwide has been adapting to the ongoing changes and challenges in recent years. New fuel regulations will significantly affect refinery operations. Catalytic reforming will continue to be an important process unit in refinery operations for gasoline production and to further link refining and petrochemical operations. The longer-term trend for catalytic reforming shows increased interest in the development of more selective and stable catalysts for converting naphtha into BTX aromatics and hydrogen alone.

ACKNOWLEDGMENT

The author acknowledges the support of the Research Institute at King Fahd University of Petroleum & Minerals, Dhahran, in publishing this chapter.

REFERENCES

1. Bell, L. Worldwide refining. Oil Gas J. **2001**, Dec. 20, 46.
2. Godwin, G.; Moser, M.; Marr, G.; Gautam, R. In *Latest Developments in CCR Platforming Catalyst Technology*, 40th International Petroleum Conference, Bratislava, September 2001.
3. Dachos, N.; Kelly, A.; Felch, D.; Reis, E. UOP platforming process. In *Handbook of Petroleum Refining Processes*; Meyers, R. ed., McGraw-Hill: New York, 2nd ed., 1997; p. 4.3.
4. Refining Handbook 2000; *Hydrocarbon Processing*; November 2000; 97.
5. Clause, O.; Dupraz, C.; Frank, J. Continuing innovation in catalytic reforming. In *NPRA Annual Meeting*, San Antonio, Texas, March 1998.
6. Le Goff, P.; Pike, M. Increasing semi-regenerative reformer performance through catalytic solutions. In *3rd European Catalyst Technology Conference*, Amsterdam, February 2002.
7. Axens process brochures for octanizing and aromizing processes, Paris, 2002.
8. Aitani, A. Catalytic reforming processes. In *Catalytic Naphtha Reforming*; Antos G. et al. Eds.; Marcel Dekker: New York, 1995; p. 409.
9. Stepanov, V.G.; Snytnikova, G.P.; Ione, K.G. In *A New Effective Process for Motor Gasoline Production over Zeolite Catalysts*, 5th European Congress on Catalysis, Limerick, September 2001.
10. Stell, J. Catalyst developments driven by clean fuel strategies. Oil Gas J. **2003**, Oct. 6, 42.

14

Control Systems for Commercial Reformers

Lee Turpin
Aspen Technology, Inc.
Bothell, Washington, U.S.A.

1 INTRODUCTION

The typical catalytic reforming process is designed to convert a low-aromatics naphtha feedstock to a stabilized reformate stream high in aromatics. Commercial reformer units consist of a reactor section and a fractionation section. The reactor section consists of a feed system, several pairs of heaters, and reactors in series followed by a flash drum. A portion of the flashed hydrogen is recycled to the feed prior to entering the first heater. The flash liquid is then sent to the fractionation section with a distillation tower frequently referred to as the stabilizer. Some units incorporate a recontactor section to increase the hydrogen purity of the net gas stream. The light ends are stripped from the flash liquid to form an off-gas and a liquefied petroleum gas (LPG) stream off of the top of the stabilizer. The stabilizer bottoms product is referred to as the reformate. Several heater reactor pairs are used in the process to maintain the reactor temperature profiles within an operating range of roughly 750–950°F (400–500°C). To achieve these relatively simple tasks in a commercial environment requires a complex set of integrated systems referred to as the management of a catalytic reformer.

Management of a commercial reformer operation has two primary and equally important objectives. One objective is to maintain a safe operation of the process, and the other objective is to run the process in a manner that meets the overall processing requirements for the facility. The safe operation of the process unit covers a variety of issues, foremost of which is the safety of the operators, maintenance workers, and other personnel within the operating facility. Included

497

in the general objective of safe operation is sustaining an operation that ensures the structural integrity of the owner's capital investment in process equipment. The processing objective includes operating at the planned conversion of the feedstock and meeting government requirements such as protection of the environment. To meet the primary and other supplementary operating objectives, a catalytic reformer is controlled by a management continuum as represented by Figure 1. Note that one end of the continuum is basic measurements of process variables such as temperature, pressure, and flow. The other end of the continuum is the economic planning function. Figure 1 is a representative drawing that shows a partial set of the control functions found in a catalytic reformer process unit. Note that there are two distinct paths through the management continuum, one for each objective; one path for safe operation of the unit and the second for control of the unit to meet processing objectives. Providing continuity from one end of the continuum to the other is a set of interlinked control systems. Some of the control systems are automated with traditional instrumentation, and some are supervisory systems implemented with the unit's operating staff. The control systems are tools designed to work together to assist in the management of the unit operation. As the catalytic reformer control systems are further integrated with technology such as regulatory controllers, advanced process control (APC), and real-time online optimization (RT-OPT), the mundane human interactions with the basic regulatory controls are reduced. However, ultimate control by humans is not eliminated. The controllers can be viewed as chained technology packages, each linked to adjacent packages to refine the overall control of the process.

All of the common control systems found in the management of catalytic reformers are discussed in this chapter, with the objective of giving an overall picture of the interlinking of these control systems. More detail is given to automated control systems that are unique to the catalytic reforming process. The fundamental control systems are the same for all catalytic reformers, regardless of mechanical configuration. In those instances where the mechanical configurations require unique control system designs, the control systems are presented as exceptions to the typical controls.

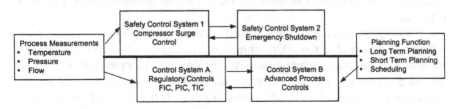

Figure 1   The management continuum.

The following section on fundamental control systems for commercial reformers presents material applicable to all reformer operations. The objective of the discussion is to present information on the different control systems incorporated into the management of a catalytic reformer. The section on APC of catalytic reformers discusses the requirements, benefits, and techniques used in the application of advanced process control to catalytic reformers. The final section on RT-OPT of catalytic reformers presents information on the types of optimization systems available, where they are applicable, and the techniques used to implement an RT-OPT into the overall control system architecture.

2 FUNDAMENTAL CONTROL SYSTEMS FOR COMMERCIAL REFORMERS

From an instrumentation and control perspective, the process unit is limited to the reactor section and primary separation section in this chapter. The reactor section consists of the feed control, feed/effluent exchangers, fired heaters, reactors, effluent cooler, product separator, and recycle compressor. For the case of continuous catalyst circulation units, the catalyst regeneration section will be covered in the APC and optimization sections. Moving from the primary instruments along the management continuum are the fundamental regulatory control systems. These systems follow two parallel paths as shown in Table 1, both taking process measurements from the primary instruments.

2.1. Basic Regulatory Control Systems

All of the control systems are built on measurements of the process conditions. Primary instrumentation in commercial catalytic reformers consists of basic temperature, pressure, and flow transmitters integrated with single-loop proportional integral (PI) and proportional integral derivative (PID) controllers. More sophisticated online analyzers are incorporated into some units.

Conventional orifice plates are used for flow measurements. Pressure is usually measured using force-type pressure sensors. Type J thermocouples are

Table 1 Fundamental Control Systems

Process control systems	Safety systems
Fundamental regulatory control system	Equipment shutdown systems
Supervision control systems	Emergency shutdown systems
Planning and scheduling function	

used for temperature measurement in cold service and type K thermocouples in hot service. Duplicate sensors are often installed to provide redundant readings for the safety systems. The standard primary sensing elements have been selected to meet three criteria: (1) producing products on specification, (2) equipment protection, and (3) personnel protection, with attention given to capital cost, reliability, and ease of maintenance. The level of accuracy is determined by the point needed to safely monitor and operate the facility.

The majority of catalytic reformer control systems include an on-line analyzer for the recycle gas hydrogen concentration. Frequently an analyzer for recycle gas moisture is also installed. Other process analyzers are used, but infrequently.

Traditional Control Systems

The traditional control systems consist of a number of individual single-task control and monitoring instruments. Control rooms have banks of instrument packages, each processing a separate control function. In some cases multiple sensor readings (such as temperatures) may be collected into a single display mechanism. Sensor readings are either transformed in the field to a usable measurement (torsion to pressure), or are collected and transformed in an instrument located in the control room (emf from a thermocouple to temperature via a Wheatstone bridge). Sensor readings that are used for feedback in a control loop are normally limited to that instrument. All of the controllers are single-input, single-output PI or PID control loops. Controllers can be cascaded from one controller to the next, but only within the limitations of PI or PID control [1].

The time to steady state (TSS) in all of the catalytic reformer control loops is sufficiently short to allow a traditional control system to support the implementation of PI and PID controllers. Some controls, such as heater excess air control, cannot be implemented safely within the confines of the basic instrumentation.

In the operation of catalytic reformers, a traditional control system is perfectly adequate for performing basic regulatory tasks. The traditional control system is not adequate for supporting more sophisticated control technology such as APC, expert advisor systems, and on-line optimization.

Distributed Control Systems

A distributed control system (DCS) performs all of the same basic functions of the traditional control system, but a DCS system processes sensor signals and handles the control functions in a slightly different manner than the traditional control system. Common to all DCS systems are four key functions. First, a single sensor reading can be shared by several controlling and monitoring instruments without deterioration of the sensor signal. Second, all sensor readings can be easily processed. This means they can be easily placed into common engineering units, bad signal analyses can be performed, and signals can be modified to constant conditions of temperature, pressure, and gravity. Third, the controllers are embedded into the DCS system and can be changed from PI to PID simply via setup specifications. The fourth difference is that the sensor signals, controller settings, and any other information can be easily shared with other assemblies in the DCS or connected to the DCS. This could be a memory device or a computer for application of more sophisticated controllers, such as APC, expert advisors, and RT-OPT.

2.2. Reformer Regulatory Control System

The regulatory control system in a commercial reformer allows the unit operators to convert the operating plan developed by management into the conversion of naphtha to reformate, LPG, and hydrogen. Typically, the management operating plan will tell the operators to run a certain feedstock, at a certain rate, and at a certain processing severity. This plan has to be integrated with safety policies, efficiency guidelines, environmental regulations, and the unit's physical limitations, and then transformed into control terms of flow rates, temperatures, and pressures.

Regulatory Controls in a Catalytic Reformer

The regulatory control system is used to control the conversion of naphtha to reformate with the required product properties and recovery, and to meet defined mechanical constraints such as heater excess air. Typically the reformate properties are research octane (RON) and vapor pressure (RVP) [2]. Figure 2 is a simplified instrument diagram of a catalytic reformer unit [3]. Only the key closed-loop controllers are shown and are represented by flags at their feedback sensor location. For simplicity, level controllers have not been included because they are not pertinent to the control of the operation of the process.

Table 2 lists the primary operating objectives of a catalytic reformer (control variables) and the controllers used to meet the processing objectives (manipulated variables). RON, or an alternate conversion target, is maintained by adjusting the reactor temperature controller setpoints [4]. In conventional operations, the control is open loop. The operators make adjustments based on feedback from laboratory analyses or to account for process disturbances such as changes in feed rate. In an APC application, the setpoints are adjusted automatically based on on-line measurement or inference of severity. Temperature controllers adjusting the fuel to the fired heater preceding each reactor maintain the reactor temperatures. Quite often the temperature controller

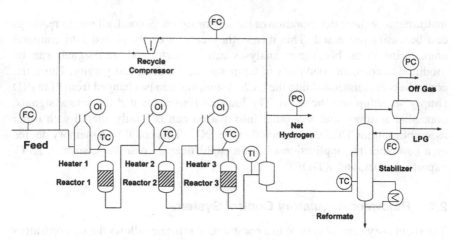

Figure 2 Simplified instrument diagram—catalytic reformer.

is cascaded to either a pressure controller or a flow controller on the fuel source. Some applications use on-line measured fuel gas heating value compensation to the fuel controller to alleviate disturbances. Heater outlet temperature setpoints are commonly adjusted simultaneously and to approximately the same value. When economics or equipment constraints dictate, the temperature profiles are skewed [5].

The driver for varying hydrogen recycle is its effect on coke production. Units in an unconstrained coke operation are run to a specified hydrogen-to-hydrocarbon ratio in an attempt to control the rate of coke laydown to an acceptable rate. Those units that are coke make-constrained maximize hydrogen

Table 2 Basic Regulatory Control: Typical Operating Variables

Control variables	Manipulated variables
Severity (usually RON)	Heater outlet (reactor inlet) temperature
Hydrogen recycle rate	Recycle flow
Feed rate	Feed flow controller
Heater excess air/firebox pressure	Stack damper/air register position
Reactor system pressure	Product separator pressure
Stabilizer pressure	Stabilizer overhead receiver pressure
Reformate RVP	Stabilizer bottoms temperature
Stabilizer overhead specification	Stabilizer reflux flow controller

recycle to the limits of the instrumentation or process equipment, with equipment constraints being the most common limit. The type of compressor and associated driver usually determines the instrumentation used for hydrogen recycle gas control. The three most common methods of instrumentation are shown in Figure 3. Depending on the design requirements, aspects of the various systems may be incorporated into an integrated design.

The reformer feed rate is usually specified by the refiner's planning department [6]. To reduce the impact of moisture absorbed in stored feed on the catalyst, reformer feed control systems are designed to maximize feed coming directly from the hydrotreater and still be on flow control. If the bottom of the hydrotreater stripper is adequately sized, the flow to the reformer can be on flow control, and the surge volume of the hydrotreater stripper is maintained by adjusting the feed rate to the hydrotreater as shown in Figure 4a. In this configuration, all of the reformer feed comes directly from the hydrotreater. A common feed line-up is presented in Figure 4b. A small part of the treated naphtha from the hydrotreater goes to storage, the bulk goes to the reformer, and the reformer feed rate is maintained by a slip stream from storage. The reformer feed tank serves as a surge drum between the hydrotreater unit and the reformer.

Heater excess air and firebox pressure are controlled simultaneously with the air registers and the stack damper(s). Closed-loop controllers must be designed in such a manner that disturbances in the unit operation that (1) increase the firing rate are used as feedforwards to open the damper prior to additional fuel being added to the system, and those that (2) decrease the firing rate are not used as feedforwards to close the damper prior to the reduction in fuel rate [7]. The burner manufacturer specifies minimal excess air rates. Recommended control schemes for fired heater excess air and firebox pressure controls are outlined in API RP-560. In most regulatory control systems on catalytic reformers, the damper position and registrars are in open-loop control with the operators making the required adjustments to meet both operating objectives and safe operation. This is done with either manually operated devices such as a chain wheel drive on a damper or a hydraulic positioner.

The reactor system pressure is simply controlled by a pressure controller on the product separator or on the product separator vent line. The excess hydrogen produced in the dehydrogenation reactions is vented to fuel or, more commonly, to a hydrogen header. In most reformer complexes the product separator pressure is set by the design and cannot be varied significantly due to a combination of factors, including feed pump head, compressor constraint issues, valve and pipe sizes, relief valve settings, stabilizer pressure, and downstream unit pressures.

Product separator temperature is usually not directly controlled. Tight control of the product separator temperature is normally not critical to the unit operation or equipment safety. Some units have temperature controllers adjusting the fin fan speed or louver positions.

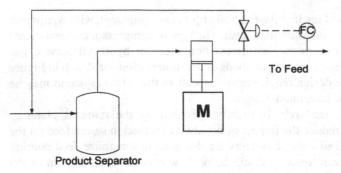

To Feed

Product Separator

(a) Reciprocating compressor with motor drive and spill-back

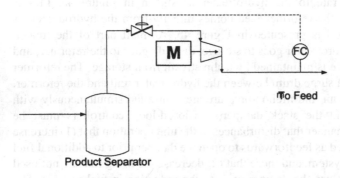

To Feed

Product Separator

(b) Centrifugal compressor with suction throttle valve

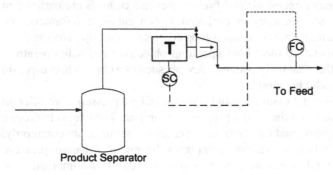

To Feed

Product Separator

(c) Centrifugal compressor with variable speed drive

Figure 3 Recycle compressor instrumentation.

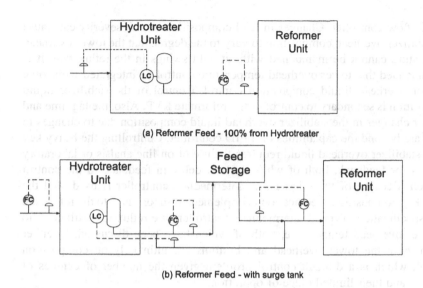

(a) Reformer Feed - 100% from Hydrotreater

(b) Reformer Feed - with surge tank

Figure 4 Reformer feed diagrams.

Stabilizer pressure is maintained using a traditional pressure controller venting the off gas either to fuel or to a light-ends recovery unit. Maintaining system pressure is critical to maintaining proper separation in the stabilizer between the designated light and heavy keys in the tower. In most reformer complexes the stabilizer pressure is set by the design and cannot be varied due to a combination of mechanical limits.

RVP is typically maintained by adjusting the bottoms temperature of the stabilizer tower and using laboratory feedback. The best point for temperature measurement to control RVP is approximately three trays from the bottom of the tower. The third tray also offers the best location for the temperature measurement used in an inferred RVP calculation.

The stabilizer reflux rate is typically set to meet one of four specifications: (1) a defined reflux rate; (2) a reflux-to-distillate ratio (reflux-to-feed is occasionally specified); (3) the tower overhead temperature; or (4) a heavy key concentration in the overhead liquid. The control of a defined reflux rate is accomplished using a standard flow controller on the reflux line. Because the liquid distillate off the stabilizer overhead receiver is on level control, the calculated reflux ratio used in reflux-to-distillate control uses a weighted average distillate rate in the denominator. (The calculation of the reflux-to-feed ratio uses the same type of weighted average feed rate because the stabilizer feed is on level control from the product separator.) The use of reflux to control the tower overhead temperature is normally done by cascading the overhead temperature to a reflux flow controller. Changes in feed composition or unit severity can cause the stabilizer overhead composition to vary to the degree that the tower overhead temperature cannot be maintained without wild swings in the reflux rate. It is recommended that tower overhead temperature control be integrated with some form of overhead liquid composition control. Control of the stabilizer liquid composition is secondary to control of the reformate RVP. Also, the lag time and TSS for changes in the stabilizer overhead liquid composition due to changes in reflux are beyond the capabilities of a PID controller. Controlling the heavy key in the stabilizer overhead liquid requires the use of on-line analsis or laboratory analysis for feedback, both of which have a delay in feedback to the control function. Because of these issues, an intermediate controller is used, and the heavy key composition specification is implemented subordinate to the reformate RVP specification. Typical intermediate controllers are reflux-to-distillate ratio or tower overhead temperature, both of which have been discussed. In either option, both the tower overhead and bottoms are ultimately on composition control, which is a difficult control problem given the number of degrees of freedom and their limited range of operation.

Auxiliary Control System

In addition to the fundamental process control system, there are a series of auxiliary control systems. These will vary from unit to unit depending on the plant mechanical configuration and process equipment used in the reformer design. The systems listed in Table 3 and discussed in this section are representative of the auxiliary control systems to control utilities or control process equipment systems. Each auxiliary control system may have a set of subsystems. For example, a waste heat boiler will have a regulatory control system for steam generation, and subsystems for injection of chemicals and boiler blowdown. Auxiliary control systems may also have their own safety systems, such as pressure relief systems.

Table 3 Example Auxiliary Control Systems

Waste heat boiler control system
Water/chloride injection system
Compressor lubrication system
Cooling tower acid/chlorine control system
Instrument air system
Fuel system

Heat to the reactor section of the reforming process is transferred in the radiant sections of the charge and interheaters. To increase the efficiency of the fired heater(s) a convection section is added with a convection duty coil for the stabilizer reboiler, an air preheater, a waste heat boiler, or some combination of these waste heat devices. The most common use of the convection heat is a waste heat boiler. A typical control strategy is shown in Figure 5.

The boiler feed water is made up to the steam drum on level control. The water-circulating loop maintains the flow at roughly five times the steam rate. The drum is on conventional pressure control. A superheater is shown in Figure 5 and may have a quench. Control of water quality is done with standard wet laboratory quantitative laboratory tests. Chemical injection and blowdown rates are adjusted manually based on routine test results and changes in rate of steam production.

The catalyst acidity is maintained by injection of water (or a compound that converts to water in the reformer environment) and an organic chloride-bearing compound that will release the chloride on the catalyst. The water and chloride agent are injected into the reactor system in very small quantities using small chemical injection pumps. These are typically plunger pumps with self-contained regulators that permit the operators to set a specified injection rate. Some of the regulators allow the settings to be adjusted through a DCS system.

Compressor lubrication systems are self-contained oil systems with a circulation pump, filters, and associated flow control devices. Subsystems include pressure relief systems and alarms systems. Signals to the emergency shutdown

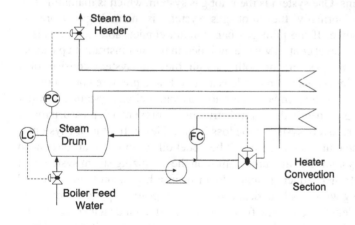

Note: Blowdown, chemical injection system, pressure relief and corrosion monitoring not shown.

Figure 5   Typical waste heat boiler controls.

system are included to shut down the compressor upon failure of the lubrication system.

The water quality in a cooling tower system is primarily maintained by injecting acid and chlorine. On-line analyzers are used to maintain pH, conductivity, and dispersions by regulating the quantity of acid and dispersant injected into the system and blowdown rate. The dead time and TSS are sufficiently long so that conventional PID control is not suitable for this application. Control systems are used that are specifically designed to handle cooling water chemistry and operate in the severe atmosphere around a cooling tower. These are usually programmable logic controllers (PLCs) installed at the cooling tower, which receive input from the analyzers and send output signals to regulatory valves and chemical injection pump flow controllers.

Instrument air systems are often located in process units or situated between units in close proximity to one another. Plant air is passed through a dryer and then reduced in pressure to approximately 30 psig. Instrumentation on the plant air system usually consists of a pressure controller sized to take wide swings in load and still maintain a stable air system pressure throughout the process unit. In some cases moisture analyzers are included in the system design. Most instrument air systems have a primary air source and a secondary air source to provide air in the situation where there has been a failure in the primary air source. This normally includes a second pressure regulator with its setpoint slightly below the primary regulator setpoint. The difference in settings must be large enough to provide for normal fluctuations in the air supply pressure.

The fuel system in a reformer unit normally consists of two or more independent systems. One system is the pilot gas system, which is maintained on pressure control. Normally the pilot gas system is natural gas from an uninterruptable source. If the pilot gas can contain condensable compounds, a knockout drum with level control will be included in the unit instrument package. Reformer heaters fire gas, oil, or both. A unit fuel gas system consists of a knockout pot with level control on condensed liquids and pressure control on the fuel gas going to the heaters. In both the pilot gas and fuel gas systems, blowout instrumentation is provided in the control structure to prevent a depressurizing of gas into the liquid recovery system upon loss in liquid level. If a fuel oil system is also included in the unit design, there will be a fuel oil system with an associated atomizing steam system. The instrumentation on the atomizing steam system will be a pressure controller reducing steam down from a higher pressure saturated steam header, along with knockout drums or steam traps to keep the system from accumulating condensed water. The fuel oil system will include a level controller to maintain a stored quantity of fuel oil on site, a pressure controller to maintain pressure control on the system, and a flow controller to maintain a circulating oil supply. Quite often the circulating oil flow is controlled simply by having a restriction orifice (RO) in the fuel oil return line. Figure 6 shows the controls on

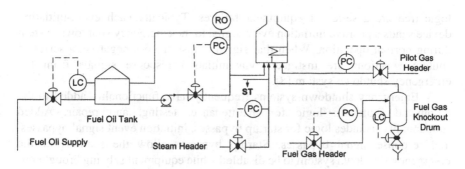

Figure 6 Fuel system controls.

an integrated fuel system. The process controllers have been omitted from the drawing for clarity. Significant control problems can exist in an operation where both fuel oil and fuel gas are being simultaneously fired to maintain a process temperature setpoint. The dynamic response to changes in the firing systems between fuel gas and fuel oil are appreciably different. One controller cannot be used for both services with adequate responsiveness. If the operation is going to switch between fuel oil and fuel gas, two separate controllers need to be utilized. The same primary sensor element can be used, but the controllers must be different. If both fuels are to be fired simultaneously, one fuel should be base loaded and the second used for control of the process temperature. Base loading oil and controlling with fuel gas is viewed by most operating companies as the preferred operation. Fuel availability issues may dictate that the fuel gas be base loaded and the fuel oil used to control the setpoint.

2.3. Emergency Shutdown Systems

The control system on a catalytic reformer that handles the shutdown of the equipment and blocking in of hydrocarbon lines in the case of catastrophic failure or when a piece of equipment has exceeded a safe limit of operation is known as the emergency shutdown system. Emergency shutdown systems are designed to minimize damage to process equipment and protect personnel; they are not used for equipment shutdown in the normal course of operation, only in emergencies. The basic emergency shutdown system is a logic system that sets a sequence of actions into place when one of a number of causes, called "initiating events," activate the system via an "initiation event signal." The resulting action(s) are triggered by "initiation response signals" generated by the emergency shutdown system. This is handled using a logic tree, with an operator-initiated emergency shutdown as one of the initiating actions. The other initiating branches in the logic tree are a series of equipment failures. Typically each event-initiating device sends a positive initiation event signal to the emergency shutdown system during normal operation. When that signal is lost or goes negative, a series of shutdown actions are instigated via initiation response signals from the emergency shutdown system [8].

Emergency shutdown systems frequently have functionality added to the automated shutdown logic for maintenance, testing, and repair. Added functionality includes logic for startup bypasses, initiation event signal bypasses, and response signal bypasses. Startup bypasses allow the sections of the emergency shutdown system to be disabled while equipment is being brought on-line. Initiation event signal bypasses cut off the signal from the various initiation event generators and send a false positive (normal operation signal) initiating event signal to the emergency shutdown logic. This allows instrumentation to be repaired without shutting down the unit. The response signal bypasses block an event-generated output signal so that testing and maintenance can be performed on various pieces of equipment, including the emergency shutdown system [9].

Emergency Shutdown System—Initiating Events

Initiating events come from three sources. The first initiating event is an operator-initiated event. An operator initiates the shutdown system in one of two cases. The first case would be when a system failure has occurred outside of the planned instrumentation. An example is processor utility line rupture. A second case in which an operator would initiate the emergency shutdown system is when there has been an instrument failure and the emergency shutdown system failed to respond to an initiating event.

The second initiating event is failure of a piece of equipment or a supporting system. For example, the feed pump to the unit may shut down due to a bearings failure. The feed flow would go to zero and the shutdown system would be initiated. An example of a piece of equipment failing due to a supporting system is a recycle compressor shutdown due to the failure of a lubricating pump. The recycle flow would go to zero and the shutdown system would be initiated.

The third source of an initiating event is the loss of a utility system such as fuel gas, steam, instrument air, or electricity.

Logic Tree for a Catalytic Reformer Emergency Shutdown System

Emergency shutdown systems can range from a simple series of integrated switches to extremely complex logic functions. As the sophistication of instrumentation and the use of computers, DCSs, and PLCs have become more prevalent, the emergency shutdown systems have become more complex. One of

Table 4 Sample Emergency Shutdown Response Matrix

	Response actions. Check indicates response				
Event initiators	Cut fuel to heaters	Close feed valve	Shut down recycle compressor	Shut valve to stabilizer	Shut net hydrogen valve
Operator initiated	✓	✓	✓	✓	✓
Feed pump shutdown	✓				
Recycle compressor shutdown	✓	✓			
Electric power failure	✓	✓		✓	✓

the most easily understood shutdown logic systems is a shutdown matrix as shown in Table 4. This emergency shutdown response matrix is abbreviated for demonstration. The event initiators are listed on one axis and the resulting actions listed on the other.

Complex Boolean logic diagrams are sometimes prepared for catalytic reformers. The complex logic diagram methodology increases the flexibility of the system. For example, if two initiation events happen simultaneously, the resulting set of responses may be different from the sum of the two responses designated in a response matrix [9].

Emergency Shutdown System and Overall Plant Safety

Design and implementation of emergency shutdown systems are regulated by governments, industry groups, instrument and safety societies, or company policies as part of a process safety management program. Life cycle management programs often require multidisciplined reviews and associated record keeping of the initial installation of control systems and all subsequent modifications. Regulations often include requirements of multilayered control and safety systems that can each function independently of one another yet work together to form an integrated system. Layered safety systems within the reformer process area are often extended beyond the battery limits to incorporate the reformer operations and safety systems as part of an extended plant safety system [8].

2.4. Process Equipment Safety Systems

Each piece of process equipment has at least one safety system associated with its operation designed to protect equipment and the personnel who operate the process units. The mechanical design of the equipment, type of equipment, and company safety policies dictate the systems to be used in a particular reforming unit. The equipment safety systems may or may not be event-initiating devices in the emergency shutdown system.

Fired Heaters

Fired heaters are typically instrumented with at least three different safety systems. One is a shutoff on the fuel system if there is a disruption to the fuel flow—typically triggered by a low flow measurement device. A second safety system is a shutoff of the fuel system if the flame goes out in the heater. Instruments located in the heater detect the presence of flame via light intensity. The third common safety system is low oxygen content in the flue gas stack or the flue gas in the transition zone between the radiant and convection sections of the heater. Upon the measurement of low oxygen content in the flue gas, the burner louvers and/or stack damper are opened to increase the oxygen entering the heater. In addition to directly measured safety variables, such as temperature and oxygen content, calculated tube skin temperatures and flux rates are used as operating constraints [10].

Recycle Compressor and Drivers

Centrifugal recycle compressors are typically instrumented with surge control systems. When applicable, a kickback valve will be opened to move the operation away from a potential surge in the compressor. Electric motor drivers are monitored for overamperage operation through time. The electrical switch gear will break the electric supply to the motor before the motor coils or windings exceed a specified temperature. Steam turbine drivers are installed with govenors.

Pressure Relief System

In a catalytic reformer, all of the piping and equipment is protected by pressure relief systems from uncontrolled process fluid expansion due to a heat source, chemical reaction, or mechanical compression. Typically these are relief valves, but they can be rupture disks, or a combination of both rupture disks and relief valves. The vessels and piping are divided into operating zones such that any area of pipe and vessels that can be blocked in and are subject to a process fluid expansion are protected by a pressure relief system.

2.5. Catalyst Control Functions

Control of the reformer catalyst is centered around two areas. The first is the catalyst environment—the managing of the chloride level and its distribution on the catalyst. The second area of catalyst control is management of the catalyst deactivation. Several performance indicators are calculated either by engineering staffs or automated systems to aid in monitoring and controlling performance of reformer catalyst.

Catalytic Reformer Performance Indicators

The following performance indicators are calculated on a scheduled or as-needed basis. These performance indicators are used in monitoring the process unit and in some cases are used as control variables in APC applications [12].

WAIT—weighted average inlet temperature = $\Sigma(T_i * RXWT_i)$
  T_i = inlet temperature to reactor (i)
  $RXWT_i$ = weight of catalyst in reactor (i)
Adjusted WAIT—WAIT adjusted to base operating conditions
ABT—average bed temperature = $\int T * d(L)$, L = 0–1
  T = temperature of catalyst at distance L through the catalyst bed
  L = fraction of the distance through the catalyst bed in the direction of hydrocarbon flow
WABT—weighted average bed temperature = $\Sigma(ABT_i * RXWT_i)$
  ABT_i = average bed temperature of reactor (i)
  $RXWT_i$ = weight of catalyst in reactor (i)
LHSV—liquid hourly space velocity
WHSV—weight hourly space velocity
H2/HC—hydrogen-to-hydrogen ratio = mole H_2 recycle/mole feed
C_1 molar ratio, C_2 molar ratio, C_3 molar ratio
C_4 isomer ratio = iC_4/nC_4
Aromatics production as weight percent of feed
Aromatics as a weight (or volume) percent of reformate (or C_5+)
MCP (methylcyclopentane) conversion = $100 * (MCP_{IN} - MCP_{OUT})/MCP_{IN}$
  MCP_{IN} = moles of MCP in the feed
  MCP_{OUT} = moles of MCP in the net reaction products
Paraffin disappearance = $100 * (PARA_{IN} - PARA_{OUT})/PARA_{IN}$
  $PARA_{IN}$ = moles of C_5+ paraffin in the feed
  $PARA_{OUT}$ = moles of C_5+ paraffin net reaction products
Aromatics selectivity = $100 *$ paraffin converted to aromatics/paraffin disappearance
RON (research octane number)
RVP (Reid vapor pressure)

Catalyst Environment

The catalyst environment refers to quantity and distribution of chloride on the catalyst and the amount of water in the system. Because chloride and water leave the unit with the reaction products, small quantities are injected to maintain the prescribed water/chloride balance. In many units catalyst samples are taken to monitor the chloride content, and in some systems correlations of the methane, ethane, and propane molar ratios are used for feedback in the control of the catalyst environment. Moisture analyzers are normally installed to measure the water content of the recycle gas to (1) provide feedback to the calculated recycle gas moisture content and (2) signal if the water content of the feed has varied [3].

Catalyst Deactivation

Catalyst deactivation is primarily a function of coke deposition on the catalyst. The catalyst poisons other than coke are managed through proper operation of the preceding naphtha hydrotreater. Controlling coke deposition is equivalent to controlling catalyst deactivation. Several methods of tracking the deactivation are employed, but all serve the same objective: to anticipate unit performance by monitoring past and present operation. In many units, two key variables are tracked: coke laydown rate (coke deposition per unit time) and weight percent carbon on catalyst. In a continuous catalyst regeneration (CCR) operation the coke on catalyst must be maintained at a level to maintain reasonable activity and proper regenerator performance. In a semiregenerative reformer operation the objective in catalyst activity management is to end a cycle of several months of operation with a specified weight percent coke on catalyst. The coke on the catalyst through a cycle follows an exponential rate of deposition. Controlling the coke laydown rate by adjustment of feed rate, feed composition, hydrogen recycle, and severity allows a processor to proceed through a cycle at a planned WAIT through the cycle. Chapters 10 and 11 discuss coking in more detail.

3 ADVANCED PROCESS CONTROL OF CATALYTIC REFORMERS

Modern APC systems use one or more multivariable controllers to control the operation of a catalytic reformer. APC assists in the control of a reformer in two areas of the operator's standard duties. First, APC adjusts the basic regulatory controllers to meet a set of operator-entered processing objectives, and second, APC monitors the unit operation and makes adjustments as needed for disturbances in the unit operation. Unlike an operator, APC applications are normally proactive and start adjustment of control variables at the time of the disturbance, before changes in measured variables can be recognized [4].

There are three fundamental differences between the way an operator performs and the way an APC application works in a catalytic reformer. First, APC applications may be used to process delayed feedback (long lag times) from process instruments and long intervals to steady state (long TSS) on a model predictive basis. However, operators need to wait well past the lag time to analyze the impact of the disturbance on the process before taking an action. Second, operators tend to make a series of directionally correct changes with necessary time intervals for the process to come to steady state, for sample collection, and for sample analysis, and eventually arrive at an appropriate operating condition. An APC application makes a more precise move in its first step; then it starts making adjustments as soon as feedback can be detected rather than waiting for the unit to come to steady state. On-line analyzers or inferred properties are used for feedback when a process variable cannot be measured directly with traditional temperature, pressure, and flow instrumentation. The third difference between an operator and APC is the way overlapping constraints are handled. An APC application recognizes at the initial time of the operator entering the new target whether or not there will be multiple constraint/target conflicts and resolves them at the time of first move of the independent variable. In some cases, multiple independent variables may be moved to resolve the processing conflicts.

APC can only manipulate those variables that are in closed-loop control under the APC system. Operators must bridge the control gap when a system is open loop. For example, if the cooling water temperature starts to increase, the flash drum temperature may increase. This in turn results in a higher molecular weight of the recycle hydrogen gas, which may cause the recycle compressor to bump a driver limit. The only solution available to the APC system may be to reduce charge rate. However, the operator may have another option, such as turning on another fan in the cooling tower.

An APC application works with three kinds of variables: manipulated variables (MVs), control variables (CVs), and feedforward variables (FFs). In general terms, the multivariable controller maintains the operation of the reformer at multiple CVs. All MVs in an APC application are variables with closed-loop control in the regulatory control system, usually temperature, flow, and pressure controllers. The CVs are most often measured variables such as instrument readings and analyzer output, but can also include calculated performance indicators or inferred properties. Table 5 is a list of the common MVs, CVs, and FFs found in a catalytic reformer APC application.

3.1. Instrumentation Requirements for APC

The instrument requirements for implementation of APC in a catalytic reformer are typically the same as for basic regulatory control. APC can control a reformer

Table 5   APC Variables List

Variable description	Type	Notes
Manipulated Variables		
Feed rate	SP	
Recycle compressor turbine speed	SP	
Charge heater outlet temp.	SP	
Interheater 1 outlet temp.	SP	
Interheater 2 outlet temp.	SP	
Interheater 3 outlet temp.	SP	
Product separator pressure	SP	
Heater burners—primary air louver position	SP	
Stabilizer feed temp.	SP	
Stabilizer reflux rate	SP	
Stabilizer tray 26 liquid temp.	SP	
Feedforward Variables		
Cooling water temp.	PV	
Control Variables		
Reformate RON	PV	Inferred Property
C_5+ in stabilizer overhead liquid	PV	GC
Reformate RVP	PV	Inferred Property
Weighted average bed temp. (WABT)	PV	Calculated
Hydrogen-to-hydrocarbon ratio (H_2HC)	PV	Calculated
Heater box pressure	PV	
Charge heater tube skin temp.	PV	Inferred
Interheater 1 tube skin temp.	PV	Inferred
Interheater 2 tube skin temp.	PV	Inferred
Interheater 3 tube skin temp.	PV	Inferred
Stabilizer reflux ratio	PV	Calculated
Heater excess air	PV	Calculated
Net gas valve position	PV	

SP, setpoint; PV, process variable.

operation more precisely than human beings, but human beings can handle poorly tuned controllers, open-loop control, and discontinuous operations.

The degrees of freedom for each unit operation must be analyzed to ensure that the problem is not overspecified and yet is completely specified. It is recommended that a pairing chart be constructed to assure that each CV has a valid function, and that each CV has at least one associated MV which when moved will significantly impact the magnitude of the value of the CV. If the number of CVs exceeds the number of MVs, all of the CVs will probably not be solved exactly. For example, if the steam rate to the stabilizer reboiler is the MV, and the bottoms temperature is a CV (process operating target) and the steam valve position is a second CV (mechanical constraint), when the bottoms temperature specification is solved (exactly) by adjusting the steam flow rate, the steam valve position will be less than its mechanical constraint (not exactly at limit). To solve the problem of multiple CVs without direct solutions, typically a linear program is included in the multivariable controller to maximize a profit function.

The multivariable controller on a reformer unit must have each contributing MV and disturbance variable properly associated with the CVs. Table 6 is a sample theoretical pairings matrix relating typical and contributing MVs and FF (disturbance variables) to the CVs of a catalytic reformer APC application. Plant tests performed at the beginning of an APC implementation project determine the actual pairing with the CVs in the controller. Basic regulatory controls must be properly paired with the correct feedback variable. If the CV does not have a paired MV that can move the control variable over the anticipated operating range, the function cannot be implemented in an APC application.

Instruments must not display hysteresis. Hysteresis is a control problem where the output from an instrument is not consistent, both in response times and in magnitude for a specific change of input, or does not give a mirror image in magnitude of response when the input is reversed. An example of hysteresis is a stack damper with positioner input signal 'X' for position 'A'. The input signal is changed from 'X' to 'Y' and the damper moves from position 'A' to 'B'. But when input signal X is reapplied, the damper moves to position 'C' rather than back to position 'A'.

3.2. APC Application in the Reactor Section

Advanced process control in the reactor section of a catalytic reformer is centered on (1) maintaining a target operating severity in the process, (2) controlling catalyst deactivation, and (3) controlling the fired heaters. In a reformer with continuous catalyst circulation, a fourth area is controlling the coke loading on the catalyst to meet the regenerator burn requirements [13], and will be covered in Section 3.4.

Reformer Severity Control with APC

Traditional reformer severity control in gasoline operations and many chemical operations is based on RON. Other severity targets, such as aromatics make, are frequently used in chemicals operations. RON and the alternative severity definitions are primarily basic expressions of conversion of paraffins to

Table 6 Example MV and CV Pairings Matrix

CVs	MVs and FFs										
	Feed rate	Comp. turbine speed	Chg. htr. outlet temp.	Heater 1 outlet temp.	Heater 2 outlet temp.	Heater 3 outlet temp.	Prod. sep. pressure	Burner louver position	Stab. feed temp.	Stab. reflux rate	Stab. tray 26 liq. temp.
Reformate RON	✓	✓					✓		✓	✓	✓
C₅ + in stab. OH	✓						✓		✓	✓	✓
Reformate RVP	✓						✓		✓	✓	
Wt. avg. bed temp.	✓	✓	✓	✓	✓	✓					
H₂HC ratio	✓	✓	✓	✓	✓	✓	✓				
Heater box pres.	✓	✓	✓	✓	✓	✓		✓			
Charge htr. TST	✓	✓	✓	✓	✓	✓		✓			
Heater 1 TST	✓	✓	✓	✓							
Heater 2 TST	✓	✓	✓		✓						
Heater 3 TST	✓	✓	✓			✓					
Stab. ref. ratio	✓	✓					✓		✓	✓	✓
Heater excess air	✓	✓		✓	✓	✓		✓			
Net gas valve pos.	✓	✓	✓	✓	✓	✓					

aromatics, which is driven by reactor temperature. In a modern APC application the CV severity can be measured or inferred and the reactor inlet temperature MVs specified directly.

Reformer Catalyst Deactivation Control with APC

Catalyst deactivation is a result of poisons in the feed and coke laydown due to unit upsets, feed composition, temperature, and hydrogen partial pressure. APC applications can be designed to handle all of these factors except unit upsets. Hydrogen partial pressure is normally adjusted by changing the hydrogen recycle rate as opposed to changing system pressure. Very few commercial reformers have been built with the capability of making significant shifts in unit pressure. In the simplest of APC applications, the hydrogen-to-hydrocarbon ratio is maintained.

In more sophisticated applications, the APC controls to a maximal coke laydown rate in mass per unit time. Figure 7a shows an idealized plot of coke deposition over a catalyst cycle. The slope of the plot at any point in time can then be taken as the coke laydown constraint in the APC application. However, the planned coke deposition plot and reality rarely match. Unplanned poisons in the feed, or unit upsets and other factors create variations in the planned operation. To account for these factors, the coke deposition plot must be updated through the cycle using either measured coke or inferred coke deposition. Some reactor sections have catalyst samplers that can be used to sample the catalyst and permit measurement of the coke. An alternative is to generate a coke vs. adjusted WAIT plot. Figure 7b is a representative plot of coke vs. temperature. As the adjusted WAIT requirement increases through the cycle, the coke deposition plot can be reconstructed to provide new coke laydown constraints.

APC of Reformer Fired Heaters

Firebox management describes the strategies that (1) provide the correct amount of heat to meet processing objectives, such that (2) heat loss to the environment is minimized, and is done in a manner that is (3) safe for workers and maintains equipment integrity. APC regulatory controls must be integrated in such a way as to provide stability and flexibility. Heat loss is a function of mechanical design, maintenance, and the excess air provided for combustion. The latter is the only parameter that can be affected by heater controls.

There are several aspects of fired heater safety. In the domain of control, this means providing mechanisms that prevent explosive mixtures from being formed in the firebox and both general and localized overfiring. Maintaining the burner vendor-prescribed excess air will ensure that all fuel is burned in the

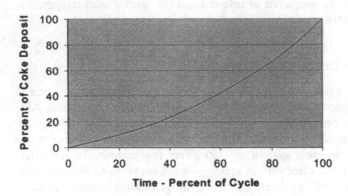

(a) Coke Disposition Over Time

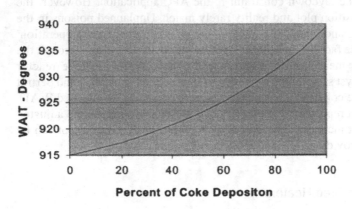

(b) Planned Adjusted WAIT Over Time

Figure 7 Coke deposition plots.

combustion zone of the burner and the resultant flame pattern is compatible with
the firebox configuration. It is also necessary to maintain the correct pressure
(draft) balance in a fired heater to ensure correct air/fuel mixing in the burner and
to minimize air leakage through the heater walls.

3.3. APC Application in the Separation Section

In the application of APC to catalytic reformer units, the area that is typically underspecified is the stabilizer operation. Sufficient degrees of freedom typically exist to allow for two tower specifications, one on the top operation and one on the reformate. Common tower top specifications include (1) total C_5+ in the overhead liquid, (2) benzene in the overhead liquid, or (3) reflux ratio. The composition specifications require either an on-line analyzer or use of an inferred property for feedback. Because of the concentrations being specified, the inferred properties are considered adequate for most operations but not precise. Inferred properties have the advantage over on-line analyzers because of reduced lag time in feedback. A calculated value is nearly instantaneous and continuous. In contrast, the output signal from an on-line gas chromatograph is incremental, typically with a 5- to 10-min gap in feedback results. To overcome this problem even when on-line analyzers are available, inferred properties are used with the analyzer providing frequent updates. This also allows the APC application to remain in operation when the analyzer fails or is down for routine maintenance. The specification on the reformate for most commercial applications is either RVP or a specific component concentration.

3.4. APC Special Case Applications

Coke Handling

In a continuous catalyst circulation reformer operation, the catalyst circulation rate can be adjusted to maintain a specified weight percent carbon on catalyst. The lag time in the response of the coke on catalyst to changes in unit operations, including changes in catalyst circulation rate, is very long. The ratio of lag time of the coke on catalyst to the next slowest variable is roughly 100 : 1 or twice the maximum recommended for an APC application. Sophisticated interlocking controllers are used to handle this problem. To implement a coke on catalyst controller, the system must be installed with a coke laydown rate calculation package. The more important the need to accurately control the coke on catalyst, the higher the requirement for an accurate coke deposition calculation package. When APC is to be utilized to operate the reformer against a regenerator coke burn constraint, very precise coke calculation is required.

Catalyst Regeneration

Catalyst regeneration systems in both cyclic units and continuous catalyst circulation units are typically controlled with PLCs. Changing the control of the catalyst regeneration to an APC application is not needed. However, using a small APC application installed in parallel to the PLC application may improve the regenerator performance at a very low capital investment.

Integrated Feed Systems

Reformer APC systems often have integrated feed systems. One type of integrated feed system is a unit with a base feed and a supplemental feed. The APC application adjusts the supplemental feed to a constraint. Typical supplemental feed constraints are coke laydown rate, maximal supplemental feed rate, maximal total feed rate, LHSV, and catalyst pinning. The second type of integrated feed system is the reformer unit with feed coming directly from the hydrotreating unit, and the feed rate to the hydrotreating unit is adjusted to maintain the charge rate to the reformer. Traditional PI controllers do not work well in these applications; heavy operator involvement is almost always required. This is a very simple problem for an APC application.

Minimization of System Pressure

The system pressure in a reformer can be minimized with an APC application. The economic driver for this is increased C_5+ yield at constant severity. There are several constraints that must be evaluated in commercial units. First, the hydrogen must be able to leave the unit. The typical constraint used in APC is the product separator pressure control valve position. To facilitate operation, the control valve trim needs to be checked to ensure tight pressure control when operating at the end of the valve curve. The second common constraint is coke make. If pressure is going to be minimized, a coke laydown rate constraint or pressure compensated hydrogen-to-hydrocarbon ratio must be used.

3.5. Inferred Properties

Inferred properties can be generated with a number of math tools ranging from simple regressions to neural networks and artificial intelligence tool kits. In all situations it is essential that process measurements be transformed to appropriate format for use in the calculation package. A simple example is the use of pressure-compensated temperature to infer the concentration of C_5+ in the stabilizer tower overhead. In many cases, an analyzer measuring one stream property can be used to infer a second stream property. For example, hydrogen purity of the recycle gas can be easily inferred from the recycle gas gravity, which is an easy and inexpensive on-line measurement. In some instances it is critical that first-principle calculations be used to calculate inferred properties. Common regression and neural network inferential tool packages can generate calculations that work well over the base set of data used for the development and work for several sets of test data, but then fail during routine application on a live process unit.

When analyzers are used for controlling the process, it is common practice to use an inferred property to control the process and have the analyzer update the inferred property package. This allows an analyzer with long delays between executions to be used in an APC application, which is better served with near-instantaneous feedback from the process. Use of a combined analyzer and inferred property package also allows the controller to continue operating if the analyzer is taken out of service for routine maintenance or fails due to a mechanical problem.

In the development of inferred properties, the engineer designing the calculation must take into consideration the characterization [11] (random or systematic) and the magnitude of the errors of the measured values being used in the calculation. The calculated expected error of the inferred property should not be more than one and a half times the direct measurement reproducibility of the inferred variable. The following summaries provide guidelines on proven methodologies for inferring various properties for catalytic reformer applications.

Research Octane Number

RON can be inferred from reactor temperatures, feed rate, recycle rate, and feed properties. It is important to update the inferred RON as catalyst activity changes, or have a function incorporated into the inferential that allows for changes in catalyst activity. One method of obtaining this information is from plant tests, which are expensive, have low reproducibility, are often limited in range of variable operation, and can rarely be done at constant catalyst activity. A second method of obtaining inferential data for RON is to use correlations developed by catalyst vendors, research centers, or other sources of large volumes of reformer operating data. The effects developed from this data will at least be directionally correct and will be close to the right magnitude if the unit for which the inferred RON calculation has an operation similar to the average of the collected data. The preferred method for determining a RON correlation is to exercise a kinetic model through the range of operation of the variables to be included in the calculation.

Reid Vapor Pressure

The inferential calculation of RVP is usually based on a modification of the Clausius–Clapeyron equation [14]. This is applicable because the vapor pressure of reformate is low and it can be assumed that the light hydrocarbons in the reformate obey the ideal gas law. These inferentials work well when the RVP is in the range of 5–10 psia, which is a common operation for most reformers. To obtain good results, the stabilizer bottom pressure must be measured rather than calculated from a top pressure and assumed pressure drop across the tower. The temperature used must be representative of the equilibrium tower bottom liquid temperature.

Stabilizer Overhead Composition

Preparing a robust inferential calculation for a component in the stabilizer overhead liquid is nearly impossible in a dynamic operating environment. The typical approach is to infer the C_5+ content of the stream using tower temperature and pressure measurements. A gross regression of operating data and laboratory measurements can be used. A slightly more rigorous approach can be used by assuming a vapor pressure for the C_4-, a second vapor pressure for the C_5+, and calculating the concentrations required to meet the process temperature and pressure. Both methods are restricted due to changes in C_4- concentration due to changes in feed composition, operating severity in the reactor system, and chloride level on the catalyst.

Feed Composition from Density

The P (paraffin), N (naphthene), and A (aromatic) distribution can be inferred from a gravity measurement. On-line liquid gravity (density) analyzers are inexpensive, have a low maintenance cost, and give a continuous analysis signal to the control system. Properly installed, the lag time is relatively small. The calculation is done by assuming a constant N/A ratio, a bulk P gravity, bulk N gravity, and bulk A gravity. From the N/A ratio a bulk NA gravity can be calculated. With a given measured gravity, the concentration of P and NA can be calculated. From the calculated NA concentration and assumed N/A ratio the N and A concentrations can be calculated. The N+2A or N+3A feed descriptors can then be calculated. This technique is restricted to operations with limited variations in feed distillation range and feeds with roughly the same N/A ratio.

Coke Laydown Rate

Coke laydown rate is a function of the coke precursors in the feed, hydrogen partial pressure, space velocity, and temperature. As with most inferentials, an equation for coke laydown rate can be regressed from operating data taken over an extended period of time and a broad range of operating conditions. Each of the variables used in the correlation must be varied over the anticipated operating range. A second way of developing a coke laydown rate equation is to use basic kinetic terms. These need to be simplified to keep the number of calculations to a minimum and the processor utilization to an acceptable level. Activation energies and frequency factor data can be obtained from pilot plant runs or literature sources [15].

Fired Heater Tubeskin Temperatures

Fired heater tubeskin temperatures can be calculated from basic radiant heat transfer equations [16]. Although the primary heat transfer mechanism is radiant heat, a small percentage of convective heat can be assumed. This can be determined from field tests. Heater-specific mechanical configuration data should be used to calculate the radiant section cold-plane effectiveness and cold-plane area. The heat transfer coefficient can be calculated for the tube walls. An effective firebox temperature can be calculated on-line from the bridgewall temperature using a bias. The flux rates used in the tubeskin temperature computation can be calculated from fireside and process data [10]. The inferred tubeskin temperatures need to be updated from measured values. Measurement is usually made with either tubeskin thermocouples or an optical pyrometer. There are three critical areas in the calculation. First is developing the thin wall resistance of the internal fluid in the tubes; the second is estimating any change in resistance on the external surface of the tubes; and the third is measurement of the bridgewall temperature. The bridgewall temperature measurement in most commercial heaters is done with a thermocouple dropped through the ceiling of the heater. The use of a shielded, high-velocity thermocouple is preferred.

Recycle H₂ Purity

Recycle gas hydrogen purity can be inferred from online recycle gas density measurement. A simple regression can be used. As with all analyzer-based inferentials, the inferred property validity should be a function of the analyzer signal validity.

3.6. On-line Analyzers

Recycle Hydrogen Purity Analyzers

Two types of analyzers are used for on-line measurement of recycle gas hydrogen purity. In some instances, on-line gas chromatographs are used. These analyzers also provide a composition analysis of the entire stream with the exception of trace compounds. Most refiners use an on-line density analyzer with inferential calculation to convert gravity to hydrogen concentration.

Recycle Hydrogen Moisture Analyzers

Two types of moisture analyzers are in common use for measuring the moisture content in the hydrogen recycle gas [17]. One analyzer uses an in situ probe with an aluminum oxide sensor. Because these analyzers are in situ installations, they cannot be calibrated on-line. Analyzer drift can be a problem. The second type of analyzer is an extractive analyzer that is based on the change in vibrational frequency of a hydroscopically sensitized quartz crystal as the moisture content of the stream varies. Because moisture analyzers are not used for direct process control, a lag time of several minutes is of no consequence.

Reformate Octane Analyzers

Several different technologies have been used to generate a RON measurement for the reformate [17]. One technology is to use an ASTM/CFR engine configured for continuous service. The analysis is not instantaneous, and a lag time is built into the system with the sample collection and preparation system. These analyzers have a tendency to drift and have a high downtime factor for routine maintenance. This method is the only direct on-line measurement of octane.

Delta knock analyzers have been developed for on-stream octane analysis. The difference in knock index is determined between the sample and a base material. The difference is then added to the known base material octane. The analysis is incremental with added lag times for stream sampling and preparation. Significant maintenance is required.

Four other technologies have been developed in which secondary properties are measured and then octane inferred from these properties:

Gas chromatographs are used to first measure the composition of the
    reformate, and the octane is calculated using octane blending factors for
    each component. The lag time for analysis in this method is long.
Near-infrared (NIR) analyzers are based on light passed through the
    process fluid [18]. The analyzers have a fast response time, high
    reliability, multiple component monitoring, multiple sample point
    analyses, and very high on-stream times. However, calibration can be
    difficult.
A flame peak displacement analyzer monitors the reactions that are
    precursors to knock. Octane can then be inferred from the peak
    displacement of a sample fluid as compared to a standard fluid.
A technology utilizing a thermal oxidation reaction of the reformate infers
    octane by simulating the reactions in a knock engine. The sample
    analysis is incremental.

In modern installations, the preponderance of installed octane analyzers will be NIR-based analyzers. Specific technology and hardware is available from a number of vendors around the world. The instantaneous and reliable analyses make these analyzers good tools for APC applications.

Stabilizer Component Analyzers

Online gas chromatographs are the primary type of analyzer used for component analysis of the stabilizer overhead liquid. The samples are intermittent, and there is a significant lag time (typically in the 10-min range) between when the sample is taken and the results reported. NIR analyzers can also be used for component analysis of the stabilizer overhead liquid.

Reformer Feed Analyzers

Reformer feed analysis is primarily handled by on-line gas chromatographs. NIR analyzers could be used. As discussed in the feed inferential section, density analyzers can be used to generate bulk paraffin, naphthene, and aromatic concentrations of the feed. The gas chromatograph results are valuable information in real-time optimization of a commercial reformer operation. However, the results significantly lag the process and will not suffice for feed disturbance measurement.

Fired Heater O₂ Analyzers

The oxygen content of the flue gas from fired heaters is measured with two different types of analyzers: extractive and in situ [17]. The in situ analyzer uses a zirconia oxygen sensor situated in a probe inserted into the fired heater. The signals are continuous, near-instantaneous, and very reliable, all of which are required for an APC application. The extractive type of oxygen analyzer uses either paramagnetic or electrochemical technology. These analyzers are not instantaneous and may not be continuous. Because of the extractive (vacuum) systems, the samples are subject to air contamination. Significant maintenance resources are spent maintaining these systems.

Utility System On-line Analyzers

In addition to the primary process analyzers, catalytic reformer units typically have several utility analyzers that may or may not be incorporated into an APC application.

Cooling water pH is monitored using an electrode.

The chemicals used in cooling water for dispersing salts and other materials are laced with various tracer compounds. Chemical vendors provide proprietary analyzer technology to measure the tracers and thus the level of dispersant chemicals in the cooling water.

Cooling water conductivity is determined on-line using an electrode to measure resistance.

Regeneration System On-line Analyzers

Analyzers are used in the regenerators of both cyclic reformers and continuous reformers to measure carbon monoxide, carbon dioxide, and oxygen. The same technologies used for the fired-heater oxygen analyzers can be used for the regeneration system gas analysis.

3.7. Benefits from Application of APC to a Catalytic Reformer

The financial benefits from application of APC to a catalytic reformer operation are centered in three areas: (1) minimizing operating costs, (2) maximizing yields, and (3) reducing variability. Table 7 summarizes the benefits for each area. Although all reformer APC applications have benefits in each of these areas, the distribution of benefits will vary from unit to unit depending on unit constraints and operating objectives. Typical benefits from application of APC on a catalytic reformer range from $0.10 to $0.25 per barrel of feed.

Table 7 Benefits from Application of APC

	Eliminate Unnecessary Operating Cost
1	Reduce heater excess air to minimal specification
2	Reduce reboiler duty to minimal reflux subject to optimal tower operation
3	Reduce compressor load
	Maximize Yield of Valuable Components
1	Reduce reactor operating pressure to mechanical constraints
2	Maximize feed rate in heater (or other) constrained operation
3	Reduce C_5+ material in the stabilizer overhead subject to optimum tower operation and process specifications
4	Maximize reformate RVP to product specifications
	Reduce Variability of Reformer Operation
1	Reduce variation of severity
2	Reduce variability of RVP of reformate
3	Reduce variability of C_5+ in stabilizer overhead liquid

Benefits from application of APC to catalytic reformers normally result in a simple payout of a year or less. But it must be noted that application of APC will not improve poor instrumentation nor significantly increase unit throughput or yields [19]. APC applications will not improve mediocre operator performance due to poor attitudes or training. To significantly increase yields and/or throughput, major mechanical modifications to or replacement of the reformer unit are usually required [20].

Benefits from Minimizing Operating Costs with APC in a Reformer Unit

Given the operating objective of processing a specified amount of feedstock to make a particular set of products, APC eliminates unnecessary operating costs while maintaining all products within specification. For example, reducing heater excess air to the minimal specification is of significant value for large heaters that are experiencing poor manual control. This is normally the case when the instrumentation is in poor condition and the heater does not have a functional oxygen analyzer. Unfortunately, benefits from the application of APC to the fired heaters in a reformer may be overwhelmed by the costs of repairing the heater structure on an old heater so that it doesn't leak oxygen, and of repairing or installing new instrumentation. Several tools are available for calculating the reduction in fuel consumption associated with the reduction in excess air in a fired heater. These include the use of a flue gas enthalpy graph or a chemical engineering flowsheet simulator.

Overrefluxing the stabilizer tower is a common waste of energy. The energy savings benefit from reduction in overrefluxing is simply the reduction of reboiler duty associated with the reduction of reflux. Once the allowable reduction in reflux is calculated, the reduction in reboiler duty can be calculated using a chemical engineering flowsheet simulator.

Compressor loads can be reduced if the recycle hydrogen is being run at an excessively high rate, or if the system includes a spill-back line that is open, or if there is an antisurge line that is open. Control of flow rate through centrifugal compressors is easily implemented in the regulatory control system and can be adjusted by APC. This is true for both electric motor-driven compressors and turbine-driven compressors. It may be necessary to justify the inclusion of a surge control system for the compressor. Reciprocating compressor flow control offers a significant challenge. The only way to reduce flow rate through the compressor (flow control through the reactors can be maintained with a spill-back system) is by adjusting the driver speed or by use of unloaders. Typically APC projects on reformers with reciprocating compressors include installing new or improved instrumentation on the compressors if the compressor work is to be reduced via APC. This additional cost must be included in the benefits calculation.

Benefits from Maximizing Yields with APC in a Reformer Unit

Significant benefits from implementation of APC in a reformer operation come from maximizing the yield of valuable components while meeting process objectives. Reducing the reactor operating pressure to mechanical constraints is applicable when the unit is not being run to maximize feed rate and is not coke laydown rate limited. Typically the product separator pressure setpoint is reduced, thus reducing the reactor pressures. The increased yield can be calculated using a kinetic model of the process. Offsetting costs for increased compressor work and increased coke laydown must be subtracted from the gross yield improvement benefit to generate the net benefit from pressure reduction.

Product rates are maximized if feed rate is maximized. The overwhelming benefit from reformer APC applications is feed maximization in a mechanical equipment-constrained operation. The two constraints most frequently encountered are heater constraints and coke laydown constraints. In most cases, increasing the feed rate has associated incremental costs such as higher coke laydown rates, lower heater efficiency, and increased pressure drop. Thus, compressor load must be taken into consideration when calculating the net benefit of increasing feed rate.

Reduction of the C_5+ material in the stabilizer overhead subject to optimal tower operation and process specifications is accomplished through better control of the tower temperature profile. The benefit can best be calculated using a chemical engineering flowsheet simulator.

Maximal reformate RVP to product specifications is accomplished by dropping additional light component into the bottoms stream of the stabilizer. In a chemicals operation where the reformate is going to an extraction process this may result in an incremental increase in benzene content but will certainly increase the raffinate load on the extractor. The additional operating cost in the extraction unit must be subtracted from the gross benefit from maximizing the RVP of the reformate. In a gasoline operation, the benefit from maximizing RVP is subject to the effect on the final blended gasoline pool. If dropping light key into the reformate adversely affects the blended gasoline pool, the associated cost must be deducted from the unit benefit.

Benefits of Reducing Variability with APC in a Reformer Unit

Reduction of variation of severity is usually a significant benefit. In a gasoline operation, the impact of reducing the variation of severity at the process unit is minimal because of the flat profit curve in most gasoline operations. The exception to this is when the severity is constrained and the plant is octane limited. The big impact of variations in reformer severity is on the variation in the blended gasoline pool RON and the resulting RON give-away. In chemicals operations the profit function with respect to severity has significant curvature. Because the function is curved, the profit gains when operating on the high side of the curve are less than the profit losses when operating on the low side of the curve. The greater the curvature of the profit curve, the greater the benefit from improved severity control. A local profit function can be developed for the stabilizer operation with profit being a function of C_5+ in the overhead. The same effect of variation of C_5+ material in the stabilizer overhead liquid exists as for variations of severity.

Reduction of variability of RVP of reformate in a gasoline operation is similar to RON variation. The effect is primarily seen on the resulting variation in the blended gasoline RVP giveaway. In a chemicals operation the RVP variations may result in benzene loss to the stabilizer overhead when the RVP is low and disruptions in the extractor operation when the RVP is high.

Calculation of Benefits in a Reformer Operation

Reformer benefits are typically calculated using one set of operating data per day over a month or more of operation. At least 30 sets of data are needed for the statistical calculations. Although laboratory analyses are frequently done two or three times per week in normal operation, the analyses need to be done on a daily basis during the month of operation to be studied. All samples need to be taken at the same time, and when the unit is in steady state. The process data used for the benefits calculations should be from the time the samples are taken. Statistical techniques are commonly used to evaluate the benefits from reduction of variability when APC is implemented on a catalytic reformer unit. More details are described elsewhere [21].

4   EXPERT ADVISOR SYSTEMS FOR CATALYTIC REFORMERS

Expert advisor systems are relatively new to industry and are slowly making inroads into commercial applications. Expert advisors are tools that are integrated among several systems to (1) advise operators of potential problems before they become major incidents and (2) offer advice and operation options to solve various operating problems [22]. Through the use of a knowledge-based system, an expert advisor predicts future events based on past similar event sequences that led to a problem or using logic patterns that predict a problem based on an analysis of several nonrelated operating parameters. Expert advisors take the knowledge stored in HAZOP documentation, PSM reports, and unit operating manuals and make it available to the operator in a timely, organized, and useful manner. The recommendations for operation changes can be for process variables that are directly accessible though a DCS system, such as temperature flow and pressure controllers, or for process variables not under conventional control, such as turning fans on and off, opening valve bypasses, or putting a standby pump into service. Expert systems use knowledge-based systems that are loosely divided into three major logic tree groups: (1) case based, (2) model based, and (3) rule based. The preponderance of commercial applications on catalytic reformer units use a rule-based system.

4.1. Expert Advisor Functions in Catalytic Reforming

As explained in the introduction, all control systems in a catalytic reformer unit fall on one of two paths along the management continuum—either the process control path or the safety path. The expert advisor is the most automated control system along the safety path. The expert advisor systems in catalytic reforming units are individually designed to meet the considerations of the mechanical design and the operating objectives of the unit. Where expert advisor systems are not installed, it is the responsibility of the operators to perform the functionality of these systems.

Four functional areas are typically addressed in expert advisor systems in all catalytic reformer units [23]. The first is to monitor catalyst performance-type issues. This includes warning operators of potential catalyst poisoning and providing them with the potential source of the problem so that corrective action can be initiated. Poisoning of the reformer catalyst can be temporary, such as from sulfur, or permanent, such as from iron deposition. Although sulfur poisoning is temporary, it can be extremely expensive in a reformer operation. This is particularly true in reformers being run to an operating constraint. Chloride and water addition rates can be monitored, along with light-ends yields, to ensure the correct acid function on the catalyst. Expert advisors are also used to advise of potential problems in the regeneration of the catalyst in cyclic and CCR units.

The second functional area is to warn of variations of feed content outside of normal operation. Particular emphasis is placed on detecting deviations of feed endpoint, which can cause accelerated catalyst deactivation. Major swings in feed composition can be detected and operators can be advised of actions that must be taken. For example, a unit may be processing a high-naphthene feed, with a corresponding high-hydrogen yield, and the operators open the net gas control valve bypass to keep the control valve in a preferred operating range. A shift from a high-naphthene feed to a low-naphthene feed may require the operators to close the bypass around the net gas control simply to maintain unit pressure. The expert advisor would prompt the operators to start closing the bypass as the feed is changing, rather than wait for the regulatory controller to be saturated with the valve fully closed and the unit still losing pressure.

Expert advisors are also used in a catalytic reformer operation to monitor equipment performance and changes in operation and to anticipate equipment failure. Although emergency shutdown systems are designed to handle equipment failure and prevent catastrophic unit failures, initiating an emergency shutdown should be avoided if at all possible. The expert advisor will aid operations by generating and advising the operators of an alternate operating scenario prior to a catastrophic event. For example, if a pending turbine-driven compressor shutdown is imminent because of a sagging steam header problem, the expert advisor would alert the operators of the problem, recommending that the feed be cut out of the unit and reactor inlet temperatures reduced prior to losing power from the turbine. Clearing the catalyst of naphtha and dropping inlet temperatures prior to losing the compressor results in minimal extraordinary coke deposition on the catalyst. If the emergency shutdown system had been activated at the time of the compressor failure, the unit would have been safely shut down, but naphtha would have been left on the catalyst at an elevated temperature and a significant extraordinary coke deposition would have occurred.

The final functional area for expert advisors is to identify control system failures and inconsistent operations. Included in this area of monitoring are failures in instruments, analyzers, and hardware as well as software errors. The potential loss of function of key equipment is monitored by the emergency shutdown system. The loss of some systems is a nuisance to operators, but is not of short-term concern and does not require immediate notification and action. Not all system failures are at the extreme, and an expert advisor can notify the operator in a timely manner and recommend corrective action. A simple example is an operator transposing the hydrogen-to-hydrocarbon ratio setpoint entry into the advanced process control system, entering 3.4 as the setpoint rather than the desired 4.3. This may be within the minimal/maximal constraint of the APC application and accepted without an advanced process control system-generated warning. The expert advisor would give a warning that the hydrogen-to-hydrocarbon ratio was outside the preferred unit operation for the given feed rate and feedstock composition.

4.2. Expert Advisor Integration into the Control System

At a minimum, a DCS system with historian, computing capability, and communication is required for implementation of an expert advisor. Signals from equipment monitoring instruments, such as vibration monitors on rotating equipment, must be available to the expert advisor. Also, all input variables from the emergency shutdown system should be available to the expert advisor. When APC systems have been implemented, the APC setpoints, constraints, and models can be included in the expert advisor.

The expert advisor will work with the information provided and its knowledge base to provide the operators with explanations on why key dependent variables in the process are changing, i.e., why is the hydrogen production increasing? When an APC application is moving a manipulated variable, the expert advisor can provide to the operator information on what constraint or target is prompting the change. In the introduction to the APC section, the need for the operator to deal with process variables not in closed-loop control was discussed. Expert advisors fill in part of the gap by first telling the operator there is a situation and then providing a recommended course of action. In the referenced example, the expert advisor would notify the operator that the molecular weight of the recycle gas was increasing, the cause is higher cooling water temperature, a compressor driver overload is anticipated, and would provide some alternatives to allowing the advanced process control function to reduce feed rate.

5 ON-LINE OPTIMIZATION OF CATALYTIC REFORMERS

On-line optimization is included in the control scenario of a catalytic reformer as a connection between the real-time unit control and the planning function [6]. The reformer on-line optimization function does not replace the APC function and should not be used when an APC function can be used to push a predefined optimal constraint such as maximizing feed rate. For on-line optimization of a catalytic reformer to be of value, one or more MVs must be free to move and have an optimal condition that is not constrained. An example is calculating the firing rate for optimal aromatics yield in a chemicals operation where octane is not required for gasoline blending. The on-line optimization of a catalytic reformer operation is a mix of process simulation and application of business processes. There are five areas of activity in the development of a real-time on-line optimization tool for a catalytic reformer: (1) definition of an economic envelope so that an optimization of the reformer process can be completed; (2) selection of a mathematical technique that will meet the calculation requirements for both the catalytic reformer process problem and the economic problem; (3) evaluation of reformer instrument and analyzer requirements from an optimization perspective; (4) consideration of the impact of the real time reformer optimization problem on the underlying reformer APC application; (5) coordination of an implementation plan through the plant that links the reformer control strategy with the planning function to deliver effective real-time optimization.

5.1. Defining the Catalytic Reformer Economic Envelope

Every reformer has an economic envelope as well as an energy envelope or a material balance envelope. The same methodology is used. The difference is that traditionally the heat and material balance envelopes are drawn concurrent to the unit's mechanical battery limits. The economic envelope may coincide with the battery limits but in many cases crosses into other process areas. Drawing the economic envelope for a reformer on-line optimization project requires an evaluation of potential sources of benefits, defining a set of economic drivers, and, finally, preparing a reformer economic statement.

Sources of Benefits

Just as the functionality of APC and on-line optimization are separate, the justifications for both applications are separate [24]. Benefits from on-line optimization come primarily from utility savings and improved composition distribution. Utility saving through optimization is normally a minor function. Reformers use utilities to make money—they have to be consumed to generate profit. Quite often the utility savings are generated by minimizing utilities while meeting a process objective, and this can almost always be efficiently done in the APC application. For example, if the process objective is to run at a given reactor inlet temperature, the optimal heater operation is going to be to run to a specified temperature at the minimal allowable excess air. This is a constrained optimization solution and does not need an on-line optimizer to determine the answer. However, if there are options on fuel gas utilization, there may be a utility optimization problem.

The preponderance of on-line optimization benefits come from correct selection of feeds to be processed and/or unit severity. On-line optimization of feed between multiple process units including one or more catalytic reformers is a major source of improved unit profitability. Table 8 lists several applications for on-line optimization in a catalytic reformer.

In those reformer units in gasoline production, where a specified octane need not be met to meet the gasoline blend recipe, unit severity can be optimized. Because of the effect of severity on C_5+ yield this can be a very lucrative

Table 8 Optimization Potential from Interunit Feed Allocation

1	Allocation of feed between two reformers
2	Incremental feed allocation between two reformers
3	Feed cutpoint definition between a reformer and an Isom unit
4	Aromatics section feed IBP definition in an ethylene plant
5	Feed IBP/EP definition in a feed rate–constrained reformer operation
6	Feed EP definition in a coke laydown rate–constrained reformer operation
7	Feed bypassing of the reformer (the bypass is a null unit)
8	Selection between incremental feeds to the reformer from different process units

optimization problem in a gasoline operation with a high-octane credit or in a plant with two reformer units that are creating a pooled reformate [26]. Simply optimizing a single reformer based on octane barrels will probably only be attractive to a plant that is selling gasoline blending components. A refinery that has a single reformer in gasoline operation and is running in naphtha balance will probably not be able to take advantage of varying severity.

Reformers operated to produce aromatic chemicals normally have the opportunity to vary severity and can make good use of on-line optimizers. The exception is when the reformer unit is being run to a mechanical constraint or a downstream unit is causing a process constraint. Chemical operations offer the most potential for use of an on-line optimizer, particularly when aromatics markets are in flux. The optimal severity will vary with the differential in value of benzene and C_8 aromatics.

Economic Drivers in a Catalytic Reformer

An economic driver in a reformer operation is an activity or condition that will significantly impact the profit function of the process. When designing an on-line optimization strategy, it is important to define those conditions that affect profit and their frequency of change [27]. If the disturbance happens on an infrequent basis, particularly if it can be foreseen, the plant planning tool will be able to define the optimal reformer operation. However, if the disturbance is frequent and without warning, the plant planning department cannot react quickly enough to effectively optimize the process. Table 9 lists various unit disturbances that can

Table 9 Potential Economic Drivers with Application to On-line Optimization

	Potential	Driver
1	High	Changes in feed composition
2	Low	Summer to winter change in gasoline demand
3	High	Changes in weather conditions
4	High	Changes in aromatics spot market prices
5	Low	Changes in mechanical equipment or process flow
6	High	Equipment failure
7	High	Downstream processing capacity
8	High	Feed processing capacity
9	High	Incremental feed availability
10	High	Unit operating severity

be economic drivers and indicates their potential application to on-line optimization.

Once variables are determined that have potential impact on the short-term operation, each variable needs to be perturbated over its range of operation and the economic consequence established. The variables that show a significant impact on the reformer economics are the economic drivers of the unit. Use of a sophisticated planning model or kinetic model with an economic function is of great value in generating economic drivers. These drivers are going to be in two forms: disturbance economic drivers and manipulated economic drivers. The disturbance drivers are those over which the optimization project cannot have control, such as weather, equipment failures, and the like. The manipulated economic drivers are those that can be controlled, such as feed rates, feed cutpoints, etc. The manipulated economic drivers may be outside of the conventional unit operating area but must be within the overall realm of operation.

Defining the Reformer Economic Envelope

When the economic drivers for the catalytic reformer have been established, the manipulated ones can be laid over the plant process flowsheet to determine the scope of the economic envelope from an economic perspective. A process review must then be made to ensure that any defined disturbance drivers are included in the scope, and the envelope is enlarged if necessary. The third step is to review the process for constraint issues that will affect the operation of the reformer with respect to the manipulated or disturbance drivers. Again, the scope of the envelope may need to be expanded. Finally, all energy and material balance streams entering and leaving the envelope must be identified. Each stream must be of a nature that can be assigned an economic cost in conventional engineering units. Again, it may be necessary to expand the envelope to place streams in measurable economic units. Once the economic envelope for the reformer unit has been drawn, an optimization statement can be prepared.

The Reformer Optimization Statement

The reformer optimization statement is a summary of the economic drivers and manipulated variables that will be adjusted to maximize the reformer economic performance. The objective of the statement is to communicate to management, operations, and programmers the intent of the real-time optimization project. An example of a statement is as follows: The feed rate and feed initial boiling point will be adjusted to maximize the unit profitability as the cost of feed varies and as the product value of benzene varies with respect to feed costs.

The profit function is then prepared by identifying all significant products and values and operating costs and values. The profit function is the sum of the value of the product streams less the operating costs including feed.

$$\text{Profit (\$)} = \Sigma(a_i{}^*P_i) - \Sigma(b_j{}^*C_j)$$

where

P = magnitude of product streams 1 through i

a = product pricing coefficient, value per flow measurement of products 1 through i

C = magnitude of cost vector 1 through j

b = cost pricing coefficient, value per vector measurement of costs 1 through j

A typical profit function is presented in tabular form in Table 10. The figure is representative of a typical operation. Product stream and operating cost stream definitions will vary from application to application.

Table 10 Typical Profit Function

Products and operating cost	Flow	Engineering units	Price coefficients	Net value
Products				
Benzene	200	mLB/h	700	140,000
Toluene	625	mLB/h	540	337,500
Mixed xylenes	770	mLB/h	600	462,000
Heavy aromatics	210	mLB/h	440	92,400
Raffinate	730	mLB/h	300	219,000
Net hydrogen	170	mLB/h	800	136,000
Fuel gas (stabilizer O/H gas)	70	mLB/h	220	15,400
LPG (stabilizer O/H liquid)	125	mLB/h	270	33,750
Waste heat steam	15	tons/h	9	135
Net Product Value ($/h)				1,436,185
Operating costs				
Base feed	2000	lb/h	450	900,000
Incremental feed 1	700	lb/h	465	325,500
Incremental feed 2	200	lb/h	480	96,000
Boiler feed water	4	mGal/h	1	4
Fuel gas	300	mmBTU/h	4	1200
Electricity	3000	mKW	7	21,000
Catalyst regeneration cost	1	mLB/h of coke	10	10
Aromatics extraction cost	1805	mLB/h of arom.	30	54,150
Misc. chemicals and catalyst	2900	lb/h of feed	1	2900
Net operating costs ($/h)				1,400,764
Reformer Profit ($/h)				35,421

Product and Cost Pricing Coefficients

Providing accurate reformer product and feed pricing coefficients is critical to optimizing the unit operation [27,28]. Defining what is not appropriate for establishing reformer price coefficients is frequently easier than defining the correct method for establishing price coefficients. For example, using average past market prices is not a good idea, as this tells the optimizer what it should have done in the past, not what is appropriate at present. Using LP shadow prices is also not appropriate. Using a tool to dissect the planning LP and determine base and incremental product prices that reflect the true economic position of products and costs in the reformer operation is the best solution. Using one of these dissection packages will allocate the appropriate profit to the reformer operation creating a gap between products and cost. This can be supplemented by adjusting specific price coefficients for short-term discrete changes in the plant operation. For example, if the benzene produced in the reformer is normally lined up to the production of cyclohexane, but because of a change in downstream operation the benzene is to be sold on the spot market, the spot market price is used for that time interval.

5.2. Mathematical Techniques for Real-Time Optimization

Different mathematical techniques are available for modeling the reformer optimization problem. The more detailed the solution method, the higher the implementation cost and the higher the fidelity of the answer. The shape of the profit curve as a function of key operating parameters will give a good estimate of the value of investing in a more rigorous and higher cost solution. If the optimization curve is relatively flat, there will be little incentive to apply on-line optimization. However, if the curve is highly pronounced, a rigorous optimization technique may be justified.

Using an LP for Real-Time Optimization

The planning workhorse in industry is an LP. A base operation and shift vectors are supplied to the LP as the reformer model representation. Using linear projections, an optimal operation, either constrained or unconstrained, can be calculated. Multiple base operations or a recursion model can be added to improve model validity. These models are not in heat or material balance and are not based on fundamental chemical engineering principles. Composite LPs can be used to span multiple process applications [27,28]. The major problem with using LPs for on-line optimization of a reformer is that the reformer operation is nonlinear. However, the predicted optimal solution will be directionally correct; and given a sufficient number of iterations, the LP optimizer will come to the correct solution. After each iteration, constraints are rechecked and limits updated. If the process and economic conditions do not change during the iterative process, the optimizer slowly creeps toward the optimal solution. Typically on-line reformer optimizers run once every 3–6 h. Assuming it will take between 6 and 10 iterations to close to solution, the process must remain steady from 1.5 to 2.5 days. The net result of using an LP for optimization of a reformer operation is a margin between the calculated and actual optimal operation. The margin degrades every time the economic or process conditions change in the reformer operation. This results in a persistent deviation between the calculated optimal operation and the true optimal operation.

Optimization Using Statistical Regression Models

Statistical regression models including neural networks are correlation models based on observation of historical operation to calculate typical process gains with changes in independent variables. In one methodology the observations are taken over a long period of time. There is high confidence that the gains will be directionally correct because the data used in the regression span a large band of operation for each independent variable. The magnitude of the gain calculation is less certain. The second method for fitting statistical regression models is to collect a large number of data over a relatively short time on only the reformer operation being optimized. Both the gain direction and magnitude are subject to error because of the narrow span of operation used in data collection. Like the LP models, the statistical regression reformer models are in neither heat nor material balance and are not based on fundamental chemical engineering principles. If the reformer economic and process conditions are not changing, the statistical regression models can be biased to match the unit operation. As with the LP modeling technique, errors in heat and material balance predictions can significantly affect the accuracy of the profit calculation used as the objective function in the optimization problem.

Optimization Using Kinetic Models

Rigorous first-order models of the reformer operation include the use of reaction kinetics to predict changes in yield as the process changes. Detailed hydraulic and heat transfer models are included as needed to model the process and associated constraints. Accuracy is dependent on the degree of modeling sophistication. Like the LP and regression models, the rigorous models are steady state, but unlike the previously discussed models, the rigorous models are in both heat and material balance. Rigorous models of the reforming process use a flowsheeting structure. The approaches to operating constraints are accurately predicted in addition to precise predictions of yields and operating costs. In addition to highly accurate bulk reformate yield predictions, kinetic models can accurately predict component yields [6,26]. This is critical in calculating the optimal operation for chemicals reformer operations, which are the mostly likely candidates for on-line optimization. Kinetic models are dependent on an accurate feed stream analysis for most advantageous calculation performance. System hydraulic calculations, heat transfer calculations, separation calculations, and power calculations are performed using conventional first-principle modeling techniques. For example, rigorous tray-to-tray distillation tower models are used for the stabilizer.

Model Solver Technology—Sequential Modular Calculations and Equation-Oriented Calculations

Catalytic reformer operations always include at least one process recycle stream, the hydrogen recycle. If a recontactor is included for hydrogen purification, multiple additional process recycles are entered into the problem. Adding heater firebox models also adds more recycle calculations. Rigorous distillation tower models inherently include multiple recycle-type calculations. Each operating constraint included in the problem will generate at least one recycle calculation. There are two fundamental mathematical approaches to modeling a catalytic reformer operation, using sequential modular calculations or open-equation modeling. Both techniques have significant advantages and disadvantages. Two significant advantages of the open-equation technique over the sequential modular technique are the simpler approach to model calibration and the faster time of execution.

In the updating of a sequential modular model, modifications of various parameters such as rate constants are done on a trial-and-error basis using a Wegstein or similar closure technique. Depending on the number of recycle loops in the problem and the degree of overlapping, this can be a very involved and lengthy process. In an open-equation solution the same parameters are found directly by simply redefining the variable specifications and solving the program in the parameterization mode. Reconciliation of several sets of reformer data to develop average parameters is standard procedure in open-equation modeling but is not practical in the application in most sequential modular calculations.

RT-OPT Frequency of Execution

The frequency of execution for a given reformer on-line optimization application is determined by the anticipated frequency and magnitude of operating changes. The operators are typically given the option to initiate an optimization run when certain changes in the process have occurred, such as variations in (1) weather, (2) feed composition, (3) feed rate, (4) unit economics, (5) equipment performance, and (6) catalyst activity.

Reformer Model Calibration

Model calibration addresses validating the data to be used in the reformer model and adjusting imperfect measured data to the perfect world of mathematics. The less sophisticated the optimization tool being utilized, the less rigorous the data validity requirement. Calibration of kinetic models require data checking beyond simple material balances to ensure that measured reaction rates are reasonable [29].

Data errors in the reforming process may be described as precision error or systematic error. Precision error is uncorrelated error and can only be reduced though improved measurement devices. Systematic error, which is a correlated offset error, is difficult, if not impossible, to detect and quantify with a single set of measurements. One simple way to discover the existence of systematic error is to perform two different tests on the same data point, e.g., measuring the gravity of a stream directly by ASTM 1298 and by calculation from a gas chromatographic analysis, or measuring recycle gas content by both on-line and off-line analyzers. If both techniques generate the same results within the precision for the test procedures, the systematic error is very low. However, if the results are different, the systematic error is significant on one or both of the test procedures. This type of comparative analysis does not solve the data error problem but can be used as justification for rejecting the data as unusable.

The measured data will never be perfect, but at some degree of accuracy it is acceptable to use in an on-line optimization model. Two methods are commonly used in a reformer optimization project to account for deviations between measured data and model entries. The first method is to use average process data and disregard the difference between the model and a single-process measurement. The second method is to apply a bias between the model and the process.

5.3.   Instrumentation and Analyzer Requirements

Modeling a process is different than operating a process, and the location of the process sensors and on-line analyzers must be reviewed from the perspective of using the readings for model validation. The precision error problems should have been addressed in the implementation of the APC, which can operate effectively with systematic error but requires minimal precision error. For model applications, the systematic errors of process variables, whether measured directly, calculated, or inferred, must be eliminated. Critical to both instruments and analyzers used in an on-line reformer optimization project is a well-coordinated maintenance program. Deviations in instrument and analyzer accuracy will distort the optimizer solution and result in the process operating at less than optimal conditions.

The systematic error in the process instruments must be reduced to the minimum achievable with standard instrumentation. For example, systematic error can be reduced in flow meters by adding temperature and pressure sensors to permit accurate temperature and pressure compensation. Orifice installations need to be checked for proper installation with respect to upstream and downstream turbulence. Orifice plates need to be precisely machined and correctly installed. Differential pressure transmitter installations should be reviewed, particularly where condensable vapor streams are being measured. Temperature sensor installations need to be checked for correct thermowell penetration, thermocouple insertion, and shielding of the head when appropriate. Use of temperatures of streams in two-phase flow should be avoided. Thermocouple installation locations need to be reviewed where flashing between two points can affect the temperature of a stream.

Analyzers used for reformer on-line optimization must have a low precision error and a low systematic error. Single-component or property analyzers are the exception because they can be easily biased to account for a systematic error. A reformer optimization project needs frequent feed analysis and the use of an on-line feed analyzer is preferred. If the optimization problem includes a component yield as a constraint, the stream with the preponderance of that component needs to be analyzed on-line. This measurement is critical as part of the model calibration and feedback.

5.4. Integration of a Reformer Optimization Project into the Control System

Successful integration of the reformer into the control system centers on three primary areas: operator training, support issues, and elimination of conservative limit creep.

Operator Training

Engineers design on-line optimization systems for catalytic reformers, but operators turn the programs on. If the operators do not understand the application, they turn the software off. Unlike the installation of APC, the operators are not as concerned initially with job security. In most cases, operators want the reformer to run at top performance and thus are predisposed to accept the on-line optimization software. The training must be centered in four areas: data entry, what the controller will adjust in the APC application, safeguards in the system, and why the optimizer makes the moves it makes. Data entry must be simple and easy to understand. The biggest problem is making sure the operators understand the engineering units used in the optimizer constraints and the economic product and cost coefficients. The operators must be thoroughly trained in which APC control functions will be adjusted by the optimizer. They must be familiar with the constraint handling in the shutdown system, the regulatory controls, and the APC and how these limits are implemented in the on-line optimizer. The integration of the on-line optimizer and the APC will have a series of safeguards to ensure that optimizer-specified setpoints do not violate the underlying controller setpoint limits. The biggest hurdle in training is instructing operators on how and why the optimizer calculates different optimal operations than the operators have done for years using rules of thumb and common practices.6pt

Support Issues with Real-Time Optimization of Reformers

Support for the on-line optimizer will come from several functions in the plant. The information technology department that has never seen the reformer unit will need to support the various interface applications between the on-line optimization computer and the computers of other functions in the plant such as the planning section.

Process engineering support is nearly a full-time project for most on-line optimization projects in a catalytic reformer. There are constant changes in the plant operating plan, maintenance schedules, process equipment line-ups, and equipment maintenance status that must be monitored. The optimizer constraints and objective function must be routinely reviewed for applicability to the present operation. Routine testing of the model performance and model updating must be performed to ensure that the modeling is being done correctly.

The planning department must adjust their activities to provide a steady set of economic product and cost price coefficients for the reformer optimizer. Procedures need to be established for the development of the price coefficients, verifying the accuracy of the price coefficients and implementation of the price coefficients into the on-line optimizer. If the optimizer is designed to vary feed rates, the planning department will need to be involved in setting limits on feed rate moves. If any of the other independent variable moves are going to affect operation outside of the reformer area, feed IBP cutpoint for example, the planning department will also need to be involved in setting these move limits.

The Danger in Being Overly Conservative in RT-OPT of Reformers

If the operators of an on-line reformer optimizer are overly cautious in setting operating constraints, the advantage of using an on-line optimizer is lost. Consider the constrained optimal operation depicted in Figure 8. The true operating constraint has been first limited by the emergency shutdown system. To avoid an unintended emergency shutdown, a second constraint was set by operations as a guideline for the regulatory control. Then a third constraint was

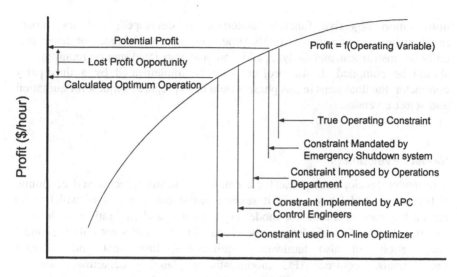

Potential Profit

Lost Profit Opportunity

Calculated Optimum Operation

Profit = f(Operating Variable)

True Operating Constraint

Constraint Mandated by
Emergency Shutdown system

Constraint Imposed by Operations
Department

Constraint Implemented by APC
Control Engineers

Constraint used in On-line Optimizer

Profit ($/hour)

Operating Variable

Figure 8 Constraint creep impact on optimization profitability.

added in the APC application by the control engineers to ensure that the operations imposed limit was not violated. Finally, the optimization engineers added a fourth constraint to protect the APC application. The result is a calculated optimal operation significantly moved from the potential optimal operation. Care must be taken to ensure that the normal variation in the unit operation does not cause the reformer unit to be shut down. A simple statistical analysis of the operation can establish the limits. Then, the constraints added by each layer of control must be checked not against the previous layer's constraint but against the true operating constraint.

5.5. Implementation Plans for Real-Time Optimization

Implementation of a real-time optimization project on a catalytic reformer can be divided into four steps: (1) project analysis, (2) model development, (3) model on-line implementation, and (4) ongoing support and evaluation. Well-defined policies and procedures are established by companies routinely installing on-line optimization applications [30].

Project Analysis

In the project evaluation phase, the economic need for on-line optimization on a specific catalytic reformer is established, an economic envelope is defined, an optimization objective function statement is developed, and the model requirements are specified. At this time, preliminary estimates of hardware, software, instrument, and analyzer additions and associated support requirements should be compiled. If the project is to be implemented by a third-party contractor, the final steps in this phase would be to prepare a request for quotation and select a vendor.

Model Development

The model development phase for a catalytic reformer process and economic optimization model has within it several milestones, the first of which is to establish a base operation for model configuration and calibration. A detailed function design of the model must be prepared that specifies not only the model specification but also hardware requirements, instrument and analyzer requirements, required APC modifications, planning department support requirements, an implementation plan, and a plan for ongoing maintenance. The model development phase includes the programming of the process configuration and the economic vectors used in the model objective function. Upon model completion, the model must be calibrated and tested. A model interface is then built for use by operations, engineering, and planning.

Model On-line Implementation

The catalytic reformer model on-line implementation should be done in two steps. First, the model is installed on-line in an open-loop mode. This initial open-loop mode is used for testing of the following:

    User interfaces (operations, engineering, and planning)
    Plant process data to model connections
    Model gains to the underlying APC controller gains
    Model for robustness in solution
    Model to controller connections

While in the open-loop mode, any document changes are completed prior to operator and support engineer training. Methodology for coordinating the input of economic values to the model must be prepared. Methodology for engineering reviews of the model performance and troubleshooting must also be prepared.

The second step in the catalytic reformer on-line implementation is to connect the model controller output targets to the APC controllers. The controller movements should be initially tightly constrained and the constraints relaxed as systems and communication links are retested.

Ongoing Support and Evaluation

Ongoing support of a catalytic reformer on-line optimization project starts with the integration of the optimization problem into the plant's work practices. An on-line optimizer requires input and evaluation from all sources along the length of the control spectrum, from the technician in the instrument shop to the chief of the planning department.

On-line catalytic reformer optimization models require frequent review by operations management and support engineering. Initially, there will be many questions as to why the optimizer pushes the reformer into a particular operating zone. There are counteracting economic drivers that the reformer model may evaluate differently, which are based on rigorous calculation rather than a human using hunches and rules of thumb. As process objectives change, the ability of the reformer model to correctly represent the new conditions should be reviewed. This is particularly true when operating changes, mechanical repairs/changes, or economic changes have been made that push the reformer operation against a different set of constraints.

Operating plants are in a constant state of flux. They are responding to pressures from government agencies, environmental groups, corporate policy initiatives, improvements in process technology, and economic pressures from supply and demand. Periodic review need to be made of the economic boundaries and objective function in an on-line catalytic reformer optimizer. When appropriate, minor and major overhauls to the reformer optimization model must be incorporated.

6 SUMMARY

Integration of the control systems into an effective management tool results in a safe and efficient operation of the catalytic reforming process [31]. This requires coordination of multiple disciplines within the planning, operations, engineering, management, and maintenance departments. Instrumentation and controls must be properly sized and designed to meet the processing objectives established by the planning function. The instrumentation must be properly installed and maintained in good working order to be effectively utilized. The operations group must implement the planned operation with the tools provided, make timely and accurate analyses of instrument and control problems and requirements, and effectively communicate those issues to the appropriate plant organization. Above all, each individual involved in the catalytic reforming unit control system must function in a manner that is consistent with safe design, maintenance, and operation of the facility.

REFERENCES

1. Murrill, P.W. *Fundamentals of Process Control Theory*; Instrument Society of America: Research Triangle Park, NC, 1981.
2. *ASTM Manual on Hydrocarbon Analysis*, 4th ed.; ASTM: Philadelphia, 1989.
3. Little, D.M. *Catalytic Reforming*; Penn Well: Tulsa, OK, 1985.
4. *DMCplus® Multivariable Control Software Training Manual*; Aspen Technology: Houston, TX 1998.
5. Rossi, R.J.; Lee, R.W.; Turpin, L.E. In *Maximizing Reformer Profitability Against Mechanical Constraints Using a Kinetic Reaction Model*; NPRA Computer Conference, November 1991.
6. Turpin, L.E. Catalytic reforming: real-time planning and optimization. Hydrocarbon Technol. Quart. **1997**, 23–30.
7. Turpin, L.E.; Korchinski, W.J. Applying advanced controls to fired heaters. Petroleum Technol. Quart. **1999**.
8. Summers, A.E. Setting the standard for safety-instrumented systems. Chem. Eng. **2000**, *207* (13), 92–94.
9. Onderdonk, J.K. Understanding shutdown systems. Chem. Eng. **1986**, *205* (7), 45–50.
10. Martin, G.R. Heat-flux imbalances in fired heaters cause operating problems. Hydrocarbon Processing **1998**, *77* (5), 103–109.
11. Dieck, R.H. *Measurement Uncertainty Methods and Applications*; Instrument Society of America: Research Triangle Park, NC, 1992.
12. McDonald, G.W.G. To judge reformer performance. Hydrocarbon Processing **1977**, *55* (6), 147–150.
13. Advanced control and information systems. Hydrocarbon Processing. Maximizing reformer profitability against mechanical constraints using a kinetic model. (L. Kane, ed.) September, **1995**, *277* (9).
14. Maxwell, J.B. *Data Book on Hydrocarbons*; Krieger: New York, 1977.
15. Turpin, L.E. *Predicting Component Yields in a Commercial Catalytic Reformer Using a Kinetic Model*, 222nd American Chemical Society National Meeting, 2001 Spring National Meeting.
16. Kern, D.Q. *Process Heat Transfer*; McGraw-Hill: New York, 1990.
17. Clevett, K.J. *Process Analyzer Technology*; Wiley-Interscience: New York, 1986.
18. McIntosh, B. Optimizing processes with on-line infrared monitoring. Control Eng. **1998**.
19. Hugo, A. Limitations of model predictive controllers. Hydrocarbon Processing **2000**, *79* (1), 83–88.
20. Gilsdorf, N.L.; Furfaro, A.P.; Rachford, R.H.; Schmidt, R.J.; York, D.L. Reforming processes for aromatics production. In *Encyclopedia of Chemical Processing and Design*; KcKetta, J.J., Ed.; Marcel Dekker: New York, 1994; Vol. 47, 92–113.
21. Martin, G.D.; Cline, R.P.; Turpin, L.E. Estimating control function benefits. Hydrocarbon Processing **1991**, *70* (6), 68–73.
22. Ayral, T.E.; Stahl, D.E.; Glidewell, M.C. Expert systems aid plant operations managers. Hydrocarbon Processing **1998**, *77* (11), 115–119.

23. Advanced control and information systems '99, catalytic reformer advisor. Hydrocarbon Processing **1999**, *78* (11), September (T. Ayral, ed.) p. 95.

24. White, D.C. Online optimization: what, where and estimating ROI. Hydrocarbon Processing **1997**, *76* (6), 43–51.

25. Turpin, L.E.; Kim, H.C.; Yoon, H.S.; Lakshmanan, A.; Park, N.S. A reformer model for a petrochemical complex. Petroleum Technology Quarterly **2001**, 133–139.

26. Yu, C.D.; Bluck, D.; Powel, R.; Turpin, L.E. *Integration of Refinery Reactors into Flowsheet Simulation*, AIChE 1997 Spring National Meeting, March 1997.

27. Verne, T.; Jerrit, B.; Bester, P. Unique approach to real-time optimization. Hydrocarbon Processing **1999**, *78* (3), 53–60.

28. Friedman, Y.Z. Closed-loop optimization update—a step closer to fulfilling the dream. Hydrocarbon Processing **2000**, *79* (1), 15–16.

29. Turpin, L.E. Reforming processes, benzene reduction. In *Encyclopedia of Chemical Processing and Design*; KcKetta, J.J., Ed.; Marcel Dekker: New York, 1994, Vol. 47, 113–133.

30. *RT-OPT* ® *Real-Time Optimization System User's Manual, Release 3*. Aspen Technology, 1998.

31. Turpin, L.E. Performance optimization through software integration. Today's Refinery **2000**, *15* (1), 9–11.

15
Modeling Catalytic Naphtha Reforming
Temperature Profile Selection and Benzene Reduction

Rafael Larraz* and Raimundo Arvelo
University of La Laguna
Laguna, Spain

1 INTRODUCTION

In 1952 CEPSA became the first European company to license UOP Platforming at its refinery on Tenerife [1]. Many things have changed since those days, but catalytic reforming of naphtha remains one of the major processes used at refineries, with the aim of increasing the octane number of naphtha or producing aromatics. Several of the reactions in the reformer, including naphthene dehydrogenation and cracking, also produce a high hydrogen content gas useful in middle distillate hydrotreating. The reformer unit is an integral part of the overall refinery or aromatics plant flow scheme and is a key contributor to profit generation.

Semiregenerative reformers typically consist of three or four adiabatic reactors loaded with a binfunctional catalyst. Reforming reactors operate at temperatures between 460°C and 540°C, and at a total pressure between 7 and 40 bar and with molar H_2/hydrocarbon ratios between 3 and 8 [2]. High H_2/hydrocarbon ratios minimize coke deposition over the catalyst. Coke deposition decreases naphtha reforming activity and shortens the run length depending on operational severity. The reformer can be optimized for product quality, product yield, and catalyst cycle life by changing the operating variables, of which the

---

*Current affiliation: CEPSA, Madrid, Spain

reactor inlet temperatures are the most widely used. The values of reactor inlet temperatures are known as the reactor temperature profile. The profile is characterized as flat if all temperatures are equal, ascending if temperature increases from the first reactor to the last, or descending if they decrease. The establishment of operational rules based on reformer operating variables helps the operating staff to improve unit yields.

Models are powerful tools, but the user has to be aware that sometimes the model does not match the plant which it is sought to simulate. Understanding the reasons can assist in using the model to the maximum of its capabilities. The reasons why simulations do not match the plant fall into three main categories [3]: (1) simulation effects or inherent errors, (2) sampling and analysis effects or measurement errors, and (3) model misapplication effects. A model must predict behavior not only within the reactor but in the auxiliary areas of the unit as well. It should also consider the complex nature of the process and the reaction that takes place during the process of reforming. Turpin [2] has given a detailed outline of the steps in catalytic reforming modeling, including the definition of the modeling objective, process identification, model selection, data collection and validation, and, finally, model calibration and verification.

The objective of this chapter is to outline a model for catalytic naphtha reforming [4,5] that can provide operational rules. The model also simulates how feed composition and operating conditions affect product compositions and yields. In addition, as practical applications of the model, a procedure for reactor temperature profile optimization is presented, as are some guidelines for improving reforming performance. Finally, an important issue for refiners, namely, reformate benzene reduction, is studied from the operational point of view.

The reforming unit model comprises a series of modules. One module covers feed composition calculation and a second is the naphtha reforming kinetic model where important issues such as reaction mechanism, reactor model, and flash drum calculations are discussed. A survey of thermodynamic equilibrium calculations is also made. A visually oriented development software (Microsoft VBA) has been used in model development allowing a user-friendly interface. The main screen snapshot is shown in Figure 1, along with the process flow diagram of the reformer modeled in this work.

2   REFORMING THERMODYNAMIC EQUILIBRIUM CALCULATIONS

Equilibrium compositions are not usually attained under reformer operation conditions with the exception of naphthenes. But the knowledge of the reaction equilibrium values for C_6 through C_{10} hydrocarbons provides upper bounds on

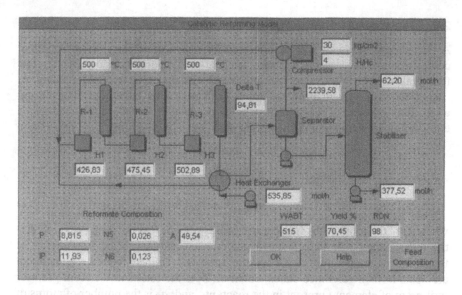

Figure 1 Reforming model main screen snapshot.

actual aromatic and individual component yields. From the information cited in [6], a number of significant results are obtained:

For all carbon numbers, equilibrium aromatics decrease with decreasing temperature and increasing hydrogen pressures.

Even at high temperatures and low hydrogen partial pressures, equilibrium aromatics are low for C_6 and C_7 hydrocarbons; the effect is more pronounced at higher pressures. At high hydrogen pressures, this phenomenon is also found for the C_8 and C_9 hydrocarbons, where aromatic formation is no longer quite as favorable as at low hydrogen pressures.

For all temperatures, hydrogen pressures, and carbon numbers, both alkylcyclohexanes and alkylcyclopentanes are essentially absent from the equilibrium mixture.

For all temperatures, hydrogen pressures, and carbon numbers, the equilibrium favors isoparaffins over normal paraffins.

The availability of an equilibrium calculation procedure gives a deep insight into the behavior of reforming reactions. Some thermodynamic equilibrium calculation procedures are discussed in the following sections.

2.1. Rigorous Methodology

A free-energy minimization method is used to determine equilibrium chemical composition of the gaseous mixture at constant temperature and pressure. The free total energy of the system is minimized by means of the free molar energy of each species, thus maintaining the number of atoms of each element. Considering a reaction or system of reactions producing products from different reactants, the equilibrium conversion of each species is obtained by minimizing the expression

$$G(n_1, n_2, \ldots, n_N) = \Sigma n_j[(\Delta G_T)_j/RT + \ln(n_j \cdot P)] \qquad (1)$$

subject to the mass balance constraints:

$$\Sigma a_{ij} \cdot n_j = b_i \ (i = 1, 2, \ldots, m); \qquad n_j \geq 0 \text{ for } \forall j \qquad (2)$$

where G is the total Gibbs free energy of the system, P is the total pressure, T is the temperature, R is the universal gas constant, $\Delta(G_T)_j$ is the free energy of formation at temperature $T(K)$ for the species j defined from its constituent elements in their reference states, cal/mol; b_i is the total sum of atom-gram or mol-gram of element i present in the reactants, and, a_{ij} is the number of atoms of element i of the reactants present in the products j. Correlation to obtain the free energy of formation, ΔG_f, for the main components of naphtha has been compiled and expressed as a function of the temperature using the following equation [7]:

$$\Delta G_f = A + BT + CT^2 \qquad (3)$$

The linear mass balance, Eq. (2), becomes [8]:

$$2n_1 + m_2 n_2 + \cdots + m_n n_n = b_1 \qquad (4)$$

$$c_1 n_1 + c_2 n_2 + \cdots + c_n n_m = b_2 \qquad (5)$$

Equation (4) is a hydrogen balance with m_i the number of hydrogen atoms in species i; Eq. (5) is a carbon balance with c_i the number of carbon atoms in species i. The ratio b_1/b_2 is the hydrogen-to-carbon ratio in the mixture.

Hydrogen partial pressure has an extremely important effect on process yields and on the rate of catalyst deactivation. Consequently, each catalyst is designed to operate at specific hydrogen pressures. Therefore, it is desirable to report the results of the idealized equilibrium calculations at constant hydrogen partial pressure so that they may be compared to actual process data.

The equilibrium compositions are determined without knowledge of the specific chemical reactions involved [9]. Due to the fact that the function described in Eq. (1) is convex, the conjugate gradient algorithm is ideally suited to solve this problem [10]. In carrying out the equilibrium calculation, care must be taken to ensure nontrivial results. Because hydrocracking reactions are the most energetically favorable reactions [11], the equilibrium compositions are calculated separately for hydrocarbon mixtures of the same carbon number. The equilibrium results are thus idealized because they are applicable only in the absence of cracking.

2.2. Shortcut Method

When it is necessary to estimate the equilibrium composition of the many complex reactions that occur during catalytic reforming of naphtha, a simple and precise method could be employed [12]. The method uses individual equilibrium compositions to derive generalized equilibrium constants for each carbon number group from six carbons up to nine carbons. This method assumes that paraffins do not change their carbon number when converted to aromatics during the catalytic reforming process. As previously, hydrocracking reactions are neglected. It is also assumed the reformate naphthenes content remains low. The key reaction of catalytic reforming, dehydrocyclization, converts paraffins to aromatics:

$$P_i \Leftrightarrow A_i + 4H_2 \qquad i = 6, 7, 8, 9$$

The equilibrium constant for this reaction is

$$K_{i,P} = (x_{i,A}/x_{i,P})p_{H2}{}^4 \tag{6}$$

where

$K_{i,P}$ = equilibrium constant for the paraffins in the y carbon number group

$x_{i,A}$ = aromatic content of the y carbon group, mole fraction of the total mixture

$x_{i,P}$ = paraffin content of the y carbon group, mole fraction of the total mixture

p_{H2} = hydrogen partial pressure, atm

If we let B symbolize the mole fraction of each carbon number group based on the total moles of the mixture, the material balance is:

$$\sum_{i=6}^{9} \quad B_i = 1 \tag{7}$$

Since naphthenes are assumed zero

$$P_i = x_{i,A} + x_{i,P} \qquad \text{and} \qquad x_{i,P} = B_i - x_{i,A} \tag{8}$$

Including this expression to the equilibrium constant equation and rearranging terms, we obtain:

$$x_{i,A} = [K_{i,P}/(p_{H2}{}^4 + K_{i,P})]B_i \tag{9}$$

Table 1 Dehydrocylization
Equilibrium Constant Coefficients
($\ln K = a(10^3/T) + b$. Range 650–
829 K)

Carbon number	a	b
C_6	-32.6	51.17
C_7	-31.24	53.36
C_8	-29.70	53.36
C_9	-28.94	53.78

Source: Ref. 12.

The total aromatic concentration of the mixture is shown by the following

$$x_{i,A} = \sum_{i=6}^{9} x_{i,A} = \sum_{i=6}^{9} [K_{i,P}/(p_{H2}^4 + K_{i,P})]B_i \qquad (10)$$

The total paraffin concentration is found by the difference with the mole fraction of each carbon number group B_i.

The equilibrium constants for each hydrocarbon type are given in Table 1, reporting temperature effects as an integrated form of the Van't Hoff equation. These data were obtained by means of equilibrium calculations of a mixture of 104 components including hydrogen. Free-energy minimization techniques were employed and substitution of the equilibrium composition for each carbon number group in the mass balance equation provides the specific equilibrium constants $K_{i,P}$.

The procedure includes some corrections in order to account for the cracking reactions, which occur during the reforming reactions. Cracking tends to increase the composition of the C_6 and C_7 hydrocarbons in the product, while the composition of the C_8 and C_9 hydrocarbons is decreased. Table 2 presents calculation results for two different operating conditions.

3 FEED CHARACTERIZATION TECHNIQUE

Advanced control strategies and optimization techniques are frequently applied to catalytic reforming in order to obtain greater economical benefits and ease unit operations. The basis for success of these strategies and techniques is the availability of on-line feed and product quality measurements. These are further described in Chapter 14. The traditional approaches employed tend to be on-line analyzers and algorithms based on quick analyses of density or distillation performed in situ. In order to obtain reformer feed composition on-line and

Table 2 Equilibrium Concentration of Aromatics

	Feed composition $B_i = x_{i,A} + x_{i,P};\ i = 6, \ldots, 9/\Sigma B_{i=6}^9 = 1$		
B_6	B_7	B_8	B_9
0.24	0.30	0.28	0.18

	Results $T = 482/530°C,\ P = 10/30\,\text{kg/cm}^2$			
T/P	$x_{6,A}$	$x_{7,A}$	$x_{8,A}$	$x_{9,A}$	x_A
482/10	0.103	0.281	0.278	0.180	0.842
482/30	0.004	0.069	0.216	0.163	0.452
530/10	0.225	0.299	0.280	0.180	0.984
530/30	0.054	0.250	0.275	0.179	0.785

Source: Ref. 12.

without time delay, an estimative naphtha characterization algorithm based on the correlation of physical properties is presented and a relatively new methodology for naphtha characterization, known as nuclear magnetic resonance (NMR), is described in brief.

3.1. Feed Characterization Algorithm

The naphtha used as catalytic reformer feedstock usually contains a mixture of paraffins, naphthenes, and aromatics in the carbon number range C_5 to C_{10}. The analysis of feed naphtha is typically reported in terms of its American Society for Testing and Materials (ASTM) distillation curve and API gravity. Since reforming reactions are described in terms of lumped chemical species, a feed characterization technique based on the correlation of physical properties with reformate boiling range and density has been implemented by Taskar and Riggs [13] in naphtha reforming modeling of industrial units. Several other methods based on neural networks exist in the literature that allow feed composition in terms of paraffins, olefins, naphtenes, and aromatics (PONA) and reformate octane number prediction [14,15]. The proposed method results exhibit errors ranging from 3% to 10%. The main steps of the procedure are described below.

The ASTM distillation data are first converted to true boiling point data. Riazi et al. [16] compile experimental data on ASTM, TBP, and EFV distillation of fractions from different sources. Using this set of data, the following equation was found to be the simple form for interconversion of distillation data.

$$t = a\theta^b S^c \tag{11}$$

where t (TBP or EFV) and θ (ASTM) are temperatures (in °R) at the same vol % vaporized, S is the specific gravity, and a, b, and c are constants. If the specific gravity is not known, Eq. (12) could be used with an average deviation of about 2%.

$$S = at_{10}{}^b t_{50}{}^c \qquad (12)$$

where t_{10} and t_{50} (°R) are temperature at 10% and 50% in ASTM, TBP, or EFV curve. The TBP curve is decomposed into a number of discrete volume fractions. According to Lion and Edmister [17], the specific gravity of each volume fraction is evaluated assuming constant UOP characterization factors.

Riazi et al. [18] have given a correlation table to predict physical properties of hydrocarbon mixtures using average boiling point and specific gravity in an empirical equation of the form

$$P = aT^b S^c \qquad (13)$$

where P is a physical property to be predicted, T is the average boiling point in °R, and S is the specific gravity at 60°F. Constants a, b, and c are shown in Table 3 for different physical properties. Prediction accuracy is reasonable, and average deviations remain under 3.5% over the boiling range of 40–450°C for the following properties: molecular weight, liquid density, liquid molar volume, critical temperature, critical pressure, critical volume, refractive index, heat of vaporization, and ideal gas heat capacity. Since the materials boil over a range of temperatures, any one average boiling point fails to correlate all of the properties. In general, the average boiling points listed are used for estimating the physical properties:

Volume average boiling point	Specific heat
Molal average boiling point	Pseudo-critical temperature
Mean average boiling point	Pseudo-critical pressure
	Molecular weight
	Heat of vaporization
	Density
	Liquid molar volume

Table 3 Correlation Constants for Equation $P = aT_b{}^b S^c$

Property	A	b	c
Mol. wt	4.5673E-3	2.1962	−1.0164
Critical temp. (°R)	24.2787	0.58848	0.3596
Critical pressure (psia)	3.12281E-9	−2.3125	2.3201
Molar volume at 20°C (cm³/gmol)	7.6211E-5	2.1262	−1.8688
Density (g/cm³)	0.982554	0.002016	1.0055
Heat of vaporizat. (Btu/lb mol)	8.48585	1.1347	0.0214
Specific heat at 0°F (Btu/lb mol)	4.0394E-7	2.6724	−2.363

P, physical property; T_b, normal boiling point (°R); S, specific gravity.
Source: Ref. 16.

Each TBP fraction is composed of a mixture of chemical species whose boiling points fall within the TBP range of the corresponding fraction. An adaptive random search (ARS) algorithm [19–22] is applied in order to find the naphtha chemical composition, in terms of paraffins, naphthenes, and aromatics, that provides the best match between predicted and observed physical properties in each volume fraction. The entire mixture composition is then calculated from the volume fractions. Olefin species are not considered, introducing an error of about 0.5% into the naphtha composition. Pure hydrocarbon properties have been extracted from the API tables [23].

The composition of a naphtha reforming feed is studied employing the outlined characterization method. Specific gravity and ASTM D86 distillation of reforming feed are used as source data, then a split into discrete fractions with gravity assay estimations is made, as shown in Tables 4 and 5. Density and molecular weight are the physical properties that have to be adjusted. Table 6 gives the predicted values for each fraction of feed naphtha. The adjusted composition of the naphtha feed is presented in Table 7, detailing each fraction composition. Assay needs are limited to a D86 ASTM and an API or specific gravity; and even this information could be avoided employing Eq. (12). Method error varies from 3% to 10% but gives enough information to predict feed quality changes and to monitor unit performance.

3.2. NMR On-line Analyzers

During an NMR analysis, a sample stream is passed through a precisely controlled magnetic field. The protons of the sample 'line up' with the homogeneous magnetic field. To take a reading, the NMR analyzer transmits

Table 4 Reformer Naphtha Feed Assay

Assay (vol %)	Naphtha feed (°C)
ASTM 0%	105
ASTM 10%	114
ASTM 30%	122
ASTM 50%	131
ASTM 70%	147
ASTM 90%	168
ASTM 95%	182
ASTM 100%	200
RON	39.5
SPGR at 15°C	0.7471

Table 5 Distillation Ranges and Density Estimation Naphtha Feed

Fraction (°C)	Vol %	Density (K_{UOP})
10–40		
40–70		
70–100	21.0	0.703
100–125	22.0	0.737
125–150	26.0	0.755
150–175	20.0	0.769
175–EP	11.0	0.787
IP–EP	100.0	0.7471[a]

Table 6 Estimated Properties Naphtha Feed

Fraction (°C)	Density ($P/N/A$)	Molecular weight ($P/N/A$)
10–40		
40–70		
70–100	0.684/0.756/0.878	96/89/80
100–125	0.702/0.770/0.866	112/102/94
125–150	0.717/0.779/0.865	129/115/104
150–175	0.730/0.793/0.865	145/130/120
175–EP	0.750/0.800/0.870	160/150/135

Table 7 Calculation Results

Fraction (°C)	Naphtha feed ($P/N/A$) (vol %)
10–40	
40–70	
70–100	85.2/13.1/1.7
100–125	59.0/30.7/10.5
125–150	53.3/29.9/16.8
150–175	55.3/28.5/16.2
175–EP	53.2/27.1/19.7
IP–EP	62.5/25.6/11.9

pulses of radiofrequency energy into the stream, which deflects the protons off their aligned axis. The amount of deflection and the subsequent recovery time vary according to molecular structure, and the NMR analyzer reads this structure by analyzing frequency signals emitted by the spinning protons. The analyzer averages multiple pulses into a spectrum that reveals chemical species and their concentrations. This spectrum can then be correlated with physical properties in addition to the chemical composition, enabling determination of multiple parameters in a single spectrum. And since NMR is not an optical technology, the analysis is essentially independent of sample state or physical condition. In NMR spectroscopy, the surrounding molecular structure determines where the species will be resolved in the spectrum. These peaks occur at known positions, and their intensity is directly proportional to concentration. These facts contribute to the high precision and accuracy of the NMR method.

In Figure 2 several groups of spectra showing different naphtha streams and the chemical makeup of the streams as observed by NMR are presented. The aromatic, olefinic, and aliphatic regions are in their own distinct areas of the spectrum. Different types of olefins can be quantified, i.e., terminal vs. internal olefin, as well as trisubstituted, disubstituted, and vinylidene olefin distributions. Monoaromatics are readily discerned from PNAs, and benzene is observed as a single resonance. The aliphatic protons are observed as peaks due to CH_3, CH_2, and CH proton types. This yields direct information about branching character and linear character, i.e., paraffin vs. isoparaffin content. The chemical detail that NMR provides will not only allow correlation to PONA-type analyses but also allow speciation of olefinic and aromatic types. In Table 8 a statistical analysis is presented [24]. This extremely precise analysis allows kinetic model parameter correlation to be made frequently and on-line.

4 NAPHTHA REFORMING KINETIC MODEL

From a fundamental point of view, kinetics contributes to a better understanding of the reaction mechanisms and of the effect of catalysts on these. From a practical point of view, accurate kinetic equations are of great importance for the reliable design and simulation of the reactors. Relatively few reaction models of the reforming process have been suggested in the literature pertaining to this field. The first significant attempt at delumping naphtha into different constituents [25] considered naphtha to consist of three basic components: paraffins, naphthenes, and aromatics. Due to its simplicity, this model is still used. Basing his work on Smith's model, Barreto et al. [27] included certain modifications concerning discrimination between the reaction rates for aromatization of five- and six-ring naphthenes, the consideration of two kinds of paraffins with different reactivities, and the inclusion of an overall hydrodealkylation reaction. Also

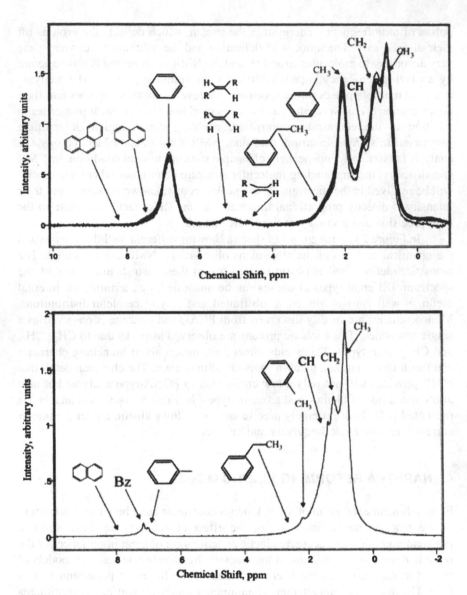

Figure 2   Naphtha composition by NMR. (a) Heavy Cat–cracked naphtha. (b) Straight-run naphtha.

Table 8 On-line Comparison of I/A Series Process NMR and Laboratory GC Analysis

Parameter	Mean conc. (wt %)	Mean dev. (wt %)	Std. dev. (wt %)	Correlation coefficient
Total normal paraffin	34.10	−0.15	0.78	0.980
Total isoparaffin	32.01	0.01	0.66	0.975
Total naphthenes	20.40	0.04	0.63	0.983
Benzene	1.99	0.01	0.11	0.993
Toluene	1.99	−0.01	0.15	0.995
Ethylbenzene	0.33	0.00	0.04	0.993
Xylenes	1.56	0.01	0.08	0.999
C_9 Aromatics	0.65	0.01	0.12	0.993
C_{10} Aromatics	0.14	0.01	0.06	0.962
Total aromatics	6.66	0.01	0.17	0.999

Source: Ref. 24.

based on Smith's model, Ansari and Tade [28] developed a multivariate control application after testing it against existing plant data. In a more extensive attempt to model reforming reactions of whole naphtha, Krane et al. [29] recognized the presence of various carbon numbers from C_6 to C_{10} as well as the difference between paraffins, naphthenes, and aromatics within each carbon number group. The model derived by Krane contained a reaction network of 20 different components. An improved model has been proposed by Aguilar and Anchyeta [30,31] where the Krane model is modified to account for temperature and pressure variations. Burnett [32] proposed a similar model. Jenkins and Stephens [33] employed first order rate equations including reversible ones to develop a kinetic model, the effect of pressure on the reaction rate is simulated by means of a pressure factor with a characteristic exponent for each particular reaction. The first mention of a different treatment for the C_5 and C_6 ring naphthenes was made by Henningsen and Bundgaard-Nielson [34] who also gave accurate values for the heat of reaction and activation energies of the reactions involved in the model.

All the models described were pseudo-homogeneous in nature. Kmak [35] presented the first model incorporating the catalytic nature of the reactions by deriving a reaction scheme with Hougen–Watson–Langmuir–Hinshelwood–type kinetics. Rate equations derived from this type explicitly account for the interaction of chemical species with the catalyst. In another notable effort, Ramage et al. [36] developed a detailed kinetic model based on extensive studies of an industrial pilot plant reactor. The model accurately simulates commercial reactor performance and has been applied routinely in Mobil refineries under the name of KINPTR [37]. New continuous regeneration reforming units have been simulated by Lee et al. [38], and Taskar has employed a rigorous kinetic model to optimize performance in an industrial reformer by studying operating modes and the influence of operational variables [13].

The Kmak model was later refined by Marin and coworkers [39], who presented the reaction network for the whole naphtha. Containing hydrocarbons in the carbon number fraction from C_5 to C_{10}, the reaction network included 23 pseudocomponents and used Hougen–Watson rate equations. Marin and Froment [40] and Van Trimpont et al. [41] also conducted separate studies on C_6 and C_7 carbon number fractions, respectively, and developed the corresponding Hougen–Watson rate equations. Various possible reaction paths and mechanisms were systematically evaluated before choosing the one that best fits the experimental data provided by a laboratory scale reactor. An extensive experimental program with various feeds led to a plausible reaction mechanism using kinetics of the Hougen–Watson type [39,42]. The reaction network involves hydrogen and 92 hydrocarbons ranging from C_1 to C_{11}, including all individual detectable hydrocarbons up to and including C_8. Between certain hydrocarbons, equilibrium is reached so that lumping is permitted. This leads to 29 lumps: hydrogen, methane, ethane, propane, butanes, pentanes, multibranched hexanes, single-branched hexanes, n-hexane, five-ring naphthenes, benzene, six-ring naphthenes, the analogues of the C_6 lumps for C_7 and for C_8 separately, and, finally, five C_5+ lumps involving C_9 to C_{11} hydrocarbons divided into isoparaffins, n-paraffins, five-ring naphthenes, six-ring naphthenes, and aromatics [43]. This definitive model has been extended as a collaboration between IFP and Froment's group, in order to cover reformer units dynamic simulations thus providing an operator training tool as well as unit performance monitoring [44].

Diffusional effects in naphtha reforming modeling are not usually treated separately. The uncertainty associated with diffusion effects in the catalyst pellets is lumped into the kinetic rate parameters [13,39]. If there are diffusional limitations inside catalyst pellets, the effective reaction rates differ from the intrinsic reaction rates at surface conditions. To solve the problem the pre-exponential factor in the rate equation is tuned as an adjustable parameter matching simulation results to reality. Coppens and Froment [43] improved reforming models by including diffusional effects into rate equations. Inside catalyst pellets, diffusion and reaction is described by the continuity equations for the M lumps:

$$\nabla \cdot N_j = \rho_s R_j \qquad j = 1, \dots, M \tag{15}$$

together with Stefan–Maxwell flux equations for multicomponent diffusion [45]:

$$-100/RT \nabla p_j = \sum_{k=1}^{M} (p_k N_j - p_j N_k)/(p_t D_{e,jk}) = N_j / D_{e,jk}$$

$$j = 1, \dots, M \tag{16}$$

In their model, the porous nature of the reforming catalyst support (γ-alumina) is approximated by a self-similar fractal structure (see Appendix at the end of the chapter). Fractal dimension, D, of the catalyst is obtained by means of small-angle X-ray scattering (SAXS) and expressions for Knudsen diffusion in terms of fractal dimension are provided [46–49]. Model results showed that the effects of the total fractal surface morphology on the product distribution and yields are significant. When the intrinsic reaction rates per unit catalyst mass are the same, a catalyst with a smooth surface, $D = 2$, yields 18.9 wt % isoparaffins and 70.1 wt % aromatics while a catalyst with a fractal surface, $D = 3$, yields 21.7 wt % isoparaffins and 66.2 wt % aromatics. This is a result of the stronger diffusional limitations found within catalysts with a fractal surface, as compared to catalysts with a smooth surface.

4.1. Reaction Mechanism

As reaction kinetics play a paramount role in catalytic reforming model performance we have employed a rigorous model [36,39] in our calculations. The naphtha used as catalytic reformer feedstock contains a mixture of more than 285 different chemical species, mainly paraffins, naphthenes, and aromatics in the carbon number range C_5 to C_{10}. This overwhelming number of chemical components requires chemical component lumping into a smaller set of kinetic lumps [50] to make the reforming reactions network tractable. Naphtha reactions in catalytic reforming are dehydrogenation of naphthenes, isomerization of paraffins and naphthenes, dehydrocyclization of paraffins, and hydrocracking of paraffins. The selectivity changes considerably between six-carbon, seven-carbon, and eight-carbon species in a given molecular class. For hydrocarbons containing eight or more carbon atoms the selectivity within a molecular class does not vary significantly due to the similarity of their aromatization equilibrium data [6]. Therefore, the C_6, C_7, and C_8+ lumps are considered plus a total C_5- lump. The reaction network and lumping scheme are shown in Figure 3.

Due to recent legislation reformate benzene content has been limited, and knowledge of the benzene formation mechanism could be an aid to finding procedures that ease the task of meeting the new specifications. The first sources of benzene in reformate that can be identified are benzene and benzene precursors from the C_6 fraction of the feed. As a result of the benzene formation mechanism [40], it can be expected that cyclohexane from the feed is almost completely dehydrogenated into benzene. To convert methylcyclopentane to benzene is much more difficult because methylcyclopentane first has to be converted into cyclohexane by functional dehydrogenation and a ring isomerization reaction. Particularly in the case of methylcyclopentane, ring isomerization is a difficult step since a more stable secondary methylcyclopentane carbenium ion has to be isomerized into a less stable primary cyclohexane carbenium ion. Nevertheless,

Carbon Number	Six Carbon Naphthenes (N6)	Five Carbon Naphthenes (N5)	Isoparaffins IP	Paraffins P	Aromatics A
C8+	Cyclohexanes	Cyclopentanes	Isoparaffins	Paraffins	Aromatics
C7	Methyl Cyclohexane	Cyclopentanes	Isoheptanes	Heptanes	Toluene
C6	Cyclohexane	Methyl Cyclopentane	Isohexanes	Hexanes	Benzene
C5-			C5- Hydrocarbons	C5- Hydrocarbons	

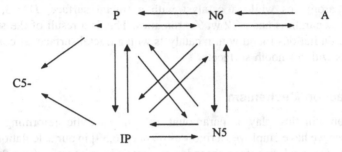

Figure 3   Lumping scheme reaction network.

once the cyclohexane intermediate is formed, it is easily dehydrogenated to benzene. Finally, the aromatization of *n*-hexane is even more difficult because *n*-hexane must be converted to cyclic compounds by slow dehydrocyclization reactions. As a rule of thumb it can be said that only about 20 wt % of *n*-hexane and 50 wt % of methylcyclopentane from the feed are converted into benzene [51]. Next to benzene precursors from the C_6 fraction of the feed, hydrocracking of higher molecular weight hydrocarbon fractions can form benzene precursors. C_7+ aromatic compounds can be dealkylated, which also can lead to the formation of benzene. A differentiation is made between C_6, C_7, and C_8+ compounds only. This is justified by the fact that in general the aromatization selectivity of paraffins and cyclopentanes increases with chain length. However, the differences in selectivity between C_8+ compounds are very small. An increase in the selectivity for hydrocracking reactions may lead to a higher benzene content of the reformate [52]. Reforming catalysts are generally known as bifunctional catalysts because the metallic and the acid sites are discrete components. They cooperate in promoting the desired overall reactions, but each appears to be responsible for certain steps. The two functions of a reforming catalyst—metallic and acid—are balanced carefully for reaction conditions. If the balance is shifted, side reactions occur. Benzene selectivity is enhanced with high acidity catalyst as the C_8+ aromatics selectivity is decreased [52]. So the catalyst chloride level has to be strictly controlled in order to avoid benzene formation due to a strong acid function.

4.2 Reactor Model

Simulation of reforming reactors which incorporates the kinetic model has been made according to the following procedure. The mass, energy, and momentum equations for an axial flow reactor are:

$$dF/dz = \rho_b \Omega R \tag{17}$$

$$dT/dz = \rho_b M_m/(G c_p) \sum_{j=1}^{NC} (-\Delta H^0{}_f)_j R_j \tag{18}$$

$$dp_t/dz = -\frac{1.5 \cdot 10^{-3}(1-\varepsilon)^2 \mu G}{\varepsilon^3 p_f (d_p)^2} - \frac{1.75 \cdot 10^{-5}(1-\varepsilon)G^2}{\varepsilon^3 p_f d_p} \tag{19}$$

Initial conditions are $F = F_o$, $T = T_o$, and $p_t = p_{to}$ at $z = 0$. The production rates R_j are obtained from:

$$R_j = \Sigma \beta_{ij} r_i \qquad j = 1, \ldots, Nc \tag{20}$$

β_{ij} is negative for reactants and positive for products of the single reaction i. The rate expressions of the individual reactions, r_i, are of the Hougen–Watson type and are shown in Table 9. For the dehydrogenation of six-ring naphthenes a first-order rate expression is used. The dehydrogenation is much faster than the former reactions and is diffusion limited, and this criteria is analogous to the approach of Marin et al. [39]. The activation energies and pre-exponential factors used are shown in Table 10 as well as the reaction heats involved. Gear's numerical integration scheme for solving stiff ordinary differential equations was employed to integrate the equations through each fixed-bed reactor. Figure 4 presents the calculation sequence, including flash drum calculations.

Table 9 Catalytic Reforming Reaction Rate Expressions

Hydrocracking	$r_{\text{Hydrocracking}} = Ae^{-E/RT}(P_A P_H)/\Gamma$
Isomerization	$r_{\text{Isomerization}} = Ae^{-E/RT}(P_A - P_B/K_{AB})/\Gamma$
Dehydrocyclization	$r_{\text{Dehydrocyclization}} = Ae^{-E/RT}(P_A P_B P_H/K_{AB})/\Gamma$

A, pre-exponential factor; E, activation energy; Γ, adsorption term; H, hydrogen; A, reactant; B, product; K_{AB}, equilibrium constant; P_i, partial pressure of component i.

Table 10 Heats of Reaction (kcal/gmol)

Reaction	C_6	C_7	C_8	C_9
$N_6 \Leftrightarrow N_5$	3.8	4.5	7.1	8.8
$N_6 + H_2 \Leftrightarrow P$	−10.6	−8.4	−7.3	−5.8
$N_5 + H_2 \Leftrightarrow P$	−14.4	−13.0	−14.4	−14.6
$N_6 \Leftrightarrow A + 3H_2$	52.3	51.5	49.9	51.0
$N_5 \Leftrightarrow A + 3H_2$	48.5	47.0	42.8	42.2
$P \Leftrightarrow IP$	−2.3	−2.3	−1.9	−2.0
$N_6 + H_2 \Leftrightarrow IP$	−12.9	−10.7	−9.1	−7.8
$N_5 + H_2 \Leftrightarrow IP$	−16.7	−15.2	−16.2	−16.6

Kinetic Rate Equation Parameters

Reaction	Pre-exponential factor			Activation energy (cal/mol)
	C_6	C_7	C_8+	
Hydrocracking	4.18 E12	1.61 E13	1.76 E13	53000
Dehydrocylization	1.0188 E11	1.37 E10	3.23 E10	45000
Dehydrogenation	1.35 E08	3.04 E08	7.24 E08	30000
Isomerization	7.72 E07	3.08 E08	3.25 E08	40000

Source: Ref. 6.

Deactivation Function

Coke deposition causes deactivation of the catalyst, resulting in a reduction of the reforming reaction rates. Coke is generated from the olefinic intermediates formed during the course of the main reforming reactions. Higher hydrogen pressures suppress the diolefin formation, thereby reducing coke formation [54]. As has been confirmed experimentally by means of temperature-programmed oxidation tests, deactivation during the first hours of the run is due to coke deposition on the metal and thereafter to deposition on the support, which produces minor deactivation [55,56]. Catalyst deactivation can be accounted for by introducing a deactivation function [40,57] ϕ for each of the so-called main reactions:

$$r_i = r^o{}_i \phi \qquad 0 < \phi < 1 \tag{21}$$

The deactivation function depends on catalyst coke content rather than run length as commonly considered. The dependence has the form

$$\phi_i = e^{-\alpha_i C_c} \tag{22}$$

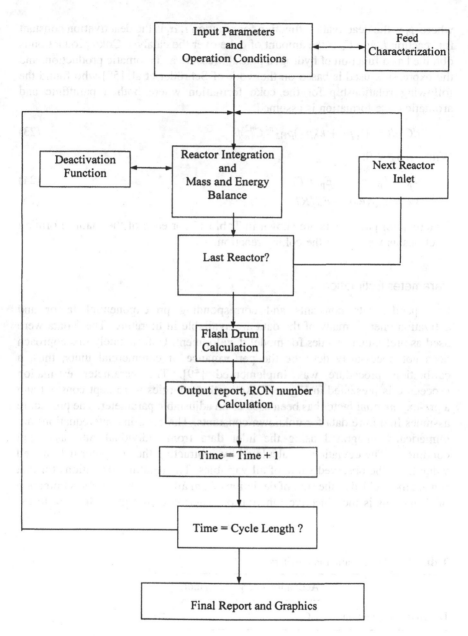

Figure 4 Model calculation flow chart.

where ϕ_i is the deactivation function for reaction i, α_i is the deactivation constant for reaction i, and C_c is the amount of coke over the catalyst. Coke production is obtained as a function of hydrogen partial pressure and aromatic production, and the expression used is based on the work of Schröder et al. [58] who found the following relationship for the coke formation where both a paraffinic and aromatic coke formation is assumed.

$$dC/dt = (k_p p_p + k_A p_A) p_{H2}{}^{n1} C^{n2} \phi \tag{23}$$

With $n_1, n_2 < 0$:

$$k_p = k_{p\infty} \exp(-E_p/RT) \tag{24}$$

$$k_A = k_{A\infty} \exp(-E_A/RT) \tag{25}$$

Deactivation parameters are shown in Table 11 for each of the main reforming reactions as well as for the coking reaction.

Parameter Estimation

For specific rate constants and corresponding pre-exponential factor and activation energy, many of the data are available in literature. These data were used as preliminary values for model development. Unfortunately this approach does not necessarily describe the performance of commercial units; thus, a calibration procedure was implemented [59]. The parameter estimation procedure is presented in Figure 5. Activation energies were kept constant and a pre-exponential factor has been used as an adjustable parameter. The procedure assumes literature data for unknown constants. The resulting rate equations are numerically integrated using the inlet data from individual runs as initial conditions. The deviation is calculated by subtracting the computer-determined value from the observed value of all variables. The resulting deviations for that run are then added to the sum of deviations from all runs. This double summation of deviations is the objective function or criterion of the fit of the data to the

Table 11 Deactivation Parameters

Deactivation function	Activation energy (cal/mol)		
	C_6	C_7	C_8+
k_P (paraffinic)	49000	90000	90000
k_A (aromatic)	14000	47000	47000

Source: Ref. 58.

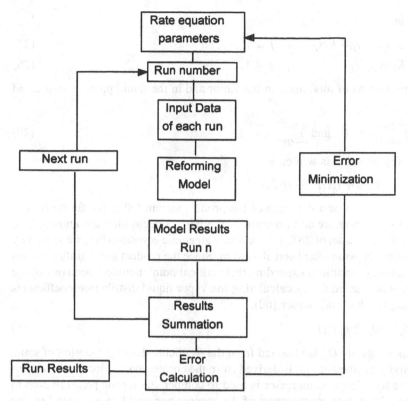

Figure 5 Parameter estimation procedure.

model. The constants are then varied one at a time to minimize this objective function, which is presented in Eq. (26):

$$\Phi(j, i) = \Sigma_{j=1}{}^{j}\Sigma_{i=1}{}^{i}\|Y_{j,i}^* - Y_{j,i}\|$$ (26)

A conjugate gradient algorithm is used in the calculations, and different initial values have been tested in order to avoid local minimums.

5 FLASH DRUM SEPARATOR MODEL

The flash drum separator, which separates the hydrogen and some lighter gases from heavier products, was modeled as an isothermal flash operation. Equilibrium is established between the vapor and liquid phase. The mass balances and vapor liquid equilibrium equations for component j of the stream are written as:

$$F_t z_j = F_v y_j + F_1 x_j \qquad j = 1, \dots, Nc \qquad (27)$$

$$K_j = y_j / x_j \qquad j = 1, \dots, Nc \qquad (28)$$

The conservation of total mass in the vapor and in the liquid phase is expressed by:

$$\sum_{j=1}^{NC} y_j = 1 \quad \text{and} \sum_{j=1}^{NC} x_j = 1 \qquad (29)$$

The energy equation is written as:

$$Q = H_v F_v + (H_1 F_1 - H_f F_t) \qquad (30)$$

When composition and flow rate of the product stream fed to the flash drum as well as the temperature and pressure of the flash condensation are known, Eqs. (27) to (30) form a set of 2NC+3 nonlinear algebraic equations that are solved by means of the Newton–Raphson algorithm. Since the product gas mainly contains components at conditions exceeding their critical point, nonideal behavior of the mixture is accounted for by calculating the vapor liquid distribution coefficients according to Chao and Seader [60]:

$$K_j = O_{jv} / (\gamma_{j1} V_{j1}) \qquad (31)$$

The vapor fugacity O_{jv} is obtained from the Redlichh–Kwong equation of state. The activity coefficient, γ_{j1}, is derived from the correlation of Flory and Huggins [61]. The recycle gas compressor is used to compensate for the pressure drop in the loop. The power consumption of the compressor could be estimated on the basis of calculated adiabatic head and flow rate of recycle gases:

$$\Delta H_{\text{adiabatic}} = RT_{\text{inlet}}(k/k - 1)\{(P_{\text{out}}/P_{\text{in}})^{k-1/k} - 1\} \qquad (32)$$

6 MODEL APPLICATIONS

Catalytic reformers are designed with operational flexibility in mind, whether for motor fuel production or to produce aromatics such as benzene, toluene, and xylenes. The variables that affect the performance of the catalyst and change the yield and quality of reformate are feedstock properties, reaction temperature, space velocity, reaction pressure, and hydrogen-to-hydrocarbon ratio [62]. The reactor inlet temperatures are invariably the most effective tool at the refiner's disposal to obtain the required reformate quality. As catalysts lose activity in operation, reactor temperature is gradually increased to maintain a constant octane number in the product reformate [63]. Since multiple factors affect reforming performance, a unit model appears as an useful tool to improve unit capabilities and extend cycle length. In the following sections, two important issues have been selected as reforming model applications. First, model performance against an industrial unit is outlined; then an analysis is made of reactor temperature profile influence over reforming performance and run length. Second, the effect of operational variables over benzene formation during naphtha reforming is studied, and some recommendations are made in order to reduce the reformate benzene content.

6.1. Reforming Model Performance

The reforming model described was used as a monitoring tool for an industrial semiregenerative reforming unit. The main operating parameters for the industrial reformer are presented in Table 12. The data employed were obtained from the control system and cover a 6-month period so catalyst deactivation could be evaluated. Reformate composition and reactor ΔT values were used to calibrate the model. Model fitting to plant data are shown in Figure 6, and model accuracy is satisfactory for the selected variables. Based on these results, one can expect a reasonable model predictive capability.

Catalyst deactivation was estimated using the deactivation function paradigm, and deactivation results for the main reforming reactions are presented in Figure 7. From the graphs obtained, a slight relative deactivation is observed for the aromatization, dehydrocyclization, and isomerization reactions ranging from 0.85 to 0.7. This supports the fact that the catalyst is still active and is at about midlife condition.

Table 12 Industrial Semiregenerative Reformer Operating Parameters

Date	31/10/2001	30/11/2001	30/12/2001	29/01/2002	26/02/2002
Flow rate (BPSD)	24034	25326	24985	25748	25867
Reactor temp.					
1 Rx	519	521	522	523	524
2 Rx	511	514	515	516	517
3 Rx	511	514	515	516	517
Pressure (kg/cm^2)	29	29	29.8	30	30
Feed composition					
Paraffins	25.8	24.7	22.4	24.0	25.3
Isoparaffins	31.5	29.6	28.9	28.1	29.6
Naphthenes	31.8	34.8	37.4	37.7	34.5
Aromatics	10.8	10.6	10.3	10.0	10.5

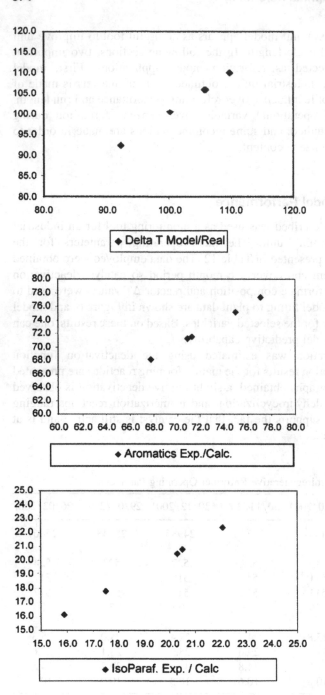

Figure 6   Comparison of industrial reformer data and model calculations for reactor ΔT, aromatics, isoparaffins, paraffins, and naphthenes.

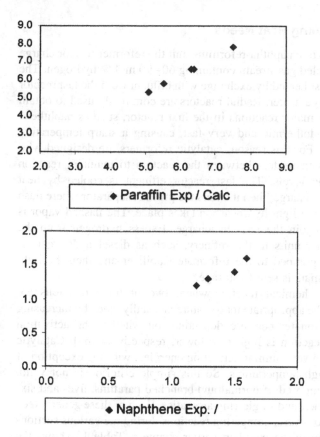

Figure 6 Continued.

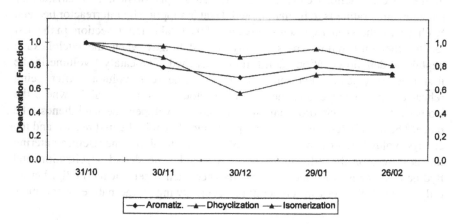

Figure 7 Deactivation function evolution with time for reforming reactions.

6.2. Naphtha Reforming Heat Needs

In a typical semiregenerative naphtha-reforming unit the reformer reactor charge is combined with a recycled gas stream containing 60–90 mol % hydrogen. The total reactor charge is first heated by exchange with effluent from the last reactor and then in the first charge heater. Radial reactors are commonly used to obtain low-pressure drops. The major reactions in the first reactor, such as naphthene dehydrogenation, are endothermic and very fast, causing a sharp temperature drop in the first reactor. For this reason, catalytic reformers are designed with multiple reactors and with heaters between the reactors to maintain reaction temperature at operation levels. The last reactor effluent is cooled by heat exchange with the reactor charge; then it enters the product separator where flash separation of hydrogen and light hydrocarbon takes place. The flashed vapor is compressed and sent to join the naphtha charge. Excess hydrogen from the separator is sent to other units in the refinery, such as diesel hydrotreating. The separator liquid is pumped to the reformate stabilizer and then the light hydrocarbons free reformate is sent to storage.

In the design of chemical reactors where two or three reactions are involved, the choice of the appropriate temperature is usually made by increasing or decreasing the reaction temperature depending on whether the activation energy of the desired reaction is higher or lower, respectively [64]. Catalytic reforming reactions have very similar activation energies, with the exception of coking which favors high temperatures. So this simple criterion is not valid. Reformer naphtha is composed of normal and branched paraffins, five- and six-membered ring naphthenes, and single-ring aromatics. Each of these general feed constituents can undergo several competing reactions during the various sections of the reaction as has been discussed in earlier chapters. Table 13 shows the influence of operating conditions on each of the main catalytic reforming reactions. The differing rates of the various reactions and the high endothermicity of the fastest reactions call for a substantial supply of heat if a satisfactory reaction temperature is to be maintained. Heat is supplied by interreactor heaters. Each reactor therefore represents only part of the total reaction section. In the first reactor the temperature drops rapidly and stabilizes when the cyclohexane naphthene dehydrogenation equilibrium is reached. The catalyst volume in the first reactor typically represents 10–15% of the total volume. After being reheated the effluent enters the second reactor, and this is mainly where the isomerization and dehydrogenation reactions of cyclopentane naphthenes take place. The drop in temperature is less pronounced than in the first reactor and the catalyst volume represents about 20–30% of the total volume. Before entering the next reactor the effluent is again reheated. The dehydrocyclization and hydrocracking reactions more or less balance each other out thermally; hence, only minor differences in temperature occur. Owing to the nature of the main reactions in each reactor and of the average level of temperature in each reactor, coke formation increases from the first to the last reactor. Recycling of hydrogen-rich gas provides sufficient hydrogen partial pressure to prevent coke forming too rapidly on the surface of the catalyst. The main variable used in the unit is the temperature of reaction, which can be adjusted by means of furnaces installed upstream of the reactors.

A steady-state reactor heat balance is usually raised to estimate process and product conditions. Kugelman [65] has derived formulas to calculate the reformer steady-state heat balance in terms of enthalpies above the reference state at both inlet and outlet conditions. The reactor fluid is treated as the sum of the following constituents: C_6+ mixture, C_5- mixture, recycle mixture, and H_2 produced via reaction. The contribution to the heat balance from the C_6+ mixture and C_5- mixture is represented by enthalpy balance equations for each of the species present in the mixtures. The recycled portion has no composition change, and the change in enthalpy contributed by the recycle depends only on the temperature difference between the outlet and the inlet. The hydrogen produced is given as a function of the C_6+ feed. In Table 14 linear equations for reforming mixture component enthalpy calculation are given as a function of temperature. Aguilar et al. [30,31] has also proposed an equation to estimate the temperature profile along the reactor, considering adiabatic regime, from an energy balance over the differential reactor control volume.

The reforming model results are presented in Figure 8 where the temperature drop and composition profiles of paraffins, naphthenes, and aromatics along the reaction system are shown. Operational conditions and feed naphtha composition used are presented in Table 15. The model reproduces typical unit behavior. Naphthene dehydrogenation results in a sharp temperature drop in the first reactor and the remaining naphthenes disappear along the reactors, increasing the aromatic content and causing minor temperature drops. Hydrocracking reaction produces an increase in C_5- hydrocarbons and a

Table 13 Catalytic Reforming Reactions Characteristics

Reaction	Heat of reaction (kcal/mol)	Equilibrium favorable conditions	Activation energy (kcal/mol)	Rate of reaction
Isomerization	2–4	T Low	20–25	Medium
Naphthene dehydrogenation	−50	High T/low P	20	High
Dehydrocylization	−60	High T/low P	35	Low
Hydrocracking	10	Complete	40–50	Low
Coke formation		Complete	30–40	Very Low

Table 14 Coefficients for the Linear Expression of the
Component j Enthalpy in the i Lump as a Function of
Temperature T. $\Delta H_j{}^i = m_j{}^i T + b_j{}^i$

j	$m_j{}^i$	b_j^i
	$i = N_6$, six-carbon-ring naphthenes	
C_6	0.762	-858.7
C_7	0.772	-900.3
C_8	0.783	-912.6
C_9	0.793	-924.4
	$i = N_5$, five-carbon-ring naphthenes	
C_6	0.732	-755.1
C_7	0.768	-816.6
C_8	0.768	-816.6
C_9	0.768	-816.6
	Paraffins	
C_6	0.793	-1082.5
C_7	0.794	-1052.7
C_8	0.792	-1015.2
C_9	0.792	-984.0
	Aromatics	
C_6	0.558	305.7
C_7	0.585	75
C_8	0.606	-75.9
C_9	0.624	-157.7
	Light hydrocarbons	
C_1	0.907	-2208.6
C_2	0.828	-1432.8
C_3	0.813	-1231.0
iC_4	0.809	-1211.1
nC_4	0.801	-1145.6
iC_5	0.797	-1130.5
nC_5	0.795	-1082.7
	Hydrogen	
H_2	3.50	-296.8

Source: Ref. 65.

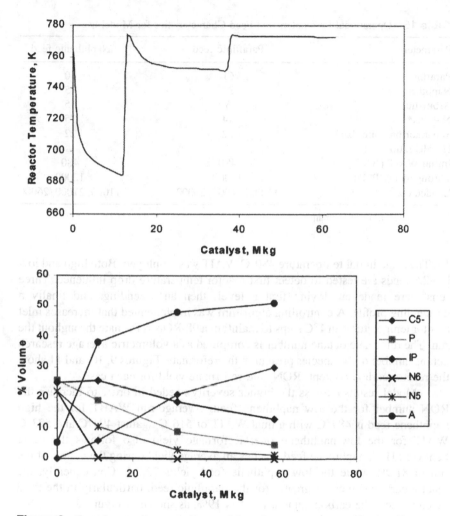

Figure 8 Reaction temperature drop and lumping group composition profiles across reactor system.

decrease in paraffin content of reformate. As the objective is to analyze unit performance with different temperature profiles, an instantaneous simulation is not adequate. More information can be inferred from longer operation times, where coke deposition over the catalyst decreases its activity and higher temperatures are needed to maintain reformate quality. The simulation has been made over a time period of 8000 h similar to a commercial unit run length and using operational conditions analogous to the ones described previously in Table

Table 15 Operational Parameters and Feed Characteristics for Model use

Parameter	Paraffinic feed	Naphthenic feed
Paraffin	61	40
Naphthene	34	45
Aromatic	5	15
N + 2A*	44	75
Separator pressure (bar)	22	22
H_2/HC ratio	4	4
Initial WAIT (°C)	480	480
Throughput (BPSD)	12000	12000
Loaded catalyst (kg)	11000/21000/26000	11000/21000/26000

*Naphthenes plus twice aromatics.

15. The same initial temperature 480°C WAIT was employed. Both high and low N+2A feeds are tested to detect first reactor temperature drop influence. Three runs were made employing first a level, then an ascending, and finally a descending profile. A controlling algorithm was implemented that increases inlet reactor temperatures in 1°C steps to maintain a 98-RON reformate throughout the run. The reformate octane number is computed as a volumetric average research octane number of the species present in the reformate. Figures 9, 10, and 11 show the reactor carbon content, RON, and reformate yield for each case.

Model results express the higher severity needed in order to increase the RON number for the low naphthene feed. Average-run WABT for the high naphthene feed is 487°C with a final WAIT of 516°C against 494°C and 523°C WAIT for the low-naphthene one. Reformate yield also follows the same behavior. High-naphthene feed reaches an average yield during the course of the run of 81.9% while the low naphthene feed yields 72.2%. Consequently, the reactor carbon content is greater for the paraffinic feed, particularly in the third reactor where the carbon content reaches 19%, as shown in Figure 9.

From the simulation results, some insights regarding reactor temperature profile are obtained. Catalyst deactivation tends to be greatest in the last reactor, and ascending profiles will tend to concentrate coke deposition in this reactor and will reduce the catalyst cycle. For units with furnace temperature limitations at the end of the cycle, starting the cycle with one of the reactors closer to its end of run condition will also limit cycle life.

High-naphthene feeds can cause large temperature drops in the first reactor such that the latter part of the bed in that reactor becomes so cold that the catalyst is quenched and no reaction occurs. Some advantage in this situation may be possible from a descending profile to keep the first reactor bed temperatures higher. A descending profile increases run length due to lower temperature of the

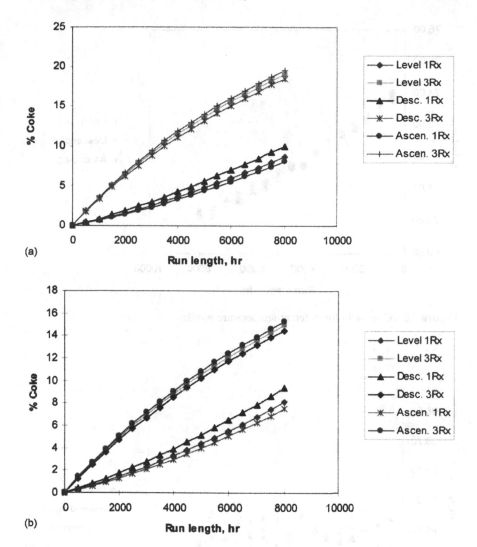

Figure 9 Reactor carbon content. (a) Paraffinic feed. (b) Naphthenic feed.

third reactor and consequently lower degree of cracking reactions. As shown in Figure 9, a 1% decrease in coke content could be obtained. A descending profile produces an increase of around 2% in first reactor coke content due to higher severity. As this reactor presents the lowest coke content, this slight increase could be tolerated so as to permit better use of the loaded catalyst.

With regard to RON and C_5+ yield, from the simulation results there does not appear to be significant advantage in shifting temperature profile. As shown in

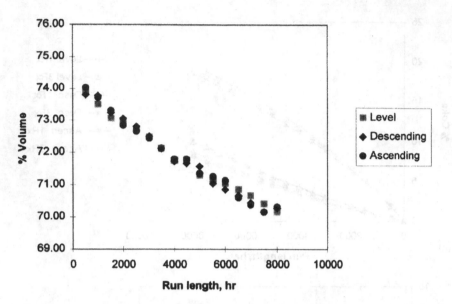

Figure 10 C_5+ yield for different temperature profiles.

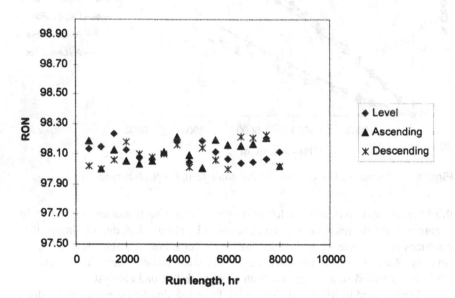

Figure 11   RON for different temperature profiles.

Figures 10 and 11, level profile results in average reactor coke contents, RON, and C_5+ yield.

The optimal reactor temperature profile may be established based not only on the reforming unit operation but on other factors due to the paramount role of reforming in the refinery operation. Temperature profile influences reformer performance, and wise use of reactor temperature profiles could extend cycle life if needed although the expected benefits may not compensate operational control and monitoring disadvantages when compared to the level profile operation.

6.3. Benzene Reduction and Reformer Operation Conditions

Refiners are confronted with new specifications for gasoline and these are expected to be followed by more stringent specifications after 2005, although timing and levels are yet to be finalized. Table 16 shows the new specifications that are in the pipeline [66]. One of the most significant changes is the restriction in benzene and aromatics. These proposed regulatory changes directly affect naphtha reforming, as the goal of this process is precisely the production of

Table 16 Evolution of European Gasoline Specifications

Property	2000	2005
TVP summer (kPa)	70	60
Sulfur (ppm)	150 max	50 max
Oxygen (wt %)	2.3 max	To be decided
Benzene (vol %)	1.0 max	1.0 max
Aromatics (vol %)	42.0 max	35.0 max
Olefins (vol %)	18.0 max	To be decided

Source: Ref. 66.

Table 17 Trends in Catalytic Reforming Catalysts

Catalyst changes	Impact
High activity	Higher throughput
High selectivity	More H_2 + aromatics
Low coke formation	Benzene reduction
High thermal stability of the carrier	Higher throughput
	Extended cycle length

Source: Ref. 67.

Table 18 Benzene Reduction Simulation: Operation Conditions and Feed Compositions, LHSV = 2

Component	Paraffinic1	Paraffinic2	Naphthenic1	Naphthenic2
P_6	11.20	15.00	12.30	8.28
P_7	13.40	27.60	15.60	8.79
P_8	14.30	18.40	8.87	8.93
P_9+	28.20	5.20	6.80	15.66
CH	1.00	2.60	2.00	4.55
MCP	1.60	3.00	1.71	2.77
N_7	4.30	8.80	16.83	14.52
N_8	4.80	10.70	11.38	10.78
N_9+	12.10	1.60	9.81	13.95
A_6	0.60	0.50	0.90	0.64
A_7	1.50	3.70	4.00	3.10
A_8	2.90	2.90	6.80	4.25
A_9+	4.10	—	3.00	3.78
Pres. (kg/cm^2)	10	15	25	
WAIT paraffinic feed (°C)	499.0	504.5	510.5	
WAIT naphthenic feed (°C)	491.2	493.8	501.6	
RON	90	94	98	

aromatics. Changes in the refinery configuration, including new units such as isomerization, naphtha splitting, and reformate benzene removal, are the first solutions to be implemented [67]. Accomplishing benzene reduction is a potential trend in reforming catalyst design, as shown in Table 17 [67 and references herein].

The benzene content in the reformate is mainly dependent on benzene precursors in the feed and reforming operation conditions [51–53,68,69]. So for a major reduction of benzene content, two basic approaches are possible: either minimize benzene formation by removing benzene precursors from the reformer feed, or fractionate out a light reformate for subsequent conversion or extraction. The use of operating conditions to reduce the benzene formation implies a low-

Table 19 Model Prediction: Aromatic Distribution at Different Operation Conditions—Paraffinic Feed

Pressure (kg/cm^2)	WAIT (°C)	C_6 (%)	C_7 (%)	C_8+ (%)
20	517	9.8	37.9	51.7
7	506	8.5	34.7	55.4

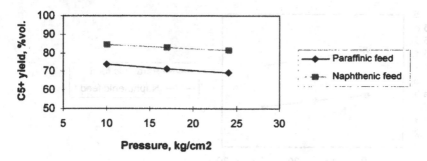

Figure 12 Influence of pressure on C_5+ yield. Model prediction.

severity operation that is frequently beyond the refiners' capabilities. Despite the low capability of operating variables for benzene reduction, the reforming model previously described has been used to study the effect of temperature and pressure as the main operating variables in the process of benzene formation.

In Table 18, the reformer feed compositions and operation conditions used in the model are shown. Naphthenic as well as paraffinic feeds are treated, with N+2A between 45 and 71. The table shows the temperatures needed to obtain the same RON for naphthenic and paraffinic feeds. Operation conditions are typical for a semiregenerative reforming unit. Table 19 shows some results obtained in reforming paraffinic feed. Conversion to aromatics increases as the pressure is reduced. Aromatic distribution shows that the products are mainly toluene and C_8 aromatics. There is still not much benzene produced, and results show a slight shift to heavier products at lower pressures. Lower pressures not only increase the conversion to aromatics but also reduce the extent of hydrocracking and hence the loss to light product formation. The general behavior is consistent with the

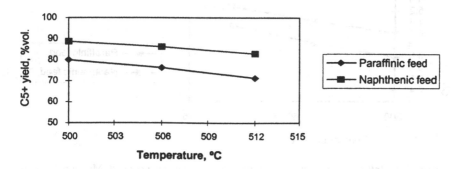

Figure 13 Influence of reactor temperature on C_5+ yield. Model prediction.

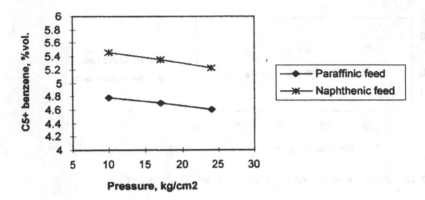

Figure 14 Influence of pressure on C_5+ benzene content. Model prediction.

fact that aromatics formation is accompanied by hydrogen production while hydrocracking consumes hydrogen.

An important issue related to operational control of benzene content is reformer C_5+ yield. Yield is also a dependent function of feed quality and operating conditions. In Figures 12 and 13, the influence of pressure and temperature on C_5+ yields is shown for naphthenic and paraffinic feeds. Temperature is varied from 490°C to 510°C at a constant pressure of 15 kg/cm². Separator pressure ranges from 10 to 25 kg/cm² for a constant severity of 98 RON. As expected yield decreases deeply with pressure for the paraffinic feed varying from 73.6 to 69.1 vol %, the naphthenic feed yield loss is lower, i.e., 84.3–81.4 vol %. Yield variation with temperature is much greater than that

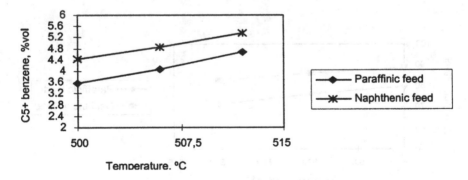

Figure 15 Influence of reactor temperature on C_5+ benzene content. Model prediction.

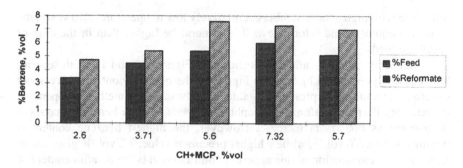

Figure 16 Influence of benzene precursors on C_5+ benzene content. Model predictions.

obtained with pressure variations and is again higher for the paraffinic feed, 79.8–71.4 vol %, while naphthenic feed shows a change from 88.5 to 82.9 vol %. The influence of pressure and temperature on reformate benzene content is presented in Figures 14 and 15. The reformate benzene content decreases slightly as the pressure increases; the values obtained are related to the variations in yield and maintain the benzene content almost constant. Naphthenic feed produces higher benzene content, from 5.45 to 5.22 vol %, while the paraffinic feed results in a 1% lower value, from 4.8 to 4.66 vol %. The yield effect previously described becomes less pronounced as temperature increases. Benzene content is higher for naphthenic feeds ranging from 4.2 to 5.3 vol %, while paraffinic feed gives 3.6 to 4.7 vol %. The effect of feed composition for naphthenic feeds gives

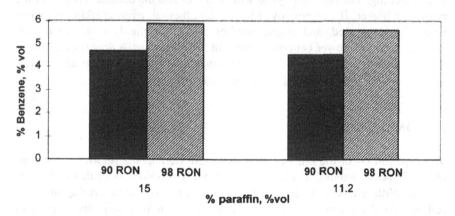

Figure 17 Influence of dehydrocyclization on C_5+ benzene content. Model predictions.

reformate with high octane numbers at relatively low temperatures. However, the benzene content of the reformate will in general be higher than in the case of paraffinic feeds.

Benzene precursor influence is shown in Figures 16 and 17, with severity 98 RON and pressure 15 kg/cm^2. In Figure 16 the benzene content with respect to reformate and feed is presented against cyclohexane and methylcyclopentane content for different paraffinic and naphthenic feeds. A clear benzene increment is detected as precursors increase. However, the highest benzene content is obtained with a 5.6 vol %, while a higher precursor value, 7.2 vol %, gives lower benzene. The explanation of this apparent contradiction is the paraffin content of each feed. The feed that produces the maximal benzene has 15 vol % of paraffins, with only 11 vol % for the second charge, and due to the high-severity conditions, dehydrocyclization reactions are surely responsible for this high benzene content. To support this argument, in Figure 17 the simulation results for the same feeds but at a lower severity, 90 RON, are presented. As can be seen in the figure, at the lower severity benzene content is reduced overall and the percentage difference for the two feeds is also reduced.

As noted, in the catalytic reforming kinetic model, dehydrocyclization reaction rates are lower than similar dehydrogenation rates and take place in the second and mainly in the third reactor. Also hydrocracking reactions that decrease reformate yield and promote heavy aromatic dealkylation reactions occur in the third reactor, so that one would expect a benzene reduction if third-reactor severity is decreased. At the same reformate RON, naphthenic feeds need a lower operating temperature, thus diminishing the benzene production due to hydrocracking and dehydrocyclization reactions. Increasing the reformer operating temperature leads to higher dehydrocyclization rates and favors hydrocracking. The reformate yield will be lower and the benzene content of the reformate higher. If one employed pressure reduction, hydrocracking reactions would be suppressed, and octane number as well as yields would be higher, accompanied by a lower benzene content in the reformate. A drawback is that at low pressures more heavy aromatic compounds and coke precursors are formed. Deactivation is therefore much faster at lower pressures.

7 CONCLUSION

The applications of the kinetic model for catalytic naphtha reforming to actual refiner issues as exemplified here demonstrate the usefulness of this tool to the refiner. With a model describing the performance of the reforming unit, the refiner can determine at least directionally, if not precisely, the impact of variations in the process conditions or feedstocks available. Additions of other refinery streams as feedstocks can be tested for impact on stability or profitability without jeopardizing the operation. As will be seen in the chapter on process control, modeling of the reformer is an important aspect in maximizing the utilization of this key refining asset.

ACKNOWLEDGMENTS

The first author thanks Dr. Perez Pascual, head of CEPSA Research Center, for his help during development of this work.

SYMBOLS

C_c	coke content, kg coke/kg cat.
c_p	specific heat, kcal/kmol°
d_p	particle diameter, cm
D_e	effective diffusivity, cm^2/s
E_i	activation energy of i, cal/mol
$F_{l,v,t}$	mixed, liquid, or vapor molar flow rate, kmol/h
G	superficial mass flow rate, kg/m^2 h
$H_{f,l,v}$	mixed, liquid, or vapor molar enthalpy, kcal/kmol
K_i	vapor liquid distribution coefficient of ith component
k_i	rate constant, units consistent with rate expression
k	adiabatic coefficient
M_m	mean molecular weight, kg/kmol
N_c	number of components
N_j	molar flux of j, mol/m^2 s
O_{jv}	vapor fugacity
P_t	total pressure, bar
p_i	partial pressure of i, atm
Q	heat balance, kcal/h
r_i	reaction rate of ith reaction, kmol/kg cat. h
r_i^0	reaction rate of ith reaction in absence of coke, kmol/kg cat. h
R_j	production rate of jth component, kmol/kg cat. h
R	gas constant
T	temperature, °C
V_{j1}	liquid fugacity of pure component j
WABT	weight average bed temperature, °C
WAIT	weight average inlet temperature, °C
x_i	mole fraction of component i in the liquid phase
Y^*	calculated parameter
Y	experimental parameter

y_i	mole fraction of component i in vapor phase
z	axial reactor coordinate, m
z_j	mole fraction of j in feedstream to the flash drum
γ	activity coefficient
$\rho_{b,f}$	bulk density of the packing, kg cat./m^3; fluid density, kg/m^3
ρ_s	catalyst pellet density, kg/m^3
Ω	cross-section of reactor, m^2
$\Delta H_f{}^0$	standard enthalpy of formation, kcal/kmol
ε	void fraction of the packing, m^3/m^3
β_{ji}	stoichiometric coefficient of component j in ith reaction
μ	viscosity of gas passing through the bed, Pa s
α	deactivation constant, kg cat./kg coke
Φ	objective function
ϕ	deactivation function

REFERENCES

1. Recasens, E. Combustibles **1955**, *78/9*, 137–148.
2. Turpin, L.E. Hydrocarbon Processing, June **1992**, 81–90.
3. Sowell, R. Hydrocarbon Processing, March **1998**, 102–107.
4. Larraz, R. Oil and Gas Eur. Mag. **2000**, *2*, 33–37.
5. Larraz, R. Honeywell Users Group Conference, Sitges, September 1998.
6. Kugelman, A.M. Hydrocarbon Processing, January **1976**, 95–102.
7. Yaws, C.L.; Chiang, P.Y. Hydrocarbon Processing, May **1988**, 91–98.
8. Storey, S.H.; Van Zeggeren, F. *The Computation of Chemical Equilibria*; Cambridge University Press: New York, 1970.
9. White, W.B.; Johnson, S.M.; Dantzig, G.B. J. Chem Phys. **1958**, *28*, 751.
10. Golfarb, D.; Lapidus, L. I&EC Fundam. February **1968**, *7* (1), 141–151.
11. Stull, D.R.; Westrum, E.F.; Sinke, G.C. *The Chemical Thermodynamics of Organic Compounds*; John Wiley and Sons: New York, 1969.
12. Radosz, M.; Krmarz, J. Hydrocarbon Processing, July **1978**, 201–203.
13. Taskar, U.; Riggs, J.H. AICHE J. **1997**, *43*, (3), 740–753.
14. Brambilla, A.; Trivella, F. Hydrocarbon Processing **1996**, *75*, (9), 61–66.
15. Twu, C.H.; Coon, J.E. Hydrocarbon Processing **1996** *75*, (2) 51–56.
16. Riazi, M.R.; Daubert, T.E. Oil and Gas J. August **1986**, *50*, 25.
17. Lion, A.K.; Edmister, W.C. Hydrocarbon Processing **1975**, *54*, (8) 119.
18. Riazi, M.R.; Daubert, T.E. Hydrocarbon Processing **1980**, *59*, (3) 115.
19. Martin, D.L.; Gaddy, J.L. AIChE Symp. Series **1982**, 99–107.
20. Luus, R.D. Can. J. Chem. Eng. **1975**, *53*, 217.
21. Gaines, L.D.; Gaddy, J.L. Ind. Eng. Chem. Proc. Des. Dev. **1976**, *15*, 206.
22. Jezowski, J.; Poplewski, G.; Madej, S.; Bochenek, R.; Jezowska, A. CHISA Congress, Praga, August 1998.
23. Rossini F.D. et al., API Project 44.

24. Lough, V. Foxboro. Personal communication, 2000.
25. Smith, R.B. Chem. Eng. Prog. **1959**, *55*, (6) 76.
26. Bommannan, D.; Strivastava, R.D.; Saraf, D.N. Can. J. Chem. Eng. **1989**, *67*, 405.
27. Barreto, G.F.; Viñas, J.M.; Gonzalez, M.G. Lat. Am. Appl. Res. **1996**, *26*, (1) 21–34.
28. Ansari, R.M.; Tade, M.O. Saudi Aramco J. Technol. Spring **1988**, 13–18.
29. Krane, H.G.; Groh, A.B.; Shulman, B.L.; Sinfelt, J.H. World Petroleum Congress, 1960.
30. Aguilar, E.; Anchyeta, J. Oil Gas J. July 25 **1994a**, 80–83.
31. Aguilar, E.; Anchyeta, J. Oil Gas J. January 31 **1994b**, 93–95.
32. Burnett, R.L.; Steinmetz, H.L.; Blue, E.M.; Noble, E.M. Petrol Chem. Am. Chem. Soc., Detroit Meeting, April, 1965.
33. Jenkins, J.H.; Stephens, T.W. Hydrocarbon Processing, November **1980**, 163–167.
34. Henningsen, J.; Bundgaard-Nielson, M. Br. Chem. Eng. **1970**, *15* (11) pp. 1433–1436.
35. Kmak, W.S. AICHE Meeting, Houston, 1972.
36. Ramage, M.P.; Graziani, K.P.; Krambeck, F.J. Chem. Eng. Sci. **1980**, *35*, 41–48.
37. Ramage, M.P.; Graziani, K.R.; Shipper, P.H.; Krambeck, F.J.; Choi, B.C. Adv. Chem. Eng. **1987**, *13*, 193.
38. Lee, J.W.; Ko, K.Y.; Jung, Y.K.; Lee, K.S. Comp. Chem. Eng. **1997**, *21*, S1105–S1110.
39. Marin, G.B.; Froment, G.F.; Lerou, J.J.; De Backer, W. EFCE **1983**, *2* (27), Paris, C117.
40. Marin, G.B.; Froment, G.F. Chem. Eng. Sci. **1982**, *37*, (5), 759.
41. Van Trimpont, P.A.; Marin, G.B.; Froment, G.F. Ind. Eng. Chem. Res. **1988**, *27*, 1.
42. Marin, G.B.; Froment, G.F. In *The Development and Use of Rate Equations for Catalytic Refinery Process*. Proceedings of the 1st Kuwait Conference on Hydro-treating Processes, Kuwait, March 5–9, 1989.
43. Coppens, M.O.; Froment, G.F. Chem. Eng. Sci. **1996**, *51*, 2283–2292.
44. Galtier, P.; Vacher, P.; Froment, G.F. Hydrocarbon Eng. March **1998**, 40–42.
45. Jackson, R. *Transport in Porous Catalyst*; Elsevier: Amsterdam, 1977.
46. Coppens, M.O.; Froment, G.F. Chem. Eng. Sci. **1995**, *50*, 1013–1026.
47. Coppens, M.O.; Froment, G.F. Chem. Eng. Sci. **1995**, *50*, 1027–1039.
48. Coppens, M.O. Catalysis Today **1999**, *53*, 225–243.
49. Coppens, M.O.; Froment, G.F. Chem. Eng. Sci. **1994**, *49*, 4897–4907.
50. Weekman, V.W. AICHE J. Monogr. Ser. **1979**, *75*, 11.
51. Ramage, M.P.; Graziani, K.R.; Shipper, P.H.; Krambeck, F.J.; Choi, B.C. Advances Catalyst **1987**, *13*, 193.
52. Van Broekhoven, E.H.; Bahlen, F.; Hallie, H. AICHE Spring Meeting, March 1990, *52*, 18–20.
53. Pollitzer, E.L.; Hayes, J.C.; Haensel, V. Adv. Chem. Ser. **1970**, *97*, 20–37.
54. Parera, J.M.; Querini, C.; Figoli, N. Appl. Cat. **1988**, *44*, L1–L8.
55. Parera, J.M.; Beltramini, J.N. J. Catal. **1988**, *112*, 357–365.
56. Mieville, R.L. J. Catal. **1986**, *100*, 482–488.
57. De Pauw, Froment, G.F. Chem. Eng. Sci. **1975**, *30*, 789–801.

58. Schröder, B.; Salzer, C.; Turek, F. Ind. Eng. Chem. **1991**, *30*, 326–330.
59. Moore, C.E. Hydrocarbon Processing, July **1991**, 92–94.
60. Chao, K.C.; Seader, J.D. AICHE J. **1961**, *7*, 598.
61. Henley, E.J.; Seader, J.D. *Equilibrium Stage Operations in Chemical Engineering*; John Wiley and Sons: London, 1981.
62. Pistorius, J.T. Oil Gas J., June **1985**, *10*, 10–14.
63. Little, D.M. *Catalytic Reforming*; PennWell: Tulsa, OK, 1983.
64. Levenspiel, O. *Chemical Reaction Engineering*; Reverte: Barcelona, 1981.
65. Kugelman, A.M. Hydrocarbon Processing, December **1973**, 67–70.
66. Genis, O.; Simpson, S.G.; Penner, D.W.; Gautam, R.; Glover, B.K. Hydrocarbon Eng. March **2000**, 87–93.
67. Martino, G. 12th International Congress on Catalyst, Stud. Surf. Sci. **2000**, *130*, 83–103.
68. Sertic-Bionda, K. Oil Gas Eur. Mag. **1997**, *3*, 35–37.
69. Ciapetta, F.G. Am. Chem. Soc., Div. Petrol. Chem. **1955**, *33*, 167.
70. Mandelbrot, B.B. *The Fractal Geometry of Nature*; Tusquet Eds.: Barcelona, 1997.
71. Pfeifer, P.; Avnir, D. J. Chem. Phys. **1984**, *80*, 4573.
72. Avnir, D.; Farin, D.; Pfeifer, P. J. Chem. Phys. **1983**, *79*, 3566–3571.
73. Avnir, D.; Farin, D.; Pfeifer, P. Nature **1984**, *308*, 261–263.
74. Neimark, A. Physica a **1992**, *191*, 258–262.
75. Guinier, A.; Fournet, G.; Walker, C.L.; Yudowitch, K.L. *Small Angle Scattering of X-Rays*; John Wiley and Sons: New York, 1955.
76. Schmidt, P.W. J. Appl. Crystallogr. **1991**, *24*, 414.

APPENDIX: FRACTALS AND SURFACES

Fractal objects are self-similar structures in which increasing magnifications reveal similar features at different length scales [70]. They show a power-law relation between a certain properties (like mass, void volume, or surface area) and length scale. These are termed mass fractal, pore fractal, and surface fractal, respectively [71]. Characterization and analysis of porous objects in terms of fractal geometry has become an intensive research area in recent years. Experiments show a large number of natural and industrial structures having either fractal surfaces or fractal pore distribution [72]. Fractals have been described as objects capable of simulating diffusion-controlled process structures as catalysts [73]. Their use enables important observations to be made when analyzing much of the adsorption data published so far for a large number of porous media, including many catalysts. The measured surface area S depends on the effective diameter δ of the sorbate molecules according to simple power law:

$$S \approx \delta^{-a} \tag{A1}$$

More precisely, when data are plotted as $\log S$ vs. $\log \delta$, a straight line can be fitted, with a slope between 0 and 1. If $a = 0$, the surface area does not depend on the size of the probe. Frequently a is found to be greater than 0, implying that specific area is meaningless without specifying the resolution δ. The so-called fractal dimension, D, expresses the space filling capacity of a fractal, while Euclidean shapes have integer dimensions (1 for a line, 2 for a surface, and 3 for a volume), a catalyst surface can have any dimension between 2 and 3, both limits included. Many fractals in nature can be very well approximated by a statistical self-similar or self-affine structure. A real object can only be a self-similar fractal, within a finite fractal scaling range, denoted with the inner and outer cutoffs, δ_{min} and δ_{max}. Informally the number N of units of size δ needed to cover a fractal object decreases with δ as:

$$N \approx \delta^{-D} \tag{A2}$$

Lengths are measured as $N\delta$, areas as $N\delta^2$, so that in equation (A1), $a = 1 - D$ and becomes

$$S \approx \delta^{2-D} \tag{A3}$$

Methods for fractal dimension determination have been described using the information of the complete adsorption isotherm of a single probe [74]. Small-angle X-ray scattering can be also used to study the mass fractal and surface morphology of materials [75,76].

Index

Printed in the United States
by Baker & Taylor Publisher Services